Amorphous and Polycrystalline
Thin-Film Silicon Science
and Technology — 2009

MATERIALS RESEARCH SOCIETY
SYMPOSIUM PROCEEDINGS VOLUME 1153

Amorphous and Polycrystalline Thin-Film Silicon Science and Technology — 2009

Symposium held April 14–17, 2009, San Francisco, California, U.S.A.

EDITORS:

Andrew Flewitt
University of Cambridge
Cambridge, United Kingdom

Jack Hou
LuxingTek Ltd.
Jhubei City, Taiwan

Arokia Nathan
University College London
London, United Kingdom

Qi Wang
National Renewable Energy Laboratory
Golden, Colorado, U.S.A.

Shuichi Uchikoga
Toshiba Corporation
Kawasaki, Japan

Materials Research Society
Warrendale, Pennsylvania

CAMBRIDGE
UNIVERSITY PRESS

University Printing House, Cambridge CB2 8BS, United Kingdom

One Liberty Plaza, 20th Floor, New York, NY 10006, USA

477 Williamstown Road, Port Melbourne, VIC 3207, Australia

314-321, 3rd Floor, Plot 3, Splendor Forum, Jasola District Centre, New Delhi - 110025, India

79 Anson Road, #06-04/06, Singapore 079906

Cambridge University Press is part of the University of Cambridge.

It furthers the University's mission by disseminating knowledge in the pursuit of education, learning and research at the highest international levels of excellence.

www.cambridge.org
Information on this title: www.cambridge.org/9781605111261

Materials Research Society
506 Keystone Drive, Warrendale, PA 15086
http://www.mrs.org

© Materials Research Society 2009

First published 2009
First paperback edition 2012

Single article reprints from this publication are available through University Microfilms Inc., 300 North Zeeb Road, Ann Arbor, MI 48106

CODEN: MRSPDH

A catalogue record for this publication is available from the British Library

ISBN 978-1-605-11126-1 Hardback
ISBN 978-1-107-40834-0 Paperback

CONTENTS

*Invited Paper

DEFECTS AND METASTABILITY

POSTER SESSION: CRYSTALLIZATION

POSTER SESSION:
NANOSTRUCTURED SILICON

NOVEL DEVICE APPLICATIONS

FILM GROWTH I

SOLAR CELLS

CRYSTALLIZATION

LIGHT TRAPPING IN SOLAR CELLS II

*Invited Paper

POSTER SESSION:
LARGE AREA AND FLEXIBLE PROCESSING

POSTER SESSION:
THIN-FILM TRANSISTORS

FILM GROWTH II

HYDROGEN IN SILICON

*Invited Paper

PREFACE

Thin-film silicon materials and their alloys underpin a diverse range of electronic systems from active matrix flat-panel displays, through solar panels for "green-power" generation, to surface micromachined MEMS devices. Furthermore, new application areas are emerging, including RFID tagging and biosensors. As a consequence, large-area electronics is currently one of the fastest growing semiconductor technologies. Thin-film silicon can possess a diverse range of structures, from being fully amorphous to fully polycrystalline, as well as allowing mixed-phase states, such as micro- and nanocrystalline silicon. Such diversity has enabled this growth of large-area electronics, but it has also introduced complexity.

Symposium A, "Amorphous and Polycrystalline Thin-Film Silicon Science and Technology — 2009," held April 14–17 at the 2009 MRS Spring Meeting in San Francisco, California, has been running annually at the MRS Spring Meeting for over 25 years. This symposium provides a unique annual forum for scientists and engineers dealing with thin-film silicon materials, and their alloys with Ge, C, N, and other elements, to discuss issues related to both fundamental materials science and applied technology. The symposium opened with a full day tutorial looking at the deposition, characterization and physics of thin-film silicon materials in the morning session, followed by an afternoon devoted to studying the devices that use thin-film silicon. This was followed over the course of the next three and a half days by 15 invited talks, 52 contributed oral presentations and 71 poster presentations. This volume therefore acts as a good overview of the fields discussed, which ranged from studies of film growth and crystallization, through investigations on materials characterization, defects, metastability and carrier transport, to reports on devices such as solar cells and thin-film transistors. The importance of developing efficient solar cells was reflected in the significant number of papers that looked at aspects of improving lifetime and efficiency, and two focus sessions were devoted to light trapping in solar cells. The drive towards ever larger and more flexible substrates was also evident.

The organizers would like to thank all those who attended the symposium and contributed such excellent presentations. Particular thanks are due to the contributors and reviewers of this proceedings volume, and to the members of the Symposium A Advisory Group for their help and support. We would like to extend special thanks to Craig Taylor for his active role in organizing and providing administrative support for these symposia for over 15 years. We are also indebted to Mary Ann Woolf for all her help, advice and support in managing the construction of the symposium and the process of producing this proceedings volume. Finally, we would like to thank the Industrial Technology Research Institute, Taiwan for their generous financial support by sponsoring this symposium.

Andrew Flewitt
Jack Hou
Arokia Nathan
Qi Wang
Shuichi Uchikoga

August 2009

MATERIALS RESEARCH SOCIETY SYMPOSIUM PROCEEDINGS

Volume 1153 — Amorphous and Polycrystalline Thin-Film Silicon Science and Technology — 2009, A. Flewitt, Q. Wang, J. Hou, S. Uchikoga, A. Nathan, 2009, ISBN 978-1-60511-126-1
Volume 1154 — Concepts in Molecular and Organic Electronics, N. Koch, E. Zojer, S.-W. Hla, X. Zhu, 2009, ISBN 978-1-60511-127-8
Volume 1155 — CMOS Gate-Stack Scaling — Materials, Interfaces and Reliability Implications, J. Butterbaugh, A. Demkov, R. Harris, W. Rachmady, B. Taylor, 2009, ISBN 978-1-60511-128-5
Volume 1156— Materials, Processes and Reliability for Advanced Interconnects for Micro- and Nanoelectronics — 2009, M. Gall, A. Grill, F. Iacopi, J. Koike, T. Usui, 2009, ISBN 978-1-60511-129-2
Volume 1157 — Science and Technology of Chemical Mechanical Planarization (CMP), A. Kumar, C.F. Higgs III, C.S. Korach, S. Balakumar, 2009, ISBN 978-1-60511-130-8
Volume 1158E —Packaging, Chip-Package Interactions and Solder Materials Challenges, P.A. Kohl, P.S. Ho, P. Thompson, R. Aschenbrenner, 2009, ISBN 978-1-60511-131-5
Volume 1159E —High-Throughput Synthesis and Measurement Methods for Rapid Optimization and Discovery of Advanced Materials, M.L. Green, I. Takeuchi, T. Chiang, J. Paul, 2009, ISBN 978-1-60511-132-2
Volume 1160 — Materials and Physics for Nonvolatile Memories, Y. Fujisaki, R. Waser, T. Li, C. Bonafos, 2009, ISBN 978-1-60511-133-9
Volume 1161E —Engineered Multiferroics — Magnetoelectric Interactions, Sensors and Devices, G. Srinivasan, M.I. Bichurin, S. Priya, N.X. Sun, 2009, ISBN 978-1-60511-134-6
Volume 1162E —High-Temperature Photonic Structures, V. Shklover, S.-Y. Lin, R. Biswas, E. Johnson, 2009, ISBN 978-1-60511-135-3
Volume 1163E —Materials Research for Terahertz Technology Development, C.E. Stutz, D. Ritchie, P. Schunemann, J. Deibel, 2009, ISBN 978-1-60511-136-0
Volume 1164 — Nuclear Radiation Detection Materials — 2009, D.L. Perry, A. Burger, L. Franks, K. Yasuda, M. Fiederle, 2009, ISBN 978-1-60511-137-7
Volume 1165 — Thin-Film Compound Semiconductor Photovoltaics — 2009, A. Yamada, C. Heske, M. Contreras, M. Igalson, S.J.C. Irvine, 2009, ISBN 978-1-60511-138-4
Volume 1166 — Materials and Devices for Thermal-to-Electric Energy Conversion, J. Yang, G.S. Nolas, K. Koumoto, Y. Grin, 2009, ISBN 978-1-60511-139-1
Volume 1167 — Compound Semiconductors for Energy Applications and Environmental Sustainability, F. Shahedipour-Sandvik, E.F. Schubert, L.D. Bell, V. Tilak, A.W. Bett, 2009, ISBN 978-1-60511-140-7
Volume 1168E —Three-Dimensional Architectures for Energy Generation and Storage, B. Dunn, G. Li, J.W. Long, E. Yablonovitch, 2009, ISBN 978-1-60511-141-4
Volume 1169E —Materials Science of Water Purification, Y. Cohen, 2009, ISBN 978-1-60511-142-1
Volume 1170E —Materials for Renewable Energy at the Society and Technology Nexus, R.T. Collins, 2009, ISBN 978-1-60511-143-8
Volume 1171E —Materials in Photocatalysis and Photoelectrochemistry for Environmental Applications and H₂ Generation, A. Braun, P.A. Alivisatos, E. Figgemeier, J.A. Turner, J. Ye, E.A. Chandler, 2009, ISBN 978-1-60511-144-5
Volume 1172E —Nanoscale Heat Transport — From Fundamentals to Devices, R. Venkatasubramanian, 2009, ISBN 978-1-60511-145-2
Volume 1173E —Electofluidic Materials and Applications — Micro/Biofluidics, Electowetting and Electrospinning, A. Steckl, Y. Nemirovsky, A. Singh, W.-C. Tian, 2009, ISBN 978-1-60511-146-9
Volume 1174 — Functional Metal-Oxide Nanostructures, J. Wu, W. Han, A. Janotti, H.-C. Kim, 2009, ISBN 978-1-60511-147-6

MATERIALS RESEARCH SOCIETY SYMPOSIUM PROCEEDINGS

Prior Materials Research Society Symposium Proceedings available by contacting Materials Research Society

Characterization

Mater. Res. Soc. Symp. Proc. Vol. 1153 © 2009 Materials Research Society   1153-A02-01

Photoluminescence Characterization of Hydrogenated Nanocrystalline/Amorphous Silicon

J.D. Fields[1], P.C. Taylor[2], J.G. Radziszewski[2], D.A. Baker[2], G. Yue[3], B. Yan[3]

[1] Materials Science Department, Colorado School of Mines, Golden, CO
[2] Physics Science Department, Colorado School of Mines, Golden, CO
[3] United Solar Ovanic LLC, Troy, MI

ABSTRACT

The photoluminescence in a nc-Si/a-Si:H mixture has been investigated at varying excitation intensities, and temperatures We have also observed changes in the luminescence spectra, which are induced by sequential annealing at temperatures below the a-Si:H crystallization temperature (~ 600°C). Two predominant luminescence peaks are observed at ~ 0.95 eV and ~ 1.30 eV, which are attributed to band tail-to-band tail transitions near the nc-Si grain boundaries and in the a-Si:H bulk, respectively. The 0.95 eV band saturates approaching 500 mW/cm^2 excitation intensity. Annealing the nc-Si/a-Si:H mixture brings out a new low energy peak, centered at ~ 0.70 eV, and which we believe to be due to oxygen defects.

INTRODUCTION

In order to optimize hydrogenated nanocrystalline/amorphous silicon (nc-Si/a-Si:H) for photovoltaic applications we must better understand its electrical and optical properties, especially those at the nc-Si/a-Si:H interfacial regions. We present data from an exploratory round of photoluminescence experiments, which shed light on the nature of the electronic structure of nc-Si/a-Si:H. Investigating the luminescence behavior of a nc-Si/a-Si:H mixture under various excitation intensity and temperature conditions reveals several differences between the electronic structure of the nc-Si phase from that of a-Si:H.

Radiative transitions signified by emission at ~ 1.3 eV occur when excited carriers recombine radiatively across band-tail states in a-Si:H [1]. In poly-Si and mixed phase silicon the large number of defects in grain boundary regions, namely variation in bond lengths and angles, causes local fluctuations in the electric potential. These effects form localized states in the band gap and give rise to band-tails [2]. Emission at 0.9 eV in poly-Si [2, 3] occurs by carrier recombination across these band-tail states, and a band observed at ~ 0.9 – 1.0 eV in μc-Si/a-Si:H systems is believed to involve similar band-tail states [4].

The decay of PL with increasing temperature occurs due to the increased probability of non-radiative recombination when carriers have sufficient (thermal) energy to hop to nearby states [1]. Dangling bonds, interfaces, and defects provide pathways for excited carriers to lower their energy without radiative recombination.

Both the luminescence decay with increasing temperature and the excitation intensity dependence of PL depend on local electronic structure. Band-tail luminescence in a-Si:H typically increases with temperature at very low temperatures, plateaus at ~ 50 K, and decreases rapidly above 80 K [1, 5]. Detection of nc-Si luminescence at room temperature and its unique luminescence decay with increasing temperature suggests fundamental differences in the local electronic structure near nc-Si grain boundaries compared with a-Si:H. Excitation intensity dependence of the PL emitted by our nc-Si/a-Si:H mixture also shows a striking difference between nc-Si grain boundaries and a-Si:H. Observing that the band attributed to nc-Si saturates with increasing excitation intensity while the a-Si:H signal intensity continues to rise over the entire range investigated allows us to speculate about the relative density of luminescent states in these regions.

Changes in the luminescence profile induced by sequential annealing provide insight regarding thermally activated processes in this system. Several authors argue that a low energy PL band centered at ~ 0.7 eV arises from oxygen defect states in silicon systems [3, 4, 6]. In every case, thermal exposure seems to play a role in determining whether this band emerges. Based on past work and our observations we argue that thermal energy provided by annealing promotes oxygen agglomeration, and that 0.7 eV PL involves donor states related to oxygen precipitates.

EXPERIMENTAL DETAILS

United Solar Ovanic made the samples used in these experiments, which contain intrinsic nc-Si/a-Si:H mixtures characteristic of high quality PV devices. The nc-Si/a-Si:H layers, ~ 1 μm thick, were grown by PECVD on stainless steel substrates and capped with a transparent conducting oxide. The nc-Si/a-Si:H mixture is believed to be ~ 50% crystalline, and contain crystallites averaging 20 nm in grain diameter.

A Coherent Inova 300 cw argon laser, and a Nicolet MagnaIR 860 spectrometer interfaced with Thermo Fisher Omnic software, produced the PL spectra. Filters transmitting in the spectral range of the sample emission reduced background noise and blocked reflected excitation radiation from the detector. Interference fringes, which arise from constructive and destructive effects in the nc-Si/a-Si:H layer, have been removed from all of the presented spectra as is customarily done in FTPL spectroscopy. An APD HC-2 helium compressor cryostat system was used to obtain the desired operating temperatures.

The temperature dependent experiments used 514.5 nm excitation, at 800 mW/cm^2, and the intensity dependent experiments used 488 nm excitation. Two intensity dependent experiments were performed at 18 K. In one case the power density was kept below 600 mW/cm^2, and in the other the power density was allowed to reach 1.2 W/cm^2. We normalized the two data sets by the luminescence detected at 240 mW/cm^2 to create a combined plot showing the PL dependence over the full intensity range investigated.

Isochronal annealing consisted of 20 minute exposures at the specified temperatures, with 20 minutes of ramp-up time, and several hours of cooling. A small passageway in the furnace wall permitted air flow at atmospheric pressure during annealing. The samples were successively annealed at 180°C, 260°C, 330°C, 400°C, 450°C, 500°C, and 550°C, and the luminescence was measured at 18 K after each anneal. SIMS characterization of several samples reveals the oxygen content in the nc-Si/a-Si:H layer before annealing and after a 350 °C exposure.

RESULTS

We attribute two strong PL bands in the unannealed nc-Si/a-Si:H mixture, centered at ~ 0.95 eV and 1.3 eV, to band tail-to-band tail transitions at nc-Si crystal surfaces and to band-tail to band-tail transitions in the bulk a-Si:H, respectively [1, 4].

The luminescence attributed to nc-Si surfaces (gain boundaries) varies with excitation intensity as $I_{nc-Si} \sim I_{ex}^{0.4}$ between 15 and 400 mW/cm^2, and saturates at power densities approaching 500 mW/cm^2. The a-Si:H luminescence increases with excitation intensity as I_{a-Si}

~ $I_{ex}^{0.8}$ between 15 mW/cm^2 and 1.2 W/cm^2 with no evidence of saturation. Figure 1 shows the normalized PL signal from a-Si:H and nc-Si phases as a function of the excitation intensity.

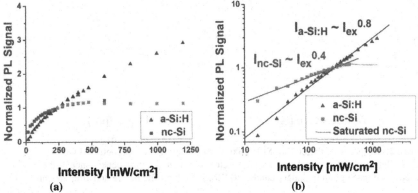

Figure 1: Photoluminescence excitation intensity dependence of the 1.3 eV and 0.95 eV bands **(a)** on linear axes **(b)** on log scale axes

Annealing this nc-Si/a-Si:H mixture at temperatures below the a-Si:H crystallization temperature (~ 600°C) produces a red-shift in the a-Si:H and nc-Si luminescence peaks. This behavior is shown in Figure 2. The a-Si:H luminescence peak shifts to 1.25 eV after 20 minutes at 260°C, and moves further to 1.15 eV after a subsequent 20 minutes at 330°C. After 400°C annealing, the luminescence from the a-Si:H regions is no longer observed, and annealing at 550°C causes all of the luminescence peaks to vanish.

After annealing at 260°C for 20 minutes, a soft shoulder becomes apparent on the low energy side of the nc-Si peak. This peak, which we label here a "defect" peak, becomes more pronounced after annealing at 400°C. Using an MCT detector, which does not distort the line-shape, this peak is centered at ~ 0.7 eV (Figure 2.b). We note that the defect peak holds its position from 18 to 130 K, while the 0.95 eV PL shifts noticeably over the same temperature range.

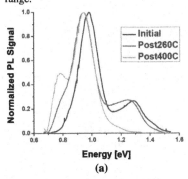

(a)

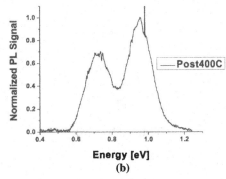

(b)

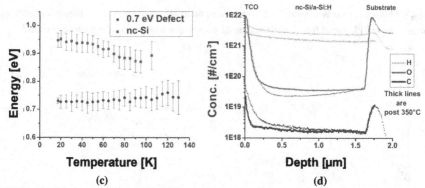

Figure 2: Response of the nc-Si/a-Si:H mixture to annealing – PL **(a)** detected with an InGaAs detector, **(b)** detected with an MCT detector, **(c)** temperature dependence of the "defect" and nc-Si grain boundary bands. **(d)** SIMS depth profile before and after annealing

The temperature dependence of the peak attributed to nc-Si grain boundary regions shows several differences from that attributed to the a-Si:H bulk regions (Figure 3). The nc-Si signal decays rapidly from 17 to 80 K, then tails off slowly as the temperature increases further with detectable luminescence at room temperature. The a-Si:H signal increases slightly with temperature initially, plateaus between 25 and 50 K, and decreases rapidly above 80 K. The a-Si:H band falls below our detection limit above about 230 K. Above 80 K, the luminescence signal attributed to nc-Si decays more slowly than that from the a-Si:H, and appears to follow a power-law decay with increasing temperature as $I_{nc-Si} \sim T^{-3}$. The a-Si:H band decays as $I_{a-Si:H} \sim T^{-6}$ above 100 K. The luminescence peaks both red-shift as temperature increases from 17 to 300 K (Figure 3.b), and in each case the shift exceeds that of the crystalline Si band gap and the mobility gap in a-Si:H [7, 8].

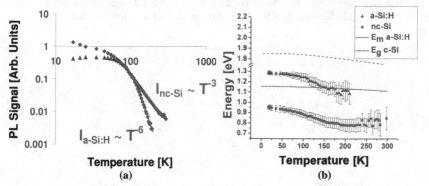

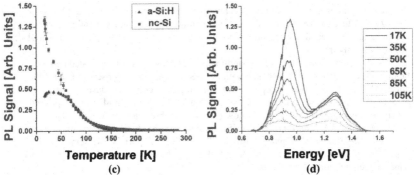

Figure 3: **(a)** PL intensity temperature dependence power laws, **(b)** red-shifts, **(c)** signal decay, and **(d)** line shapes from the nc-Si/a-Si:H mixture at various temperatures

DISCUSSION

To interpret the saturation of the 0.95 eV band which occurs for excitation intensities approaching 500 mW/cm^2, one must consider the limiting quantities which determine photoluminescence intensity. Here we present the following relationship [1]:

$$I_{PL} = \frac{N_{occ}}{\tau} \quad (1)$$

In equation 1, N_{occ} is the number of occupied states which give rise to radiative recombination, and τ is the recombination lifetime.

The number of nc-Si surface band-tail states which we believe give rise to the observed 0.95 eV luminescence is finite, limiting N_{occ}. When these band tail states fill, additional carriers are unable to thermalize into them and are more likely to find pathways for nonradiative recombination [1]. When the rate of carrier generation approaches the radiative recombination rate, the corresponding luminescence saturates. Amorphous silicon band tail-to-band tail luminescence saturates by this mechanism at excitation intensities higher than those we investigated [1]. A rough calculation, assuming the relative density of nc-Si surface band tail states scales directly with the volume fraction of the grain boundary regions in the material, suggests that nc-Si surface band tail states make up only ~ 1% of the total density of states. This limited number of states is consistent with the fact that the nc-Si surface band-tail to band-tail luminescence saturates at lower excitation intensity than the luminescence band in a-Si:H.

The 0.9 eV band attributed to band tail-to-band tail transitions at nc-Si grain boundaries has previously been observed in polycrystalline Si to follow a power law dependence as $I_{0.9 \, eV} \sim I_{ex}^{0.6}$ [2]. Our 0.9 eV band was less sensitive, varying with intensity only by $I_{0.9 \, eV} \sim I_{ex}^{0.4}$. The luminescence attributed to a-Si:H shows the strongest dependence on excitation intensity.

Transitions involving band tail states have a unique temperature dependent signature. As the temperature increases, carriers trapped in localized states are able to thermalize either to higher energy localized states or to the mobility edge. This enhanced diffusion allows carriers to find non-radiative recombination pathways more easily. In addition, the depopulation of shallow states reduces the number of transitions which give

7

rise to the higher energy part of the band tail-to-band tail luminescence band [2]. Since the 0.7 eV defect band does not shift with increasing temperature, this band is unlikely due to band-tail to band-tail transitions [4]. In fact, at least one of the carriers involved is probably deeply trapped.

The emergence of the defect band in all of the Si systems investigated appears to be thermally activated and associated with the c-Si phase. Tajima observed 0.7 eV PL in annealed CZ-grown c-Si and proposes that oxygen precipitates form during annealing and create oxygen donor states [6]. In the μc-Si/a-Si:H experiment [4] films deposited at substrate temperatures above 300°C show 0.7 eV PL, while films grown on cooler substrates do not. Furthermore, of the films grown at high substrate temperature, only those with significant μc-Si phase formation show 0.7 eV PL [4]. We observe the defect band in nc-Si/a-Si:H only after annealing, and higher temperature exposure causes this band to become more pronounced.

To monitor oxygen levels in our samples, we subjected several to SIMS analysis (Figure 2.d). The oxygen content in the middle of the nc-Si/a-Si:H layer, initially ~ 2 x 10^{19} /cm^3, almost doubles in response to a 350°C anneal. The SIMS depth profile shows that the oxygen concentration decreases moving deeper into the nc-Si/a-Si:H layer, which suggests that the transparent conducting oxide (TCO) provides additional oxygen to nc-Si/a-Si:H during annealing. However, it remains unclear whether additional oxygen from the TCO causes the formation the defect giving the observed 0.7 eV luminescence, or whether the nc-Si/a-Si:H mixture develops this defect state given thermal energy regardless of the presence of a TCO. Diffuse oxygen incorporated during deposition of the nc-Si/a-Si:H may be enough to form the oxygen agglomerate donor states described by Tajima [6]. Whatever the source of the oxygen in the nc-Si/a-Si:H mixture, the SIMS data supports the conclusion that the emergence of 0.7 eV luminescence in nc-Si/a-Si:H is thermally activated, and related to oxygen defects.

ACKNOWLEDGEMENTS

This research was partially supported by a DOE grant through United Solar Ovanic under the Solar America Initiative Program Contract, No. DE-FC36-07 GO 17053, by an NSF grant, DMR-0073004, and by an NSF cooperative agreement through the Renewable Energy MRSEC at Colorado School of Mines. Evans Analytical Group is recognized for performing the SIMS characterization.

REFERENCES

1. R. A. Street, *Hydrogenated Amorphous Silicon*, Cambridge University Press, (1991)
2. A.U. Savchouk, S. Ostapenko, G. Nowak, J. Lagowski, and L. Jastrzebski, *Appl. Phys. Lett.* 67 (1) (1995)
3. S.S. Ostapenko, A.U. Savchouk, G. Nowak, J. Lagowski, and L. Jastrzebski, *Mat. Sci. Forum* Vols. 196 – 201 (1995) pp 1897 - 1902
4. T. Merdzhanova, R. Carius, S. Klein, F. Finger, and D. Dimova-Malinovska, *Thin Solid Films*, 511 – 512 (2006) 394 – 398
5. T. Muschik and R. Schwarz, *J. Non-Cryst. Sol.* 164-166 (1993) 619
6. M. Tajima, *J. Cryst. Growth* 103 (1990) 1-7
7. J.I. Pankove, *Optical Processes in Semiconductors*, Dover Publications, Inc.(1971)
8. A. Yamaguchi, T. Tada, K. Murayama, and T. Ninomiya, *Solid State Communications*, Vol. 71, No. 4, pp 233 - 236 (1989)

Mater. Res. Soc. Symp. Proc. Vol. 1153 © 2009 Materials Research Society   1153-A02-02

Infrared photoconductivity in heavily nitrogen doped a-Si:H

David J. Shelton[1], James. C. Ginn[1], Kevin R. Coffey[2], and Glenn D. Boreman[1]

[1] CREOL, University of Central Florida, Orlando, FL 32816, USA
[2] AMPAC, University of Central Florida, Orlando, FL 32816, USA

ABSTRACT

High frequency steady-state photoconductivity in nitrogen doped hydrogenated amorphous silicon (a-Si:H-N) films has been demonstrated at infrared (IR) frequencies of 650 to 2000 cm^{-1}. This allows IR photoconductivity to be excited using a simple thermal source. In order to produce high frequency photoconductivity effects, the plasma frequency must be increased to the desired device operation frequency or higher as described by the Drude model. IR ellipsometry was used to measure the steady-state permittivity of the a-Si:H-N films as a function of pump illumination intensity. The largest permittivity change was found to be $\Delta\varepsilon_r = 2$ resulting from a photo-carrier concentration on the order of 10^{22} cm^{-3}. IR photoconductivity is shown to be limited by the effective electron mobility.

INTRODUCTION

Thin film systems with IR conductivity or permittivity that may be actively tuned with the application of a DC electric field, have been of interest for some time to IR designers. As an alternative, photoconductive devices have been proposed for active IR systems. The carrier concentration can be actively changed by illuminating a-Si:H with source energy above the band gap and thus out of the IR band. This illumination results in the generation of electron-hole pairs, and a sufficient density of these carriers will result in a change in the material's permittivity in the IR frequency range. Thus, by varying out-of-band pump power, an active IR system may be achieved.

Photoconductive elements have been used for optically generated grid arrays and as switches for reconfigurable antennas at 40 GHz [1]. In these low frequency designs high resistivity Si wafers have been used as the photoconducting elements. Due to the nanoscale size of IR systems patterned a-Si:H thin films must be used for photoconducting elements, and a higher carrier concentration is required for a contrast in permittivity. The generated electron-hole pairs form a pseudo-metallic plasma with behavior described by the Drude model. Eq. 1 gives the permittivity of the photoconductive semiconductor as the difference between the dark permittivity $\varepsilon_L(\omega)$ and a photo-plasma term

$$\varepsilon_r(\omega) = \varepsilon_L(\omega) - \frac{\omega_p^2}{\omega^2 - \frac{1}{\tau^2}} \times \left(1 + \frac{i}{\omega\tau}\right) \tag{1}$$

where ω_p is the plasma frequency, ω is the IR radiation frequency, and τ is the electronic relaxation time [2]. The plasma frequency depends upon the photo-carrier density in equation 2

$$\omega_p{}^2 = \frac{q^2}{\varepsilon_0} \times \frac{n_{ilum}}{m*} \tag{2}$$

where q is the charge on the electron, ε_0 is the permittivity of free space, and $m*$ is the effective mass of the photo-carrier, and n_{ilum} is the photo-carrier density as a function of power from the thermal pump source. n_{ilum} should be greater than 10^{20} cm^{-3} for significant IR photoconductivity to occur.

THEORY

The photo-carrier density, n_{ilum}, depends on the photon density, $G(P)$, and the recombination time, t, for electron hole hairs as shown in equation 3.

$$n_{ilum} = G(P) \times t \tag{3}$$

The recombination time t is the average time required for electron-hole pairs to recombine thus eliminating the photo-carrier. For intrinsic a-Si:H t is on the order of a microsecond which means that for significant IR photoconductivity to occur $G(P)$ needs to be on the order of 10^{26} s^{-1}cm^{-3}. By comparison a focused spot from a 100 W thermal source generates a $G(P)$ on the order of 10^{23} s^{-1}cm^{-3}. Larger $G(P)$ values may be obtained using a pulsed source such as a strobe light, but this would result in non-steady-state photoconductivity. In order to make IR photoconductivity accessible with a simple thermal source, a slow t is required on the order of millisecond or longer.

Figure 1 shows density of state functions versus energy for intrinsic and n type a-Si:H.

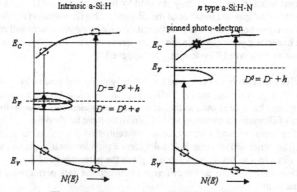

Figure 1. Density of states ($N(E)$) for a-Si:H.

In both cases E_C and E_V represent the energy of the mobility edge of the conduction and valence bands respectively while E_F is the Fermi energy. Near E_F in the forbidden gap of intrinsic a-Si:H are neutral dangling-bond states that may combine with electrons and holes to produce the charged dangling-bond defect reactions shown in the left half of Fig. 1. The production of charged dangling-bonds results in microsecond recombination times for intrinsic a-Si:H [3]. With the addition of n type donors in the right half of Fig. 1 the Fermi energy increases, and this results in charged dangling bonds which quickly recombine with hole states. This results in a defect reaction that creates a neutralized dangling bond. An excess concentration of electrons is now present at the conduction level, and thus the photo-electrons may be said to be 'pinned' and left without hole states to recombine with. Doping with P has been observed to retard the recombination process in a-Si:H resulting in a high steady-state concentration of photo-carriers [4-5]. In order to pin electron-hole pairs the added impurity needs to be chosen such that the Fermi level of a-Si:H is increased. Nitrogen impurities can be expected to have a similar effect on the electronic structure as other group V elements, but even longer recombination times have been observed than in the case of P doping. For nitrogen concentrations in the range of 10^{20} to 10^{21} cm^{-3} the recombination time was measured to slow to between 10 and 100 ms [6] – enough to produce steady-state IR photoconductivity from a thermal source. This a-Si:H alloy is referred to as a-Si:H-N.

EXPERIMENT

Ion-assisted electron-beam evaporation was selected as the method for depositing a-Si:H-N thin films. Although evaporation is a less common method for a-Si:H deposition, it has been shown to successfully yield dense a-Si:H films suitable for photovoltaic devices [7]. Nitrogen is also added to the a-Si:H films via the evaporator's ion source. Partial pressures of H$_2$ at a 25 sccm flow rate and a 99 % Ar balanced N$_2$ at 1 sccm were pumped into the chamber and ionized by a tungsten filament. The resulting plasma was maintained at a steady discharge current of 0.5 A. Simultaneously to the plasma formation Si was evaporated onto the sample at a rate of 0.5 nm/s.

The primary figure of merit for the a-Si:H-N films is the change in permittivity at IR wavelengths as measured by ellipsometry. In these experiments a J.A. Woollam Co. IR-VASE (variable angle spectroscopic ellipsometer) was used under standard *ex-situ* conditions. A test fixture was constructed to take ellipsometry data under illumination from a 12 V, 100 W quartz-halogen incandescent source. A high-pass filter is placed in front of the thermal source in the test fixture to block IR radiation so that the pump illumination is entirely out of band. The sample was constantly illuminated for 20 minutes during data collection.

The hydrogen concentration and Si-H bonding related microstructure may be evaluated using FTIR (fourier transform IR) spectroscopy. Nitrogen concentrations were measured directly using SIMS. The band gap energy, E_g, was measured using visible ellipsometry (J.A. Woollam Co. V-VASE), and E_g was found by fitting to a Lorentz-Tauc oscillator.

RESULTS AND DISCUSSION

Measured results are shown in Table 1 for three films selected because of their large photoconductive response. For the deposition parameters used the hydrogen concentration was in the desired range at around 10 at. %, and the nitrogen concentration was sufficient to slow the recombination time to around 100 ms based on comparison to reference 6. The dark (n_{dark}) and illuminated carrier concentrations (n_{ilum}) were based on comparing the generation rate and recombination time following equation 3, and then to DC secondary photoconductivity measurements. Due to the presence of localized states as shown in figure 1, the mobility is not constant for all states since carriers may exist on either side of the mobility edge. Equations 1-2 were used to determine an effective mobility for IR frequency transport by comparing the change in permittivity under illumination from the IR ellipsometry measurement to n_{ilum}. It should not be assumed that the effective mobility for IR frequency transport is the same as the DC transport mobility. Even for metal films it has been shown that there is an effective mobility change at IR frequencies [8]. The effective IR mobility is about an order of magnitude smaller than the typical mobility for pure a-Si:H. Measurements of the DC carrier transport in references 10 and 6 were used to find n_{dark} and n_{ilum}. DC carrier transport properties of a-Si:H-N were found to be similar to reference 6, and from comparison to reference 10, it can be seen that the carrier concentration is much higher than in pure a-Si:H.

Table I. Measured properties of a-Si:H-N with literature values for comparison. N_H and N_N are hydrogen and nitrogen concentrations, E_g is the band-gap energy, n_{dark} and n_{ilum} are carrier concentrations under illumination, and μ^* is the effective IR mobility.

Sample	N_H, at. %	N_N, at.%	Eg, eV	n_{dark}, cm^{-3}	n_{ilum}, cm^{-3}	μ^*, cm^2/V-s
hn02	20	2	1.6	5×10^{18}	2×10^{21}	0.76
hn03	6.6	1.2	1.4	4×10^{18}	5×10^{21}	0.56
hn12	28	2.8	1.62	2×10^{18}	2×10^{22}	0.18
a-Si:H [9]	10	0		2×10^{8}	2×10^{13}	
a-Si:H-N [6]	10	1.1		8×10^{18}	8×10^{21}	

Based on the high carrier concentrations and small effective mobility, equation 1 results in a nearly constant shift in permittivity. Due to the small mobility, the $1/\tau^2$ term is large compared to the frequency, so the change in permittivity is approximately equal to the squared product of the plasma frequency and relaxation time τ. Some spectral shifting of absorption features is also present as shown in Figs. 2-4. In figure 2 ellipsometry data of sample hn03 shows a change in permittivity of $\Delta\varepsilon_r = 1.3$ to 2 across much of the measured spectrum. This is due to increased loss from the photo-carriers which can be seen by comparing the imaginary permittivity functions. Sensitivity to some loss features such as the Si-H stretching mode at 630 cm-1 begin to be screened by photo-carriers and no longer appear in the spectrum. This screening effect also occurs in sample hn02 which has a similar change in permittivity as shown in figure 3.

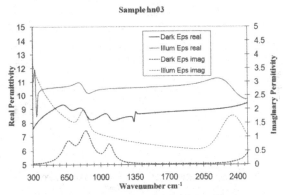

Figure 2. Real and imaginary portions of permittivity from ellipsometry measurements of sample hn03 in dark and illuminated state

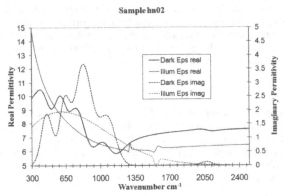

Figure 3. Real and imaginary portions of permittivity from ellipsometry measurements of sample hn02 in dark and illuminated state

Sample hn02 has both higher nitrogen and hydrogen concentrations than hn03, and as a result it also has stronger absorption features. This leads to a more pronounced photo-carrier screening effect. The change in permittivity under illumination was similar away from resonant features, and at frequencies greater than 1700 cm^{-1} $\Delta\varepsilon_r = 1.5$. The change in permittivity achieved in samples hn02 and hn03 are the highest achieved in these experiments. Figure 4 shows results from sample hn12 which are typical of most the films produced.

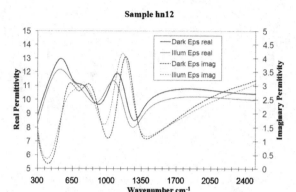

Figure 4. Real and imaginary portions of permittivity from ellipsometry measurements of sample hn12 in dark and illuminated state

The IR effective mobility of sample hn12 is smaller than hn02 and hn03 which results in a smaller change in permittivity where $\Delta\varepsilon_r = 0.9$ away from resonance. IR frequency electron mobility is consistently the limiting factor in the permittivity change under illumination. Some photo-carrier screening is still present in sample hn12 as can be seen by the shifting of features in the imaginary part of permittivity, but the effect is smaller due to the limited mobility in hn12 compared to hn02.

CONCLUSION

Retarded electron-hole recombination in a-Si:H-N leads to high carrier concentrations suitable to create a change in permittivity under illumination at IR frequencies. This permittivity change is limited by the effective electron mobility at IR frequencies and is approximately equal to the product of the plasma frequency and the relaxation time. The largest permittivity change was found to be $\Delta\varepsilon_r = 2$ and occurs over the spectral range from 650 to 2000 cm^{-1}.

REFERENCES

1. D.S. Lockyer, J.C. Vardaxoglou, *IEEE Trans. MTT* **47**, 1391, (1999)
2. C.H. Lee, P.S. Mak, A.P. Dephonzo, *IEEE J. Quantum Electron.* **16**, 277 (1980)
3. F. Vaillant, D. Jousse, *Phys. Rev. B* **34**, 4088 (1986)
4. R.A. Street, J. Zesch, M.J. Thompson, *Appl. Phys. Lett.* **43**, 672 (1983)
5. R.A. Street, D.K. Biegelsen, R.L. Weisfield, *Physical Review B* **30**, 5861 (1984)
6. A. Morimoto, M. Matsumoto, M. Yoshita, M. Kumeda, T. Shimizu, *Applied Physics Letters* **59**, 2130 (1991)
7. H. Rinnert, M. Vergnat, G. Marchal, *Journal of Applied Physics* **83**, 1103 (1998)
8. S.R. Nagel, S.E. Schnatterly, *Physical Review B* **9**, 1299 (1974)
9. W. Beyer, B. Hoheisel, *Solid State Communications* **47**, 573 (1983)

Mater. Res. Soc. Symp. Proc. Vol. 1153 © 2009 Materials Research Society   1153-A02-03

Decomposition of Mixed Phase Silicon Raman Spectra

M. Ledinský, J. Stuchlík, A. Vetushka, A. Fejfar and J. Kočka

Institute of Physics, Academy of Sciences of the Czech Republic
Cukrovarnická 10, 162 53 Prague 6, Czech Republic

ABSTRACT

Series of Raman spectra were measured for microcrystalline silicon thin film with variable crystallinity. Five sets of Raman spectra (corresponding to excitations at 325 nm, 442 nm, 514.5 nm, 632.8 nm and 785 nm wavelengths) were subjected to factor analysis which showed that each set of spectra consisted of just two independent spectral components. Decomposition of the measured Raman spectra into the amorphous and the microcrystalline components is illustrated for 514.5 nm and 632.8 nm excitations. Effect of the light scattering on absolute intensity of Raman spectra was identified even for excitation wavelength highly absorbed in the mixed phase silicon layers.

INTRODUCTION

Crystalline volume fraction is an important structural property of thin silicon films for solar cells, which influences absorption in the active layer, open circuit voltage V_{OC} and therefore the final efficiency of the photovoltaic cells. Crystallinity is usually determined from the Raman scattering spectra which consist of a broad amorphous band centered at 480 cm^{-1} and a sharp microcrystalline peak situated around 520 cm^{-1}. Evaluation of crystallinity [1] requires a procedure for decomposition of the spectra into the amorphous and microcrystalline components.

In the article [2] we used factor analysis to demonstrate the presence of just two spectrally independent components in mixed phase silicon Raman scattering spectra for excitations 325 nm, 514.5 nm, 632.8 nm and 785 nm. We have also shown that the most frequently used decomposition procedure of Raman spectra into three Gaussian bands has no physical foundation and may lead to inaccurate results. In agreement with [3] we have concluded that the mixed phase silicon Raman spectra have to be decomposed directly into contributions of completely amorphous and fully microcrystalline phases.

In this paper we extend the number of analyzed excitation sources by HeCd 442 nm laser and we describe in detail the procedure of decomposition of the measured Raman spectra based on factor analysis.

EXPERIMENT

All Raman spectra used in this work were measured on a 1 μm thick microcrystalline layer prepared by plasma enhanced chemical vapor deposition. A 4 mm wide permanent magnet placed under the substrate [4, 5] locally changed the nearby plasma and thus influenced the nucleation density and crystallinity of the layer. Raman scattering spectra were measured at 13 predefined positions where the film structure changed from amorphous to microcrystalline. The spectra were recorded using excitation wavelengths 325 nm, 442 nm, 514.5 nm, 632.8 nm and 785 nm. Detailed description of the film deposition conditions and Raman spectra measurements are given in [4, 5].

RESULTS and DISCUSSION

Figure 1. shows a representative set of Raman spectra measured on the sample with changing crystallinity. Baseline correction was applied to all measured spectra and afterwards they were rescaled to the same intensity at 470 cm^{-1}.

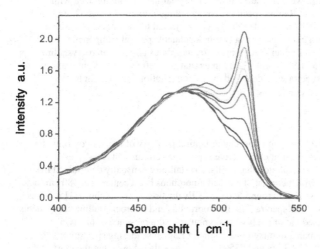

Figure 1. Raman spectra of microcrystalline silicon measured on the sample with changing crystallinity, excited by Ar^{+} ion laser at 514.5 nm. All spectra were baseline-corrected and rescaled (in order to show that all spectra have the same shape in the spectral range below 470 cm^{-1}).

In this region the contribution of microcrystalline phase is negligible and the coincidence of all spectra shows that the amorphous component in all spectra has the same spectral shape. For our special sample we can expect that the properties of amorphous and microcrystalline phases remain unchanged and hence also the shapes of amorphous and microcrystalline bands are invariant with the crystallinity.

This assumption was proved by factor analysis. All five spectra sets (for each of the excitation wavelengths) were subjected to the factor analysis. The left part of the figure 2. shows the Cattel scree plot for all five data sets. The dependence shows that the spectral dimension is 2, including the new data set excited by 442 nm laser. In other words, each set of spectra may by expressed as a linear combination of just two subspectra, all other subspectra represent the noise signal only. The same fact is illustrated in the right side of the figure 2. which shows three subspectra with the highest singular values. The first two spectral components contain spectroscopic information, while the third subspectrum represents just the noise of the measurement.

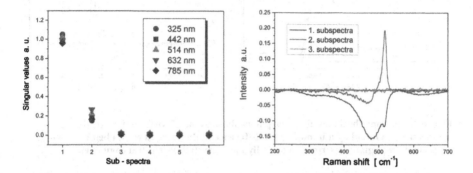

Figure 2. Left side: Cattel scree plot, i.e., the dependence of the singular values W$_j$ on the number of subspectra for five sets of the spectra measured with different excitation wavelengths. Right side: first three subspetra with the highest singular values illustrating the fact that the spectral dimension equals 2 for all five data sets.

Due to the mathematical background of factor analysis, the subspectra have no direct physical meaning (both subspectra have negative values at some spectral regions, which is not possible in case of Raman spectra). The first step of the factor analysis procedure, see [2] for details, is looking for the spectral shape which is the best representation of all the measured spectra, therefore the first subspectrum corresponds to the spectra average. The second subspectrum describes differences of the measured spectra and the first subspectrum and its shape is strongly connected with the change of the crystallinity.

We already noted that the spectral dimension of all data sets is 2. Let us now address the question whether the Raman spectra of fully amorphous and fully microcrystalline silicon can be used as a suitable spectral base. In our case we can directly measure the spectra of fully amorphous silicon. Unfortunatelly, the Raman spectra of microcrystalline films always contain some contribution of amorphous component and so it is not possible to obtain the Raman spectrum of fully microcrystalline component by direct measurement. Instead we have subtracted the amorphous Raman spectrum from the 1st and 2nd subspectra. Subtraction was made manually by adjusting the intensity of the amorphous spectrum to minimize the signal in the region below 470 cm^{-1}. The results of this procedure are shown in figure 3. for the 1st and 2nd subspectra for excitation at 514.5 nm.

17

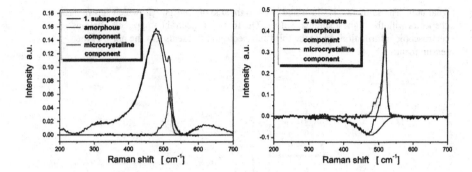

Figure 3. The contribution of the amorphous phase to the 1st (left) and 2nd (right) subspectra was found by minimizing the difference in the spectral region below 470 cm^{-1}. The remaining signal is ascribed to the fully microcrystalline spectral component.

In this way we obtained two independent Raman components for all five data sets. These components are shown in the figure 4. for excitations 514.5 nm and 632.8 nm. The spectra obtained from the 1st and 2nd subspectra are almost identical, therefore we can conclude that these spectra correspond to the Raman spectrum of fully microcrystalline phase. Moreover, this confirms that the Raman spectra of amorphous and microcrystalline phases are suitable spectral base, i.e., all measured spectra may be decomposed into amorphous and microcrystalline bands.

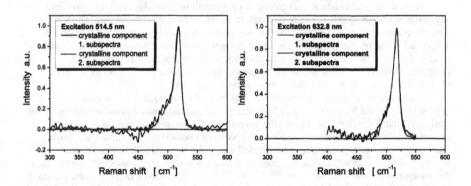

Figure 4. Comparison of the spectral shape of the fully microcrystalline bands evaluated independently from 1st and 2nd subspectra for 514.5 nm (left) and 632.8 nm (right) excitation wavelengths.

In order to evaluate the crystallinity, it is necessary to integrate the intensities of the amorphous and microcrystalline components in the decomposed Raman spectra. Integral intensities of amorphous and microcrystalline components for excitations 514.5 nm and 632.8 nm are plotted in figure 5. The x axes show a distance from the magnet edge which describes the effect of the magnetic field on the discharge plasma. With increasing distance the crystallinity of the layer is decreasing [4, 5].

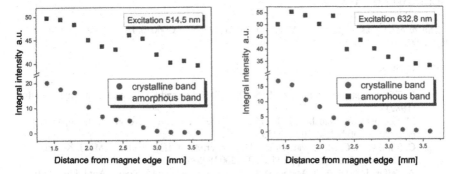

Figure 5. Integrated intensities of the microcrystalline and amorphous components in Raman spectra measured at 514.5 nm and 632.8 nm as a function of the distance from magnet edge (decreasing crystallinity). All values were calculated using the decomposition procedure described above.

Integrated intensity of the microcrystalline component is decreasing with decreasing crystallinity. On the other hand integrated intensity of the amorphous component has the same trend, in spite of the fact that the active volume of the amorphous phase is increasing with the decreasing crystallinity. In [2] we observed the same tendency of integral intensities for 785 nm excitation and we interpreted it as the effect of light scattering on the microcrystalline grains and corresponding light trapping in the silicon film. In case of 514.5 nm and 632.8 nm the absorption depth is smaller than the thickness of the layer and the light trapping does not play such an important role. This effect is for 514.5 nm and 632.8 nm excitation wavelengths given mainly by the light scattering on the microcrystalline grains. The scattered excitation light penetrates into the film under effective angle. Therefore the back scattered Raman photons cover a shorter distance in absorbing layer and total integrated Raman signal increases. Consequently the integrated Raman signal from amorphous band is decreasing with decreasing crystallinity (decreasing roughness and light scattering on the layer).

CONCLUSIONS

We have confirmed that the Raman spectra of mixed phase silicon consist of just two independent spectral components also for the 442 nm excitation. Moreover, we have demonstrated that the spectra corresponding to Raman scattering of fully amorphous and fully microcrystalline phases may be used as suitable spectral base for decomposing Raman spectra of mixed phase silicon. The observed counterintuitive increase of the integral intensity of the amorphous component in Raman spectra with increasing crystallinity was interpreted as an effect of light scattering on microcrystalline grains.

ACKNOWLEDGMENTS

This research was supported by AV0Z 10100521, LC510, LC06040, IAA100100902 and KAN400100701 projects.

REFERENCES

1. R. Tsu, J. Gonzalez-Hernandez, S. S. Chao, S. C. Lee and K. Tanaka: Appl. Phys. lett. 40 (1982) 534.
2. M. Ledinský, A. Vetushka, J. Stuchlík, T. Mates, A. Fejfar, J. Kočka and J. Štěpánek: J. Non-Cryst. solids 354 (2008) 2253.
3. C. Smit, R. A. C. M. M. van Swaaij, H. Donker, A. M. H. N. Petit, W. M. M. Kessels, M. C. M. van de Sanden, J. Appl. Phys. 94 (2003) 3582.
4. A. Fejfar, J. Stuchlík, T. Mates, M. Ledinský, S. Honda and J. Kočka: Appl. Phys. Lett. 87 (2005) 011901
5. M. Ledinský, L. Fekete, J. Stuchlík, T. Mates, A. Fejfar and J. Kočka: J. Non-Cryst. solids 352 (2006) 1209.

Light Trapping in Solar Cells I

Mater. Res. Soc. Symp. Proc. Vol. 1153 © 2009 Materials Research Society   1153-A03-02

Simulation of Plasmonic Crystal Enhancement of
Thin Film Solar Cell Absorption

R. Biswas[1,2], D. Zhou[2], L. Garcia[2]
[1]Department of Physics and Astronomy, Ames Laboratory, Iowa State University, Ames IA 50011
[2]Microelectronics Research Center and Department of Electrical and Computer Engineering, Iowa State University, Ames IA 50011

ABSTRACT

Light management and enhanced photon harvesting are critical areas for improving efficiency of thin film solar cells. Red and near infrared photons with energies just above the band edge have large absorption lengths in amorphous silicon and cannot be efficiently collected. We previously demonstrated that a photonic crystal back reflector involving a periodically patterned ZnO layer can enhance absorption of band edge photons. We propose and design alternative new plasmonic crystal structures that enhance absorption in thin film solar cell structures. These plasmonic crystals consist of a periodically patterned metal back reflector with a periodic array of holes An amorphous/nanocrystalline silicon layer resides on top of this plasmonic crystal followed by a standard anti-reflecting coating. We have found plasmonic crystal structures enhance average photon absorption by more than 10%, and by more than a factor of 10 at wavelengths just above the band edge, and should lead to improved cell efficiency. The plasmonic crystal diffracts band edge photons within the absorber layer, increasing their path length and dwell time. In addition there is concentration of light within the plasmonic crystal. Design simulations are performed with rigorous scattering matrix simulations where both polarizations of light are accounted for.

INTRODUCTION

Photovoltaics and solar cells have been an active area for research and development, driven by the world's constantly increasing demand for power. Amorphous silicon (a-Si:H) is among the most developed material for thin film solar cells.

Light trapping is the standard technique for improving the thin film solar cell efficiencies and harvesting the spectrum of incoming sunlight. The conventional light trapping schemes unitize a random textured Ag/ZnO back reflector that scatters light within the absorber layer and increases the optical path length of solar photons [1]. However, those metallic back reflectors of silver coated with ZnO, suffers from intrinsic losses from surface plasmon modes generated at the granular metal-dielectric interface [2]. Periodic metallic gratings were also used to improve absorption of polymer based thin film solar cells [3]. Recently, we developed a light trapping scheme for a-Si:H thin film solar cells, where the back reflector was replaced by two dimensional photonic crystal on top of distributed Bragg reflector (DBR) [4]. Photonic crystals have been a major scientific revolution in manipulating and guiding light in novel ways [5]. The advantage of photonic crystals is to introduce diffraction, where the photon momentum (**k**) can be scattered away from the specular direction with ($\mathbf{k}^{\|} = \mathbf{k}_i^{\|} + \mathbf{G}$), where **G** is a reciprocal lattice vector and $\mathbf{k}_i$ is the incident wave-vector. The photonic crystal diffracts photons through oblique angles in the absorber layer, thereby increasing the path length and dwell time of photons.

Here we discuss a different approach to improving light trapping using metallic photonic crystals (or plasmonic crystals) rather than the dielectric photonic crystal used previously. The dielectric DBR is replaced by flat silver mirror to reflect light specularly. The plasmonic crystal resides on this mirror. The plasmonic crystals can both diffract photons within the absorber layer and concentrate light to high intensities in regions of the cell. The diffraction mechanism of photonic crystals still applies here.

DESIGN

The typical thickness of a-Si:H solar cells is 250-500 nm and is limited by the minority carrier diffusion length. The photon absorption length (L_d) of a-Si:H with bandgap (E_g) of 1.6 eV is shown in Fig. 1 [6]. For wavelengths $\lambda > 600$ nm, the absorption length exceeds 0.5 μm and approaches 100 μm near the band edge ($\lambda_g = 775$ nm). It is extremely difficult to harvest these photons with a 500 nm absorber layer. Harvesting of the long wavelength photons is critical for improving short circuit currents (J_{sc}) and cell efficiencies.

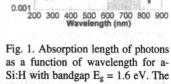

Fig. 1. Absorption length of photons as a function of wavelength for a-Si:H with bandgap $E_g = 1.6$ eV. The band edge wavelength is indicated by the arrow.

In the plasmonic crystal enhanced solar configuration (Fig. 2), we have 1) a top indium tin oxide (ITO) layer serving as antireflective coating and top contact (thickness d_0), 2) the absorber layer (thickness d_1), 3) the back reflector with silver plasmonic crystal structures. A thin layer of ZnO is deposited on silver conformally. We discuss two different plasmonic back reflector structures here:

i) The first structure consists of an Ag back reflector that has been patterned with a periodic array of holes. The depth of the holes is d_2 and they are filled with a-Si:H. The array of holes forms triangular lattice two dimensional metallic photonic crystal with lattice constant of 'a' (insert of Fig 2a).

ii) The second structure consists of conical protrusions of Ag on a base planar layer of Ag. In simulations the cones have flat tops. The height of the cones is d_2 and the space between the cones are filled with a-Si:H. The array of cones forms triangular lattice two dimensional metallic photonic crystal with lattice constant of a (Fig 2b).

The deposition of thin layer of ZnO between amorphous silicon and silver is to prevent diffusion of amorphous silicon into silver and make the interface smoother with less defects. Since we are considering a very thin ZnO layer (with thickness much smaller than the a-Si:H absorber layer or the wavelength of light), as a first approximation we do not include the ZnO layer in the scattering matrix simulations.

a)   b)

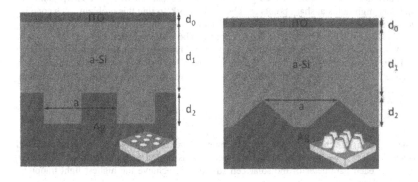

Fig 2. a) Schematics of silver back reflector with periodic hole array. b) Schematics of silver back reflector with periodic conical protrusions.

SIMULATION METHODS:

We simulate solar cell structures with a rigorous scattering matrix (S-matrix) method [7, 8], where Maxwell's equations are solved in Fourier space and the electric/magnetic fields are expanded in Bloch waves. The structure is divided into slices along z direction. In each slice, the dielectric function $\varepsilon(r)$ is a periodic function of x, y only and independent of z. Hence the dielectric function and its inverse are a Fourier expansion with coefficients $\varepsilon(G)$ or $\varepsilon^{-1}(G)$. A transfer matrix M in each layer can be calculated and diagonalized to obtain the eigenmodes within each layer for both polarizations [7, 8]. The continuity of the parallel components of **E** and **H** at each interface leads to the scattering matrices S_i of each layer from which we obtain the scattering matrix S for the entire structure. Using the S-matrix, we can simulate the reflection, transmission and absorption [8] for the whole structure. Since the solutions of Maxwell's equations are independent for each frequency, the computational algorithm has been parallelized where each frequency is simulated on a separate processor.

In the individual layers, realistic frequency dependent dielectric functions are used to include absorption and dispersion. We ignored the absorption and dispersion in ITO (that are appreciable below 400 nm) and assume a refractive index of 1.95. For an a-Si:H absorber with bandgap of 1.6 eV, we used the frequency dependent dielectric functions determined from spectroscopic ellipsometry for a-Si:H and analytically continued to the infrared by Ferlauto et al [6]. We used the experimental frequency dependent dielectric functions for Ag to account for absorption and dispersion.

In the case of periodic hole array, the division of layers is straight forward. However, the division with no dielectric constant variation along z direction can not be naturally done on the cone shaped gratings with flat tops. We avoid the sharp point so that very high fields at sharp points are absent. To work around this problem, each cone is approximated with a stack of 6 cylindrical disks with the same height and decreasing radii. With sufficiently large number of disks, a cone can be well simulated.

We used an absorber layer thickness of 500 nm, typical for single junction p-i-n solar cells. In this paper we ignore the absorption in the p-layer that typically reduces the blue response. The calculated total absorption in the i-layer is weighted by the AM 1.5 solar spectrum [9] and integrated from 280 nm (λ_{min}) to 775 nm (λ_g) to obtain the average absorption <A>

$$< A >= \int_{\lambda_{min}}^{\lambda_g} A(\lambda)\frac{dI}{d\lambda}d\lambda \qquad (1)$$

where dI/dλ is the incident solar radiation intensity per unit wavelength. We use average absorption <A> weighted by the solar spectrum as a figure of merit to systemically optimize each parameter of the solar cell structure to achieve the highest light trapping enhancement.

RESULTS

The thickness of the ITO layer is assumed to be 65 nm from previous simulations [5]. By systematically varying parameters of the plasmonic crystal, we can design back reflectors to maximize the average absorption of the solar cells. We started with back reflector with an array of holes and used R/a = 0.25 and hole depth d_2 of 200 nm to explore the dependence of average absorption on the lattice constant (Fig. 3a). The average absorption has strong dependence on the lattice constant. For a=700-800 nm, the average absorption is maximized. With lattice constant of 700 nm, the hole depth and R/a ratio are varied with the other parameters fixed. The average absorption variation (Fig. 3b) is optimized for a hole radius R/a=0.25 and a hole depth near 250 nm.

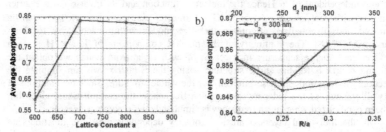

Fig. 3. a) The variation of average absorption with lattice constant 'a' for metallic hole array with depth of 200 nm and R/a = 0.25. b) The variation of average absorption with hole diameter for hole depth of 300 nm and the variation with hole depth for R/a = 0.25. The lattice constant is fixed at 700 nm.

The design parameters of the back reflector with periodic cone protrusions (Fig. 2b) can be optimized in the similar fashion. It is found that the best absorption enhancement can be achieved with cone protrusions nearly touching each other (R/a ~ 0.5). The absorption of the solar cells with plasmonic crystal structures are compared with solar cell with same absorber and flat silver reflector (Fig. 4). Most of the enhanced absorption occurs near the band edge (600 – 775 nm), where photons have long absorption lengths. The plasmonic crystal generates modes of diffraction at these wavelengths, effectively increasing the path length or dwell time. Below 600 nm, the photonic crystal has little effect, since photons have absorption lengths smaller than the film thickness and are effectively absorbed within the a-Si:H absorber layer, without reaching the back surface. The fall-off in the absorption at short wavelengths is due to the anti-reflection layer being optimized for the green region of the spectrum.

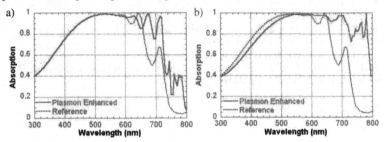

Fig 4. a) The absorption enhancement with a metallic hole array with respect to the reference cell with a flat silver reflector and anti-reflective coating. b) The absorption enhancement with metallic conical array with respect to the reference cell with a flat silver reflector and anti-reflective coating.

In experimental solar cells high absorption of the p-layer also decreases the absorption of blue photons. It is well known that the patterned back reflector will lead to conformal lattices in the top layer of the solar cell including at the a-Si:H/ITO interface. Preliminary calculations indicate that conformal patterns decrease the absorption enhancement.

CONCLUSIONS

Metallic photonic crystal back reflectors generate significant increase of absorption and succeed in harvesting red and near infrared photons in amorphous silicon solar cells.

ACKNOWLEDGEMENTS

We thank Vikram Dalal for many suggestions. We acknowledge support from the National Science Foundation under grant ECS-06013177 at Iowa State University and the Iowa Powerfund. The Ames Laboratory is operated for the Department of Energy by Iowa State University under contract No. DE-AC0207CH11385.

REFERENCES

[1] B. Yan, J. M. Owens, C. Jiang, J. Yang and S. Guha, Mater. Res. Soc. Symp. Proc. **862**, A23.3.1 (2005).

[2] J. Springer, A. Poruba, L. Mullerova, M. Vanecek, O. Kluth and B. Rech, J. Appl. Phys. **95**, 1427 (2004).

[3] K. Tvingstedt, N-K. Persson, O. Inganas, A Rahachou and I. V. Zozoulenko, Appl. Phys. Lett. **91**, 113514 (2007).

[4] D. Zhou and R. Biswas, J. Appl. Phys. **103**, 093102 (2008).

[5] J. D. Joannopoulos, R. D. Meade and J. N. Winn, Photonic Crystals, Princeton, NJ: Princeton University Press, 1995.

[6] A. S. Ferlauto, G. M. Ferreira, J. M. Pearce, C. R. Wronski, R. W. Collins, X. Deng and G. Ganguly, J. Appl. Phys. **92**, 2424 (2002).

[7] R. Biswas, C.G. Ding, I. Puscasu, M. Pralle, M. McNeal, J. Daly, A. Greenwald and E. Johnson, Phys. Rev. B. **74**, 045107 (2006).

[8] Z. Y. Li and L. L. Lin, Phys. Rev. E. **67**, 046607 (2003).

[9] ASTMG173-03, Standard Tables for Reference Solar Spectral Irradiances, West Conshohocken, PA: ASTM International, 2005.

Mater. Res. Soc. Symp. Proc. Vol. 1153 © 2009 Materials Research Society   1153-A03-03

Light Trapping in Thin-Film μc-Si:H Solar Cells Using Self-Ordered 2D Grating Reflector

Hitoshi Sai[1], Yoshiaki Kanamori[2], and Michio Kondo[1]
[1]Research Center for Photovoltaics (RCPV), National Institute of Advanced Industrial Science and Technology (AIST), Central 2, 1-1-1 Umezono, Tsukuba, Ibaraki, 305-8568, Japan
[2] Graduate School of Engineering, Tohoku University, Aoba 6-6-01, Aramaki, Aobo-ku, Sendai, 980-8579, Japan

ABSTRACT

Effect of back reflectors on light trapping in μc-Si:H cells has been investigated with self-ordered Al substrates obtained by anodic oxidation. With increasing the period of the patterned substrates from 0 to 1.1 μm, 1-μm-thick μc-Si:H cells on the patterned substrates have shown a significant enhancement of spectral response in the near infrared region, giving an increment of the short circuit current density from 18 to 24 mA/cm^2. This enhanced light trapping effect are attributed to the improved reflectivity of the rear side and effective light scattering at the front side, as well as light scattering at the rear side.

INTRODUCTION

Thin-film Si solar cells with tandem structures using hydrogenated amorphous Si (a-Si:H) and hydrogenated microcrystalline Si (μc-Si:H) are promising candidates for future low-cost and high-efficiency solar cells [1,2]. In the high-efficiency tandem cells, μc-Si:H plays a key role as an absorber material in the bottom or middle cell. For μc-Si:H cells, light trapping is crucial to obtain a high current density, as the absorption coefficient of μc-Si:H is small in the near infrared (NIR) region. Theoretically, by ideal light trapping, the optical path length in a material with a refractive index of n is enhanced by a factor of $4n^2$ [3], and the amount of photons absorbed in the material can be increased drastically.

To date, various light trapping structures have been researched and developed by many researchers to enhance light trapping in thin-film Si solar cells [4-17]. Randomly textured substrates have been realized by using SnO$_2$ [5,6], ZnO [7-9,14,15], glass [10], Ag [11], and plastic films [12]. In fact, they can improve the spectral response in longer wavelengths, and some of them have been already applied in commercial solar cells. When incorporated into μc-Si:H solar cells, such textured surfaces significantly increase the optical path length of the incident light, due to (i) light scattering on textured surface at oblique angles and (ii) enhanced total internal reflection at the interface within the cells. In addition to random textures, periodically textured substrates like surface gratings have also been investigated for light trapping in thin-film Si solar cells [13-17]. Calculated results have shown that a significant enhancement of the photon absorption in thin-film Si cells is expected by applying optically functional periodic structures [13-17]. Actually, the effectiveness of surface gratings on light trapping has been demonstrated in μc-Si:H cells [14,15], although the current gain by the gratings has not exceed the gain obtained by random textures.

In general, μc-Si:H cells require a light trapping structure with a larger feature size compared with a-Si:H cells, because μc-Si:H can absorb photons with longer wavelengths than a-Si:H. In superstrate-type *p-i-n* μc-Si:H cells, it has been reported that textured ZnO films with crater-like textures with a typical diameter of 0.5 – 2 μm, which are fabricated by sputtering and post wet etching, show a better light trapping effect [7]. On the other hand, in substrate-type n-i-p μc-Si:H cells, textured SnO₂ with a typical lateral size of 0.2 – 0.3 μm show a significant increase in the spectral response in the NIR region [6]. As another approach, textured ZnO films with a grain size of 1.3 μm, which was fabricated by electroplating, has been applied for n-i-p μc-Si:H cells and given a high NIR response [9].

As mentioned above, light trapping in μc-Si:H solar cells has been extensively researched by many groups. Nevertheless, the best light-trapping structure for thin-film μc-Si:H cells is still ambiguous. One big problem is that, in the previous reports, each textured substrate is fabricated by totally different approach including chemical vapor deposition (CVD), sputtering, wet etching, electroplating and reactive ion etching. This means that the morphology and the feature size of these textured substrates are quite different. Therefore, it is quite difficult to systematically compare the light-trapping effect of the textured substrates previously reported, and to deduce the optimized parameter for realizing better light trapping.

Recently, we have reported unique periodically textured back surface reflectors (BSR) fabricated by using anodic oxidation of Al [18]. In that paper, we have shown that BSR with an appropriate dimple size of ~ 0.9 μm significantly improves the spectral response of substrate-type μc-Si:H cells in the NIR region. An advantage of our approach is the availability of systematic control of the dimple size, namely, period, by simply changing the applied voltage and electrolytes during anodic oxidation [19-21]. Using this approach, self-ordered dimple patterns with a wide range of periods can be prepared on Al substrates with high uniformity as well as large substrate sizes. In this paper, we report the structural and optical properties of patterned Al substrates with a wide range of periods, and their effects on the light trapping in substrate-type n-i-p μc-Si:H single cells. The relationship between the period of substrates and the light trapping effect in μc-Si:H cells is presented, and the optimum size is determined. In addition, it is shown that angular reflection property of substrates is well correlated with the light trapping effect in the cells. Light trapping effect by natural textures grown during μc-Si:H deposition is also discussed.

Generally, efficiency of solar cells is given by multiplying the open circuit voltage V_{OC}, the short circuit current density J_{SC}, and the fill factor FF. To boost the cell efficiency, all the parameters should be improved. However, light trapping textures for increasing J_{SC} often induce the defects in μc-Si:H layers and deteriorate V_{OC} and FF of the cells [6]. In fact, we have experienced a decrease in V_{OC} and FF by applying textured substrates. However, the purpose of this work is to examine the light trapping effect of textured substrates. Therefore, in this paper, we will focus only on J_{SC} and spectral response of solar cells, and not discuss the other cell parameters. The other cell parameters can be referred in our previous paper, and in some cases, our patterned substrates give a rise of efficiency as well as J_{SC}, compared with conventional textured substrates [18].

OPTICAL SIMULATION

According to grating theories, a surface grating structure with a period of $\Lambda > 0.3$ μm is required to generate diffracted light inside μc-Si:H in the NIR region. This means that BSRs with a grating structure with $\Lambda > 0.3$ μm can reflect the incident light obliquely, resulting in a longer optical path length in the μc-Si:H cells. To estimate the diffraction behavior of BSRs with grating structures quantitatively, we have performed numerical simulations based on rigorous coupled-wave analysis (RCWA) [22] using the optical constants of Ag [23] and crystalline Si [24].

Figure 1 shows diffraction behaviors of Ag BSRs with grating structures embedded in Si as schematically shown in the inset. In Figs. 1 (a) – (c), backward diffraction intensities of two-dimensional (2D) Ag gratings with three different periods, Λ = 0.3, 0.6 and 0.9 µm, are plotted as functions of wavelength and reflection angle. For simplicity, here we assume that the gratings have a 2D symmetrical pyramid structure with a constant aspect ratio, h/Λ = 0.25, where h denotes the grating height. In these figures, the gray scale indicates the light intensity reflected at each angle. The pairs of integers in the brackets denote the corresponding diffraction orders for 2D gratings, (m, n), where m and n denote the diffraction orders for the x and y-directions. Each line in these figures is corresponding to each diffraction order in this system. Note that the specular reflection component is denoted by (0, 0). It can be seen in Figs. 1 (a) – (c) that the shortest wavelength where non-zero diffractions occur is shifted to longer wavelengths with increasing Λ. It is confirmed that a period of $\Lambda > 0.3$ µm is required to generate diffracted light inside Si in the NIR region, as expected from grating theories. It is also found that the number of diffraction lines increases with increasing Λ. This result indicates that a BSR with longer Λ can effectively diffract incident light at larger angles not only for normal incidence but also for obliquely incident light. In addition, by increasing Λ, the intensity of the specular reflection is

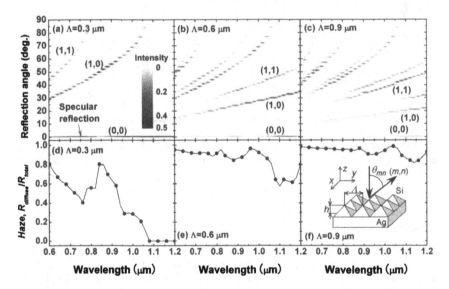

FIG. 1. (a) – (c) calculated diffraction behaviors and (d) – (f) the corresponding spectral hazes of 2D pyramidal Ag gratings embedded in Si as functions of wavelengths for Λ = 0.3, 0.6 and 0.9 µm. The pairs of integers in the brackets denote the diffraction order (m, n). In this calculation, a constant aspect ratio of h/Λ = 0.25 and normal incidence are assumed.

substantially suppressed, which is crucial for better light trapping. The reduction of specular reflection by increasing Λ is more clearly found in Figs. 1 (d) – (f). In these figures, haze ratios $R_{diffuse}/R_{total}$, where $R_{diffuse}$ and R_{total} denotes diffuse reflectivity and total reflectivity, are plotted as functions of wavelengths. In this calculation, $R_{diffuse}$ was obtained by summing all non-zero diffraction components. It is found from these figures that the haze ratio in the NIR region can be effectively increased by increasing Λ. We can expect that more than 90 % of the incident light is reflected diffusively in Si by lengthening Λ.

Figure 2 shows that the angular distribution of reflectivity of 2D Ag gratings with Λ = 0.3, 0.6, and 0.9 μm for two different NIR wavelengths, (a) λ = 0.8 μm and (b) λ = 1.2 μm. As can be seen in Fig. 2, by increasing Λ from 0.3 to 0.9 μm, reflectivity at the normal or small angles is suppressed to less than 10% together with the enhancement of reflectivity at larger angles. From the results shown in Figs. 1 and 2, it is confirmed that gratings with $\Lambda \sim 0.9$ μm are preferable for light trapping in thin-film Si solar cells.

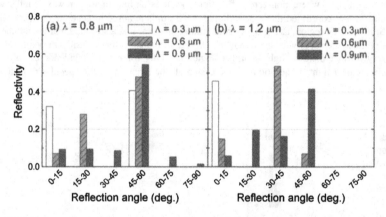

FIG. 2. Angular reflectivity of 2D Ag gratings embedded in Si with Λ = 0.3, 0.6 and 0.9 μm at the wavelengths of (a) λ = 0.8 μm and (b) λ = 1.2 μm. The parameters used in this calculation are the same as in Fig. 1.

EXPERIMENT

Substrate fabrication

BSRs with a 2D periodic dimple pattern were fabricated by the following procedure. Electrochemically polished Al sheets are first anodically oxidized at a constant voltage in an electrolyte solution [25-27]. By the anodization, an Al_2O_3 film with pore arrays is formed on the Al surfaces. During the oxidation, a self-ordered dimple pattern is simultaneously formed on the Al surface beneath the Al_2O_3 film. The Al_2O_3 film is easily etched off with a mixture of chromium and phosphoric acid solutions, leaving a bare patterned Al substrate.

A unique feature of self-ordered porous alumna films is that the period, Λ, is roughly proportional to the applied voltage during the anodic oxidation. To control Λ from 0.1 to over 1 µm, anodic oxidations were performed using three different acid solutions, namely, oxalic acid ((COOH)$_2$), phosphoric acid (H$_3$PO$_4$), and citric acid (C$_6$H$_8$O$_7$) under the different applied voltages from 40 to 450 V.

Solar cell fabrication

Substrate-type n-i-p µc-Si:H cells with an active area of 1 cm^2 were fabricated on the patterned Al substrates coated with a Ag/ZnO stacked film, by conventional plasma-enhanced chemical vapor deposition (PECVD). The structure of the solar cell consists of Al substrate/Ag(200 nm)/ZnO(40 nm)/µc-Si:H n-i-p layers/ In$_2$O$_3$:Sn (ITO, 75 nm)/Ag grid. The µc-Si:H i-layers with a thickness of t_i = 1 µm were deposited on the substrates using a SiH$_4$ and H$_2$ gas mixture (SiH$_4$/H$_2$ = 10.5/380 sccm) at a substrate temperature of 170°C, a pressure of 1.5 torr, and a RF power density of 0.04 W/cm^2. For comparison, n-i-p µc-Si:H cells were also fabricated on flat-glass/Ag/ZnO and textured-SnO$_2$(Asahi-U)/Ag/ZnO substrates [7] in the same manner.

Characterization

Surface morphology of Al substrates and solar cells was observed with a scanning electron microscope (SEM) and an atomic force microscope (AFM). Total and diffuse reflectivity of the samples was measured by using a spectrometer (PerkinElmer, Lambda950) with an integral sphere (Labsphere, 150mm RSA ASSY). Angle-dependent reflectivity of the substrates was characterized by using a photodiode which can rotate around the samples illuminated by a laser beam. The performance of the cells was evaluated by measuring current-voltage characteristics and quantum efficiency spectra under AM 1.5G 100 mW/cm^2 illumination.

RESULTS AND DISCUSSION

Structural and optical properties of patterned substrates

Figure 3 shows SEM images of Al surfaces obtained after the anodic oxidation process at (a) 40 V, (b) 120 V and (c) 370 V. Also, Fig. 3 (d) shows a SEM image of a commercially available Asahi-U type textured SnO$_2$ substrate, which is used as a reference textured substrate in this study. It can be seen in Figs. 3 (a) – (c) that self-ordered dimple patterns with different periods are uniformly formed on Al substrates. The patterns are almost arranged in a hexagonal array, and each dimple has a smooth concave shape. The obtained average periods for Figs. 3(a) – (c) are Λ ~ 0.1, 0.3, and 0.9 µm, respectively. The aspect ratio of these dimples, namely, the ratio of the peak height to Λ, is about 0.2. The typical RMS roughness of the patterned Al substrates varies from 5 nm to 70 nm by increasing Λ from 0.1 µm to 0.9 µm.

Figure 4 shows (a) R_{total} and (b) haze spectra of the patterned Al substrates with a Ag coating. As can be seen in Fig. 4 (a), all the samples show high R_{total} over 96% in the NIR region. However, an absorption peak at a wavelength of λ ~ 0.35 µm and its absorption tail are also observed in Fig. 4 (a). The intensity of this peak is sharply increased by increasing Λ from 0.1 to 0.3 µm, and then decreased by increasing Λ > 0.3 µm. This tendency is also valid for the absorption tail in the visible (VIS) and NIR regions. As a result, the R_{total} of the Al substrates is slightly increased in the VIS and NIR regions with increasing Λ from 0.3 to 0.9 µm. For the

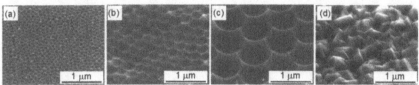

FIG. 3.SEM images of patterned Al substrates fabricated by anodic oxidation process under the applied voltages of (a) 40 V, (b) 120 V, (c) 370 V and (d) Asahi-U type textured SnO_2 on glass. The average periods of the patterned Al substrates (a) – (c) are Λ = 0.1 μm, 0.3 μm and 0.9 μm, respectively.

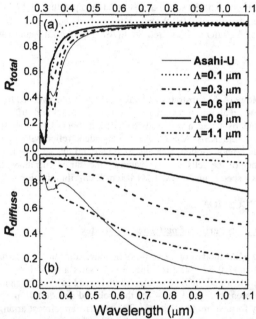

FIG. 4. (a) Total reflectivity R_{total} and (b) haze $R_{diffuse}/R_{total}$ spectra of the patterned Al substrates and Asahi-U type textured SnO_2 on glass. All samples are coated with a Ag film.

substrates with Λ = 0.9 and 1.1 μm, R_{total} is almost identical. This absorption is attributed to localized surface plasmon (LSP) excited in Ag nanostructures, which is often observed in Ag surface with roughness [28]. In other words, the intensity of LSP is related to a degree of nano-roughness. Therefore, the results in Fig. 4 (a) indicate that the higher reflectivity of the Al substrates with larger Λ is attributed to the less nano-roughness resulting in the weaker LSP absorption. It should be remarked that absorption by surface plasmon polariton supported by periodic structures as observed in perfect surface gratings [29] is not observed in our substrates.

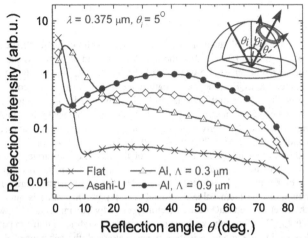

FIG. 5. Angular distribution of the reflected light intensities from the patterned Al and Asahi-U substrates with a Ag/ZnO coating measured at $\lambda = 0.375$ μm and $\theta_i = 5°$ as shown in the inset.

This is due to insufficient long-range periodicity in our substrates, as often observed in self-ordered structures fabricated by anodic oxidation. On the other hand, it is clearly shown in Fig. 4 (b) that the haze increases monotonically with increasing Λ from 0.1 to 1.1 μm in the VIS and NIR regions. For the substrates with $\Lambda > 0.9$ μm, more than 90% of visible light is reflected diffusively. This property is preferable to increase the optical path length in thin μc-Si:H layers. This behavior quantitatively agrees with the calculated haze spectra shown in Figs. 1 (f) – (h), although Fig. 4(b) was obtained for the air/Ag interfaces.

Figure 5 shows the angular distribution of the reflected light intensities from the patterned Al and Asahi-U substrates with a Ag/ZnO coating. In this experiment, a laser beam with $\lambda = 0.375$ μm was used as a light source and the incident angle was set near the normal ($\theta_i = 5°$) as shown in the inset. The wavelength of $\lambda = 0.375$ μm corresponds to NIR light ($\lambda \sim 1.3$ μm) when propagating in μc-Si:H films with a refractive index of ~ 3.5. Thus, by this configuration, we can estimate the optical behavior at the rear side of μc-Si:H solar cells in the NIR region, where light trapping is crucial. It can be seen from Fig. 5 that the patterned Al substrate reflects the incident light at larger angles more effectively with increasing Λ. At the same time, the intensity of specular reflection decreases significantly. The patterned Al substrate with $\Lambda = 0.9$ μm exhibits stronger reflection at larger angles compared with the Asahi-U substrate. However, any specific diffraction peaks are not observed on these spectra because of the insufficient long-range periodicity in our substrates. From this result, it is confirmed that the patterned substrates with larger Λ reflect the incident light at oblique angles quite effectively. This property is useful to increase the optical path length within the i-layer of μc-Si:H solar cells and enhance the total internal reflection at the interface between the μc-Si:H film and the transparent conducting oxide (TCO) top electrode.

Solar cells

Figure 6 shows the external quantum efficiency (EQE) spectra of the μc-Si:H cells with an i-layer thickness of t_i = 1.0 μm fabricated on the flat, Asahi-U, and patterned Al substrates with Λ = 0.1 – 0.9 μm measured under AM1.5 white bias light. It can be seen in Fig. 6 that the patterned Al substrates enhance the EQE in the wavelength range from 0.6 to 1.1 μm compared with the flat substrate. In addition, this EQE gain is gradually improved by increasing Λ, as expected from Fig. 5. The patterned Al substrate with Λ = 0.9 μm shows higher EQE than Asahi-U type randomly textured substrate. The corresponding J_{SC} of the flat, Asahi-U and patterned Al with Λ = 0.1, 0.3, 0.45 and 0.9 μm are 18.6, 22.5, 19.6, 21.0, 22.8, and 24.3 mA/cm^2, respectively.

Figure 7 shows the relationship between Λ of the patterned substrates and short circuit current densities J_{SC} of the n-i-p μc-Si:H solar cells with t_i = 1 μm. In this figure, J_{SC} were calculated from EQE curves. It is clearly found that the J_{SC} is significantly improved by increasing Λ, having a maximum value at around Λ ~ 1 μm. This value is quite close to the feature size of the electroplated ZnO back reflector reported by Toyama et al. [14], which gives a significantly high J_{SC}. Unfortunately, we have not succeeded to fabricate uniform patterns with Λ longer than 1.2 μm using the anodic oxidation process, although the behavior of EQE for much longer Λ is of interest.

In Figure 8, EQE, internal quantum efficiency (IQE), and R_{total} spectra of the μc-Si:H cells fabricated on the patterned Al substrate with Λ = 0.9 μm are compared with those of the cell on the Asahi-U substrate. As a reference, R_{total} of a crystalline Si (c-Si) wafer coated with an ITO film is also plotted in this figure. Interestingly, R_{total} of the both solar cells are almost the same, in spite of the clear difference in EQE. As a result, the cell on the patterned Al substrate exhibits a significant increase in IQE in the wavelengths longer than 0.6 μm compared with that of the Asahi-U substrate. A higher IQE means that photons absorbed in the cell are more effectively absorbed within the active layer and contribute to generate photo-carriers. In other words, this result indicates that the improved light trapping structure reduces the parasitic absorption in the cell. However, in the μc-Si:H cells, there remains a large reflection loss in the NIR region, compared with the thick c-Si with an ITO anti-reflection coating. This reveals that light trapping in μc-Si:H cells can be improved further in this region.

The main factors of the parasitic absorption in our solar cells are the p- and n-layers, the front ITO layer, and the BSR. In this experiment, only the BSR is different among these factors. Therefore, an increase of reflectivity at the BSR is a possible reason for this IQE gain. In addition, the amount of the parasitic absorption can be varied by changing the optical path length inside the cell. If incident light is effectively scattered at the front side of the cell, it has a higher possibility to be absorbed within the active layer before reaching the absorbing BSR. In the same manner, if the light reflected at the BSR back to the cell is effectively scattered, it has higher possibility to be absorbed within the active layer before reaching the absorbing front TCO. Thus, the improved BSR structure can offer the three beneficial effects: (i) an increase of reflectivity of the BSR, (ii) effective light scattering at the front side, and (iii) effective light scattering at the rear side. The effect (i) can be confirmed in Fig. 4 (a), showing that R_{total} of the patterned substrate with Λ = 0.9 μm is slightly higher than that of Ag-coated Asahi-U in the VIS – NIR region, although the data were obtained in air atmosphere. The effect (iii) is obvious from Fig. 5. To check the effect (ii), the angular reflection intensity of the cell surfaces was measured.

36

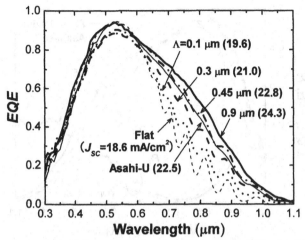

FIG. 6. External quantum efficiency (EQE) spectra of the *n-i-p* μc-Si:H solar cells with $t_i = 1$ μm fabricated on a flat, Asahi-U and the patterned Al substrates with Λ = 0.1, 0.3, 0.45 and 0.9 μm measured under white bias light. The values in the brackets show the corresponding short circuit current densities.

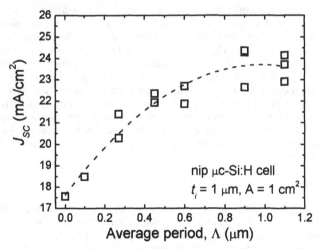

FIG. 7. J_{SC} of the *n-i-p* μc-Si:H solar cells with $t_i = 1$ μm fabricated on a the patterned Al substrates as a function of Λ. These values were calculated from EQE curves. The line is the guide to the eyes.

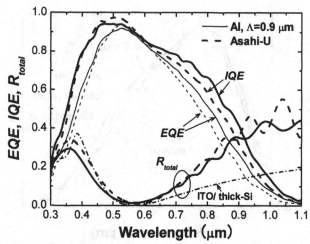

FIG. 8. EQE, IQE and R_{total} spectra of the n-i-p μc-Si:H solar cells with t_i = 1 μm fabricated on Asahi-U and the patterned Al substrate with Λ = 0.9 μm measured under white bias light and zero bias voltage. As a reference, R_{total} of c-Si wafer coated with an ITO film is also plotted.

Figure 9 shows the angular reflection intensity of the n-i-p μc-Si:H solar cells measured by the same experimental setup used in Fig. 5. In this figure, reflection intensities at 40° are plotted as functions of Λ. It should be noted that the reflection from the cells is caused only by the front reflection because the incident beam with λ = 0.375 μm cannot reach the rear surface. Obviously, the reflection of the cells shows the similar behavior with that of the substrates. This result indicates that the light scattering property of substrates is replicated to the cells fabricated on them, and light trapping effect is enhanced by the textures on the front side as well as on the rear side. However, in the case of longer Λ > 0.5 μm, reflection intensity of the cells shows the lower values than those of the substrates. This result reveals that light scattering property at the front side of the cells is decreased as compared with the substrates. It is expected that the results obtained here are valid in the NIR region where light trapping is crucial, although the absolute value must be different.

CONCLUSIONS

In this work, effect of patterned back reflectors on light trapping in μc-Si:H cells has been investigated with self-ordered Al substrates obtained by anodic oxidation. With increasing the period of the patterned substrates from 0 to 1.1 μm, 1-μm-thick μc-Si:H cells on the patterned substrates have shown a significant enhancement of spectral response in the NIR region, giving an increment of J_{SC} from 18 to 24 mA/cm^2. The optimum period of patterned substrates is determined to be around 1 μm for 1-μm-thick μc-Si:H cells. This enhanced light trapping effect are attributed to the improved reflectivity of the rear side and effective light scattering at the front side, as well as light scattering at the rear side. It is expected that independent control of the

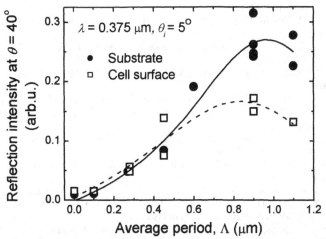

FIG. 9. Angular reflection intensity of the patterned Al substrates and the *n-i-p* μc-Si:H solar cells with $t_i = 1$ μm measured at 40° as a function of Λ. The lines are the guides to the eyes.

surface morphology of the front and the back sides yields an additional gain in the NIR response. Optimization of the texture shape will also improve the NIR response, although we have treated a fixed shape in this study. In addition, there remains a room for optimization of the aspect ratio. Needless to say, development of improved reflector materials including low-absorption TCO has been also a big challenge to enhance the light trapping in thin-film Si solar cells.

ACKNOWLEDGMENTS

The authors thank Dr. Koida and Dr. Turkevych of AIST for their help in this work. This work was supported by New Energy and Industrial Technology Development Organization (NEDO) under Ministry of Economy, Trade and Industry (METI), Japan. A part of the calculated results in this research was obtained by using the supercomputing resources at the Information Synergy Center, Tohoku University.

REFERENCES

1. J. Meier, S. Dubail, R. Platz, P. Torres, U. Kroll, J.A. Anna Selvan, N. Pellaton Vaucher, Ch. Hof, D. Fischer, H. Keppner, R. Flückiger, A. Shah,V. Shklover, K.-D. Ufert, Sol. Energy Mater. Sol. Cells **49**, 35 (1997).
2. K. Yamamoto, T. Suzuki, M. Yoshimi, A. Nakajima, Jpn. J. Appl. Phys. **36**, L569 (1997).
3. E. Yablonovich, J. Opt. Soc. Am. **72**, 899 (1982).
4. K. Yamamoto, M. Yoshimi, Y. Tawada, Y. Okamoto, A. Nakajima, S. Igari, Appl. Phys. A **69**, 179 (1999).
5. K. Sato, Y. Gotoh, Y. Wakayama, Y. Hayashi, K. Adachi, H. Nishimura, Rep. Res. Lab.

Asahi Glass Co., Ltd. **42**, 129 (1992).
6. T. Matsui, M. Tsukiji, H. Saika, T. Toyama, H. Okamoto, J. Non-Cryst. Solids **299–302**, 1152 (2002).
7. M. Berginski, J. Hüpkes, M. Schulte, G. Schöpe, H. Steibig, B. Rech, M. Wuttig, J. Appl. Phys. **101**, 074903 (2007).
8. S. Fäy, U. Kroll, C. Bucher, E. Vallat-Sauvian, A. Shah, Sol. Energy Mater. Sol. Cells **86**, 385 (2005).
9. N. Toyama, R. Hayashi, Y. Sonoda, M. Iwata, Y. Yamamoto, H. Otoshi, K. Saito, K. Ogawa, *Proceedings of 3rd World Conference on Photovoltaic Energy Conversion* (IEEE, New York, 2003) pp.1601.
10. K. Niira, H. Senta, H. Hakuma, M. Komoda, H. Okui, K. Fukui, H. Arimune, K. Shirasawa, Sol. Energy Mater. Sol. Cells **74**, 247 (2002).
11. A. Takano, M. Uno, M. Tanda, S. Iwasaki, H. Tanaka, J. Yasuda, T. Kamoshita, Jpn. J. Appl. Phys. **43**, L277 (2004).
12. V. Terrazzoni Daudrix, J. Guillet, F. Freitas, A. Shah, C. Ballif, P. Winkler, M. Ferreloc, S. Benagli, X. Niquille, D. Fischer, and R. Morf, Prog. Photovolt. **14**, 485 (2006).
13. C. Heine, R.H. Morf, Appl. Optics 34 2476 (1995).
14. H. Stiebig, C. Haase, C. Zahren, B. Rech, N. Senoussaoui, J. Non-Crystalline Solids **352**, 1949 (2006).
15. H. Stiebig, N. Senoussaoui, C. Zahren, C. Haase, J. Müller, Prog. Photovol. **14**, 13 (2006).
16. P. Bermel, C. Luo,L. Zeng, L.C. Kimerling, J.D. Joannopoulos, Optics Express **15**, 16987 (2007)
17. J.G. Mutitu, S. Shi, C. Chen, T. Creazzo, A. Barnett, C. Honsberg, D.W. Prather, Optics Express **16**, 15238 (2008).
18. H. Sai, H. Fujiwara, M. Kondo, Y. Kanamori, Appl. Phys. Lett. **93**, 143501 (2008).
19. A. P. Li, F. Müller, A. Birner, K. Nielsch, U. Gösele, J. Appl. Phys. **84**, 6023 (1998).
20. W. Lee, R. Ji, U. Gösele, K. Nielsch, Nature Mater. **5**, 741 (2006).
21. S. Z. Chu, K. Wada, S. Inoue, M. Isogai, Y. Katsuta, A. Yasumori J. Electrochem. Soc. **153**, B384 (2006).
22. M. G. Moharam, Proc. SPIE **883**, 8 (1988).
23. D. W. Lynch and D. W. Hunter, *Optical Constants of Solids*, edited by E.D. Palik, (Academic Press, Boston, 1985) pp. 350-357.
24. D. E. Asnes, *Properties of Crystalline Silicon*, edited by R. Hull (INSPEC, London, 1999), pp. 683–690.
25. A. P. Li, F. Müller, A. Birner, K. Nielsch, U. Gösele, J. Appl. Phys. **84**, 6023 (1998).
26. W. Lee, R. Ji, U. Gösele, K. Nielsch, Nature Mater. **5**, 741 (2006).
27. S. Z. Chu, K. Wada, S. Inoue, M. Isogai, Y. Katsuta, A. Yasumori J. Electrochem. Soc. **153**, B384 (2006).
28. J. Springer, A. Poruda, L. Müllerova, M. Vanecek, O. Kluth, B. Rech, J. Appl. Phys. **95**, 1427 (2004).
29. F.-J. Haug, T. Söderström, O. Cubero, V. Terrazzoni-Daudrix, C. Ballif, J. Appl. Phys. **104**, 064509 (2008).

Mater. Res. Soc. Symp. Proc. Vol. 1153 © 2009 Materials Research Society   1153-A03-05

Photonic crystal back reflector in thin-film silicon solar cells

O. Isabella[1], B. Lipovšek[2], J. Krč[2], and M. Zeman[1]

[1] Delft University of Technology, EEC Unit / DIMES, 2600 GB Delft, The Netherlands
[2] University of Ljubljana, Faculty of Electrical Engineering, SI-1000 Ljubljana, Slovenia

ABSTRACT

One-dimensional photonic crystals having desired broad region of high reflectance (R) were fabricated by alternating the deposition of amorphous silicon and amorphous silicon nitride layers. The effect of the deposition temperature and angle of incidence on the optical properties of photonic crystals deposited on glass substrate was determined and an excellent matching was found with the simulated results. The broad region of high R of photonic crystals deposited on flat and textured ZnO:Al substrates decreases when compared to the R of photonic crystals deposited on glass. The performance of amorphous silicon solar cells with 1-D photonic crystals integrated as the back reflector was evaluated. The external quantum efficiency measurement demonstrated that the solar cells with the photonic crystals back reflector had an enhanced response in the long wavelength region (above 550 nm) compared to the cells with the Ag reflector.

INTRODUCTION

For obtaining high conversion efficiency of thin-film solar cells the proper light management inside the solar cell structures is of great importance. In today's thin-film solar cells light management is accomplished by implementing light-trapping techniques that are based on the introduction of surface-textured substrates and the use of special layers called back reflectors. Reflection at the textured back contact is a critical issue, since conventional metal reflectors (Ag, Al) suffer from undesired plasmon absorption, limiting the long wavelength response of the solar cell [1]. Recently, novel concepts for wavelength-selective manipulation of the reflection and transmission at a particular interface inside a solar cell have been investigated, such as the use of Photonic Crystals (PCs) [2]. PCs in the role of a Distributed Bragg Reflector (DBR) have attracted attention as back reflectors in thin-film silicon solar cell [3, 4] or as intermediate layer in double-junction silicon-based solar cell [5]. PCs with the desired wavelength-selective behavior can be designed by using optical simulations that help to tune the thickness of layers, the number of alternating pairs, and the combination of refractive indexes [2].

In this contribution we investigate the optical properties of one dimensional (1-D) PCs based on alternating layers of hydrogenated amorphous silicon (a-Si:H) and hydrogenated amorphous silicon nitride (a-SiN$_x$:H). The total reflectance of the PCs deposited at different substrate temperatures and on different substrate carriers was measured under various angles of incidence. The experimental results of wavelength-selective high reflectance were compared to the simulation results and an excellent matching has been obtained. The 1-D PC in the role of DBR was applied at the rear side of a single-junction a-Si:H solar cell. An enhancement of the long wavelength EQE (above 550 nm) was observed in comparison with cells with a metal reflector.

THEORY

1-D PC is a multilayer structure in which two layers with different optical properties (refractive indexes) are periodically alternated (Fig. 1). When light propagates through such a structure, constructive and deconstructive interferences arise, resulting in the wavelength-selective reflectance or transmittance behavior. Using the 1-D PC as a DBR, high reflectance (close to 100 %) can be achieved for a broad region of wavelengths around the Bragg's wavelength (λ_B). We refer to this band of high reflectance as the photonic band-gap of the 1-D PC ($\Delta\lambda$), since the transmittance is almost completely suppressed in the $\Delta\lambda$ range. Around λ_B, the reflectance $R(\lambda_B)$ is maximized, even though high reflectance, close to $R(\lambda_B)$, is present in the entire $\Delta\lambda$. When λ_B is chosen for maximizing the reflective behavior of a 1-D PC, the characteristic parameters of the 1-D PC $R(\lambda_B)$ and $\Delta\lambda$ can be calculated as follows:

$$\lambda_B = 4d_L n_L = 4d_H n_H \tag{1}$$

$$R(\lambda_B) = \left[\frac{n_0(n_L)^{2M} - n_s(n_H)^{2M}}{n_0(n_L)^{2M} + n_s(n_H)^{2M}} \right]^2 \tag{2}$$

$$\Delta\lambda \approx \frac{2\lambda_B \Delta n_B}{\pi < n_{GROUP} >} \tag{3}$$

where n_H and n_L are the high/low refractive indexes of the alternating layers, respectively, d_H and d_L are their thicknesses, n_0 and n_s are the refractive indexes of the incident medium and of the substrate, respectively, M is the number of stacked pairs, $\Delta n_B = |n_L - n_H|$ is the refractive index step and $<n_{GROUP}>$ is the spatial average of the group index along the direction of incoming light [6]. We use the characteristic parameters calculated in equations (2-3) to describe the quality of the 1-D PC and compare them with the results of the computer simulations [7].

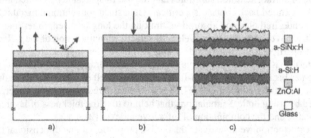

Figure 1. Schematic of the 1-D PC deposited on glass (a), on glass / flat ZnO:Al (b), and on glass / rough ZnO:Al (c). 1-D PC on glass is also characterized for different angled illumination.

EXPERIMENT

The investigated 1-D PC consists of three pairs of alternating a-Si:H / a-SiN$_x$:H layers deposited in a PE-CVD reactor. The used substrates were either glass (Fig. 1-a) or glass coated with ZnO:Al. The ZnO:Al layers were magnetron-sputtered and used as flat (as deposited) or rough (wet-etched) substrate carriers (Fig. 1-b and Fig. 1-c, respectively). The measurement of the optical properties of the 1-D PC was carried out using the Perkin-Elmer Lambda 950 spectro-photometer. The Absolute Reflectance / Transmittance Analyzer (ARTA) accessory was com-bined with the spectro-photometer to test the reflectance of the 1-D PC for different angles of incidence.

The 1-D PCs were implemented as back reflectors in a-Si:H solar cells. The schematic structure of the solar cells consists of glass coated with flat (Fig. 2-a) or rough (Fig. 2-b) front ZnO:Al layers, a typical single-junction thin-film silicon p-i-n solar cell, a ZnO:Al film used as back electrode, and three different back reflectors, namely ZnO:Al / air, ZnO:Al / Ag, and ZnO:Al / PC. The EQE of the solar cells with different back reflectors was measured.

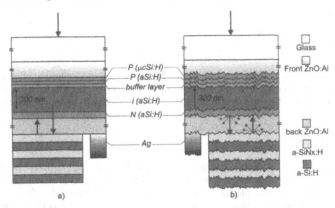

Figure 2. Schematic structure of the thin-film a-Si:H solar cells with 1-D PC as the back reflec-tor. The solar cells are deposited on the flat front ZnO:Al (a) and the rough front ZnO:Al (b).

RESULTS AND DISCUSSION

Effect of deposition temperature of 1-D PCs

We investigated the effect of the deposition temperature of the 1-D PC on glass. In figure 3 the result of measured and simulated reflectance of the 1-D PCs is presented. Already with three pairs of layers, broad high reflectance region can be achieved. As the temperature of the deposition decreases, the photonic band-gap of the 1-D PC blue-shifts almost rigidly. Indeed, both PCs present $\Delta\lambda \sim 450$ nm and $R(\lambda_B) > 0.94$. Simulations indicated and the measurements confirmed that this shift is related to the change in the thickness of the PC layers when the depo-sition temperature changes, whereas all other deposition parameters remain the same.

Effect of angle of incidence

Since the light scattering in solar cells changes the directions of light, we studied the in-fluence of different angles of incidence on the optical properties of the 1-D PC. In figure 4 the total reflectance of the 1-D PC on glass under three angles of incidence is shown. Increasing the angle of incidence, the $\Delta\lambda$ shrinks and consequently the Bragg's wavelength λ_B shifts towards shorter wavelengths. Experimental results match the simulated predictions very well (shown only for p-polarized light).

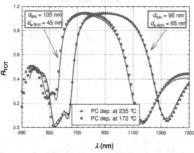 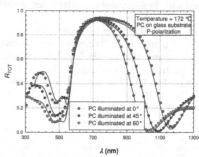

Figure 3. Temperature effect on a-Si:H / a-SiN$_x$:H based PC. Solid curves are simulations, symbol-dotted curves are measurements.

Figure 4. Angled illumination effect on a-Si:H / a-SiN$_x$:H based PC. Solid curves are simulations, circle-dotted curves are measurements.

Substrate effect

Since the 1-D PC is deposited in the solar cells on flat and rough ZnO:Al, also the use of substrate carriers different from bare glass (Fig. 1-b and Fig. 1-c) was studied. In figure 5 the total reflectance of 1-D PC deposited on flat and rough ZnO:Al is plotted. Using ZnO:Al as flat or rough substrate, $\Delta\lambda$ and $R(\lambda_B)$ become lower with respect to the 1-D PC deposited on glass.

Simulations of 1-D PCs indicate that the optical properties of a substrate, on which the PC is deposited, have minor effect on the reflective behavior, unless the substrate is highly reflective in the entire wavelength spectrum. Although this is not the case for ZnO:Al, such a substrate can affect the optical properties of the layers grown on the top of it.

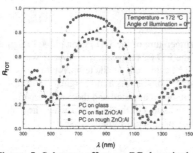 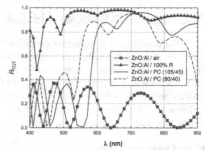

Figure 5. Substrate effect on PC deposited on glass (circles), flat ZnO:Al (triangles), and rough ZnO:Al (squares) at 172 °C.

Figure 6. Simulation of the internal reflectance of the interface n / back reflector inside the solar cell (p-i-n configuration).

PC on Solar cells

When designing the 1-D PC as a back reflector we considered that for a-Si:H solar cells high reflectance should be achieved in the wavelength region 600-800 nm. Simulations were used to calculate the internal reflectance at the interface n-layer / back reflector inside the p-i-n solar cell. The results of the calculations are shown in figure 6. Considering the PC so far discussed (temperature of deposition = 172°C, $d_{a\text{-SiNx:H}}$ = 105 nm, and $d_{a\text{-Si:H}}$ = 45 nm), the reflectance rises just after 550 nm. The reflective behavior of the ZnO:Al / PC stack can be eventually shifted before by changing the thicknesses of the layers involved in the 1-D PC.

A step further in the design of the 1-D PC was the simulation of realistic reflecting structures placed at the back side of a p-i-n solar cell endowed with flat interfaces (Fig. 2-a). The calculated EQEs are presented in figure 7. In case of ZnO:Al / Ag reflector, ideal interface between TCO and Ag was considered (without interfacial layers that might decrease the reflectance in realistic case). It's noticeable that in case of ZnO:Al / PC, starting from 550 nm, the peaks of the EQE are shifted to larger wavelengths with respect to those of the ZnO:Al / Ag structure, correspondingly with the plot drawn in figure 6.

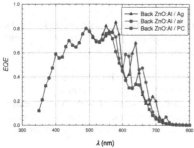 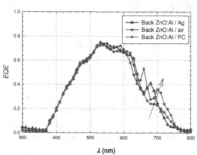

Figure 7. Simulated EQE of solar cells on flat front ZnO:Al stacked with different back reflectors.

Figure 8. Experimental EQE of solar cells on flat front ZnO:Al stacked with different back reflectors.

In figure 8 and in figure 9 the measured EQE plots of actual solar cells, according to the schematic in figure 2, are presented. In both cases the quantum efficiency turns out slightly smaller than the predictions of the simulator because of processing issue. Additionally, at shorter wavelength a higher absorption in actual ZnO:Al may occur.

In the long-wavelength region, whereas the influence of the back reflectors is higher, the EQE increases as predicted by simulations. When different back reflectors are used and considering ZnO:Al / air back reflector as reference, the trend of increasing is clarified by the arrows in the plots. As the enhancements can be recognized in the wavelength range 550-800 nm, we convoluted in that range the EQE data points with the incident spectrum AM 1.5 in order to calculate the J_{SC} achieved by the tested solar cells. Assuming 0% as reference percentage increasing of J_{SC} obtained by the ZnO:Al / air back reflector, the percentage increasing when using different substrates and different back reflectors is shown in figure 10. In case of textured metallic back reflector, surface plasmon absorption starts to take place, decreasing the improvement of J_{SC}, compared to flat case. This problem in case of ZnO:Al / PC back reflector is not present.

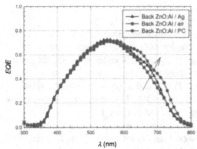

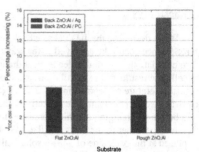

Figure 9. Experimental EQE of solar cells on rough front ZnO:Al stacked with different back reflectors.

Figure 10. J_{SC} percentage increasing in the range 550-800 nm. 0% line in the figure is the ZnO:Al / air back reflector (reference).

CONCLUSIONS

1-D PC was fabricated using amorphous silicon and amorphous silicon nitride layers. The effects of the deposition temperature and angle of incidence on the optical properties were investigated. The $\Delta\lambda$ and the $R(\lambda_B)$ expressed in the measurements and prettily matched by simulations validate the quality of the structure realized.

The 1-D PC was also tested in real thin-film silicon solar cells. Both flat and surface-textured front ZnO:Al layers were used as carrier for solar cells equipped with three different back reflectors. A percentage increasing up to 14.9% in the J_{SC} is supplied by the back reflector ZnO:Al / PC with respect to the ZnO:Al / air back reflector (reference). The absence of absorption losses makes the combination ZnO:Al / PC a potential viable candidate for back reflectors.

ACKNOWLEDGEMENTS

This work was carried out with a subsidy of the Dutch Ministry of Economic Affairs under EOS-LT program (project number EOSLT04029).

REFERENCES

1. J. Springer, A. Poruba, L. Mullerova, M. Vanecek, O. Kluth, and B. Rech, J. Appl. Phys. 95, 1427 (2004).
2. J. Krc, M. Zeman, A. Campa, F. Smole, and M. Topic, Mater. Res. Soc. Symp. Proc. Vol. 910, 0910-A25-01.
3. P. Bermel, C. Luo, L. Zeng, L. C. Kimerling, and J. D. Joannopoulos, Optics Express Vol. 15, No. 25, pp. 16987-17000 (2007).
4. M. Zeman and J. Krc, Journal of Material Research, **23** (4), 889 (2008).
5. A. Bielawny et al., Physica Status Solidi - Applications and Materials Science Vol. 205 Issue: 12, Special Issue: Sp. Iss. SI, Pages: 2796-2810 Published: DEC 2008.
6. R. Michalzik and K.J. Ebeling, Vertical-Cavity Surface-Emitting Laser Devices, pp. 53–98. Berlin: Springer-Verlag, 2003.
7. J. Krc, F. Smole, and M. Topic, Prog. Photovolt. Res. Appl., **11** (2003) 15.

Defects and Metastability

Mater. Res. Soc. Symp. Proc. Vol. 1153 © 2009 Materials Research Society   1153-A04-03

Amorphous semiconductors studied by first-principles simulations: structure and electronic properties

Karol Jarolimek[1,2] and Robert A. de Groot[2] and Gilles A. de Wijs[2] and Miro Zeman[1]
[1]Electrical Energy Conversion Unit/DIMES, Delft University of Technology, P.O. Box 5053, 2600 GB Delft, The Netherlands
[2]Electronic Structure of Materials, Institute for Molecules and Materials, Radboud University Nijmegen, Heyendaalseweg 135 , 6525 AJ Nijmegen, The Netherlands

ABSTRACT

A simulation study of the electronic structure of two amorphous semiconductors with great technological importance, namely hydrogenated amorphous silicon (a-Si:H) and hydrogenated amorphous silicon nitride (a-SiN:H), is presented. The simulations were based on density functional theory (DFT) which provides accurate description of interactions between the atoms for a wide range of chemical environments. Our model structures were prepared by a cooling-from-liquid approach. We found that the cooling rate during the thermalization process has a considerable impact on the quality of the resulting models. A rate of 0.023 K/fs proved to be sufficient to prepare models with low defect concentrations. To our knowledge we present for the first time calculations that are entirely based on the first-principles and produce defect-free models of both a-Si:H and a-SiN:H.

Although creating models without any defects is important, on the other hand a small number of defects present in the models can give valuable information about the structure and electronic properties of defects in a-Si:H and a-SiN:H. The presence of both the dangling bond and the floating bond was observed. Structural defects were related to electronic defect states within the band gap. In a-SiN:H the Si-Si bonds induce states at the valence and conduction band edges, thus decreasing the band gap energy. This finding is in agreement with measurements of the optical band gap, where increasing the nitrogen content increases the band gap.

INTRODUCTION

Amorphous semiconductors are materials of a great technological interest. Hydrogenated amorphous silicon (a-Si:H) deposited by the plasma-enhanced chemical vapor deposition (PECVD) technique is widely used in large area electronic devices such as solar cells and liquid-crystal displays [1]. Thin films of hydrogenated amorphous silicon nitride (a-SiN:H) grown by PECVD are routinely used as insulation in integrated circuits [2,3]. Other applications include optical waveguides [4], nonvolatile memory devices [5] and passivation layers [6].

There have been numerous theoretical studies on amorphous semiconductors in the past. The models of their atomic structures have been prepared with methods ranging from classical potentials to the state-of-the-art *ab initio* methods. Atomic structures are often prepared by cooling from the liquid using a molecular dynamics (MD) simulation [7-12]. This approach was criticized for generating a-Si:H models with a high defect concentration that was unrealistic [13]. A device quality a-Si:H material has a defect concentration of in the range of 10^{16} to 10^{17} cm^{-3} [1]. This implies that a simulation cell containing 10^7 atoms should have one defect. Thus model structures containing hundreds of atoms should be essentially free of defects. We find that the cooling rate has a considerable impact on the amorphous structure in our MD simulation and that a rate of ~ 0.023 K/fs is sufficient to prepare defect-free structures.

THEORY

We carried out a MD study of a-Si:H and a-SiN:H. The calculation is referred to as first-principles, since the interaction between the atoms is treated quantum mechanically without adjustable parameters. The forces acting on the atoms are obtained by solving the quantum mechanical ground state within the density functional theory (DFT) approximation. This method is more accurate and is limited to small-size systems and shorter simulation times as compared to classical model potential MD. All calculations were performed with the VASP package [14,15]. The exchange and correlation functional is evaluated in the generalized gradient approximation (GGA) [16]. The electron-ion interactions are described using the projector augmented wave method (PAW) [17,18]. We used the standard PAW potentials, which are a part of the VASP package, with kinetic energy cut-offs of 200 eV and 300 eV for a-Si:H and a-SiN:H, respectively. During the MD the Brillouin zone was sampled with Γ point only. For the density of states calculations the Brillouin zone was sampled with a 5×5×5 Monkhorst-Pack mesh [19]. The time step of 1 fs was used for both materials.

The amorphous model structures were prepared by cooling from the liquid phase. We have used compositions of $Si_{64}H_8$ and $Si_{38}N_{38}H_{34}$, and a density of 2.2 g/cm^3 and 2.0 g/cm^3 for a-Si:H and a-SiN:H respectively. First the atoms were placed into the simulation cell with velocities initialized randomly according to the Maxwell-Boltzmann distribution, corresponding to a temperature of 300 K. Next the cells were heated up to 2370 K and 3060 K for the a-Si:H and a-SiN:H, respectively. At these temperatures the systems are in a liquid state, since the mean squared displacement of the atoms increases linearly in time. The systems were further evolved for approximately 10 ps in order to loose any memory of the initial structure. Subsequently, they were cooled down to 300 K. Radial distributions at room temperature were obtained by averaging over a 0.5 ps time interval (i.e. over 500 configurations). The electronic properties were calculated on relaxed models. We prepared 5 a-Si:H models (A, B, C, D, E) and 3 a-SiN:H models (A', B', C').

RESULTS

Structural properties

The atomic structure of amorphous materials is not random, it exhibits short-range order. A common way of characterizing the short-range order is through the use of radial distributions. The distribution gives the probability of finding two atoms at a certain distance apart, relative to a system with atoms positioned randomly in space. Experimental radial distributions can be obtained from the structure factors (measured with neutron or X-ray diffraction) using a Fourier transform. In figure 1 we compare the calculated Si-Si radial distribution and structure factor with the measurement results of Bellissent et al. [20]. The overall agreement is very good. The discrepancy between the calculated and measured peak height (in the radial distribution) can be attributed to insufficient experimental resolution. In figure 1 we plot partial radial distributions of a-SiN:H. To our knowledge experimental partial radial distributions for a-SiN:H of this composition are not available. The Si-Si and N-N radial distributions contain small first peaks. These turn out to be due to "square structures" present in silicon nitride of this particular composition. The "square structures" are made of two silicon atoms in the opposing corners of a square and of two nitrogen atoms in the remaining corners. Although the Si-Si and N-N atoms in the square structure are relatively close to each other (and within the defined cut-off distances) they are *not* bonded. The Si-Si first peak is much more profound than the N-N one, because also

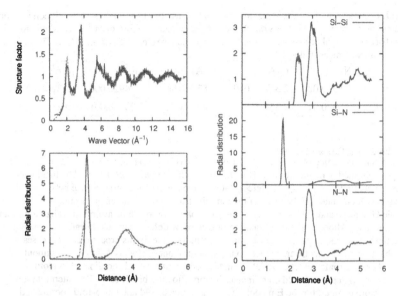

Figure 1. Left panel: Si-Si structure factor (top) and radial distribution (bottom) of a-Si:H. Calculated curves are averages over 5 models (in solid red line), experimental curves by Bellissent et al.[20] in green dashed line. Right panel: Calculated partial radial distributions of a-SiN:H. All curves refer to systems at 300 K.

a number of Si-Si bonds is present in our model. There are, however, no N-N bonds formed, most likely due to then composition used, which is nitrogen deficient as compared to the stochiometric silicon nitride. The second peak in the Si-Si partial radial distribution at ~3 Å is due to silicon atoms that are bonded to a common nitrogen atom. Similarly the second peak in the N-N partial radial distribution (at ~2.9 Å) is due to two nitrogen atoms bonded to a common silicon atom. The Si-N radial distribution has a very well defined first neighbor peak with a small standard deviation that reflects the stiffness of the Si-N bond. Partial radial distributions containing hydrogen have in general a more simple interpretation. The position of the first peak marks the length of the Si-H or N-H bond. The bond lengths are almost the same one would find in the calculated geometries of the silane and ammonia molecules. Hydrogen molecules are found in both materials under study.

A more quantitative characterization of the amorphous network is given in Table I. Here we limit ourselves to a-Si:H only, since the experimental record on a-SiN:H is rather sparse. We compare our calculated values to a high resolution measurement of Laaziri et al. [21] on pure amorphous silicon. The mean first and second neighbor distances coincide to within a few hundreds of an Angström. From the deviations we notice that our models contain slightly more static disorder.

Table I. Comparison of experimental and calculated structural parameters of the silicon network: coordination number N_c (Si atoms only), mean first/second neighbor distance r_1/r_2, deviation of the first/second neighbor distance σ_1/σ_2. Calculated values refer to a system at 0 K. The data of Laaziri et al. was measured at 10 K [21].

	N_c	r_1 (Å)	r_2 (Å)	σ_1(Å)	σ_2(Å)
this work	3.89±0.02	2.371±0.010	3.835±0.019	0.049±0.006	0.326±0.027
Laaziri et al.	3.88±0.01	2.351±0.001	3.80±0.01	0.031±0.001	0.23

Electronic structure and defects

The electronic structure of a-Si:H consists of relatively flat bands (see peaks in figure 2). The average calculated band gap for a-Si:H is 0.92 eV with a deviation of 0.1 eV. The band gap values are obtained by *leaving out* defect states inside the gap. We discriminate between tail states and defects states on the basis of the participation ratio of a given electronic state (see for example G. Kresse and J. Hafner [22]). The participation ratio is a measure of localization of an electronic state. Although the identification of electronic defects can be arbitrary to some extend, the participation ratio imposes at least some quantitative rule to this process. In most cases we were able to map the electronic defects to the structural ones. Models A and C are without structural or electronic defects. Models B and D contain one 3-fold coordinated Si atom (dangling bond) and one 5-fold coordinated Si atom (floating bond) each. The interpretation becomes more difficult for the E model; there is also one 3-fold and one 5-fold coordinated silicon, but these are bonded to each other. The defect states of the E model are thus localized on both defective atoms.

The average band gap of a-SiN:H is 2.42 eV with a deviation of 0.14 eV. Model B' is completely defect-free. Models A' and C' contain one 2-fold coordinated N atom each. Additionally model A' contains a 4-fold coordinated N atom and model C' contains a 3-fold coordinated silicon. All of these structural defects give rise to gap states. In addition to the defect states, we find that Si–Si bonds give rise to states near the band gap edges. This is in agreement with measurements of the optical gap with varying composition of a-SiN$_x$:H. The band gap value increases with increasing nitrogen content or decreasing number of Si–Si bonds. Interestingly, nitrogen atoms that are 3-fold coordinated do contribute to defect states, but only near the valence band. These states have a π like character.

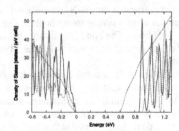

Figure 2. Calculated density of states of a-Si:H for the model A (red solid line), model C (green dashed line) and crystalline silicon (blue dotted line). The c-Si DOS is scaled by a factor of 100.

Table II. Number of defects found in models prepared with the fast, normal and slow cooling rates. Si3 denotes a silicon atom that has only 3 neighbors and is thus under-coordinated.

$Si_{64}H_8$	Si3	Si5	$Si_{38}N_{38}H_{34}$	Si3	Si5	N2	N4
fast	0	2	fast	5	0	2	1
normal	3	1	normal	4	0	1	1
slow(average)	0.6	0.6	slow(average)	0.3	0	0.7	0.3

Table III. Structural properties of the small models and the big model. The mean first and second neighbor distances are r_1 and r_2 respectively. The corresponding standard deviations are σ_1 and σ_2. E_g denotes the DFT band gap. All parameter refer to systems at 0 K.

	N_c	r_1 (Å)	r_2 (Å)	σ_1(Å)	σ_2(Å)	E_g(eV)
$Si_{64}H_8$	3.89±0.02	2.371±0.010	3.835±0.019	0.049±0.006	0.326±0.027	0.92±0.10
$Si_{216}H_{27}$	3.87	2.367	3.833	0.041	0.303	0.90

Effect of the cooling rate

It is reasonable to assume that the quality of the resulting model structure depends on how fast the liquid is quenched. To investigate this dependence we employed three different cooling rates: 1.380 K/fs, 0.138 K/fs and 0.023 K/fs. We refer to these rates as fast, normal and slow respectively. In general slower cooling results in structures with less static disorder, i.e. the deviation of the first and second neighbor distances σ_1 and σ_2 become smaller. The number of structural defects decreases considerably with the slower cooling (see table II). In both materials we have succeeded in preparing models without any structural nor electronic defects. In a-SiN:H the number of "square structures" seems to be almost unaffected by the cooling rate. The number of Si-Si bonds, however changes from 22 to approximately 16 in the fast and slow cooling rates respectively. The incorporation of hydrogen into the silicon-nitrogen network improves with the slower cooling rate and thus the number of H_2 molecules becomes smaller.

Convergence in the cell size, a-Si:H

Since *ab initio* calculations are more demanding than classical model potential calculations they are limited to smaller system sizes. To check how our results depend on the cell size, we generated a bigger model structure ($Si_{216}H_{27}$). The bigger model was prepared with the same computational settings using an identical cooling rate of 0.023 K/fs. The calculated Si-Si radial distributions are in close agreement for both system sizes. A more thorough comparison is presented in table III. The values of r_1 and r_2 in the big model lie within the error bars obtained from the small models. The big model has less static disorder, as the values of σ_1 and σ_2 lie just below the error bars of the small models. The calculated band gap of the big model is 0.90 eV, a value close to the average band gap of the small models. To conclude, the big model does not differ substantially from the small ones. The smaller static disorder might, however suggest that the periodic boundary conditions impose non-negligible restrictions on the amorphous structure.

CONCLUSIONS

We presented a first-principles study of the electronic structure for two amorphous semiconductors; a-Si:H and a-SiN:H. The DFT molecular dynamics approach with a slow cooling rate resulted in realistic atomic-scale models. The cooling rate has a considerable impact on the quality of the resulting model structures. From the present experimental knowledge (scattering data up to at least 40 Å^{-1} is necessary [21]), we can conclude that cooling rates of approximately 0.023 K/fs generate representative models of the real structures. The theoretical structures contain slightly more disorder. This method can be applied to compound amorphous materials consisting of several chemical species without the necessity of developing and testing classical potentials.

Atomic scale models of disordered semiconductors can reveal information that is difficult to obtain experimentally. In particular, we have learned that both under- and over-coordinated atoms are responsible for electronic defect states in the gap. In the case of a-SiN:H, in addition to silicon based defects we find 2-fold and 4-fold coordinated nitrogen atoms. Increasing nitrogen content in the model structure decreases the number of Si-Si bonds resulting in less electronic states near the band edges and thus the band gap of a-SiN:H opens up.

ACKNOWLEDGMENTS

This work was carried out with a subsidy of the Dutch Ministry of Economic Affairs under EOS-LT program (project number EOSLT02028) and is a part of the research program of the Stichting voor Fundamenteel Onderzoek der Materie (FOM). FOM is financially supported by the Nederlandse Organisatie voor Wetenschappelijk Onderzoek (NWO).

REFERENCES

1. R.A. Street, Hydrogenated amorphous silicon, (Cambridge University Press, New York, 1991) p. 363.
2. M. Gupta, V.K. Rathi, R. Thangaraj and O.P. Agnihotri, Thin Solid Films 204, 77(1991).
3. M.C. Hugon, F. Delmotte, B. Agius and J.L Courant, J. Vac. Sci. Technol. A 15, 3143 (1997).
4. W. Stutius and W. Streifer, Applied Optics 16, 3218 (1977).
5. Y. Yang and M. H. White, Solid-State Electronics 44, 949 (2000).
6. C.J. Dell'oca and M.L. Barry, Solid-State Electronics 15, 659 (1972).
7. F. Buda, G.L. Chiarotti, R. Car, and M. Parrinello, Phys. Rev. B 44, 5908 (1991).
8. P.A. Fedders and D.A. Drabold, Phys. Rev. B 47, 13277 (1993).
9. B. Tuttle and J.B. Adams, Phys. Rev. B 53, 16265 (1996).
10. G.R. Gupte, R. Prasad, V. Kumar, and G.L. Chiarotti, Bull. Mater. Sci. 20, 429 (1997).
11. M. Tosolini, L. Colombo, and M. Peressi, Phys. Rev. B 69, 075301 (2004).
12. B.C. Pan and R. Biswas, J. Appl. Phys. 96, 6247 (2004).
13. M. Durandurdu, D.A. Drabold and N. Mousseau, Phys. Rev. B 62, 15307 (2000).
14. G. Kresse and J. Hafner, Phys. Rev. B 47, 558 (1993).
15. G. Kresse and J. Furthmüller, Phys. Rev. B 54, 11169 (1996).
16. J.P. Perdew, J.A. Chevary, S.H. Vosko et al., Phys. Rev. B 46, 6671 (1992).
17. G. Kresse and D. Joubert, Phys. Rev. B 59, 1758 (1999).
18. P.E. Blöchl, Phys. Rev. B 50, 17953 (1994).
19. H.J. Monkhorst and J.D. Pack, Phys. Rev. B 13, 5188 (1976).
20. R. Bellissent, A. Menelle, W.S. Howells et al., Physica B 156, 217 (1989).
21. K. Laaziri, S. Kycia, S. Roorda et al., Phys. Rev. B 60, 13520 (1999).
22. G. Kresse and J. Hafner, Phys. Rev. B 55, 7539 (1997).

**Poster Session:
Crystallization**

Mater. Res. Soc. Symp. Proc. Vol. 1153 © 2009 Materials Research Society   1153-A05-01

High Temperature Post-Deposition Annealing Studies of Layer-by-Layer (LBL) Deposited Hydrogenated Amorphous Silicon Films

Goh Boon Tong, Siti Meriam Ab. Gani, Muhamad Rasat Muhamad and Saadah Abdul Rahman
Low Dimensional Materials Research Center, Department of Physics, University of Malaya, 50603 Kuala Lumpur, Malaysia

ABSTRACT

High temperature post-deposition annealing studies were done on hydrogenated amorphous silicon thin films deposited by plasma-enhanced chemical vapour deposition (PECVD) using the layer-by-layer (LBL) deposition technique. The films were annealed at temperatures of 400 °C, 600 °C, 800 °C and 1000 °C in ambient nitrogen for one hour. Auger electron spectroscopy (AES) depth profiling results showed that high concentration of O atoms were present at the substrate/film interface and at film surface. Very low concentration of O atoms was present separating silicon layers at regular intervals from the film surface and the substrate due to the nature of the LBL deposition and these silicon oxide layers were stable to high annealing temperature. Reflectance spectroscopy measurements showed that the onset of transformation from amorphous to crystalline phase in the LBL a-Si:H film structure started when annealed at temperature of 600 °C but the X-ray diffraction (XRD) and Raman scattering spectroscopy showed that this transition only started at 800°C. The films were polycrystalline with very small grains when annealed at 800 °C and 1000 °C. Fourier transform infrared spectroscopy (FTIR), measurements showed that hydrogen was completely evolved from the film at the on-set of crystallization when annealed at 800 °C. The edge of the reflectance fringes shifted to longer wavelength decrease in hydrogen content but shifted to shorter wavelength with increase in crystallinity.

INTRODUCTION

Transformation of hydrogenated amorphous silicon (a-Si:H) from amorphous meta-stable phase to stable crystalline phase has been the subject of interest for many years. The interest was mainly due to its potential applications in thin film transistors and solar cells as the polycrystalline Si films formed when annealed at high temperature had higher mobility, higher conductivity and larger breakdown voltages at metal-oxide-semiconductor interface compared to the amorphous Si films [1, 2]. The transformation from amorphous to crystalline phase has been monitored by studying the effects of thermal annealing on the various properties of a-Si:H films [3]. Monitoring the crystallization of a-Si:H films grown by layer-by-layer (LBL) deposition technique using our home-built radio-frequency plasma enhanced chemical vapour deposition (rf-PECVD) system as a result of high temperature annealing is the focus of this work. The LBL deposition technique involved periodic interruptions of the deposition process whereby the plasma discharge of silane diluted in hydrogen was stopped for a fixed period of time during which the growth surface was treated with hydrogen plasma discharge. The effects of high temperature annealing on the optical, structural and chemical bonding properties of a-Si:H films prepared by LBL deposition technique are studied in this work.

EXPERIMENT

The hydrogenated amorphous silicon (a-Si:H) films in this work were deposited on p-type (111) crystal silicon (c-Si) wafer substrates using a home-built rf PECVD (13.56 MHz) system from the discharge of SiH_4 and H_2 gas mixture. The details of the deposition system have been described elsewhere [4]. The reactor was capacitively coupled by two parallel electrodes with the electrodes distance and area of 5 cm and 28 cm^2 respectively. In this work, the LBL process was performed by periodically alternating of the deposition of Si:H layer for 5 minutes with the hydrogen plasma treatment of the growth surface for 3 minutes. This process was repeated for 4 cycles where 4 layers of Si:H film were deposited intermittently with hydrogen plasma treatment of the growth surface resulting in a total deposition time of 32 minutes. This LBL mode was done by switching off the silane source at the end of each cycle of Si:H layer deposition followed by the hydrogen plasma treatment of the deposited surface. Hydrogen plasma cleaning of the chamber and substrates was done for 10 minutes prior to the deposition process. The power for generating the hydrogen plasma cleaning was fixed at 40 W. The deposition pressure and deposition temperature were maintained at 0.8 mbar and 100 °C respectively. The rf power was fixed at 20 W which corresponded to a power density of 700 mW/cm^2. The SiH_4 and H_2 flow-rates were fixed at 5 sccm and 20 sccm respectively producing a H_2 to SiH_4 flow-rate ratio of 4. In this work, a set of films consisting of the as-deposited film and films annealed at temperatures of 400 °C, 600 °C, 800 °C and 1000 °C were studied. The films were annealed for one hour at these temperatures using a conventional furnace (Carbolyte CFM 12/1) in ambient nitrogen gas flow.

Auger electron spectroscopy (AES) was used for elemental composition depth profiling studies from the surface of the film to the substrate using a JEOL JAMP-9500F field emission Auger microprobe. The etch-rate used for the depth profiling measurement was 0.74 nm s^{-1}. XRD measurements were done using the SIEMENS D5000 X-ray diffractometer. The average grain size of the crystallites in the films was estimated from full-width at half-maximum (FWHM) of (111) peaks using Scherrer's formula [5]. The Raman spectra of the films were recorded using a Horiba Jobin Yvon 800 UV Micro-Raman Spectrometer with and CCD detector. The power of the Ar^+ laser used was 20 mW and the excitation wavelength selected for this measurement was 514.5 nm. The laser was focused onto a spot of 1 um in diameter onto the sample to collect the backscattered light from the samples for Raman measurement. The Raman spectra obtained were deconvoluted into three peaks located at around 480 cm^{-1}, 500 cm^{-1} and 520 cm^{-1} which correspond to the TO mode of amorphous component, grain boundaries and crystalline component, respectively. The crystalline volume fraction, X_C is determined from the integrated intensities of the peaks corresponding to amorphous component, grain boundaries and crystalline component and the grain size was calculated from the shift of the crystalline peak as described by Daxing Han et al [6]. The reflectance spectra were obtained using a JASCO V570 ultra-violet visible near-infrared (UV-VIS-NIR) spectrophotometer and from the spectra, the increase in structural order of the film structure was monitored. FTIR spectra of the films were measured using a Perkin-Elmer System 2000 FTIR spectrometer in the range of 4000 cm^{-1} – 400 cm^{-1}. The vibrational properties and the hydrogen content were determined from FTIR spectra.

RESULTS AND DISCUSSIONS

Figure 1(a) shows the depth profiling trends of atomic concentration of oxygen (O) and silicon (Si) atoms derived from Auger Electron Spectra (AES) of the LBL deposited films annealed at different temperatures. The results showed that the Si atom was the highest elemental component with uniform distribution within the film structure. The Si atom concentration dropped sharply while the O atom concentration increased sharply at the film/c-Si substrate interface confirming presence of SiO$_x$ at the interface. The O atom concentration was also high at the surface of the film but was significantly low within the film structure where the O atoms layers were separated from each other by layers of Si atoms at five regular intervals which was one extra from the number of cycles used for the LBL deposition. These showed that the film consisted of Si layers separated by layers of Si rich SiO$_x$ layers formed either during film deposition or the H plasma surface treatment of the LBL deposition process. The variations of the fraction of O atoms at the substrate/film interface and at the surface the film along with the duration of the depth profiling time with annealing temperature are plotted in Fig. 1(b). The depth profiling time can be estimated to be proportional to the film thickness. The variation of the fraction of O atoms at the film/substrate interface with annealing temperature was similar to the trend produced by the depth profiling time indicating that thicker films deposited by the LBL technique have higher O content at the interface. Thicker films had a more relaxed film structure as a result of more dominant biaxial tensile stress between atoms in the film compared to the compressive stress. The thicker film allowed more diffusion of O atoms to the film/substrate interface. The O content at the surface of the film decreased slightly to a minimum when annealed at 400 °C and increased with further increase in annealing temperature. O atoms from the atmosphere diffused into the film at the surface during annealing through low density regions and satisfied dangling bonds left by broken Si-H bonds leading to the formation of Si-O bonds with increase in annealing temperature above 400 °C. This has also been reported by other workers [7, 8].

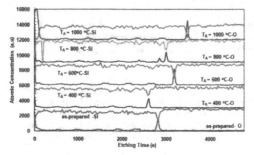

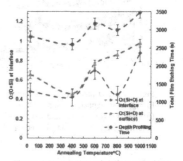

Figure 1(a) AES atomic concentration of oxygen (O) and silicon (Si) atoms depth profile of Si:H films prepared by LBL deposition technique annealed at different temperatures.

Figure 1(b) Fraction of O atoms in Si:H films at the film/substrate interface and at the surface along with the depth profiling time versus annealing temperature.

The XRD patterns plotted in Fig 2 showed that the films were amorphous in structure, as-prepared, and when annealed at 400 °C and 600 °C. Formation of crystalline structures in the film surfaced when the films were annealed at temperatures at 800 °C from the appearance XRD peaks at 2θ angles of 28.4 °, 47.3 ° and 56.1 ° representing the c-Si (111), (220) and (311) orientation planes respectively. The grain size as calculated from the FWHM of the XRD (111) peak and Scherrer's formula increased from 19 nm to 23 nm when annealed at temperatures of 800 °C to 1000 °C respectively.

Figure 3 shows the Raman scattering spectra of the films prepared by LBL deposition technique for as-prepared films, the films annealed at annealing temperatures of 400 °C,

600 °C, 800 °C and 1000 °C. The as-prepared film and the films annealed at temperatures of 400 °C and 600 °C were amorphous as indicated by the significant presence of the broad amorphous Si peak at 480 cm^{-1}. For the films annealed at temperatures of 800 °C and 1000 °C, the appearance of a significantly sharp crystalline Si peak at 520 cm^{-1} with a significant shoulder at 500 cm^{-1} indicated that the films consisted of very small crystalline grains separated by very thin amorphous grain boundaries. The results obtained from the Raman spectra confirmed the presence of crystalline grains in the films annealed at 800 °C and 1000 °C. The crystalline grain size, D_R and crystalline volume fraction, X_C as determined by the Raman spectra also increased with increase in annealing temperature from 800 °C to 1000 °C. D_R increased from 5.9 nm to 9.6 nm while the crystalline volume fraction increased from 37 % to 66 %. Since the amorphous component was still present in both films, it can be indicated that the films consisted of nano-sized crystalline grains embedded within the amorphous matrix of the film structure and the films were not totally polycrystalline in structure. The crystallite size calculated from the XRD peaks were three times larger than the crystallite size measured from the Raman crystal Si peak. This may be due the difference in sensitivity of the measurement techniques.

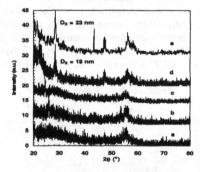

Figure 2 XRD patterns of LBL deposited Si:H films, (a) as-deposited, annealed at (b) 400 °C, (b) 600 °C, (c) 800 °C and (d) 1000 °C. The crystalline grain size, D_X is indicated on the plots.

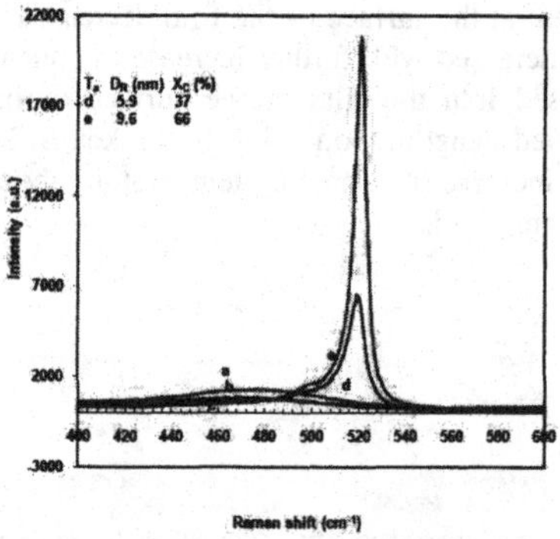

Figure 3 Raman spectra of LBL deposited Si:H films, (a) as-deposited, annealed at (b) 400 °C, (b) 600 °C, (c) 800 °C and (d) 1000 °C. The crystalline grain size, D_R and the crystalline volume fraction, X_C are indicated on the plots.

Fig. 4 shows the reflectance spectra for the as-prepared annealed at different annealing temperatures. The reflectance fringes for the as-prepared film ended at ~570 nm and the smooth feature in the ultra-violet (UV) region indicate the amorphous nature of the film [9]. When annealed at 400 °C, and 600 °C, the edge of the interference fringes shifted towards longer wavelength of 626 nm and 692 nm respectively and the amplitude of the fringes decreased. This might be due to evolution of hydrogen from the film resulting in the decrease in the optical band gap [10]. The edge of the interference fringes is shifted to shorter wavelengths of 640 nm and 598 nm when annealed at 800 °C and 1000 °C respectively and the amplitude of the interference fringes were also significantly reduced. B. Roy et al [3] contributed the shift of the edge of the interference fringes towards shorter wavelength for films with high crystallinity to the narrowing of optical band gap and increasing absorption. However, for these films, the shift was also coupled with reduction in reflectance intensity which Stradins et al [9] related to the formation of mixed crystalline and amorphous grains in the film which resulted in maximum scattering of light. Since the grain sizes, as calculated from XRD Si (111) diffraction peak can categorize the grains to be nano-sized grains and crystalline volume fraction were 37 % and 66 % for the films annealed at 800 °C and 1000 °C, the blue shift maybe the result of a mixed phase structure of nano-crystalline Si grains embedded within the amorphous matrix of the film structure. The transformation to the crystalline of the film structure in these films was also evidenced from the appearance of two clear peaks in the ultra-violet region and these results were consistent with the appearance of XRD peaks in these films. These two peaks even started to appear when the film was annealed at 600 °C marking the onset of increase in structural order in the amorphous film structure. The higher intensity of these peaks in the UV region for the film annealed at 1000 °C showed that higher crystalline volume fraction and larger crystalline grains increased the intensity of the UV peaks of the reflectance spectrum of the Si:H film. These results suggested that the reflectance spectra were very sensitive to the increase in structural order of the film.

Fig 5 presents the FTIR spectra of the films prepared by LBL deposition technique, as-prepared and when annealed at different annealing temperatures. These spectra showed significant presence of Si-H bonds in the film from the appearance of absorption peaks at around 640 cm^{-1}, 880 cm^{-1} and 2000 – 2090 cm^{-1} which corresponded to Si-H wagging, (Si-H$_2$)$_n$ bending and Si-H/Si-H$_2$ stretching bands respectively [11]. These absorption bands decreased significantly when the film was annealed at 600 °C. This indicated that H atoms were released from Si-H bonds when annealed at these temperatures. When annealed at temperatures of 800 °C and 1000 °C, the absorption bands representing the Si-H wagging, (Si-H$_2$)$_n$ bending and Si-H/Si-H$_2$ stretching bands completely disappeared. The hydrogen content as calculated from the integrated intensity of the Si-H wagging band at 640 cm^{-1} [12] decreased from 29 % for the as-prepared film to 9 % and 0.2 % for the film annealed at 400 °C and 600 °C respectively. This confirmed that the red-shift of the edge of interference fringes in the reflectance spectra of these films were due to significant decrease in H content in the film.

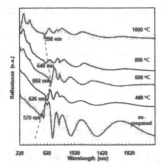

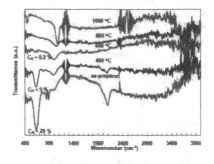

Figure 4 Reflectance spectra in the ultra-violet to near-infrared region of Si:H films prepared by LBL deposition technique, as-prepared and annealed at different temperatures. The edge of the interference fringes are indicated on the plots.

Figure 5 FTIR spectra of Si:H films prepared by LBL deposition technique annealed at different temperatures. The H content, C$_H$ in the films are indicated on the plots.

CONCLUSIONS

The effects of high temperature post-deposition annealing at temperatures of 400 °C, 600 °C, 800 °C and 1000 °C on the optical, structural and surface morphology of LBL deposited a-Si:H films have been studied. The AES profiling results showed that O atoms are abundant at the Si:H/c-Si substrate interface and at the film surface. LBL deposition formed alternate layers of Si:H films and highly Si rich SiO_x films and annealing at high temperatures had no observable effect on the SiO_x layers. The XRD and Raman spectroscopy results consistently showed that annealing of a-Si:H film prepared by LBL deposition technique only resulted in the onset of transformation of amorphous phase to crystalline phase when annealed at 800 °C for one hour which was high compared to films prepared by continuous deposition. The increase in structural order and the formation of crystalline structures in the film produced two peaks in the UV region of the reflectance spectra. These peaks increased in amplitude with increase in crystalline volume fraction and crystalline grain size. These results showed that the reflectance spectra of a film could also be used to monitor the onset of crystallinity in amorphous films and was sensitive to structural ordering in the film structure. The films annealed at 800 °C and 1000 °C are of mixed phase with nano-crystalline grains embedded within the amorphous matrix of the films. The nano-crystalline grains and crystalline volume fraction increased with increase in annealing temperature.

ACKNOWLEDGMENTS

This work was supported by the Ministry of Higher Education of Malaysia under Fundamental Research Grant Scheme (FP008/2008C, FP016/2008C) and University of Malaya Short-Term Grant (FS289/2008C, FS299/2008C).

REFERENCES

1. K. Ono, S. Oikawa, N. Konishi and K. Miyata, *Jpn. J. Appl. Phys.* **29**, 2705 (1990).
2. A.M. Funde, Nabeel Ali Bakr, D.K. Kamble, R.R. Hawaldar, D.P. Amalnerkar and S.R. Jadkar, *Sol. Energy Mater. & Sol. Cells* **92**, 1217 (2008).
3. B. Roy, A.H. Mahan, Q. Wang, R. Reedy, D.W. Readey and D.S. Ginley, *Thin Solid Films* **516**, 6517 (2008).
4. Goh Boon Tong, and Saadah Abdul Rahman, in: Proceedings of the IEEE International Conference on Semiconductor Electronics 2006 (ICSE 2006), Kuala Lumpur, Malaysia, 29th Nov. – 1st Dec., 2006, pp. 472 – 476.
5. H.P. Klung and L.E. Alexander: *X-ray Diffraction Procedures* (Wiley, New York, 1974).
6. Daxing Han, J.D. Lorentzen, J. Weinberg-Wolf, L.E. McNeil and Qi Wang, *J. Appl. Phys.* **94** (5), 2930 (2003).
7. D.K Basa, G. Ambrosone, U. Coscia and A. Setaro, *Appl. Surf. Sci.* **255**, 5528 (2009).
8. R.J. Prado, T.F. D' Addio, M.C.A. Fantini, I. Pereyra and A.M. Flank, *J. Non-Cryst. Solid* **330**, 196 (2003).
9. G. Harbeke, E. Meier, J.R. Sandercock, M. Tgetgel, M.T. Duffy and R.A. Soltis, *RCA Review* **44**, 19 (1983).
10. P. Stradins, D. Yaung, H.M.Branz, M. Page and Q. Wang, *Mater. Res. Soc. Symp. Proc.* **862**, 227 (2005).
11. G. Lucovsky, R.J. Nemanich and J.C. Knights, *Phys. Rev. B* **19**, 2064 (1979).
12. A.A. Langford, M.L. Fleet, B.P. Nelson, W.A. Lanford and N. Maley, *Phys. Rev. B* **45**, 367 (1992).

Mater. Res. Soc. Symp. Proc. Vol. 1153 © 2009 Materials Research Society   1153-A05-03

Modeling the Grain Size Distribution During Solid Phase Crystallization of Silicon

Andreas Bill[1], Anthony V. Teran[1] and Ralf B. Bergmann[2]
[1]Department of Physics & Astronomy, California State University Long Beach, 1250 Bellflower Blvd., Long Beach, CA 90840, U.S.A.
[2]Bremen Institute for Applied Beam Technology (BIAS), Klagenfurter Str. 2, 28359 Bremen, Germany

ABSTRACT

We analyze the grain size distribution during solid phase crystallization of Silicon thin films. We use a model developed recently that offers analytical expressions for the time-evolution of the grain size distribution during crystallization of a d-dimensional solid. Contrary to the usual fit of the experimental results with a lognormal distribution, the theory describes the data from basic physical principles such as nucleation and growth processes. The theory allows for a good description of the grain size distribution except for early stages of crystallization. The latter case is expected and discussed. An important outcome of the model is that the distribution at full crystallization is determined by the time-dependence of the nucleation and growth rates of grains. In the case under consideration, the theory leads to an analytical expression that has the form of a lognormal-type distribution for the fully crystallized sample.

INTRODUCTION

In applications one targets specific functionalities of a material, which are generally defined by the microstructure of the solid. More often than not one relies on empirical knowledge to determine the growth conditions for which the desired microstructure is obtained. This is for example the case in the crystallization of an amorphous solid such as Silicon. It is of interest to be able to determine the microstructure from a more fundamental point of view.

One important tool to characterize the micromorphology of a solid during its crystallization is the time-dependent grain size distribution (GSD) $N(g,t)$, where g defines some physical quantity that measures the size of the grain (diameter of the grain, number of atoms, volume, etc.) [1-6]. N counts the number of grains that have a certain size g at time t during crystallization. Recently we developed a theory describing the evolution of the GSD during the crystallization of a d-dimensional solid ($d = 2$ for a thin film, $d = 3$ for a bulk solid) [5,6]. The theory considers random nucleation and growth (RNG) processes, but the model might be applicable to a variety of growth phenomena both in and outside the realm of solid-state physics.

We apply the theory [5,6] to solid phase crystallization of a Si thin film [1-4]. Within the RNG model nuclei are first formed at random in the sample. These nuclei grow while others are created. This leads to a time dependent GSD, which can for example be observed by Transmission Electron Microscopy (TEM). The procedure employed for the evaluation of the cross-sectional TEM samples is described in [3]. In the present application we compare the theoretically determined GSD $N(g,t)$ to the measured one. Here g describes the diameter of a sphere, the volume of which equals the volume of the observed grain. It is worth pointing out that in the experiment we have a constant film thickness. The grain growth is isotropic at first in all three dimensions of space until the diameter of the grains reaches the film thickness. From that point on, the largest grains continue to grow laterally. In the thin films analyzed below, a certain number of grains have a diameter that is larger than the film thickness of about 1 μm. Hence, the growth is actually a mix between 2D and 3D growth. Since the majority of grains have a size smaller than the quoted thickness, we assume that the grain size distribution will mainly reflect a three-dimensional growth. We analyze histograms of the size distribution obtained at five different times ranging from early to the final stages of crystallization.

THEORY

The general formulation of the theory [5,6] gives the grain size distribution for grains modeled by d-dimensional ellipsoids. In the present study we limit our discussion to three-dimensional spheres of diameter g, as the experimentally obtained histograms are represented in terms of this quantity. The theory relies on two well-established models. The first is the random nucleation and growth model described above. This model leads the GSD $N(g,t)$ to satisfy the following partial differential equation for three-dimensional growth of grains

$$\frac{\partial N(g,t)}{\partial t} + \frac{v(t)}{g^2}\frac{\partial}{\partial g}\left(g^2 N\right) = I(t)\delta(g - g_c).$$ (1)

The right hand side is the source term, where g_c is the diameter of the nucleus, the smallest stable grain formed in the system. The crystallization process is determined in terms of two time-dependent rates, the nucleation rate $I(t)$ and the growth rate $v(t)$. The second model, the Kolmogorov-Avrami-Mehl-Johnson (KAMJ) model [7-9], provides explicit expressions for these rates. These rates contain an exponential that gives the fraction of material available at time t for further nucleation and growth. For the present work these rates have the form [5,6]

$$I(t) = I_0 \exp\left[-\left(\frac{t-t_0}{t_{cl}}\right)^4\right] \times \Theta\left(\frac{t-t_0}{t_{cl}}\right), \quad v(t) = v_0 \exp\left[-\left(\frac{t-t_0}{t_{cv}}\right)\right] \times \Theta\left(\frac{t-t_0}{t_{cv}}\right),$$ (2)

where t_{cl}, t_{cv} are the critical times for the decay of the nucleation and growth rates, respectively. The theory actually essentially depends on the ratio $t_r^2 = t_{cv}/t_{cl}$ rather than on the individual critical times [5,6]. In Eq.(2) t_0 is the incubation time and $\Theta(t)$ is the Heaviside function. As stated in Eq.(2) we assume that the KAMJ model provides the nucleation rate $I(t)$. By analogy, we assume that the growth rate has similar time dependence, though with different critical time and power law of the time-dependence in the exponential.

Solving analytically Eq.(1) with the rates given by Eq.(2) we obtain

$$N(g,t) = \frac{I_0}{v_0}\frac{1}{\pi g_c^2}\left(\frac{g_c}{g}\right)^2 \frac{1}{\alpha(g,t)}\exp\left\{\left[t_r^2 \ln \alpha(g,t)\right]^4\right\} \times \left[\Theta\left(\frac{g-g_c}{g_\infty - g_c}\right) - \Theta\left(\frac{g-g_{max}(t)}{g_\infty - g_c}\right)\right]$$ (3)

with

64

$$\alpha(g,t) = \frac{g - g_c}{g_\infty - g_c} + \exp\left(-v_0 \frac{t - t_0}{g_\infty - g_c}\right), \tag{4}$$

and

$$g_{max}(t) = g_c + \left(g_\infty - g_c\right)\left[1 - \exp\left(-v_0 \frac{t - t_0}{g_\infty - g_c}\right)\right], \qquad g_\infty = g_c + t_{cv}v_0, \tag{5}$$

The latter two expressions give the diameter of the largest grain found in the sample at time t, $g_{max}(t)$, and at full crystallization, $g_\infty = g_{max}(t \to \infty)$.

In the figures of the following section the histograms are represented in such a way that the area of the bins sum up to one (normalized). To compare the theory with these histograms we normalize the above expression by dividing Eq.(3) by

$$N(t) = \int_0^\infty dg N(g,t) = \int_{g_c}^{g_{max}(t)} dg N(g,t). \tag{6}$$

Since the constant prefactor $I_0/v_0\pi$ appears both in Eq. (3) and (6) I_0 cancels out from the normalized GSD.

APPLICATION TO SOLID PHASE CRYSTALLIZATION OF SILICON

Fig. 1 displays the experimental histograms together with the theoretical curve obtained from Eqs.(1-6) for $t = 15, 9$, and 5h. Choices of parameters are found in table I.

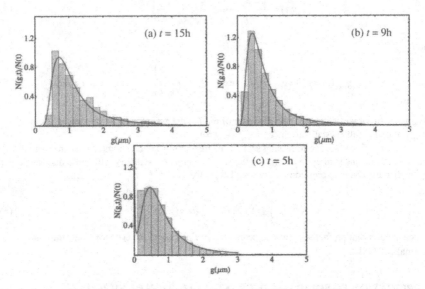

Figure 1. Histogram of the grain size distribution obtained at three different times t during solid-phase crystallization of a Si film [1-4]. Overlaid are the theoretical curves discussed in the text.

The histograms were obtained for solid phase crystallization of Si thin film grown from TEM [1-4]. Films are deposited by low-pressure chemical vapor deposition and are then thermally crystallized. Details of the fabrication and crystallization process are described in Ref. [2]. The GSD was determined for five samples at $t = 2.5, 5, 9, 15$ and 24h (the first and the latter data set are not shown here). The samples at 9, 15, and 24h are fully crystallized thin films ($X = 100\%$.) Table I summarizes the experimental data and the parameters used in the theory.

The theory requires the knowledge of the following quantities: t_0, t_r, v_0, g_∞, and g_c. As mentioned earlier the normalized expression, the ratio of Eq.(3) and (6), does not depend on I_0. From the experimental data we have (see Refs.[1,3]) $t_0 = (1.5 \pm 0.5)$h, $v_0 = (1.72 \pm 0.34)\mu$m/h, $g_c = 2.2$nm. The maximal grain size at full crystallization, g_∞, used for the theory was obtained from the average of g_{max} measured at $t = 9, 15, 24$ h. Furthermore, from Ref.[5] we take $t_{cl}/t_{cv} = 1.74 \pm 0.02$ which leads to the value $t_r = 0.76$. Guided by previous work done on a fully crystallized sample [5] we search for the best values of the parameters within their error margins to plot the figures above. The values of the parameters are found in table I.

Table I. Compilation of experimental data and values of the parameters used in Eqs. (3,4) for Figs. 1a-c. Experimental: Crystallization time t, crystallized fraction X, largest observed grain size g_{max}. Theory parameters: maximal grain size at full crystallization g_∞, growth rate v_0, incubation time t_0, and critical time ratio t_r (the choice of parameters is described in the text.)

Experiment			Parameters used in the theory			
Time (h)	Fraction X (%)	g_{max} (μm)	g_∞ (μm)	v_0 (μm/h)	t_0 (h)	t_r
15	100	4.9	4.3	1.72	1.4	0.78
9	100	3.6	3.7	1.72	1	0.76
5	20	4.0	3.4	2.06	1.4	0.9

DISCUSSION

We note first that the values of the parameters chosen for the theoretical curves of Fig.1 and summarized in table I are within the standard deviation of the experimentally determined quantities. We have chosen values such that the reduced χ^2 defined by

$$\chi^2 = \frac{\sum_{i=1}^{M}[y_i(t) - N(g_i,(t))]^2}{M-p},$$ (7)

is minimized within the range of standard deviations. M and p are the number of measured points (number of bins of the histogram) and the number of free parameters in the model (here $p = 2$), y_i are the experimental values of the normalized GSD (the heights of the bins) and $N(g_i,t)$ is the theoretically determined GSD for the grain diameters of the histogram.

Consider first the fully crystallized samples ($X = 100\%$) obtained for $t = 15, 9h$. There is excellent agreement between the data and the theory. In the present analysis, the latter has the form of a lognormal type distribution, as seen from Eq.(3) at $t \to \infty$. It is noteworthy that with the parameters of table I, the theory predicts that the final crystallization stage is reached after about 8 hours. This agrees well with the experimental observation. The theoretical expression at $t \to \infty$ offers a better description of the fully crystallized sample than even a fit of the data with the standard lognormal distribution. We stress, however, that the approach of the present paper is not a simple fit of the data. Instead, the theory gives an expression for the GSD that is derived from basic principles, in terms of physical parameters and as a function of time. We also point out that Eq.(3) is not a pure lognormal distribution; there is a different power law in the exponential term and there are cutoffs at small and large grain diameters. These differences with respect to the lognormal distribution are discussed in Ref. [5,6].

We now consider the case of a partially crystallized sample ($t = 5h$). There is good overall agreement with the histogram. There are, however, discrepancies at small grain sizes. The three lowest bins of the histogram have the largest number of grains. On the other hand, the theory displays a dip and a very sharp peak close to the critical grain size g_c. The latter comes from the nucleation term I_0 in Eq.(1), whereas the broader peak at larger values of g results from the growth term v_0. The discrepancy between experiment and theory may be related to the assumption of a Dirac delta distribution to model nucleation (right hand side of Eq. (1)). This term assumes that only nuclei with a diameter that is exactly g_c can form, which is clearly too restrictive. It is expected that replacing the Dirac delta function by a Gaussian leads to a broader nucleation peak and the filling of the dip seen at $t = 5h$ (see Ref.[1]). This is also consistent with the fact that to the authors' knowledge the nucleation and growth peaks have never been resolved.

We also attempted a description of the t=2.5h data [4]. While the parameters obtained for fractions of about 20% and above give a good description of the data, we were not able to obtain a satisfying fit for $X = 2.8\%$ ($t = 2.5h$). We attribute this to the model used here. Indeed, the nucleation term is dominant at early stages of crystallization. As explained above, that term is a simplified expression and is not expected to reproduce accurately the early stage [10]. Another reason for the discrepancy may be related to the assumed functional form of the growth rate [5,6]. While the nucleation rate $I(t)$ has a time-dependence provided by the KAMJ model, the expression for the growth does not rely on a microscopic derivation. It may be that the model proposed here does not capture the full time-dependence of the growth rate. It is possible to obtain a good fit of the data by allowing the parameters of the model to vary freely (in particular t_r and v_0). However, to reach agreement between theory and the experiment at $X = 2.8\%$ the parameters would have to be given unphysical values. This emphasizes the statement made above, that the theory is not a mere fit of data, but results from a theory motivated by the physics of random nucleation and growth crystallization processes [5,6].

CONCLUSIONS

The theory developed in Refs.[5,6] was applied to the time-evolution of the grain size distribution in solid-phase crystallization of Si thin films [1-4]. The theory was not only used to study fully crystallized material, but also to analyze partial crystallization. Overall the theory provides a good description of the observations. For samples with a fraction of crystallized material that exceeds about $X = 20\%$ the theory agrees well with the data for reasonable values of the parameters. For small fractions X of crystallized material on the other hand, the theory cannot reproduce the data with reasonable parameter values. The discrepancy was expected since early stages of crystallization strongly depend on the source term in Eq.(1), which does not accurately describe such regime. This fact emphasizes that the theory is not a mere fit to the data, but offers a description that relies on basic physical processes [5,6]. In some instances such as the one considered in the present article the theory shows that the grain size distribution evolves to become lognormal-like at full crystallization. Finally, given the general terms in which the model was developed it is expected that the theory is applicable to a wider range of phenomena in physics, chemistry, or even biology or economics.

ACKNOWLEDGMENTS

A.B. and A.V.T. gratefully acknowledge support from the Research Corporation and SCAC at CSU Long Beach.

68

REFERENCES

1. R.B. Bergmann, F.G. Shi, H.J. Queisser and J. Krinke, Appl. Surf. Sci. **123-124**, 376 (1998).
2. R.B. Bergmann, G. Oswald, M. Albrecht and V. Gross, Solar Energy Materials and Solar Cells **46**, 147 (1997).
3. R. B. Bergmann, J. Krinke, H. P. Strunk, and J. H. Werner, Mat. Res. Soc. Symp. Proc. **467**, 352 (1997).
4. R.B. Bergmann and F. Shi, Phys. Rev. Lett. **80**, 1011 (1998).
5. R.B. Bergmann and A. Bill, J. Cryst. Growth **310**, 3135 (2008).
6. A. Teran, R.B. Bergmann and A. Bill, unpublished.
7. A.N. Kolmogorov, Akad. Nauk SSSR, Izv. Ser. Matem. **1**, 355 (1937).
8. M. Avrami, J. Chem. Phys. **7**, 1103 (1939); *ibid.,* **8**, 212 (1940).
9. W. Johnson and R. Mehl, Trans AIME **135**, 416 (1939); W. Anderson and R. Mehl, *ibid.,* **161**, 140 (1945).
10. F. Shi and J.H. Seinfeld, J. Mater. Res. **6**, 2091 (1991); *ibid.* Mat. Chem. Phys. **37**, 1 (1994).

Mater. Res. Soc. Symp. Proc. Vol. 1153 © 2009 Materials Research Society   1153-A05-04

Polysilicon Films Formed on Metal Sheets by Aluminium Induced Crystallization of Amorphous Silicon: Barrier Effect

P. Pathi, Ö. Tüzün and A. Slaoui

Institut d'Électronique du Solide et des Systèmes, UMR 7163 CNRS-UdS, 23 rue du Loess, F-67037 Strasbourg Cedex 2, France

ABSTRACT

Polycrystalline silicon (pc-Si) thin films have been synthesized by aluminium induced crystallization (AIC) of amorphous silicon (a-Si) at low temperatures (≤500°C) on flexible metallic substrates for the first time. Different diffusion barrier layers were used to prepare stress free pc-Si films as well as to evaluate the effective barrier against substrate impurity diffusion. The layers of aluminum (Al) and then amorphous silicon with the thickness of 0.27 μm and 0.37 μm were deposited on barrier coated metal sheets by means of an electron beam evaporation and PECVD, respectively. The bi-layers were annealed in a tube furnace at different temperatures (400-500°C) under nitrogen flow for different time periods (1-10hours). The degree of crystallinity of the as-grown layers was monitored by micro-Raman and reflectance spectroscopies. Structure, surface morphology and impurity analysis were carried out by X-ray diffraction, scanning electron microscopy (SEM) and EDAX, respectively. The X-ray diffraction measurements were used to determine the orientation of grains. The results show that the AIC films on metal sheets are polycrystalline and the grains oriented in (100) direction preferentially. However, the properties of AIC films are highly sensitive to the surface roughness.

INTRODUCTION

Among the alternative approaches for efficient thin film solar cells, polycrystalline silicon (pc-Si) thin films on foreign substrates (such as ceramic, graphite, glass) seem a very promising candidate due to their low cost and high efficiency potential [1]. Particularly, the use of flexible substrates paves a way for a new trend in the development and application of solar cells due to their unique advantages such as light weight and flexibility in different sorts like building integration and orienting on uneven surfaces. Cr free steel foils are more effective in terms of economic production of solar cell modules. In addition, it could be possible to exploit the inherent properties of metals like optical reflection and electrical conductivity in the solar cells made on metallic substrates. However, various key points need to be addressed when the metallic sheets are used as substrates, particularly with temperature processing. Some of the main issues are; diffusion of substrate elements like Fe, Ni, etc., which is detrimental for the life time of the charge carriers and hence the device efficiency, mechanical integrity of barrier layer with substrate as well as silicon films due to the mismatch of coefficient of thermal expansion and corrosion. The migration of impurities can be avoided by incorporating an intermediate layer called, "diffusion barrier". The barrier performance against the migration of impurities depends on its physical characteristics like density, defects, etc. The mechanical integrity of the substrate, barrier layer and active layer is highly important in the course of manufacturing and operational life-time of the solar cells made on these substrates. Also, it is possible for the barrier layer to delaminate from the substrate. The reasons for the delamination could be manifold and are not known until now [2,3]. It could be mainly originated from the mismatch of thermal expansion coefficients and interfacial reactions take place when annealed at higher temperatures.

The aluminium induced crystallization (AIC) method, which is simple, low-temperature and short-time process, has been successfully used to produce large grain and defect free polycrystalline silicon (pc-Si) thin films on glass and ceramic substrates [4]. In this work, an attempt has been made for the first time to synthesize poly-Si films by AIC technique on metallic substrates. The metallic substrates were coated with different diffusion barrier layers and their effect on the AIC process was studied. The AIC grown pc-Si films have been exploited to study the barrier layer performance in terms of diffusion characteristics and the mechanical integrity of barrier layer with the substrate and pc-Si films by making use of EDAX coupled with SEM analysis of AIC grown films.

EXPERIMENTAL

The metallic substrates used in this study are made of Fe-Ni, which is cheaper compared to the stainless steel, and also these substrates do not receive any preconditioning step prior to the deposition of diffusion barrier layers. The substrates were cleaned ultrasonically using organic solvents in order to remove oil residues from rolling. Different type of dielectric and conducting barrier layers have been used as barrier layers (BL). The dielectric layers of 1 μm thickness have been deposited by spinning the FOx-25 solution (from DOW-CORNING) at 2000 rpm and baked at 90 °C followed by curing at 800 °C. The conducting layers of TiN (0.1 μm) and TaN (1 μm) were coated on the metallic substrate by radio frequency sputtering and cathodic arc deposition, respectively. Later, aluminium (Al) and amorphous silicon (a-Si) films with a thickness of 270 nm and 370 nm have been deposited by e-beam evaporation and electron cyclotron resonance plasma enhanced chemical vapour deposition (ECR-PECVD) methods, respectively. The Al layers were exposed to ambient air for 1 week prior to a-Si deposition, in order to achieve an AlO_x permeable membrane. The substrate/BL/Al/a-Si structures were isothermally annealed at different temperatures in the range of 400-500 °C for different annealing periods of 1 to 10 h under nitrogen flow. The annealing temperatures were chosen under the eutectic temperature of Al/Si system (T_{eu}=577 °C). After the completion of layers exchange process, the residual Al+Si precipitates on the top surface has been removed by chemical etching using the solutions of nitric acid, hydrofluoric acid and deionised water in the ratio of 70.5:1.5:28.

The structural, surface and crystallographic analysis of as-prepared AIC layers were investigated by micro-Raman spectroscopy, UV/VIS/NIR reflection spectroscopy, optical microscopy and scanning electron microscopy combined with an X-ray energy analyser. The crystal orientation was determined from X-ray diffraction studies.

RESULTS AND DISCUSSION

The AIC process has been carried out at different annealing temperatures and time periods on metallic substrates coated with different barrier layers. These critical processing parameters were optimised as shown in table I to form continuous polycrystalline silicon (pc-Si) layers with lower stress. The surface roughness of the barrier layer coated metallic substrates was analyzed by interference microscopy. The optimised experimental conditions varied according to the roughness of the barrier layer. The analysis of the results shows that the a-Si has been fully crystallized at 500 °C for 3 h in case of FOx-25 coating as a barrier layer, while crystallisation velocity is low in case of TiN (100 nm) barrier layer (450 °C for 8 h). This is due to the changes in the interface structure of Al/a-Si system resulted from the barrier surface structure. It indicates that the crystallization velocity can be improved by forming a smooth interface between the Al/a-Si system via the surface modification of the barrier layer.

Table I. The surface roughness of barrier layers and optimised AIC processing conditions.

Barrier layer (BL)	BL surface roughness (nm)	AIC Annealing temp./time
FOx-25	311	500 °C/3h
TaN	487	475 °C/4h
TiN	723	450 °C/8h

Figure 1 shows the reflectance spectra of crystallised Si films prepared on different diffusion barriers at the optimised conditions. The reflectance spectra of AIC grown films shows characteristic peaks of crystalline silicon in 365 nm (E_1) and 275 nm (E_2) wavelengths, which are related to direct optical transitions and absent in the reflectance spectrum of amorphous films. The intensity of these peaks varies depending on the surface structure of the barrier layer. Though the characteristic reflectance peaks of AIC pc-Si films are symmetric, the shapes of these peaks have the differences when compared to that of single crystalline silicon (c-Si), particularly in the lower wavelength region. This is probably due to the roughness of the metallic substrates. In the case of the films deposited on foreign substrates, surface roughness has a considerable effect on the UV reflectance [5]. However, the effect of surface roughness is more for the films processed on TiN coated substrates due to their higher surface roughness. Therefore, the intensity of the reflection peak is lower than that of the AIC films processed on thick barrier coated substrates.

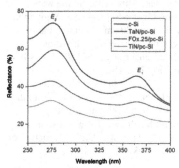

Figure 1. Reflectance spectra of AIC pc-Si samples formed on metallic substrates with different diffusion barrier layers.

The Raman spectra of AIC grown pc-Si samples at the optimised conditions are shown in figure 2. The symmetric behaviour of the Raman peaks resemble to that of c-Si, while a-Si films annealed at similar conditions without Al could not exhibit resonant Raman peak correspond to c-Si and exhibiting the spectrum similar to that of a-Si (not shown here). From this figure, it can be observed that the films showed a well defined Raman peak around 520 cm^{-1}, which is the characteristic peak position for c-Si, without showing any low energy tails in the range, 400-500 cm^{-1}, which implies that the AIC processed samples were completely crystallized into polycrystalline phase. Though there is no indication of amorphous phase, broadening of the Raman peak has been observed for films deposited on TiN coated substrates. The full width at half maximum (FWHM) for the pc-Si films on TaN is found to be lower, of the order of 7.9 cm^{-1} (error ~ ± 0.1 cm^{-1}), while it is higher (~ 9.1 cm^{-1}) for films

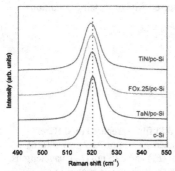

Figure 2. Raman spectra of AIC grown *pc*-Si films on different barrier layer coated metallic substrates.

processed on TiN coated substrates. The observed peak broadening is due to random nucleation of silicon that results from the higher surface roughness and hence the formation high density of smaller grains. The narrowing of the Raman peak on TaN coated substrates indicates the deterioration of short range order. The evaluated FWHM values, as given in table II, nearly match with the data reported for the AIC films processed on glass substrates [6]. However, the FWHM of experimental films differs from that of *c*-Si (~ 5 cm^{-1}), which can be attributed to grain size or strain induced effects [7]. The *pc*-Si films showed varying magnitude of stress that depends on the barrier layer, which was estimated from the wave number shift of the peak compared to the Raman line of stress free *c*-Si [8,9] and are given in table. II. The shift of peaks through the lower wave number from 520 cm^{-1} indicates the presence of tensile stress for all *pc*-Si films. However, it can be observed that the films processed on TiN coated substrates have higher magnitude of tensile stress of 150 MPa, while those on FOx-25 and TaN coated substrates showed nearly 50 MPa. This implies that the barrier layer is playing a critical role by changing the stress generated in the *pc*-Si films. Generally, the magnitude of stress comprises of internal and thermal stress that arise from the substrate/poly-Si interface, grain boundaries and dopants [10]. In the present case, the mismatch of thermal expansion coefficients of AIC grown *pc*-Si film and metal substrate as well as the defects at the grain boundaries, the degree of which varies with surface structure of the barrier, have considerable effect on shifting of the Raman peak.

Table II. The evaluated FWHM of the Raman peak and stress present in AIC grown films with respect to the barrier layer.

Barrier layer (BL)	FWHM (cm^{-1})	Stress (MPa)
Fox-25	8.3	75
TaN	7.9	50
TiN	9.1	150

The texture of the crystallized films has been evaluated using X-ray diffraction (XRD) measurements and was quantified by evaluating the crystallographic orientation factors, Θ_{hkl} with respect to the randomly oriented powder sample [11]. The orientation factors of different orientations for the films processed on TaN coated substrates are shown in figure 3, which shows that the films are highly textured in (100) orientation. E. Pihan et al. [12] observed similar results for the AIC grown *pc*-Si films on alumina and glass substrates.

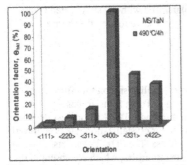

Figure 3. Orientation factors deduced from X-ray diffraction pattern of AIC *pc*-Si films synthesised at different conditions.

The optical and structural analyses of the AIC *pc*-Si films reveal the effective conversion of *a*-Si into *pc*-Si on all of the barrier metallic substrates. As the substrates are metallic substrates, it is important to study the diffusion characteristics and mechanical integrity of the barrier layers, which is shown in figure 4. In this figure, the typical SEM pictures of the layers grown at the optimised conditions with different barrier layers along with EDX spectra are shown. The surface of the layer seems to be quite uniform in the case of TaN and TiN coated substrates. The EDX analysis of these layers showed only Si and weak Al signal without any substrate metal impurities, which represents the stability of diffusion barrier at such temperatures (450-500 °C) used to carry out AIC process. It is interesting to note that though the thickness of TiN barrier is low (0.1 μm), it is effective in terms of mechanical integrity as well as against the impurity diffusion as far as the AIC annealing temperature is concerned. In the case of FOx-25 coated substrates, the surface seems to be a cracked structure. The elemental analysis of the crack free region showed only Si and Al, while that of the cracked region showed Fe, Ni and their oxides. However, these cracks are not visible after the conversion of FOx-25 coating into solid film. In general, superficial native oxide layers are responsible for the propagation of cracks [13]. It can be postulated that an ultra-thin superficial oxide layer could be formed on the substrate surface during the conversion of FOx-25 coating. It could be possible that the initiation of cracks in this brittle

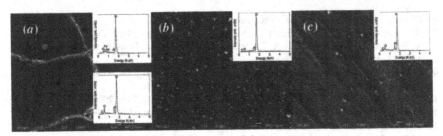

Figure 4. SEM and EDX profiles of AIC grown *pc*-Si films on (*a*) FOx.25 (*b*) TaN and (*c*) TiN coated metal substrates

and ultrathin oxide layer are likely to propagate during the AIC process to the adjacent stack layers due to interfacial stress concentrated at the crack location. Therefore, the pre-existing defects in the ultrathin layers are likely to be responsible for the failure mechanism of FOx-25 coated barriers. However, the substrate surface is well planarised with the spin coated layers and reduced the surface flaws of metallic substrates when compared to the other barriers deposited by physical methods.

CONCLUSIONS

A successful attempt has been made to prepare polycrystalline silicon thin films using AIC process on flexible Fe-Ni foil substrates for the first time. These films are highly crystallized on all barrier layer coated substrates, used in this study. The AIC process depends strongly on the surface structure of the barrier, where the faster crystallization rate is achieved on FOx-25 coated substrates due to its lower surface roughness. The FOx-25 barriers failed by triggering the cracks. The cracked region showed substrate impurities, which are absent in the crack free region. The diffusion of substrate impurities suppressed successfully by introducing TiN and TaN barriers and these barrier layers showed an excellent mechanical integrity with substrate and polysilicon films

ACKNOWLEDGEMENTS

The authors would like to thank Dr. Reydet from IMPHYS for providing the Fe-Ni foils, S. Roques, S. Schmitt, N. Zimmermann and F. Antoni from InESS for their valuable contributions. Also, Dr. C. Ducros are Dr. N. Baclet from CEA-LITEN greatly acknowledged for providing the TaN layers. This work is funded by the French National Research Agency (ANR) under project, CRISILAL.

REFERENCES

1. A. Slaoui, P. Siffert, in "Silicon: Evolution and Future of a Technology", Springer Verlag Ed., edited by P. Siffert, E.F. Krimmel, (2004), pp.45.
2. H. Gleskova, I. Cheng, S. Wagner, J.C. Sturm and Z. Suo, Sol. Energy, 80, 687 (2006).
3. J.W. Hutchinson and Z. Suo, Adv. Appl. Mech., 29, 63 (1992).
4. A. Slaoui, E. Pihan and A. Focsa, Solar Energy Materials and Solar Cells, 90, 1542 (2006).
5. G. Harbeke, Polycrystalline semiconductors, Springer Series in Solid-State Science, 57, 156 (1985).
6. G. Ekanayake and H.S. Reehal, Vacuum, 81, 272 (2006).
7. K. Kitahara, A. Moritani, A. Hara and M. Okabe, Jpn. J. Appl. Phys., 38, L1312 (1999).
8. V. Paillard, P. Puech, and M. A. Laguna, P. Temple-Boyer, B. de Mauduit, Appl. Phys. Lett. 73, 1718 (1998).
9. O. Tuzun, J.M. Auger, I. Gordon, A. Focsa, P.C. Montgomery, C. Maurice, A. Slaoui, G. Beaucarne, J. Poortmans, Thin Solid Films, 516, 6882 (2008).
10. P. Lengsfeld, and N.H. Nickel, J. Non-Cryst. Solids, 299–302, 778 (2002).
11. P. Joubert, B. Loisel, Y. Chouan, L. Haji, J. Electrochem. Soc., 134, 2541 (1987).
12. E. Pihan, A. Slaoui, C. Maurice, Journal of Crystal Growth 305, 88 (2007).
13. M.T.A. Saif, S. Zhang, A. Haque, K.J. Hsia, Acta. Mater., 50, 2779 (2002).

Mater. Res. Soc. Symp. Proc. Vol. 1153 © 2009 Materials Research Society   1153-A05-05

Low-Temperature Fabrication of a Crystallized Si Film Deposited on a Glass Substrate Using an Yttria-Stabilized Zirconia Seed Layer

Sukreen Hana Herman and Susumu Horita
School of Materials Science, Japan Advanced Institute of Science and Technology,
1-1, Asahidai, Nomi, Ishikawa, 923-1292, Japan

ABSTRACT

By using yttria-stabilized zirconia (YSZ) layer as a crystallized stimulation layer, a silicon film deposited by electron-beam evaporation was crystallized directly during the deposition at 320 ~ 430 °C. The crystallization of the Si film was stimulated and induced by the YSZ layer, as confirmed by transmission electron microscopy. The yttria content and the surface treatment of the YSZ layer are the main factors to determine the crystallization of the Si film.

INTRODUCTION

Investigations on crystalline silicon (c-Si) thin films, particularly polycrystalline Si (poly-Si) on glass or plastic substrates have been done intensively due to its superior characteristics compared with amorphous silicon (a-Si). Various methods to crystallize the a-Si have been proposed to obtain c-Si films at low temperature, such as solid phase crystallization (SPC) [1,2], metal induced crystallization (MIC) [3,4], laser annealing (LA) [5,6] and others. However, each method has its own drawbacks. SPC method needs high process temperatures and long annealing times. MIC method has provided a lower temperature process and shorter annealing time than SPC, but the poly-Si film is subject to high leakage current caused by the remnant metals as impurities. LA method, on the other hand, is a promising method with the lowest process temperature and the largest grain size compared to the above methods. However, the poly-Si film has a serious problem of large surface roughness. In order to obtain c-Si films without impurity and surface roughness at low temperature, we have proposed a new method of low-temperature growth of a poly-Si film on a quartz substrate, in which a poly-YSZ (yttria-stabilized zirconia : $(ZrO_2)_{1-x}(Y_2O_3)_x$) film is used as a seed layer to cover the whole surface of the amorphous substrate, as shown in the schematic diagram of Fig. 1. YSZ is a chemically and thermally stable material and is suitable for the seed layer owing to the small lattice mismatch with Si of only about 5%. In fact, it was reported that a YSZ film can be grown heteroepitaxially on an Si substrate [7].

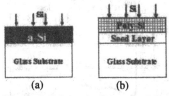

(a)   (b)

Figure 1. Schematic diagrams of the Si films deposited on the glass substrates by (a) the conventional direct deposition method and (b) the seed layer method.

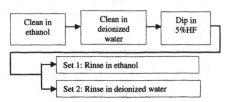

Figure 2. Two kinds of YSZ layer surface treatment prior to Si film deposition.

We have reported that the crystallized Si film was obtained on the YSZ seed layer at 515 °C, and the Si film deposited directly on the glass substrate at the same deposition temperature was amorphous [8]. Further, we have reported that the Si film deposited on the HF-etched YSZ that was rinsed with ethanol solution was crystallized at a lower deposition temperature of 430 °C, but that on the YSZ layer that was rinsed with deionized water (DIW) was amorphous [9]. Recently, we found that the Si film on the YSZ layer that was rinsed with DIW also crystallized, only if the YSZ layer has a high yttria content. In this paper, we show the relation of the chemical composition of the YSZ layer with the surface treatment and the crystallization of the Si film.

EXPERIMENTAL METHODOLOGY

A 70 nm-thick poly-YSZ seed layer was deposited by magnetron reactive sputtering on a quartz substrate. The yttria content of the YSZ layer was varied from 1.0 to 11.4 mol%. The seed layer was chemically cleaned with the two kinds of the cleaning steps shown in Fig. 2 prior to Si film deposition. The first one was to rinse the sample with ethanol (Set 1) after dipping it into 5%HF solution to remove the contaminated and damaged surface layer of the as-deposited YSZ layer, and the second one is to rinse with deionized water (DIW) (Set 2). A 60 nm-thick Si film was deposited directly on the YSZ layer by electron beam evaporation method with the deposition temperature varied from of 320 to 430 °C with the deposition rate of 1 nm/min, under the pressure of about 10^{-6} Pa. The surface of the YSZ layer that was treated chemically was characterized using x-ray photoelectron spectroscopy (XPS) with the peaks calibrated based on the carbon C $1s$ (284.6 eV) peak. The crystallization of the Si film was evaluated by the Raman spectroscopy and the film structure was characterized by high resolution transmission electron microscopy (HR-TEM) and scanning electron microscopy (SEM).

RESULTS AND DISCUSSIONS

Figure 3 shows the Raman spectra of the Si films deposited at 430 °C. Fig. 3(a)-(i) and 3(a)-(ii) are from the Si film deposited on the YSZ layer that was not dipped in the HF solution, and that of the Si film deposited directly on the glass substrate without the YSZ layer, respectively. It can be seen clearly that for both spectra, only a broad amorphous peak at around 480 cm^{-1} can be observed, meaning that neither of the Si films are crystallized. Fig. 3(b)-(i) and

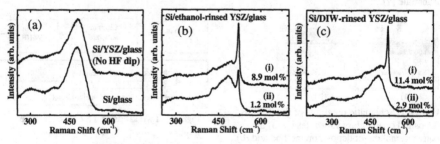

Figure 3. Raman spectra of the Si films deposited at 430 °C.

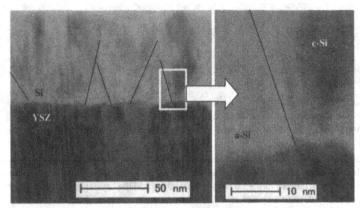

Figure 4. Microstructure of the Si film deposited at 430 °C on DIW-rinsed YSZ layer. The black lines are the eye-guide for the c-Si and a-Si regions. Right figure is the enlarged part of the region surrounded by the white square in the left figure.

(ii) are the spectra of the Si films deposited on the ethanol-rinsed YSZ layer having high and low yttria contents, respectively. Even though both the spectra in 3(b) show sharp c-Si peak at 517 cm^{-1} indicating that both the Si films are crystallized, the peak of the Si film on the YSZ with higher yttria content is much higher and sharper, which shows that the crystalline quality of the Si film on the higher yttria content is better than the lower one. Fig. 3(c) shows the spectra of the Si films on the DIW-rinsed YSZ having higher and lower yttria content. It can be seen clearly that only the Si film 3(c)-(i) on the DIW-rinsed YSZ layer with higher yttria content is crystallized, but that on the lower one is amorphous. The results in Fig. 3 show that to crystallize, the Si films needs YSZ layers, and the YSZ layers should be dipped in the HF solution. Further, we found that the yttria content, and the rinsing method of the YSZ layer have some effect on the crystallization of the Si film.

We observed the microstructure of the Si film of the spectrum Fig. 3(c)-(i), which is shown in Fig. 4, where the right-hand side figure is the enlarged image of the region framed in the white square of the left one. It can be seen that the Si grains grow from the interface of Si/YSZ in the form of V-shape towards the surface of the Si film. From the enlarged figure, we can see that the crystallization growth starts locally from the YSZ layer without any transition amorphous layer. This suggests that Si nucleation occurs on the YSZ layer surface and the crystallization of the Si is stimulated and induced by the YSZ layer during the deposition of the Si film. However, amorphous areas also exist, which correspond to the broad peak at around 480 cm^{-1} of the spectrum of Fig. 3(c)-(i).

Fig. 5 (a), (b), and (c) are the SEM images of the Si films deposited on the ethanol-rinsed YSZ layers at 320, 350 and 430 °C, respectively. The Si films were Secco-etched before the SEM observations in order to remove the amorphous components. From these figures we can confirm that the crystallization of the Si film on the ethanol-rinsed YSZ occured at a temperature as low as 320 °C. The c-Si of the film deposited at higher temperature was denser than those of lower temperature, in which, the Si film deposited at 430 °C was crystallized thoroughly, but the void regions due to the removal of a-Si by the Secco etching were observed on those deposited at

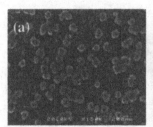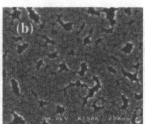

Figure 5. SEM images of the Secco-etched Si film deposited at (a) 320, (b) 350, and (c) 430 °C on YSZ layers.

350 and 320 °C. The grain size varied from 20 to 40 nm. We find that the grain size of the Si film deposited at lower temperature was more uniform than the higher temperature. This may be due to the difference in the rate of the nucleation and the crystal growth velocity of Si that depends on deposition temperature.

In order to investigate the relation of the yttria content and the rinsing method, we measured the chemical states of the treated YSZ surfaces by XPS. The XPS survey spectra of the surfaces of the as-deposited, DIW-rinsed, and ethanol-rinsed YSZ layers are shown in Fig. 6 (a)-(i), (ii), and (iii), respectively. The fluorine F 1s peak which is not observed on the as-deposited YSZ surface can be seen clearly on the other spectra, which shows that the F atoms were introduced on the YSZ surface during the HF dipping. The adsorbed F atoms were not totally removed by the rinsing process. By measuring the narrow scans of zirconium Zr 3d and yttrium Y 3d for the treated YSZ surfaces, we found that the shapes of the Zr 3d peaks were the same as the as-deposited ones, suggesting no change of the Zr chemical state after the surface treatment was observed. On the other hand, the shape of the Y 3d peaks are very different from one another as shown in Fig. 6(b). Fig. 6(b) shows the Y 3d peak of the YSZ layers; (i) is for the as-deposited, (ii) is for the DIW-rinsed and (iii) is for the ethanol-rinsed surface. As can be seen in this figure, the shape of the Y 3d for the ethanol-rinsed YSZ layer is different from those of the

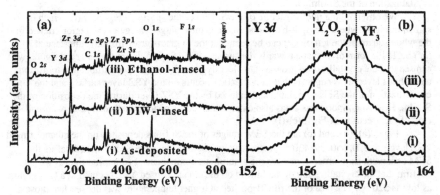

Figure 6. XPS survey spectra (a), and Y 3d peaks (b) of the YSZ layers. The dotted lines in (b) are the literature values for the binding energies of Y_2O_3 and YF_3 components.

Table 1. Relation between the F/(Y+Zr) ratio and the Y$_2$O$_3$ content of the YSZ layers with various surface treatments.

	DIW-rinsed		Ethanol-rinsed		HF-dipped (no rinse)	
Y$_2$O$_3$ content (mol%)	3.9	7.5	2.9	7.5	2.9	6.9
F/(Y+Zr) ratio	0.3	0.5	0.4	1.1	0.6	1.4

as-deposited and DIW-rinsed layers, which are mainly composed of the Y$_2$O$_3$ components. As for the peak of the ethanol-rinsed layer, besides the Y$_2$O$_3$ components, a shoulder on the higher binding energies can be observed, which corresponds well to YF$_3$ component. This means that the adsorbed F atoms during the surface treatment are bonded with Y atoms of the YSZ layer. The YF$_3$ components were removed easily by the DIW, but hardly by ethanol during the rinsing process, based on the results in Fig. 6(b). It is supposed that the remaining F atoms on the DIW-rinsed layers as indicated by the F 1s peak in the survey spectrum of Fig. 6 (a)-(ii) may be bonded with oxide Y. That is, F-Y-O compounds are formed on the YSZ surface, which were not easily recognized in the XPS measurement.

DISCUSSIONS

Considering the crystallization of the Si films and the XPS results, we speculate that when the YSZ layer was dipped in the HF solution during the surface treatment, the contaminants and the damaged components on the surface of the as-deposited YSZ layer were etched, resulting in a clean surface and exposing the ordered sites on the YSZ surface. We calculated the integrated intensity ratios of the F 1s peak with the total integrated intensities of yttrium Y 3d and zirconium Zr 3d peaks, [F/(Y+Zr)] for a few YSZ samples with various yttria contents, shown in Table 1. As can be seen from the table, the F ratio of the HF-etched surface with the high yttria content is the highest, 1.4, and that of the DIW-rinsed surface with the low yttria content is the lowest, 0.3. Further, the F ratio of the surface with the higher yttria content is larger than that in the lower content in all cases of the surface treatment. We can say from these results, that the F amount on the YSZ surface is strongly dependent not only on the final rinsing method, but also on the yttria content of the YSZ layer. The adsorbed F atoms on the treated YSZ layers protect the clean HF-etched YSZ surface from contamination during the preparation environment prior to the Si film deposition. The higher yttria content of the YSZ layer induces higher adsorption of the F atoms on the surface, thus better protection of the surface. The lack of F atoms on the DIW-rinsed with low yttria content allowed the surface to be contaminated, which subsequently disturbed the formation of the Si nuclei and also the crystallization.

Besides the role of the yttria content in the induction of the F atoms adsorption to protect the YSZ surface, we also assume that the yttria content also influences the surface crystalline quality of the YSZ layer. It is known that the crystal structure [10] and the lattice constant [11] of the YSZ layer were affected by the yttria content. In our work, we have reported that the crystalline quality of the partially-stabilized YSZ (yttria content: ~ 5 mol%) increases with the increasing yttria content [12]. Thus there are possibilities that the surface crystalline quality of the YSZ layer with high yttria content is better than that with the low content, which can prepare better ordered sites for the nucleation of the Si film and induce its crystallization. The relation between the crystalline quality of the YSZ layer and the Si film crystallization will be reported in the near future.

CONCLUSIONS

We have succeeded in the crystallization of the Si film on glass substrate directly during the deposition by using the YSZ thin film as the crystallization stimulation layer at the deposition temperature of 320 ~ 430 °C. The crystallization of the Si film on the YSZ layer was confirmed by the HR-TEM images. The yttria content and the surface treatment of the YSZ layer prior to the Si film deposition were two important factors to determine the crystallization of the Si film. The YSZ layers need to be treated by HF solution in order to remove the damage and contaminated components of the as-deposited layer. The F atoms adsorbed on the YSZ layer surface during the HF dipping may protect the surface from contamination during the preparation and transportation environment. We also speculate that the surface crystalline quality of the YSZ layer having high yttria content is better than that having low yttria content, which can prepare better ordered sites for the nucleation of the Si film and induce its crystallization.

REFERENCES

1. T. Matsuyama, K. Wakisaka, M. Kameda, M. Tanaka, T. Matsuoka, S. Tsuda, S. Nakano, Y. Kishi, Y. Kuwano, *Jpn. J. Appl. Phys.* **29**, 2327 (1990)
2. T. W. Little, K. Takahara, H. Koike, T. Nakazawa, I. Yudasaka, H. Ohshima, *Jpn. J. Appl. Phys.* **30**, 3724 (1991)
3. S. F. Gong, H. T. G. Hentzell, A. E. Robertsson, L. Hultman, S. E. Hornstrom, G. Radnoczi, *J. Appl. Phys.* **62**, 3726 (1987)
4. S. Y. Yoon, K. H. Kim, C. O. Kim, J. Y. Oh, J. Jang, *J. Appl. Phys.* **82**, 5865 (1997)
5. T. H. Ihn, B. I. Lee, S. Ki Joo, Y. C. Jeon, *Jpn. J. Appl. Phys.* **36**, 5029 (1997)
6. Y. G. Goon, M. S. Kim, G. B. Kim, S. K. Joo, *IEEE Electron Device Lett.* **24**, 649 (2003)
7. D. Pribat, L. M. Mercandalli, M. Croset, D. Dieumegard, *Mater. Lett.* **2**, 524 (1984)
8. S. Horita, K. Kanazawa, K. Nishioka, K. Higashimine, M. Koyano, *Mater. Res. Soc. Symp. Proc.* **910**, 0910-A21-17 (2006)
9. S. Horita and H. Sukreen, *Appl. Phys. Express* **2**, 041201 (2009)
10. S. G. Wu, H. Y. Zhang, G. I. Tian, Z. I. Xia, J. D. Shao, Z. X. Fan, *Appl. Surf. Sci.* **253**, 1561 (2006)
11. T. V. Chusovitina, V. M. Ust'yantsev, M. G. Tretnikova, Y. S. Toropov, *Refract. Ind. Ceram* **28**, 13 (1987)
12. S. Hana, K. Nishioka, S. Horita, *Thin Solid Films* (2009), doi: 10.1016/j.tsf.2009.03.035 – Article in press

Mater. Res. Soc. Symp. Proc. Vol. 1153 © 2009 Materials Research Society   1153-A05-09

Improving Silicon Crystallinity by Grain Reorientation Annealing

K.L. Saenger, J.P. de Souza, D. Inns, K.E. Fogel, and D.K. Sadana
IBM Semiconductor Research and Development Center
Research Division, T. J. Watson Research Center,
Yorktown Heights, NY 10598

ABSTRACT

Demand for high efficiency, low-cost solar cells has led to strong interest in post-deposition processing techniques that can improve the crystallinity of thick (1 to 40 μm) silicon films deposited at high growth rates. Here we describe a high temperature grain reorientation annealing process that enables the conversion of polycrystalline silicon (poly-Si) into a single crystal material having the orientation of an underlying single crystal Si seed layer. Poly-Si films of thickness 0.5 to 1.0 μm were deposited by low pressure chemical vapor deposition (LPCVD) on substrates comprising a surface thermal oxide or a 100-oriented single crystal silicon-on-insulator (SOI) layer. After annealing at 1300 °C for 1 hour, poly-Si on oxide shows very significant grain growth, as expected. In contrast, the poly-Si deposited on SOI showed no grain boundaries after annealing, transforming into a single crystal material with a fairly high density of stacking faults. Possible uses and drawbacks of this approach for solar cell applications will be discussed.

INTRODUCTION

Cost-effective methods for forming thick (1 to 40 μm) layers of single crystal silicon are of interest for applications that include both high efficiency solar cells and thick silicon-on-insulator (SOI) substrates for high power device applications. However, existing growth methods have potential drawbacks [1]. Epitaxial growth on single crystal Si (c-Si) seed layers, typically performed at 800-900 °C, is relatively slow and expensive, and requires careful cleaning of the initial Si substrate surface to ensure good epitaxy. High temperature (~1100 °C) chemical vapor deposition (CVD) processes that can deposit epitaxial Si at a rate of 1 μm/min (for example, atmospheric pressure iodine vapor transport (APIVT) processes) offer promise, but are less well established. Methods utilizing physical vapor deposition of amorphous Si (a-Si) followed by solid phase epitaxy (SPE) at temperatures in the range 550-600 °C are potentially low cost, but share the disadvantage of requiring a clean c-Si/a-Si interface for good epitaxy and present the additional concern that the crystallization of very thick a-Si layers will proceed by a combination of the SPE and undesired spontaneous, randomly nucleated crystallization to form poly-Si.

Processes for the bonding and transfer of thick c-Si layers have drawbacks as well: donor wafer bonding to a handle wafer followed by donor wafer etchback sacrifices the entire donor wafer, and the hydrogen ion implantation processes used for SmartCut™-type splittings are typically restricted to relatively shallow depths (a few hundred nm at most).

In this paper we investigate the possibility of creating thick single crystal Si layers by performing a grain reorientation anneal on a thick layer of poly-Si deposited onto a thin single crystal Si seed layer, a process shown schematically in Fig. 1. The present experiments were motivated by the thought that such a process might provide Si layers with single crystal silicon quality at poly-Si/aSi costs, and by recent findings showing that crystalline Si regions of one orientation in contact with a bulk single crystal substrate of a different orientation can be induced by high temperature (~1300 °C) annealing to take on the orientation of the bulk substrate. In particular, it had been found that 110-oriented Si regions embedded in a 100-oriented Si wafer can be converted to a 100 orientation (a clearly undesirable effect to be avoided when fabricating hybrid orientation substrates) [2] and that poly-Si in structures of the form Si(100, substrate)/SiO$_2$(3-4 nm)/poly-Si(160 nm) can convert to the orientation of the 100-oriented substrate after dissolution of the interfacial thermal oxide [3].

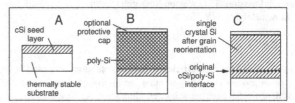

Fig. 1. A schematic of the grain reorientation annealing process.

We will first show that poly-Si layers on thin (160 nm) 100-oriented SOI layers do indeed take on the orientation of the underlying SOI layer (rather than, for example, turning the SOI layer into poly-Si), while control samples of poly-Si on thermal oxide merely show very substantial grain growth. We will then discuss some uses and drawbacks of this approach for solar cell applications.

EXPERIMENT

Polycrystalline Si layers of 500 and 1000 nm in thicknesses were deposited by low pressure CVD (LPCVD) (SiH$_4$, ~625 °C) on SC1/SC2-cleaned [4] SOI substrates comprising a 160 nm thick 100-oriented Si layer on a 145 nm thick buried oxide layer. The thicker (1000 nm) poly-Si samples were deposited by two sequential 500 nm depositions with an air break in-between. Additional samples with the same thicknesses of poly-Si were deposited in the same way on a 200 nm thermal oxide layer for comparison. A protective cap of low-temperature oxide 200 nm in thickness was deposited by CVD (SiH$_4$/O$_2$, 400 °C) on all samples prior to an anneal at 1300 °C for 1 hour in an Ar ambient, and removed after annealing with dilute aqueous HF.

The need for protective capping layers and/or low (~1%) levels of oxygen in the annealing ambient has been discussed in Ref. 3. When no capping layers are used, trace (<<1%) amounts of oxygen in the annealing ambient can cause surface pitting, the result of runaway SiO formation and desorption that can occur at ambient oxygen levels below that needed to keep the Si surface covered with a thermally stable (and less volatile) layer of SiO$_2$. Protective capping layers can eliminate the need for oxygen in the ambient, but can also be useful for reducing the amount of Si consumed by oxidation when oxygen is present in the annealing ambient.

Samples were examined by x-ray diffraction (XRD) [θ-2θ scans using a Bragg-Brentano geometry and Cu Kα radiation at 1.5418 Å] and scanning electron microscopy (SEM) before and after annealing. Before SEM, the samples were coated with Cr, cleaved, and then Secco-etched [5] to highlight grain boundaries and defects.

RESULTS

Progress of poly-Si grain growth and/or grain reorientation was qualitatively assessed from the intensities and shapes of the Si XRD peaks before and after annealing, as shown in Fig. 2. Peaks not coinciding with the 004 substrate peak (at 69°, not shown), i.e., the Si 111, 220, and 311 peaks at 28°, 47°, and 56°, are due to poly-Si grains. In the poly-Si on oxide samples (Figs. 2E - 2H), annealing narrows the poly-Si peaks and preferentially enhances Si 111, with Si 111 peak intensities higher by about a factor of 4 for the 500 nm film and a factor of 6 for the 1000 nm film relative to the as-deposited films. In contrast, the poly-Si XRD peaks for the 500 nm poly-Si on SOI sample (Fig. 2B) completely disappear after annealing, suggesting a complete conversion to the 100 orientation of the seed layer. Small poly-Si peaks still remain after annealing for the 1000 nm poly-Si on SOI sample (Fig. 2D), but these are lower in intensity by about a factor of 10 than the corresponding 1000 nm thick sample on oxide.

SEM images for the 1000 nm thick poly-Si sample on oxide before and after annealing at 1300 °C for 1 hour are shown in Fig. 3. The grain structure before annealing (Fig. 3A) indicates the presence of two separate fine-grained poly-Si layers, an expected result given that the second layer of poly-Si was deposited on the native oxide layer formed during the air break rather than on a clean Si surface. After annealing, the material is still polycrystalline (consistent with the XRD data), but there is a dramatic increase in Si grain size. More strikingly, there is no sign of original poly-Si/poly-Si interface between the first and second layers of poly-Si, indicating complete dissolution of the interfacial native oxide.

SEM images for the 500 nm thick poly-Si sample on SOI are shown in Fig. 4. Before annealing (Fig. 4A), there is a clear interface between the SOI layer and the fine-grained poly-Si layer, again due to the presence of native oxide on the SOI. After annealing (Fig. 4B), this

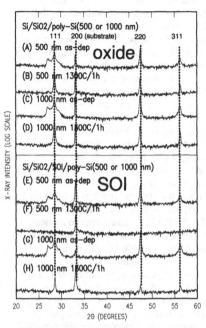

Fig. 2. XRD results for poly-Si layers on SOI (A-D) or oxide (E-H) before and after annealing at 1300 °C for 1 hour.

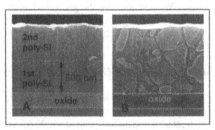

Fig. 3. SEM images for the 1000 nm thick poly-Si sample on oxide before (A) and after (B) annealing at 1300 °C for 1 hour.

interface has disappeared, once again indicating a dissolution of the oxide. In contrast to the grain growth seen in the poly-Si on oxide samples of Fig. 3B, the poly-Si layer on SOI appears to have converted into single-crystal Si, though with a fairly high density of stacking faults.

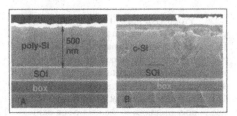

Fig. 4. SEM images for the 500 nm thick poly-Si sample on SOI before (A) and after (B) annealing at 1300 °C for 1 hour.

SEM images for the 1000 nm thick poly-Si sample are shown in Fig. 5. Before annealing (Fig. 5A), both the SOI/poly-Si and poly-Si/poly-Si interfaces are visible, as expected. After annealing (Fig. 5B), these interfaces have again disappeared, but residual poly-Si grains can still be seen in the second (upper) poly-Si layer in many areas of the samples. Qualitatively it would appear that the conversion is only 70-80% complete, from which it can be deduced that thicker films take longer to reorient and that the reorientation rate at 1300 °C might be about 700-800 nm/hour.

The optimum conditions for grain reorientation annealing will depend on the initial poly-Si thickness and grain size. Ordinarily it would be preferred that the poly-Si layers be annealed until grain reorientation is complete; however, partial or incomplete grain reorientation may be a

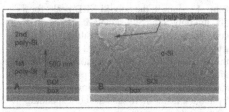

Fig. 5. SEM images for the 1000 nm thick poly-Si sample on SOI before (A) and after (B) annealing at 1300 °C for 1 hour.

satisfactory outcome if one prefers less aggressive (e.g., lower temperature or shorter duration) annealing. Suitable annealing conditions are expected to have durations of 1 to 100 hours, temperatures in the range 1250 to 1330 °C, and annealing ambients of Ar or Ar/O$_2$(1-5%).

We note also that the grain reorientation annealing process described here may also be used to convert initially amorphous Si layers into single crystal Si. In this case, the a-Si layers would crystallize into poly-Si while ramping up to the temperature of the grain reorientation anneal.

INTEGRATION SCHEMES AND DISCUSSION

In this section we discuss some solar cell structures and process flows that are, in principle, compatible with the above-described grain reorientation annealing process. Given the high rates of dopant diffusion at ~1300 °C, any grain reorientation annealing steps should occur early in process, before dopant redistribution (of localized emitter regions, for example) might be a problem.

Compatible integration schemes generally fall into two categories: those in which top surface interdigitated doped regions (and contacts) are formed last, and those in which the grain-reoriented poly-Si layer is transferred to another substrate after the anneal. Fig. 6 shows one possible interdigitated-contacts-last process flow for the case of a seed layer substrate comprising an SOI layer disposed upon a buried oxide layer on an opaque base substrate. An example of such a substrate would be a conventional SOI wafer, in which the base substrate would be a crystalline Si wafer. Because the base substrate is opaque, the top contacts will be front contacts, and will shadow some of the incident light.

Seed layer substrates comprising SOI layers disposed on thermally stable, transparent substrates (such as fused silica or quartz) would be preferable to conventional SOI wafers because light could enter the cell through the transparent substrate, leaving the same interdigitated diffusions and contacts on the cell's back surface. However, such substrates are not readily available and might be expected to be even more expensive than standard SOI substrates.

While layer transfer approaches add additional complexity, overall costs may be lower (providing a means can be found to refurbish the seed layer substrate for multiple uses) since the thermal stability requirements for the final substrate are much reduced. Figure 7 shows

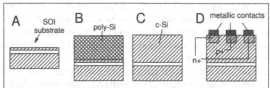

Fig. 6. Schematic cross-section view of steps for fabricating an interdigitated-contacts-last solar cell: a starting SOI substrate is selected (A); a poly-Si layer is deposited on the SOI layer (B); the structure is annealed to effect the poly-Si grain reorientation (C); and interdigitated diffusions/contacts are formed to make a solar cell (D).

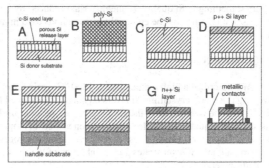

Fig. 7. Schematic cross-section view of steps for fabricating a solar cell incorporating a transferred layer of grain-reoriented poly-Si: a starting substrate comprising a seed layer and a porous Si release layer is selected (A); a poly-Si layer is deposited on the Si seed layer (B); the structure is annealed to effect the poly-Si grain reorientation (C); a surface p++ Si layer is formed on the new c-Si layer (D); the structure is inverted and bonded to a carrier substrate (E); the donor substrate is separated from the new c-Si layer (F), a surface n+ Si layer is formed on the exposed Si surface (G); and top and bottom metallic contacts are formed to make a solar cell (H).

an example of a process flow utilizing layer transfer, where the converted poly-Si layer is transferred from a donor substrate to a handle substrate after the grain reorientation. Many variations of this basic method are possible. For example, the handle substrate could be transparent and the diffusions and contacts could be interdigitated and on the top surface of converted poly-Si layer.

One concern with the transfer process of Fig. 7 is that mechanically weak porous Si release layers might not function as well after high temperature annealing due to alterations in pore microstructure. While transferring the seed layer Si/poly-Si bilayer couple to a temporary thermally stable substrate or support before annealing is possible, such an approach would add considerable complexity.

Given these limitations, it is worth remembering the two advantages that grain reorientation annealing processes have relative to other methods of forming thick single crystal Si layers: (i) poly-Si deposition can be performed as a batch process (e.g., in a tube furnace as opposed to single wafer tooling), and (ii) the Si surface on which the poly-Si is deposited does not have to be carefully cleaned to remove all traces of native oxide. On the other hand, one gives up something in Si quality, as the material produced by this method is unlikely to be as perfect as that obtained by the epitaxial CVD methods used in the semiconductor device industry.

CONCLUSIONS

We described a high temperature (~1300 °C) grain reorientation annealing process for converting thick polycrystalline Si layers on single crystal seed layers into thick single crystal Si layers having the orientation of the seed layer. This process *might* allow production of thick Si films having the quality of single crystal silicon at a cost comparable to that for poly-Si deposition. However, the usefulness of such an annealing process may be limited, due to the expense of suitable SOI substrates and the absence of low cost techniques for layer bonding/transfer.

ACKNOWLEDGMENTS

The Microelectronics Research Laboratory staff is thanked for their contributions to sample preparation.

REFERENCES

1. G. Beaucarne et al, *Thin Solid Films*, **511-512**, 533 (2006).
2. K.L. Saenger et al, *Mat. Res. Soc. Symp. Proc.* **913**, D1.1 (2006).
3. K.L. Saenger et al, *J. Electrochemical Soc.*, **155**, H80 (2008).
4. W. Kern, *J. Electrochem. Soc.*, **137**, 1887 (1990).
5. F. Secco d'Aragona, *J. Electrochem. Soc.* **119**, 948 (1972).

Poster Session:
Nanostructured Silicon

Mater. Res. Soc. Symp. Proc. Vol. 1153 © 2009 Materials Research Society   1153-A06-02

Optimization of p-type Nanocrystalline Silicon Thin Films for Solar Cells and Photodiodes

Y. Vygranenko[1], E. Fathi[2], A. Sazonov[2], M. Vieira[1], G. Heiler[3], T. Tredwell[3], A. Nathan[4]

[1]Electronics Telecommunications and Computer Engineering, ISEL, Lisbon, 1950-062, Portugal
[2]Electrical and Computer Engineering, University of Waterloo, Waterloo, N2L 3G1, Canada
[3]Carestream Health Inc., Rochester, NY, 14652-3487, USA
[4]London Centre for Nanotechnology, UCL, London, WC1H 0AH, United Kingdom

ABSTRACT

We report on structural, electronic, and optical properties of boron-doped, hydrogenated nanocrystalline silicon (nc-Si:H) thin films deposited by plasma-enhanced chemical vapor deposition (PECVD) at a substrate temperature of 150°C. Film properties were studied as a function of trimethylboron-to-silane ratio and film thickness. The film thickness was varied in the range from 14 to 100 nm. The conductivity of 60 nm thick films reached a peak value of 0.07 S/cm at a doping ratio of 1%. As a result of amorphization of the film structure, which was indicated by Raman spectra measurements, any further increase in doping reduced conductivity. We also observed an abrupt increase in conductivity with increasing film thickness ascribed to a percolation cluster composed of silicon nanocrystallites. The absorption loss of 25% at a wavelength of 400 nm was measured for the films with optimized conductivity deposited on glass and glass/ZnO:Al substrates. A low-leakage, blue-enhanced p-i-n photodiode with an nc-Si p-layer was also fabricated and characterized.

INTRODUCTION

Boron-doped hydrogenated nanocrystalline silicon (nc-Si:H) is an attractive material for large-area electronics and photovoltaic applications. It has advantages over hydrogenated amorphous silicon (a-Si:H) and silicon carbide with respect to higher conductivity and lower optical absorption in the visible range [1]. Thin p-type nc-Si:H films are used as the window layer in p-i-n solar cells with an a-Si:H or nc-Si:H intrinsic layer [2, 3]. To achieve a low absorption loss in the p-layer, the deposition conditions have to be optimized specifically for the thin (<20 nm) layers to reduce the thickness of the amorphous incubation phase. Another technological issue is that the growth mechanism of nc-Si:H strongly depends on the substrate material and surface conditions thus limiting the choice of transparent conducting oxides for junction cells in a superstrate configuration [4]. Given that nc-Si:H is a heterogeneous material of complex microstructure, the fraction of the crystalline material, the crystallite size, the grain boundaries and voids play a significant role in determining the electronic and optical film properties. These parameters may also have an effect on device performance.

In this paper, we report on electronic, structural, and optical properties of p-type nc-Si thin films deposited by conventional (13.56 MHz) plasma-enhanced chemical vapor deposition (PECVD) and their application for a-Si:H p-i-n photodiodes.

EXPERIMENT

Film and Device Deposition

Two series of boron-doped nc-Si:H films were prepared to study their structural and electronic properties. Then, the p-i-n structures with an optimized nc-Si:H p-layer were fabricated. The films and devices were deposited at 150 °C onto Corning 1737 glass substrates using a multichamber 13.56 MHz PECVD system, manufactured by MVSystems Inc.

The trimethylboron (B(CH$_3$)$_3$) (TMB), diluted in hydrogen to concentration of 1%, was used as the doping gas. The first film series (*doping series*) was deposited at different TMB-to-SiH$_4$ flow ratios in the range from 0.2 to 1.5%, but keeping the film thickness of about 60 nm. The second film series (*thickness series*) was deposited at [TMB] / [SiH$_4$] = 1%, but the thickness of the films was varied from 14 to 100 nm. The RF power density, pressure, and hydrogen dilution ratio, [H$_2$] / ([H$_2$] + [SiH$_4$]) × 100 %, were fixed at 9 mW/cm^2, 900 mTorr, and 99%, respectively.

The p-i-n diodes were fabricated by the following deposition sequence. First, a 65 nm thick ZnO:Al antireflection coating with a sheet resistance of ~160 Ω/sq. was sputtered on a 0.5 mm thick glass substrate. Then, a p-i-n stack was deposited over the substrate. Finally, a ~300 nm thick Al film was sputtered and patterned to form 5 × 5 mm^2 top electrodes.

Two types of devices were fabricated. The first sample was a p-i-n structure with a 20 nm thick nc-Si:H p-layer. The second sample was a p-δ_p-i-n structure with an 12 nm thick nc-Si:H p-layer and 8 nm thick a-SiC:H δ_p-layer. The thicknesses of the i- and n-layers were 500 and 20 nm, respectively. For both samples the p-layers were deposited at the same gas mixture of [TMB] / [SiH$_4$] = 1%. The p-type a-SiC:H was deposited using a SiH$_4$ + H$_2$ + CH$_4$ + TMB gas mixture with a hydrogen dilution ratio of 75%, [CH$_4$] / [SiH$_4$] = 1, and [TMB] / [SiH$_4$] = 1%. The intrinsic a-Si:H was deposited using a SiH$_4$+H$_2$ gas mixture with a hydrogen dilution ratio of 75%. The n-type a-Si:H material was produced by the addition of 1% phosphine (PH$_3$). The pressure and RF power density were 600 mTorr and 22 mW/cm^2, respectively.

Characterization Techniques

Raman spectroscopy was employed to estimate the crystalline fraction (X_C) of the films deposited on glass. Raman spectra were measured in the backscattering geometry using a Renishaw micro-Raman spectrometer with a 488 nm excitation laser line. Transmission and reflection spectra of nc-Si:H films were measured using a UV-visible 2501PC Shimadzu spectrophotometer.

Samples for conductivity measurement were prepared by sputtering coplanar Al electrodes through a shadow mask. A Dektak 8 surface profiler was used for film thickness measurements. Dark conductivity of the films and current-voltage characteristics of the photodiodes were measured at room temperature using a Keithley 4200-SCS semiconductor characterization system.

The spectral response measurements were performed with a PC controlled setup based on an Oriel 77 200 grating monochromator, a Stanford Research System SR540 light chopper, and a SR530 DSP lock-in amplifier. The system was calibrated in the spectral range of 300–1100 nm using a Newport 818-UV detector.

DISCUSSION

Film Properties

Figure 1 shows the Raman spectra of films deposited at a doping ratio [TMB] / [SiH$_4$] ranging from 0.25 to 1.3%. The Raman spectra show a broad shoulder associated with the amorphous phase, and asymmetric band centered at about 517 cm^{-1}, originating from the nanocrystalline phase in the film [5]. A good curve fit was achieved with four Gaussian peaks centered at 440 cm^{-1}, 480 cm^{-1}, 507-511 cm^{-1}, and 514-517 cm^{-1}, corresponding to the longitudinal optical (LO) and transverse optical (TO) phonon modes of the amorphous fraction and optical vibrational modes of Si nanocrystals, respectively [6].

The crystalline volume fraction, X_C, was determined using the relation

$$X_C = I_c / (I_c + \eta I_a), \tag{1}$$

where I_a and I_c are the integrated intensities of the peaks centered at 480 cm^{-1}, and at 507-517 cm^{-1}, respectively [7]. The ratio of the back-scattering cross-sections, η, was chosen to be 0.8 [8]. With increasing doping ratio the evaluated crystalline volume fraction decreases from 61% to 35%. A similar result has been reported for diborane (B$_2$H$_6$) doped nc-Si:H films deposited by PECVD [9].

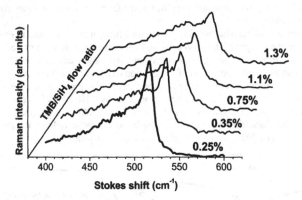

Figure 1. Series of Raman spectra of nc-Si:H films for increasing values of the TMB / SiH$_4$ flow ratio.

Figure 2 shows the dark conductivity of nc-Si:H films as a function of the TMB / SiH$_4$ flow ratio. The film conductivity increases with increasing doping ratio, reaches a value of ~0.073 S/cm at about 1%, and then, gradually decreases down to 0.002 S/cm at 1.5%. The conductivity reduction is likely related to an amorphization of the film structure observed in the series of Raman spectra.

Figure 3 shows a variation of dark conductivity for the thickness film series. It is seen that the thin (<25 nm) films exhibit low conductivity, comparable to that of doped amorphous silicon. For thicker films, the conductivity increases by 2-3 orders of magnitude in the thickness

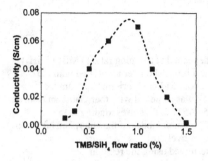
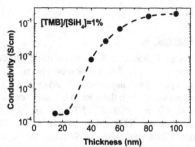

Figure 2. Conductivity of nc-Si:H films as a function of the TMB/SiH₄ flow ratio.

Figure 3. Variation of conductivity with thickness of nc-Si:H films at [TMB] / [SiH₄] = 1%.

range from 40 to 80 nm and then, tends to saturate. For a 100 nm thick film, the conductivity reaches 0.2 S/cm. A similar dependence of conductivity on film thickness has been reported for undoped and phosphine-doped nc-Si:H films deposited by PECVD [10]. In Ref. 10, the observed abrupt change in conductivity is explained in terms of percolation theory by destruction of a percolation cluster composed of nanocrystallites as the layer thickness becomes comparable to the size of a crystallite.

To ensure the nanocrystalline film growth on a foreign substrate, and to evaluate the absorption loss in the *p*-layer of the *p-i-n* cells, optical measurements were performed on the 22 nm thick films deposited onto bare glass substrate and onto glass substrate with ZnO:Al coating.

Figure 4 shows the transmission (T), reflection (R), and absorption (A=1-T-R) spectra of these samples. The transmittance/reflectance curves of the film grown on glass are comparable to the spectra of thin (~20 nm) nanocrystalline silicon films by VHF PECVD reported elsewhere [11]. The second sample is more transparent because the ZnO:Al coating reduces reflectance down to ~20%. The absorption spectra of the samples look similar yielding the absorption loss of 25% at a wavelength of 400 nm, i.e., both samples are nanocrystalline.

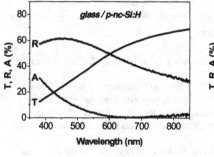
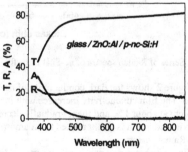

(a)   (b)

Figure 4. Transmission, reflection, and absorption spectra of thin (22 nm) nc-Si:H film on (a) glass substrate, and (b) glass with ZnO:Al coating.

Photodiode Characteristics

Figure 5 shows the typical quasi-static current-voltage characteristics of the *p-i-n* and *p-δ_p-i-n* structures. In order to minimize the transient current induced by the trapped charge in the i-layer, the sweep delay was set to 20 s and the bias voltage was varied at 25 mV increments.

For the *p-i-n* structure, the log(current) versus voltage plot is linear in the forward biasing range of 0.4–0.62 V. The diode ideality factor (n) and the saturation current density (J_o) are determined to be 1.63 and 8.6 pA/cm², respectively. The *p-δ_p-i-n* structure shows an exponential dependence of the forward current over five orders of magnitude in the biasing range of 0.2–0.6 V, yielding $n = 1.57$ and $J_o = 440$ fA/cm².

The incorporation of the a-SiC:H δ_p-layer improves not only the values of n and J_o, but it also efficiently suppresses the leakage current. The reverse dark current of the *p-i-n* structure saturates at about 100 nA/cm². The *p-δ_p-i-n* structure shows a significantly lower leakage in the low bias range. Here, the current density of 16 pA/cm² at −1 V is comparable to that reported for state-of-the-art a-Si:H photodiodes [12]. At reverse biases higher than 2 V, when the δ_p-i-layers are fully depleted and the depletion region expands into the nc-Si:H p-layer, the leakage current increases near exponentially, reaching the value of 4 nA/cm² at −5 V.

Figure 6 shows the spectral response characteristics of the *p-i-n* and *p-δ_p-i-n* structures under short circuit conditions. The external quantum efficiency of the *p-δ_p-i-n* structure reaches a peak value of 84% at a 520 nm wavelength. Its blue response is mainly limited by the absorption losses in the *p-δ_p*-layers. The deterioration in the short-wavelength response of the *p-i-n* structure is attributed to the recombination losses at the heterojunction p-nc-Si:H/i-a-Si:H interface.

Thus, the incorporation of the δ_p-layer is crucial for low-leakage, blue-enhanced *p-δ_p-i-n* photodiodes. The role of the buffer layer is discussed in prior reports on the development of a-Si:H *p-i-n* solar cells with a nc-Si:H p-layer [13]. In the photovoltaic mode, the buffer effectively blocks electron back diffusion due to the large band offset between the a-Si:H *i*-layer and the nc-Si:H *p*-layer and reduces the recombination loss at the front contact [14].

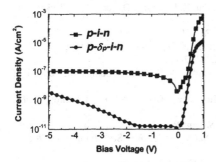

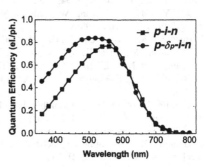

Figure 5. Dark current-voltage characteristics of the *p-i-n* and *p-δ_p-i-n* structures.

Figure 6. Spectral response of the *p-i-n* and *p-δ_p-i-n* structures.

CONCLUSIONS

The influence of boron doping and layer thickness on the structural and electronic properties of nc-Si:H films grown by PECVD has been systematically studied. The film crystallinity and conductivity were found to be tailored by controlling the TMB-to-SiH$_4$ ratio. For 60 nm thick films, the conductivity of 0.07 S/cm was achieved at a doping ratio of ~1%. It was also found that conductivity of the thin (<25 nm) film was limited by a charge-carrier transport through an a-Si:H network, while in the thicker layers, the observed conductivity enhancement was ascribed to the formation of a percolation cluster composed of Si nanocrystallites. By employing the p$^+$ nc-Si:H as a window layer, complete *p-i-n* structures were fabricated. Low leakage current and enhanced sensitivity in the UV/blue range were achieved by incorporating an a-SiC buffer between the *p*- and *i*-layers.

ACKNOWLEDGMENTS

The authors are grateful to the Portuguese Foundation of Science and Technology through fellowship BPD20264/2004 for financial support of this research.

REFERENCES

1. H. Chen, M.H. Gullanar, W.Z. Shen, *Journal of Crystal Growth* **260**, 91(2004).
2. A.V. Shah, J. Meier, E. Vallat-Sauvain, N. Wyrsch, U. Kroll, C. Droz, U. Graf, *Sol. Energy Mater. Sol. Cells* **78**, 469 (2003).
3. M. Kondo and A. Matsuda, "Low-temperature fabrication of nanocrystalline-silicon solar cells," *Thin-film solar cells*, ed. Y. Hamakawa (Springer, 2004) pp.139-148.
4. T. Fujibayashi, M. Kondo, *J. Appl. Phys.* **99**, 043703 (2006).
5. C. Smit, R.A.C.M.M. van Swaaij, H. Donker, A.M.H.N. Petit, W.M.M. Kessels and M.C.M. van de Sanden, *J. Appl. Phys.* **94**, 3582 (2003).
6. H. Xia, Y. L. He, L. C. Wang, W. Zhang, X.N. Liu, X. K. Zhang, D. Feng, H. E. Jackson, *J. Appl. Phys.* **78**, 6705 (1995).
7. E. Bustarret, M.A. Hachicha, M. Brunel, *Appl. Phys. Lett.* **52**, 1675 (1988).
8. A. T. Vautsas, M.K. Hatalis, J.B. Boyce, A. Chiang, *J. Appl. Phys.* **78**, 6999 (1995).
9. R. Saleh, N. H. Nickel, *Thin Solid Films* **427** (2003) 266.
10. V. G. Golubev, L. E. Morozova, A. B. Pevtsov, and N. A. Feoktistov, *Semiconductors* **33**, 66 (1999).
11. A. Gordijn, J. Loffler, W.M. Arnoldbik, F.D. Tichelaar, J.K. Rath, and R.E.I. Schropp, *Sol. Energy Mater. Sol. Cells* **87**, 445 (2005).
12. J.A. Theil, in *Amorphous and Nanocrystalline Silicon-Based Films,* Edited by J. R. Abelson, G. Ganguly, H. Matsumura, J. Robertson, E. A. Schiff (Mater. Res. Soc. Symp. Proc. 762, San Francisco, CA, 2003) A.21.4.1.
13. J.K. Rath, R.E.I. Schropp, *Sol. Energy Mater. Sol. Cells* **53**, 189 (1998).
14. N. Palit and P. Chatterjee, *J. Appl. Phys.* **86**, 6879 (1999).

Poster Session:
Solar Cells

Mater. Res. Soc. Symp. Proc. Vol. 1153 © 2009 Materials Research Society   1153-A07-01

High Efficiency Large Area a-Si:H and a-SiGe:H Multi-Junction Solar Cells Using MVHF at High Deposition Rate

Xixiang Xu[1], Dave Beglau[1], Scott Ehlert[1], Yang Li[1], Tining Su[1],Guozhen Yue[1], Baojie Yan[1], Ken Lord[1], Arindam Banerjee[1], Jeff Yang[1], Subhendu Guha[1], Peter G. Hugger[2], and J. David Cohen[2]
[1]United Solar Ovonic LLC, 1100 West Maple Road, Troy, MI, 48084, U.S.A.
[2]Department of Physics, University of Oregon, Eugene, OR, 97403, U.S.A.

ABSTRACT

We have developed high efficiency large area a-Si:H and a-SiGe:H multi-junction solar cells using a Modified Very High Frequency (MVHF) glow discharge process. We conducted a comparative study for different cell structures, and compared the initial and stable performance and light-induced degradation of solar cells made using MVHF and RF techniques. Besides high efficiency, the MVHF cells also demonstrate superior light stability, showing <10% degradation after 1000 hour of one-sun light soaking at 50 °C. We also studied light-induced defect level and hydrogen evolution characteristics of MVHF deposited a-SiGe:H films and compared them with the RF deposited films.

INTRODUCTION

RF glow discharge has been widely used to deposit a-Si:H and a-SiGe:H films and thin film solar cells. One serious limitation of this method is that for high quality materials, the deposition rate is limited to ~3 Å/s. High deposition rates usually lower the material and cell quality. The poor quality is attributed to high defect density in the intrinsic material, arising from di- and poly-hydride bonds in the film [1-3]. a-Si:H and a-SiGe:H based solar cells made by RF at high rate exhibit low efficiency and poor light stability.

In order to circumvent the problems associated with RF at high rate, we have developed a Modified Very High Frequency (MVHF) glow discharge technique to deposit good quality a-Si:H and a-SiGe:H thin films and solar cells using a deposition rate 2-4 times that of a typical RF deposition. The MVHF-deposited cells demonstrate high efficiency and superior light stability. Such high rates for RF deposition would adversely affect film quality, solar cell performance, and light-induced degradation. In this paper, we discuss the results of multi-junction solar cells based on a-Si:H and a-SiGe:H prepared using the high-rate MVHF deposition technique.

EXPERIMENTAL

Two large-area multi-chamber batch deposition machines were used to fabricate the various multi-junction a-Si:H and a-SiGe:H based solar cells in this study. One machine has two chambers for the intrinsic layer deposition, one powered by MVHF and the other by the conventional RF. Both machines use RF for the doped layers. The solar cell structures investigated in the current work include double-junction a-Si:H/a-SiGe:H stack, triple-junction a-

Si:H/a-Si:H/a-SiGe:H and a-Si:H/a-SiGe:H/a-SiGe:H stacks deposited on a bi-layer back reflector (Ag/ZnO or Al/ZnO) coated stainless steel (SS) substrate, as shown in Fig.1. The substrate size is 15"x14". The deposition rate and film thickness uniformity were optimized by process parameters such as cathode-to-substrate spacing, input power density, gas mixture, and total gas pressure. The deposition rate for the intrinsic layer was 1~3 Å/s for the RF-produced films and 5~12 Å/s for the MVHF-produced films. The typical Ge content is ~15-20% for a-SiGe:H middle cells and ~25-40% for a-SiGe:H bottom cells.

For small-area cells, indium-tin-oxide (ITO) and metal contact were deposited to form cells with an active area of 0.25 cm^2. For large area (>450 cm^2) encapsulated cells, grid wires are used to collect current. Current-density versus voltage (J-V) and quantum efficiency (QE) measurements were performed for solar cell characterization. The light-soaking experiments were conducted under an open-circuit condition with 100mW/cm^2 white light at 50 °C for 1000 hours.

RESULTS AND DISCUSSION

Simulation & cathode design

We designed and developed MVHF deposition hardware including cathode and gas distribution system. The design is based on modeling studies to obtain good film thickness uniformity over large areas (15"x14"). We have optimized the deposition parameters and attained high rate deposited a-SiGe:H films exhibiting hydrogen concentration and defect density similar to or slightly lower than their RF counterparts prepared at lower rate (≤3 Å/s).

One major challenge for the MVHF plasma process is to obtain spatial uniformity over a large area. We first conducted computer simulation of MVHF electric field distributions over a large-area (15"x14") cathode, testing different cathode structures and MVHF application techniques. We then designed new MVHF cathodes that resulted in electric field uniformity better than ±5% over the entire area. We also designed a gas distribution assembly to attain a desirable gas flow pattern in the high-flow regime, and a heating assembly to obtain uniform heating over the deposition area. The film thickness uniformity for both a-Si:H and a-SiGe:H deposited using MVHF plasma shows good agreement with that predicted from the simulation.

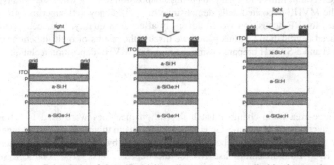

Figure 1. Schematic of three multi-junction structures studied in this work.

Multi-junction cell structures and initial performance

We use the a-Si:H/a-SiGe:H double-junction structure to develop the MVHF deposition, because the simpler double-junction cell provides a quick turnaround for various experimental conditions yet still contains a-Si:H and a-SiGe:H component cells, as well as a tunnel junction. With optimization of both the MVHF process and cell design, the MVHF double-junction cells deposited at a rate >2 times that of RF turn out to yield high efficiency, comparable to the best RF a-Si:H/a-SiGe:H/a-SiGe:H triple-junction structure. We obtained >12% initial small active-area (~0.25 cm^2) cell efficiency and 10.7% large aperture-area (~450 cm^2) encapsulated cell efficiency on Ag/ZnO (BR) coated stainless substrate.

With optimized a-Si:H and a-SiGe:H materials and double-junction cells, we started a comparative study on two triple-junction cell structures of a-Si:H/a-Si:H/a-SiGe:H and a-Si:H/a-SiGe:H/a-SiGe:H. For each cell structure, systematic experiments were conducted to obtain the optimal cell performance using small cells cut from the center of the 15"x14" deposition. Characteristics of the two encapsulated large-area triple-junction and one double-junction cell are summarized in Table I. A large-area all RF triple-junction cell deposited in a roll-to-roll machine is also included as a reference. The 4 cells yield similar initial efficiencies, 10.7-10.8%, regardless of the deposition technique (MVHF vs. RF), rate, and cell structure.

Table I. Characteristics of three different multi-junction structures deposited by MVHF. An RF cell is also included for comparison. All cells were encapsulated and measured at 25 °C under AM1.5 insolation in their initial state. The a-Si:H/a-Si:H/a-SiGe:H cell was deposited on Al/ZnO BR and the other three cells on Ag/ZnO BR. The thicknesses for groups A, B, C, and D are ~0.8 μm, ~0.3 μm, ~0.5 μm, and ~0.5 μm, respectively.

Cell Structure	Back reflector	Process	Aperture area (cm^2)	P_{max} (W)	V_{oc} (V)	FF	Efficiency (%)
A:a-Si:H/a-Si:H/a-SiGe:H	SS/Al/ZnO	VHF	464	5.00	2.60	0.68	10.8
B:a-Si:H/a-SiGe:H	SS/Ag/ZnO	VHF	464	4.96	1.62	0.66	10.7
C:a-Si:H/a-SiGe:H/a-SiGe:H	SS/Ag/ZnO	VHF	464	4.96	2.35	0.68	10.7
D:a-Si:H/a-SiGe:H/a-SiGe:H	SS/Ag/ZnO	RF	807	8.65	2.23	0.66	10.7

Light-induced Stability

Though all the above three MVHF deposited structures show comparable initial efficiencies, they resulted in quite different resistance to light-induced degradation. As illustrated in Fig. 2, the MVHF a-Si:H/a-SiGe:H double-junction (B) and MVHF a-Si:H/a-SiGe:H/a-SiGe:H triple-junction (C) structures are more stable than the MVHF a-Si:H/a-Si:H/a-SiGe:H triple-junction (A) structure, showing ~10% degradation after 1000 hours of light soaking as compared to more than 20% degradation for structure (A). The larger light-induced degradation for structure (A) is attributed to the thick a-Si:H intrinsic layer used in the middle cell. The initial and light soaked cell characteristics and their degradation are shown in Table II. For the same cell structures (C) and (D), the MVHF cell (C) is more stable than the RF cell (D), 8.5% vs. 14% degradation after 1000 hours of light soaking. For the best MVHF cell structure (C), the stable large-area aperture efficiency is 9.8%, approaching 10%.

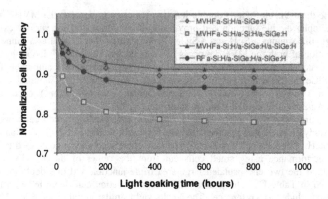

Figure 2. Normalized solar cell efficiency versus light soaking time.

Table II. Initial and light soaked J-V characteristics of the same 4 multi-junction cells in Table I. The degradation (%) is also given for each cell.

Cell Structure	Process	Aperture area (cm^2)	Light soak time (hour)	P$_{max}$ (W)	V$_{oc}$ (V)	FF	Efficiency (%)
A: a-Si:H/a-Si:H/a-SiGe:H	MVHF	464	0	5.00	2.60	0.68	10.8
			1000	3.88	2.48	0.58	8.4
			Degradation (%)	22%	5%	14%	22%
B: a-Si:H/a-SiGe:H	MVHF	464	0	4.96	1.62	0.66	10.7
			1000	4.40	1.59	0.62	9.5
			Degradation (%)	11%	2%	5%	11%
C: a-Si:H/a-SiGe:H/a-SiGe:H	MVHF	464	0	4.96	2.35	0.68	10.7
			1000	4.54	2.29	0.64	9.8
			Degradation (%)	9%	3%	6%	8.5%
D: a-Si:H/a-SiGe:H/a-SiGe:H	RF	807	0	8.65	2.23	0.66	10.7
			1000	7.44	2.21	0.58	9.2
			Degradation (%)	14%	1%	12%	14%

a-SiGe:H film study

We also conducted materials studies to understand the mechanism of better stability for the MVHF a-SiGe:H solar cells. Here we present the results of hydrogen evolution experiments and light-induced defect level analysis by drive-level capacitance profiling (DLCP) to compare MVHF and RF deposited a-SiGe:H materials.

Hydrogen evolution is a convenient method to evaluate microstructures of hydrogen incorporated into a-Si:H or a-SiGe:H based films. Estimated total hydrogen concentration and H$_2$ evolution peak shape and location are indicative of the film quality [5]. Hydrogen evolution spectra are shown in Fig. 3 for three a-SiGe:H thin film samples with similar bandgap and Ge concentration. Two are RF deposited, with 1Å/s and 10Å/s deposition rate, respectively, while the third is a MVHF deposited film at 10Å/s. Though all three samples show similar shape in the evolution spectrum, having a broad peak between 350 and 550 °C, there are two significant differences between the 10Å/s RF and MVHF films in comparison with the 1Å/s RF film. The first is that the 10Å/s MVHF and the 1Å/s RF films have similar amount of hydrogen in the film, 10.4% versus 11.7%, while the 10Å/s RF film has a larger amount of hydrogen in the film, 17.2%. The second is the hydrogen evolution peak for the 10Å/s MVHF film shifts toward higher temperature by ~15°C compared to the 10Å/s RF film, indicating stronger hydrogen bonds for the MVHF film.

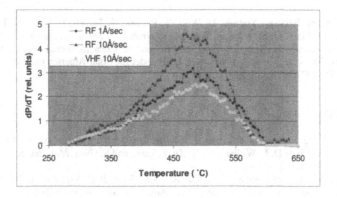

Figure 3. Hydrogen evolution spectrum of three a-SiGe:H films. Two films were deposited by RF at 1Å/s and 10Å/s, while a third sample is MVHF deposited at 10Å/s.

The electronic properties of the a-SiGe:H i-layers were also characterized by drive-level capacitance profiling (DLCP) in both the annealed and light-degraded state. The latter was obtained using a 610nm filtered ELH light for 100 hours at an intensity of 200mW/cm^2. Both mid-gap alloys with optical gaps near 1.6eV, and narrow-gap alloys with optical gaps near 1.5eV were characterized [4].

DLCP measurements for the light-degraded mid-gap RF deposited samples indicated that defect densities varied from below 10^{16} cm^{-3} to nearly 2×10^{16} cm^{-3} as the growth rate was increased from 1Å/s to 5Å/s. The defect densities for the MVHF deposited samples were at least a factor of 2 lower in the higher growth rate regime. For the light-degraded narrow gap alloys, defect densities varied from roughly 2×10^{16} cm^{-3} for the 1Å/s RF deposited layers to nearly 10^{17} cm^{-3} as the RF growth rate was increased to 5Å or 10Å/s. In contrast, the MVHF deposited narrow gap alloys at 5Å/s or 10Å/s indicated defect densities below 5×10^{16} cm^{-3}. It is clear that with the same bandgap MVHF deposited a-SiGe:H films have superior stability against light-induced degradation than the RF deposited films, which is consistent with solar cell light-soaking results.

CONCLUSION

We have developed a MVHF high rate a-Si:H and a-SiGe:H process, which yields comparable initial cell performance for multi-junction structures but superior light soaking stability compared to the RF deposited structures. The stable large-area aperture efficiency for the best encapsulated MVHF multi-junction structures is nearly 10%. Hydrogen evolution and light-induced defect studies indicate that the MVHF high rate a-SiGe:H films have superior light-induced stability than the RF films.

ACKNOWLEDGEMENTS

The authors thank K. Younan, D. Wolf, T. Palmer, N. Jackett, L. Sivec, B. Hang, R. Capangpangan, J. Piner, G. St. John, A. Webster, J. Wrobel, B. Seiler, G. Pietka, C. Worrel, J. Zhang, S. Almutawalli, D. Tran, Y. Zhou, S. Liu, and E. Chen for sample preparation and measurements. The work was supported by US DOE under the Solar America Initiative Program Contract No. DE-FC36-07 GO 17053.

REFERENCES

[1] S. Guha, J. Yang, S. Jones, Y. Chen, and D. Williamson, *Appl. Phys. Lett.* **61**, 1444 (1992).
[2] S. J. Jones, Y. Chen, D. L. Williamson, X. Xu, J. Yang, and S. Guha, *Mat. Res. Soc. Symp. Proc.* Vol. 297, p. 815 (1993).
[3] G. Yue, B. Yan, J. Yang, and S. Guha, *Mat. Res. Soc. Symp. Proc.* Vol. **989**, p.359 (2007).
[4] A. Triska, D. Dennison and H. Fritzsche, Bull. Am. Phys. Soc. **20** (1975) 392.
[5] P. Hugger, J. Lee, J. D. Cohen, X. Xu, B. Yan, J. Yang, and S. Guha, to be presented in 2009 *MRS Spring Meeting, Symp. A.*

Mater. Res. Soc. Symp. Proc. Vol. 1153 © 2009 Materials Research Society   1153-A07-03

Improvement of Quantum Efficiency of Amorphous Silicon Thin Film Solar Cells by Using Nanoporous PMMA Antireflection Coating

Liang Fang, Jong San Im, Sang Il Park and Koseng Su Lim
Division of Electrical Engineering, School of Electrical Engineering and Computer Science,
KAIST, 335 Gwahangno, Yuseong-gu, Daejeon 305-701, Republic of Korea

ABSTRACT

The enhancement of optical transmittance at the air/glass interface of amorphous silicon thin film solar cells was shown by application of a nanoporous polymethyl methacrylate (PMMA) antireflection (AR) coating. The PMMA coating was prepared by spin coating of PMMA solution in chloroform in the presence of a small amount of nonane. Because of the difference of the vapor pressure of chloroform and nonane, phase separated structure formed after complete evaporation of both of them during spin coating process. The Corning 1737 glass with the AR coating has high transmittance near 95% from 450-1100nm wavelengths. The amorphous silicon solar cells with the nanoporous PMMA AR coating realize an improvement in quantum efficiency (QE) up to 4% in 450-650nm spectral regions.

Introduction

Recently, antireflection coatings have been widely employed in optical and optoelectronic fields, and most importantly, for solar cells application. Antireflection (AR) coatings play a vital role in a wide variety of optical technologies by enhancing light transmittance at interface. For glass and common plastics, refractive index (n) is within the range of 1.45-1.7. As a result, reflection forms 4% to more than 6.5% of normally incident light from air-medium interface [1]. Until now antireflection coatings for solar cells application have been concentrated on inorganic material. These thin dielectric films with a low refractive index can enhance transmission via the destructive interference of the reflected light at the air/film and the film/glass interfaces.

With the current trend of technology moving towards optically transparent polymer media and coatings, nanoporous polymer AR coatings have attracted wide attention. Because polymer materials have relatively simple and economical fabrication process compared to inorganic materials, they also can be easily coated on a large area and flexible substrate. In the case of a single layer AR coating, two criteria must be satisfied: a film thickness must be a quarter of a reference wavelength in the optical medium, and the refractive index n_f of the film material must be satisfied by the equation $n_0/n_f = n_f/n_s$, here n_0, and n_s are the refractive indices of air, and substrate, respectively. Film thickness can be easily met. However, the refractive index of the film poses a problem, because the n_s of Corning 1737 glass is around 1.52. Based on above equation, the ideal n_f of an AR film should be 1.23. But the smallest n_f of available dielectric material is 1.38 of MgF_2. Therefore, any dense film can't satisfy the ideal value. Instead of a homogeneous layer, a nanoporous film can be used. When the pore size is much smaller than the visible wavelengths, the effective refractive index (n_e) can be estimated by using the following equation $n_e^2 = n_f^2 \times (1-\phi) + \phi$ [2]. Here n_e, n_f, and ϕ are the effective refractive indices of porous film, dense film, and porosity, respectively. The n_e of the nanoporous polymer is given by an average over the film, and the challenge is to control the volume ratio of the pores to achieve the suitable refractive index needed for AR coating. Not surprisingly, porous surfaces and materials have been extensively explored for antireflection applications, such as Plasma etching [3], phase separation of blend polymer [4], and breath figure technique [5]. These coatings show efficient antireflection properties, however, they are not well suitable for amorphous silicon (a-Si:H) solar cells application. Ex situ coating step is preferred because the a-Si:H solar cells with good quality were usually deposited by plasma enhanced chemical vapor deposition (PECVD) with a substrate temperature more than 200 °C. But the glass temperature (T_g) of polymer materials is lower than 200°C. Furthermore, polymer will cause contamination to PECVD system.

In this paper, we presented a nanoporous PMMA antireflection coating by spin coating process based on Kim's research [6]. The antireflection film was prepared by a spin coating of a polymer/solvent/nonsolvent ternary system which consists of polymethyl methacrylate (PMMA) solution in chloroform in the presence of a small amount of nonane. During the spin coating, a dense skin layer was formed because of rapid solvent evaporation, and a phase separated structure formed below this layer. The phase separated structure below the dense skin layer became a porous layer after the complete evaporation of both chloroform and nonane. The nanoporous PMMA AR coating exhibited excellent transmittance enhancement from 400-1100nm wavelengths. Amorphous silicon solar cells prepared by our group with the AR coating showed light absorption enhancement from 400-800nm wavelengths, and realized an improvement in quantum efficiency (QE) up to 4% in 450- 650nm spectral regions.

Experimental

Amorphous silicon solar cells were prepared using a four-reaction-chamber system, which consists of a three-photo-CVD, one-PECVD, and a load–lock chamber. Amorphous silicon solar cells were deposited on SnO_2-coated glass substrate with optimized qualities [7], with a cell structure of glass/SnO_2/p-a-SiC:H/buffer/i-a-Si:H/n-μc-Si:H/Al.

Corning 1737 glass was used as a substrate, PMMA with a weight-average molecular weight

(M_w) of 350 000 was purchased from Aldrich Company. Chloroform was used as a solvent, nonane was used as a nonsolvent. Various compositions of solutions were spin coated on the glass substrate to measure transmittance. The film thickness and surface morphology of the spin coated PMMA films were measured by a field emission scan electron microscopy (FE-SEM) (HITACHI S-4800). Light transmittance was measured by a UV-Vis-NIR spectrophotometer (V-570 JSACO). After fabrication of the solar cells, a nanoporous PMMA AR coating was spin coated on the glass side of the cells. Current density vs. voltage (J-V) characteristic was measured under AM1.5 conditions at 100mW/cm^2. Quantum efficiency (QE) measurement was performed on a constant energy spectroscope (JASSO, YQ-250MW-GD) in the range of 400-1100nm. As reference, both the J-V characteristic and QE measurement for each cell on each sample were measured prior to spin coating of a nanoporous PMMA AR film. Following these initial measurement, the nanoporous PMMA AR film was coated, and the J-V characteristic and QE were measured again.

Results and discussion

Figure 1(a) shows the influence of spin coating speed on transmittance. The transmittance of the glass with a nanoporous PMMA AR coating increases in certain spectral region compared to that of the glass. With increasing the spin coating speed, the maximum point of the transmittance shifts to blue wavelength region. At 9000 rpm, the transmittance has a broad band characteristic from 400-1100nm wavelengths, and with a maximum point around 550nm. Figure 1(b) shows the influence of volume amount of nonane on transmittance.

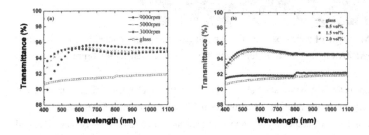

Figure 1. (a) Transmittance of glass spin-coated with PMMA solution (0.009 g/cm^3) in chloroform with 1.5 vol % of nonane at three rotating speed (9000, 5000, 3000rpm) (b) Transmittance of glass spin-coated with PMMA solution (0.009 g/cm^3) in chloroform with various amount of nonane at 9000rpm

With 0.5 vol % of nonane, the transmittance increases a little compared to that of the uncoated glass. At 1.5 vol % of nonane, the transmittance of the coated glass increases significantly. However, with further increase of nonane (2 vol %), the transmittance is smaller than that prepared with 1.5 vol % of nonane. With nonane more than 3 vol %, the film shows a milky color, which proofs the presence of the scattering of incident light. The optimized nanoporous AR coating enhances the transmittance about 4% from 450nm to 1100nm spectral regions.

Figure 2 shows SEM images of cross section, and top surface of the optimized nanoporous PMMA AR coating. There are some pores on the upper surface. The size of the pores seen in the top and inner layers was no more than 200nm, which was much smaller than micro size observed in conventional membranes. Two reasons might be considered. The first is the film thickness. The films prepared in this study have thicknesses around 50-80nm (depending on the spin coating speed, time, and amount of PMMA in chloroform solution). Therefore, the size of the phase separated domain in the conformed geometry becomes smaller. The second is the kinetic effect resulting from fast drying and solidification due to spin coating. The growth of a phase separated domain needs a mobility of polymer chains, since drying of solvent and nonsolvent during spin coating caused rapid solidification. The phase separated domain is limited to 200nm.

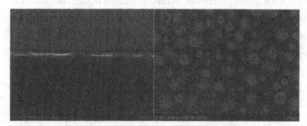

Figure 2. SEM images of a cross section (a), top image (b) of a spin-coated PMMA film from 0.009 g/cm^3 PMMA solution in chloroform with 1.5 vol% of nonane at 9000rpm

The nanoporous PMMA AR coating shows above mentioned characteristic because the vapor pressure of nonane is lower than that of chloroform. The boiling points of chloroform and nonane are 61 and 151°C, respectively. The vapor pressures at 25°C of these two are 26.2 and 0.57 kpa, respectively. Chloroform evaporates much fast from the surface of the solution upon spin coating. Then the surface of the solution can easily be solidified compared with inner parts of the solution. On the other hand, the nonane at the inner layer increases with time. The solution at the inner layer then becomes a phase separated structure consisting of a polymer rich phase and nonane rich phase. The latter becomes porous after complete evaporation of chloroform and nonane. During the phase separated process in our study, the inner layer of PMMA film forms a quasi-graded distributed structure. The structure shows a progressive narrowing of the pores radius with increasing depth towards the glass substrate. Based on the characteristic matrix theory [1], the nanoporous PMMA AR coatings with uniformity distributed pores from bottom to top show an improvement of transmittance within a certain wavelength. However, the AR coating developed by us has broadband antireflection characteristics not only at visible but also near infrared wavelengths, which is similar to the characteristic of graded index of refractive. The enhancement of the transmittance with the nanoporous PMMA AR coating is relatively small compared with that of fully graded films because of not fully graded index characteristic and relatively high refractive index of the upper layer. However, the dense upper layer can act as a simple protective layer.

Figure 3 shows the QEs of amorphous p-i-n solar cells fabricated with and without a nanoporous PMMA AR coating. The QE could be noticeably improved in the spectral region 450-650nm. The short circuit current density deduced from the QE spectrum of the solar cells with and without a nanoporous PMMA coating is 13.15mA/cm^2 and 13.63mA/cm^2, respectively. The limited improvement of the current density is due to relatively small difference of the refractive index between air and glass (1.0/1.52) compared to air and silicon (1.0/3.5).

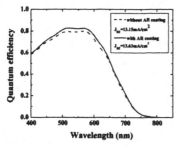

Figure 3. QEs of the amorphous single-junction solar cells with and without nanoporous PMMA AR coatings.

Conclusions

We have attained a nanoporous PMMA AR coating with a spin coating process. The AR coating enhances the transmittance at both visible and near infrared wavelengths. Applying the nanoporous PMMA AR coating on amorphous silicon solar cells realizes an improvement in QE up to 4% from 450-650nm spectral regions. Compared to traditional inorganic AR coating, we have provided a simple, flexible, and economical fabrication process suitable for AR coating of thin film silicon solar cells.

Acknowledgements

The authors would like to thank Dr. Cui Liqiang for helpful discussions regarding the phase separated process during spin coating, and acknowledge research support from Korea Science and Engineering Foundation (KOSEF) grant funded by the Korea government (MEST) (2005-211-D00247) and (R11-2008-058-00000-0)

References

[1] H. A. Macleod, Thin-Film Optical Filters, Macmillan Pub. 1986.

[2] H. Hattori, Advanced Materials **13**, 51 (2001).

[3] U. Schulz1, P. Munzert1, R. Leitel. Wendling, N. Kaiser1, A. Tünnermann, Optical Express **15**, 13108 (2007).

[4] S. Walheim, E. Schäffer, J. Mlynek, U. Steiner, Science **283**, 520 (1999).

[5] M. Park, J. Kim, Langmuir **21**, 11404 (2005).

[6] M. Park, Y. Lee, J. Kim, Chem.Mater. **17**, 3944 (2005).

[7] J. Kwak, Ph.D. Thesis, KAIST, Republic of Korea, 2007.

Mater. Res. Soc. Symp. Proc. Vol. 1153 © 2009 Materials Research Society   1153-A07-05

Hydrogenated amorphous silicon based solar cells: optimization formalism and numerical algorithm

A.I. Shkrebtii[1,*], Yu.V.Kryuchenko[1, 2], A.V.Sachenko[2], I.O.Sokolovskyi[1, 2], and F.Gaspari[1]

[1] University of Ontario Institute of Technology, Oshawa, ON, L1H 7L7, Canada
[2] V. Lashkarev Institute of Semiconductor Physics NAS, Kiev, 03028, Ukraine

ABSTRACT

Thin film hydrogenated amorphous silicon (a-Si:H) is widely used in photovoltaics. In order to get the best possible performance of the a-Si:H solar cells it is important to optimize the amorphous film and solar cells in terms their parameters such as mobility gap, p-, i- and n-layer doping levels, electron and hole lifetime and their mobilities, resistance of p-, i- and n-layers, contact grid geometry and parameters of the transparent conducting and antireflecting layers, and others. To maximize thin a-Si:H film based solar cell performance we have developed a general numerical formalism of photoconversion, which takes into account all the above parameters for the optimization. Application of the formalism is demonstrated for typical a-Si:H based solar cells before Staebler-Wronski (SW) light soaking effect. This general formalism is not limited to a-Si:H based systems only, and it can be applied to other types of solar cells as well.

INTRODUCTION

Amorphous silicon based solar cells are very promising because of low production cost, possibility of covering large uneven areas, and sufficiently high efficiency[1]. In order to get the best performance of the a-Si:H solar cells it is important first to produce a high quality and stable amorphous film and p-i-n junction. Secondly, optimization of the solar cells is a crucial part of the solar cell design. A few approaches to optimize performance of various types of solar cells are available, both analytical and numerical[2]. Analytical models have the advantage of being physically intuitive and of providing the possibility of quick estimation of photo-conversion efficiency. Previously, we have developed an analytical three-dimensional photoconversion model and applied it to optimize the amorphous Si based solar cell[3]. However, to further advance the solar cell optimization, more detailed approaches, numerical rather than analytical, are required.

A few problems have to be overcome to create efficient a-Si:H solar cells. The main problem related with a-Si:H based devices is the material stability due to the formation of metastable defects in a-Si:H that reduces the performance of these devices. Staebler and Wronski[4] found that defects can be created by illuminating a-Si:H, which decreases dark and photo conductivity. Several models have been proposed to explain the mechanism of the Staebler-Wronski (SW) effect, but no consensus has yet been reached. Microscopic mechanisms of hydrogen rebonding and diffusion are important to understand this detrimental effect. To access such processes microscopically, we theoretically investigated hydrogen behavior in a-Si:H using extensive first-principles finite temperature molecular dynamics [5-7] combined with dielectric function calculations [8] and its experimental ellipsometry measurements in the far infrared spectral range[9] as well as electronic properties[10] (both the results from [9] and [10] are presented at the 2009 MRS Spring Meeting). Such combined study allowed us to gain access at a microscopic level to a-Si:H stability related to the hydrogen diffusion, rebonding and other types of the material instability.

In this paper we outline the main features of our numerical optimization formalism and demonstrate its applications. Details of the formalism will be presented elsewhere[11].

THEORY

We consider a conventional solar cell of a sandwich type consisting of a $p - i - n$ structure between a frontal grid and rear electrodes (see figure 1). The frontal collecting grid electrode consists of parallel metal fingers, connected to each other through two conductive bus bars. The grid is placed on top of a conductive p – type ITO film. The area between the fingers is filled with SiO_2 film. A highly doped p – layer is in contact with the frontal electrode on the top and the i – layer on the bottom. A highly doped n – layer is located between the i – layer and the uniform Al rear reflective electrode (figure 1).

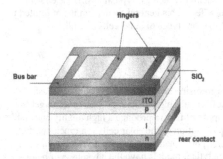

Figure 1. Schematic view of the solar cell under consideration. It consists of the $p - i - n$ structure between frontal grids and rear reflecting metal contact. The frontal collecting grid electrode contains parallel metal fingers connected to each other through two conductive bus bars. The frontal grid is filled with SiO_2 film placed on top of transparent ITO layer.

The frontal grid and rear electrode collect the photocurrent. The ITO film (i) serves as a wideband window; (ii) together with SiO_2 film it provides transparency for solar illumination; and (iii) collects photo-current between the metal fingers of the grid electrode. Highly doped p – and n – layers create a rectifying barrier that separate electrons and holes, while electron-hole pair generation takes place mainly in i – layer.

We numerically optimized the solar cell and the main features of the optimization model are as follows: (i) The short circuit current (density) J_{sc}, open circuit voltage, filling factor, optimal current and voltage, as well as the solar cell efficiency in a-Si:H p-i-n sandwich are calculated using *diffusion* theory. Diffusion theory is most appropriate to describe transport of excess electrons and holes in amorphous semiconductors. Both the black body solar radiation and experimental solar cell spectrum at the Earth's surface can be used as the numerical input of the program. (ii) The contribution of the space charge region (SCR) at the *p-i(n)*-junction to the solar cell characteristics has been taken into account by calculating band bending and adjusting currents and excess electron-hole densities at the SCR boundaries. (iii) Multiple light reflection in technological layers at the solar cell front surface (protective SiO_2 layer and conducting *p*-type ITO layer) including multiple reflection from metallic finger electrodes (for oblique light incidence) have also been considered to describe transmission of the light into the active a-Si:H region of solar cell. A generation function inside the a-Si:H region also accounts for multiple reflections from rare contact and front surface sublayers. (iv) Local orientation of the solar cell at a definite geographical point, changes in the angles of light incidence during time of year and daytime have been taken into account; it allows calculating and optimizing mean solar cell efficiency during the long-term period (*e.g.*, one year). Finally, both vertical and horizontal orientations of finger electrodes in the plane of the solar cell have been considered.

An extensive report outlining the theoretical approach is currently under preparation [11]; we demonstrate here the application of the formalism for a few typical a-Si:H based solar cells.

OPTIMIZATION RESULTS

Dependence of collected short circuit currents (figure 2) and energy conversion efficiencies (figure 3) versus the *i*-layer thickness *d* of a-Si:H solar cell is calculated considering AM1.5G (*i.e.*, global) illumination condition. Reflection coefficients from rear contact are the following: R_d equals to 1.0 for the curves 1 and 1′, 0.8 for 2 and 2′, 0.6 for 3 and 3′, and 0.4 for 4 and 4′ curves respectively.

The set of curves 1 – 4 corresponds to the case when the energy spectral region $E > E_G$ of AM1.5G solar spectrum is effective in power generation ($E_G \cong 1.75$ eV for hydrogen content C_H=10%, the value chosen for the calculations). The set of curves 1′ – 4′ corresponds to the case when a wider spectral region $E > 1.55$ eV contributes to the energy conversion. Note that for a-Si:H the mobility gap E_G can be close to 1.55 eV only for a small hydrogen content (for amorphous Si without hydrogen $E_G \cong 1.58$ eV).

The dependences, shown in figures 2 and 3, have been calculated for two sets of carrier lifetimes in the *i*-layer. In the first case these lifetimes are τ_{ei}=10⁻⁷ s and τ_{hi}=10⁻⁷ s, that is the same as τ_{ep} and τ_{hp} in the doped p^+-layer of an a-Si:H solar cell. In the second case τ_{ei}=τ_{hi}=10⁻⁶ s, which is an order of magnitude larger than in p^+-layer.

The current and efficiency dependences have been obtained for the case of p^+- layer doping with a total acceptor density n_A=10¹⁹ cm⁻³. Parameters of the acceptor energy states distribution within the mobility gap, $N_A(E) = \dfrac{n_A}{\sigma_A \sqrt{2\pi}} \exp\left[-\dfrac{(E - E_A)^2}{2\sigma_A^2} \right]$, are the following: $E_A = 0.2$ eV and $\sigma_A = 0.1$ eV, while the *i*-layer was considered to be undoped, that is n_D=0.

Other parameters were assigned basically standard values, and include operating temperature *T*=300 K; thickness of the p^+-layer of a-Si:H, d_p=0.25 μm; thickness of the top SiO_2 layer, d_{SiO2}=0.09 μm; thickness of the collecting ITO layer, d_{ITO}=1.5 μm; thickness of the technological *n*(*n*+) a-Si:H layer at rear contact, d_n=0.02 μm; relative metallization of front surface with collecting metallic fingers, *m*=0.05; width of metallic fingers, L_F=0.02 μm; hole concentration in ITO layer, n_h^{ITO}=10¹⁹ cm⁻³; saturation current density in *p-i*-junction diode characteristics of a-Si:H solar cell, j_S=10⁻¹² A/cm²; nonideality factor, *r*=1.5; mobility of photoelectrons in p^+-

layer, μ_e^p=2 cm^2/(V·s); hole mobility in p^+-layer, μ_h^p=0.3 cm^2/(V·s); electron mobility in i-layer, μ_e^i=2 cm^2/(V·s); mobility of photoexcited holes in i-layer, μ_h^i=0.3 cm^2/(V·s); hole mobility in ITO-layer, μ_h^{ITO} = 25 cm^2/(V·s); lifetime of photoelectrons in p^+-layer, τ_{ep} = 10^{-7} s; lifetime of holes in p^+-layer, τ_{hp} = 10^{-7} s; surface recombination rate of electrons at front surface, S_0 = 100 cm/s; surface recombination rate of holes at rear surface, S_d= 10 cm/s; rate of tail states density decrease in the exponential distribution $N_{ct}(E) = N_{ct0} \exp[(E - E_g)/E_{c0})]$ of tail states inside the band gap at conduction band edge, E_{c0} = 25 meV; rate of tail states density decrease in the exponential distribution $N_{vt}(E,T) = N_{vt0} \exp[-E/E_{v0}(T)]$ of tail states inside the band gap at valence band edge at T=300 K, E_{v0}(300 K)= 45 meV; correlation energy for charged deep defect states occupied by two electrons, u=0.2 eV; electron effective mass in a-Si:H conduction band, m_e=2.78 m_0; hole effective mass in a-Si:H valence band, m_h=2.34 m_0.

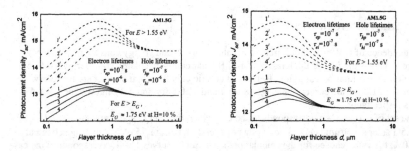

Figure 2. Collected short circuit currents versus thickness d of i-layer. In the left panel electron lifetime τ_{ei}=τ_{hi} =10^{-6} s, while for the right panel τ_{ei}=10^{-7} s and τ_{hi}=10^{-7} s.

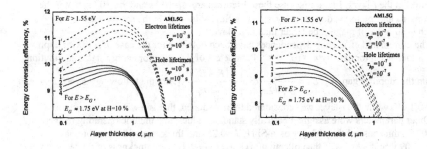

Figure 3. Energy conversion efficiencies versus thickness d of i-layer. In the left panel electron lifetime τ_{ei}=τ_{hi} =10^{-6} s, while for the right panel τ_{ei}=10^{-7} s and τ_{hi}=10^{-7} s.

As it is clear from the above, the program considers all the main parameters of the solar cells and allows for a wide range of parameters modification to maximise the solar cell efficiency. It is important to stress that the above characteristics of a-Si:H solar cells have been calculated for the material before light soaking effect. After a-Si:H light induced degradation the efficiency of the solar cell goes down. In this case an accurate optimization is even more important.

CONCLUSIONS

Optimization of the a-Si:H solar cells has been performed based on the experimental diffusion coefficients, mobilities, parameter of the *p-i-n* structures and defect distribution inside the gap. Diffusion theory of the photoconversion has been considered. The effect of reduction of the photocurrent due to both minority and majority carrier diffusion toward the collector for the short wavelength part of the solar spectra has been included. Both the black body solar radiation and the experimental solar cell spectrum at the Earth's surface can be used as the numerical program input. The contribution of the space charge region (SCR) at the *p-i(n)*-junction to the solar cell characteristics has been taken into account by calculating band bending and adjusting currents and excess electron-hole densities at the SCR boundaries. The parameter of the front - collecting metal grid was optimized with respect to the distributed resistance of the emitter and shadowing effects by the metal strips. Photoconversion has been maximized in terms of the mobility gap, modification of the light absorption coefficient and related diffusion length. Multiple light reflection in technological layers at solar cell front surface (protective silicon oxide layer and conducting p-type ITO layer) including multiple reflection from metallic finger electrodes (for oblique light incidence) has also been considered. A generation function inside the a-Si:H region also accounts for multiple reflections from rare contact and front surface sublayers. Local orientation of the solar cell in a definite geographical point, changes in the angles of light incidence during time of year and daytime have been taken into account, as well as possible changes in the orientation of finger electrodes in the plane of solar cell. This allows for optimization of the mean solar cell efficiency during long-term periods (e.g., one year).

The type of solar cell considered here is SiO_2/ITO/*p-i-n* Si:H/Al, although the approach proposed can be applied to model other types of solar cells. The corresponding executable files of the algorithm will be available on request.

ACKNOWLEDGMENTS

This research was supported by the Centre for Materials and Manufacturing/Ontario Centres of Excellence (OCE/CMM) "Sonus/PV Photovoltaic Highway Traffic Noise Barrier" project, Discovery Grants from the Natural Sciences and Engineering Research Council of Canada (NSERC) and the Shared Hierarchical Academic Research Computing Network (SHARCNET).

---

* Corresponding author, e-mail address: Anatoli.Chkrebtii@uoit.ca

REFERENCES

1. R. A. Street, *Hydrogenated Amorphous Silicon* (Cambridge Univ. Press, Cambridge, UK, 1991).

2. X. Deng and E. A. Schiff, *"Amorphous Silicon Based Solar Cells"*, in *Handbook of Photovoltaic Science and Engineering*, edited by A. Luque and S. Hegedus (John Wiley & Sons, Chichester, 2003), p. 505 – 565.

3. A.V. Sachenko, A.I. Shkrebtii, F. Gaspari, N. Kherani, and A. Kazakevitch, in *IEPIOPTICS-9: Proceedings of the 39th course of the International School of Solid State Physics, Erice, Italy, 20-26 July 2006*, edited by A. Cricenti (World Scientific, 2008), p. 76.

4. D. L. Staebler and C. R. Wronski, Appl. Phys. Lett. **31**, 292 (1977).

5. I. M. Kupchak, F. Gaspari, A. I. Shkrebtii, and J. M. Perz, J. Appl. Phys. **104**, 123525-1 (2008).

6. I. M. Kupchak, A. I. Shkrebtii, and F. Gaspari (private communication).

7. I. M. Kupchak, A. I. Shkrebtii, and F. Gaspari, Phys. Rev. B, 2009, accepted.

8. A. I. Shkrebtii, Y. V. Kryuchenko, I. M. Kupchak, F. Gaspari, A. V. Sachenko, I. O. Sokolovsky, and A. Kazakevitch, in *Proceedings of 33rd IEEE Photovoltaic Specialists Conference, San Diego, USA, May 11–16, 2008*, 2009, p. 470-1.

9. F. Gaspari, A. I. Shkrebtii, S. Mohammed, T. Kosteski, K. Leong, and A. Fuchser, *Proceedings 2009 MRS Spring meeting, San Francisco, April 2009, submitted.*

10. A. I. Shkrebtii, I. M. Kupchak, and F. Gaspari, *Proceedings 2009 MRS Spring meeting, San Francisco, April 2009, submitted.*

11. Y. V. Kryuchenko, A. I. Shkrebtii, A. V. Sachenko, and F. Gaspari, *Research Report on a-Si:H based solar cell optimization*, 2009, in preparation.

Mater. Res. Soc. Symp. Proc. Vol. 1153 © 2009 Materials Research Society   1153-A07-08

Observation of the Evolution of Etch Features on Polycrystalline ZnO:Al Thin-Films

Jorj Owen, Jürgen Hüpkes, and Eerke Bunte
IEF5-Photovoltaik, Forschungszentrum Jülich GmbH, D-52425 Jülich, Germany

ABSTRACT

The transparent conducting oxide (TCO) ZnO:Al is often used as the window layer and a source of light trapping in thin-film silicon solar cells. Light scattering in sputtered zinc oxide is achieved by wet chemical etching, which results in craters distributed randomly over the ZnO surface. To gain a better understanding of the etching process on ZnO thin films, a method for atomic force microscope (AFM) realignment between etching steps is developed. Using this method, the evolution of the HCl etch on a polycrystalline ZnO thin-film is observed. Results showed that this observation method did not modify the etching behavior, nor did stopping and restarting the etching change the points of attack, indicating that the points of HCl attack are built into the films as they are grown. Additionally, we investigated the evolution of the HCl etch on a ZnO surface previously etched in KOH, and found that the etch sites for both the acidic and basic solution are identical. We conclude that "peculiar" defects, which induce accelerated etching, are built into the film during growth, and that these defects can extend part or all the way though the thin-film in a similar way as screw dislocations in single crystalline ZnO.

INTRODUCTION

ZnO is a wide bandgap semiconductor that can be doped to act as a transparent conducting oxide (TCO) for use in both inorganic and organic photovoltaic devices [1, 2, 3]. In Si thin-film solar cells, the ZnO layer is also often also used as a source of light trapping [1, 3]. The texturization of polycrystalline ZnO films can be done through various means, such as wet chemical etching in HCl, or as grown in low pressure chemical vapor deposition (LPCVD), for example [3].

While the etching process for the ZnO single crystals is well understood, there is still no model for polycrystalline ZnO [4, 5, 6]. Despite there being no model, trends have been observed [6, 7]. For example, using a series of reactively mid-frequency (MF)-sputtered ZnO films, Hüpkes et. al. showed that even though the resultant structure from the various chemical (both acidic and basic) and physical etching methods vary, the density of the points of attack decreased for all etch types as the oxygen partial pressure was decreased [6]. From this they concluded that the number of points of attack is built into the material itself.

To verify these observations and better understand the nature of the ZnO etching process, we developed a method for reproducibly aligning and imaging a sample between short etching steps using an atomic force microscope (AFM) in non-contact mode, which minimizes modifications made to the surface. Using this stepwise imaging process, we have been able to observe the evolution of various etches on a ZnO surface indirectly. This paper contains two examples of applications of this method. First, the evolution of the standard HCl etch on the surface of a polycrystalline ZnO thin-film is explored. Second, we investigate the observations of Hüpkes et. al. further.

The discussion will focus on three major areas: first, the applicability and limitations of this alignment method; second, the evolution of the HCl etch on polycrystalline ZnO; and third, the evolution of an acidic etch on a sample previously etched in a base.

EXPERIMENT

Polycrystalline ZnO:Al thin-films were prepared by depositing approximately 800 nm of ZnO:Al on a cleaned Corning glass (Eagle 2000) substrate using radio-frequency (RF)-sputtering (VISS 300, VAAT) from a ceramic target consisting of $ZnO:Al_2O_3$ with 1 % wt. Al_2O_3. The deposition was carried out at a substrate temperature of 300°C, discharge power of 1.5 kW, and 0.1 Pa deposition pressure of pure argon. More details on this and similar films are described by Berginski et. al. [7].

To position the sample on the µm-scale between the multiple etching steps reproducibly, four progressively smaller alignment methods were used as indicated in Figure 1: a) large scale manual marks were made using a diamond tip for the mm-scale alignment of the sample. b) A photolithographically prepared grid was used to align the sample within ±30 µm. The grid consisted of thermally evaporated silver dots of approximately 3-4 µm in diameter with a spacing of 8 µm and a thickness of 50 nm. Defects were chosen as markers and areas to scan since they were uniquely spread across the ZnO surface and provided large available areas to scan. c) A large AFM (SIS nanostation 300) scan was used to further localize (within 1 µm) the defect of interest in the Ag grid. d) The area of interest, which in this case was approximately $12*12$ µm^2, was aligned accurately along the silver dots and scanned.

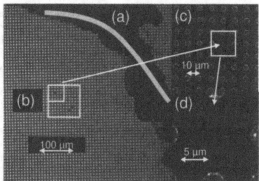

Figure 1. (a) Optical image of a manual mark made using a diamond tip (highlighted). (b) Optical image of defects in the Ag grid (large square), and the area analyzed (small square). (c) Larger AFM scan of the area analyzed. (d) Smaller scan of only the area analyzed.

We applied this stepwise etch-imaging process to track the evolution of the HCl (0.5%) etch on a polycrystalline ZnO thin film with incremental etching times of either 2 or 4 seconds. The method was also used to image a ZnO sample after a 400 seconds etch in KOH (30%), followed by incremental 5 or 10 second etches in the HCl solution. All etches were performed at room temperature. Following all etching steps, the process was stopped by rinsing the sample with deionized (DI) water and dried with nitrogen.

To compare line scans and estimate the remaining ZnO, the AFM data was shifted by setting the glass level to zero in the sample when the etch had progressed far enough to expose it. For the other line scans the maximum AFM value was assumed to fit to the maximum thickness measured by surface profiler (Dektak 3030). This approach fit quite well as the AFM and surface profiler thickness measurements had an error of less than 30 nm.

RESULTS AND DISCUSSION

Strengths and weaknesses of the alignment method

Figure 2 (a) and (b) show SEM images of the resultant HCl etch structure after 30 seconds without and with the Ag grid, respectively. Note that despite the additional process steps required to deposit the Ag grid the resultant etch features appear unchanged. In contrast, realignment in an SEM, which may be faster and easier, contaminates the surface by depositing carbon during the imaging process; this acts as a barrier in subsequent ZnO etching steps. Due to the partly isotropic nature of the HCl etch on ZnO, the ZnO under the Ag grids was also etched causing many of the markers to wash away after repeated etches. Figure 2 (c) and (d) show AFM images in the same location after two and ten cumulative seconds of etching, respectively. Notice that all eight Ag grid points are still present after the single etching step, while only half remain after five etching steps, which provides further evidence that the presence of an Ag grid does not modify the HCl etch. Despite the loss of some of the Ag markers, repositioning the sample in the AFM was still possible using the remaining Ag dots and larger manual marks. After approximately five etch steps, the surface developed unique features that could be identified and remained fairly consistent between etching steps. Once the surface has evolved to this point, the Ag grid is helpful but no longer necessary for realignment. In this case, realignment could alternatively be done using a modified keystone technique [8]. The intact grid is, however, necessary for realignment during the initial etching steps.

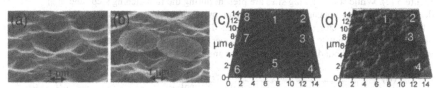

Figure 2. SEM images of an HCl etched ZnO thin-film without (a) and with (b) Ag grid. 15x15 µm scans after a cumulative etching time of 2 seconds (c) and 10 seconds (d).

While the etch features remained unchanged, it is important to note that the etch rates of the samples in this experiment are faster than those observed in the static etching in HCl 0.5% solution. This is most likely due to the dynamic nature of these very short etches, which prevents the shielding of the ZnO by used acid solution. We have observed that samples are etched more quickly when they are kept in motion or placed in a sonicator (not shown). The etch rates observed in this work were a factor of about two larger than similar samples which were statically etched. The higher etch rate observed in these samples is also partly due to the etching that occurs during the transfer of the sample from the etching bath to DI water rinse; this error in actual etching time is compounded with each additional etching step.

HCl etch evolution

Figure 3 shows AFM images taken at the same location on a ZnO surface at various steps in the etching process. Notice that even though the etching process was stopped and restarted many times, the same points of attack were picked in subsequent etching steps.

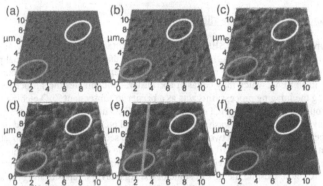

Figure 3. 12x12 μm^2 AFM images taken at cumulative etch times of 2, 4, 8, 12, 16, and 30 seconds shown in (a), (b), (c), (d), (e), and (f), respectively. The line in (e) indicates the area of the line scan analysis.

This is even more evident in the evolution of single line scans at the same location, as shown in Figure 4 (a). The location of the lines relative to the total scanned area is indicated in Figure 3 (e). While some etching sites are present during the first etching steps and then disappear, and others appear only later, most are present throughout the entire etching process. After 18 seconds of etching, the glass was reached at some points (see Figure 4 (a, first dashed line)), while other points still had thicknesses over 660 nm, showing that the etching rate can vary by a factor of about 4.5 between etch sites and other locations on the film surface.

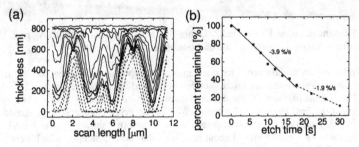

Figure 4. (a) Line scans at approximately the same location throughout the etching process. (b) Percentage of the sample remaining as a function of the cumulative etch time. In both (a) and (b) solid and dashed lines indicate that the glass was not or was observed, respectively.

Further AFM data analysis yielded the fraction of the original ZnO material remaining, depicted in Figure 4 (b) as a function of the cumulative etch time. A linear fit matches the data very well up to 18 seconds, showing that the ZnO was etched at a constant rate of 3.9 %/s. The deviation to another linear fit of 1.9 %/s at longer times can be attributed to a couple of factors. First, 18 seconds is the point at which the glass first appeared, exposing less ZnO to the HCl solution than during earlier etching steps. Second, after the glass appeared, longer etching steps were taken (4 rather than 2 seconds). This longer etching time may allow larger shielding effects.

KOH followed by HCl etches

To study the development of etch sites further we now apply this method to etching in acidic and basic solutions. AFM images taken at the same location on a polycrystalline ZnO:Al surface after a 400 second KOH etch (Figure 5 (a)), and subsequent HCl etches (Figure 5 (b)-(d)) are shown in Figure 5. While the surface structures from the KOH etch are softer, the density of the etching points is similar to the ones observed in Figure 3, which agrees with previous conclusions [6].

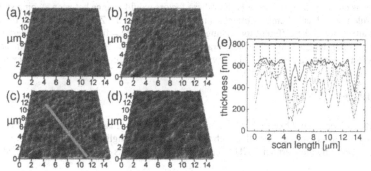

Figure 5. 15x15 µm AFM images and line scans at approximately the same points of the ZnO surface after a 400 second etch in KOH (a, e solid), and additional 5 (b, e dash), 10 (c, e dot), and 20 (d, e dash dot) cumulative seconds etch in HCl. The line in (c) indicates the approximate location of the line scans (e). The thick line in (e) approximates the film's original thickness as determined by surface profiler.

From AFM data we extracted the line traces for the different etching steps, see Figure 5 (e). Notice that while the points of attack become more pronounced in the HCl etch, the points of attack are in general already present in the KOH etch (see markers in Figure 5 (e)). Since ZnO is an amphoteric oxide, it is not surprising that it etches in both basic and acidic solutions. It is interesting, however, that these solutions attack the same points, while in single crystals, the etching of O- and Zn-terminated surfaces are prohibited in bases and acids, respectively, and is only possible at defects [4].

From the evolution of the HCl etch we can conclude that the points of attack are built into the film during growth. If the points of attack were due to the state of the sample in the HCl solution or particular adsorbates on the ZnO surface, subsequent etching steps would attack different points. This conclusion is supported by the observation that KOH and HCl attack the same points. We further conclude that the etch attack is induced by a structural defect rather than a local chemical property like polarity, which would act differently in acids and bases.

These polycrystalline ZnO:Al films have an approximate grain size between 50 and 200 nm, and thus have a much higher density of defects than etching points. To explain the etching behavior of ZnO we assume the presence of certain "peculiar" defects that etch more rapidly in acidic and basic solutions. These defects often reach all the way through the film, but they appear to also have shorter range effects: etching sites may begin at the surface or somewhere in the bulk of the ZnO film, and may extend part or all the way through the film.

CONCLUSIONS

Using an AFM repositioning method we have been able to image the same location between small etching steps. With this technique, the evolution of the HCl etch on polycrystalline ZnO, and the relation between etching in acidic and basic solutions were studied. Specifically, we observed that stopping and restarting the etching process did not change the points of attack, and that both acidic and basic solutions attack the same points. From these observations we ruled out the status of the solution close to the film surface and adsorbed molecules as etching catalysts, since both of these factors would change between etching steps. We also ruled out a local chemistry effect such as the polarity of the ZnO crystallites, as the points of attack would be different for etching in acidic and basic solutions. We concluded that there are peculiar defects that induce accelerated etching built into the film during growth, and that these defects can extent part or all the way though the thin-film in a similar way to screw dislocations in single crystalline ZnO.

ACKNOWLEDGMENTS

The authors would like to thank Janine Worbs, Hilde Siekmann, and Hans Peter Bochem for their technical assistance, as well as Rebecca Owen for her English grammar corrections. Financial support by the German BMU (contract no. 0327693A) is gratefully acknowledged.

REFERENCES

1. O. Kluth, B. Rech, L. Houben, S. Wieder, G. Schöpe, C. Beneking, H. Wagner, A. Löffl, H.W. Schock, Thin Solid Films **351**, 247 (1999).
2. J. Owen, M. S. Son, K.-H. Yoo, B. D. Ahn, S. Y. Lee, Appl. Phys. Lett. **90**, 033512 (2007).
3. S. Faÿ, U. Kroll, C. Bucher, E. Vallat-Sauvain, A. Shah, Sol. Energy Mater. & Sol. Cells **86**, 385 (2005).
4. A. N. Mariano and R. E. Hanneman, J. Appl. Phys. **34**, 384 (1963).
5. G. Heiland, P. Kunstmann, Surf. Sci. 13, 72 (1969).
6. J. Hüpkes, J. Müller, and B. Rech, in *Transparent Conductive Zinc Oxide*, edited by K. Ellmer, A. Klein, and B. Rech (Springer, Berlin Heidelberg, 2008) p. 359.
7. M. Berginski, J. Hüpkes, M. Schulte, G. Schöpe, H. Stiebig, B. Rech, M. Wuttig, J. Appl. Phys. **101**, 1911 (2007).
8. M. Su, Z. Pan, V. P. Dravid, J. Microsc. **216**, 194 (2004).

Mater. Res. Soc. Symp. Proc. Vol. 1153 © 2009 Materials Research Society   1153-A07-09

Erik V. Johnson, Ka-Hyun.Kim, and Pere Roca i Cabarrocas
LPICM, CNRS, École Polytechnique, 91128 Palaiseau, France

ABSTRACT

The efficiencies of hydrogenated polymorphous silicon (pm-Si:H) solar cells have been previously demonstrated to show superior stability under light-soaking. This stability arises due to the fact that the decrease they show in fill factor (FF) is partially offset by an accompanying increase in open circuit voltage (V_{OC}). Recently, high-deposition rate (9Å/s) pm-Si:H material deposited by standard RF-PECVD at 13.56MHz has been investigated as the intrinsic layer in photovoltaic modules as it has shown excellent electronic properties. The degradation behaviour of these high-deposition rate cells, however, differs significantly from that of lower deposition rate material. In particular, no beneficial increase in V_{OC} is observed during light soaking. We investigate the degradation dynamics of solar cells made from this high growth rate material using a Variable Illumination Method (VIM) during light soaking to quantify the changes to these high-rate cells during light-soaking and directly contrast them with those of low-rate (1.5Å/s) cells. In particular, we discuss the importance of bulk recombination effects vs interface quality changes, as well as the dynamics of changes in V_{OC}.

INTRODUCTION

In an industrial environment where equipment depreciation is a significant portion of the cost of thin-film photovoltaic devices, an effective strategy in reducing the cost-per watt of thin-film solar cells is to increase the deposition rate while maintaining the same cell efficiency. The application of RF-PECVD at 13.56MHz to this strategy through the high-rate (HR) deposition of hydrogenated polymorphous silicon (pm-Si:H) has recently been demonstrated to result in device quality material [1]. In general, pm-Si:H is grown from hydrogen-diluted silane at relatively high (>1Torr) gas pressures during deposition – conditions that result in the presence of a small volume fraction of nanocrystallites embedded in the amorphous matrix. This material differs from protocrystalline or transition material [2] in that it does not undergo a structural change once a certain thickness has been reached, but rather maintains constant structural and electronic properties for layers of increasing thickness [3]. When pm-Si:H is applied as the i-layer in PIN structures, this has resulted in 10x10cm mini-modules with initial efficiencies up to 9%, even when the i-layer was deposited at 8-9Å/s [4,5]. Devices made from this high-deposition rate pm-Si:H material, however, display significant differences under light-soaking (LS) when compared to devices made from lower deposition rate (1.5Å/s) pm-Si:H material. In this work, we develop a tool to examine these differences – an in-situ Light-Soaking and Variable Intensity Method (LS-VIM) setup – and to compare changes for two sets of devices grown at these significantly different deposition rates.

EXPERIMENT

Cell deposition conditions

The cells were deposited in the ARCAM reactor, a multi-plasma, mono-vacuum RF-PECVD reactor with an excitation frequency of 13.56MHz [6]. Results for two sets of cells, with i-layers deposited at 1.5Å/s and 8Å/s, are presented in this work, and are denoted as "low-rate" (LR) and "high-rate" (HR), respectively. A summary of the deposition conditions is presented in Table 1. As well, the interelectrode distance was decreased in the HR case to reduce the formation of powder. After deposition, the cell areas ($0.126cm^2$) were defined by evaporation of aluminum back-contacts, and the devices were annealed for 30 mins at 125°C in ambient atmosphere.

Table I. Deposition conditions for pm-Si:H pins

Layer↓ Cell →	Low Deposition rate cells	High Deposition rate cells
Back reflector	Aluminium (200nm)	
n-layer	P-doped a-Si:H at 0.5 Å/s (150Å)	
i – layer	**1.5Å/s**	**8Å/s**
	Flow H_2:SiH_4=200:12	Flow H_2:SiH_4=200:30
	P_{RF}=35 mW/cm^2	P_{RF}=140 mW/cm^2
	Pressure=1.8 Torr	Pressure=3 Torr
	(2500Å)	(2700Å)
p-layer	B-doped a-SiC:H at 0.5 A/s (180 Å)	
Substrate	ASAHI-U	

In-situ Variable Intensity Method Measurements

The Variable Intensity Method [7] is an effective procedure to extract additional, detailed information about the operating characteristics of photovoltaic devices. In this method, Current-Density-Voltage (JV) curves are measured for a wide variety of illumination conditions, ranging from dark conditions to multiple suns. In the present study, conditions from dark to three suns have been used. By measuring the devices under multiple illumination conditions, the true balance between parallel resistance and recombination, as well as between series resistance vs diode characteristics can be determined [4]. Similar information can be extracted concerning V_{OC} and FF limitations, not to mention the cell's performance under varying light conditions (as is necessary to accurately calculate annual energy output). The JV curves in each case were acquired going from forward bias to reverse bias, and the full curve from +1.4V to -1.4V was acquired in ~10s.

DISCUSSION

Comparison of Degradation for Cells Deposited at 1.5Å/s and 8Å/s

The two types of cells in this study were each light soaked using a Xenon arc-lamp under open-circuit conditions at ambient temperature (25°C) for 800-900 minutes, during which time, at certain intervals, a VIM measurement was taken. Table 2 displays the photovoltaic characteristics at a light intensity corresponding to AM1.5 for the cells both before and after light soaking. The J_{SC} for these cells is quite low, as the p-layer thickness was greater than normal. An important result from Table 2 is that for the high deposition rate cell, the FF *increases* during light soaking. The LS-VIM tool gives us the opportunity to examine this unusual effect in detail during the light-soaking process.

The precise dynamics of the changes that occur during light soaking were monitored, and the temporal changes in V_{OC} are presented Figure 1a and 1b for the low-rate and high-rate cell, respectively, under AM1.5 conditions. The dynamics for the *FF* are presented in Figure 2. Also shown on these graphs are the dynamics for "yellow" light (with a high-pass, Kapton filter) and for blue-green light (notch filter at 500nm) at similar photon fluxes. As previously observed, an increase in V_{OC} was noted for the LR-pm-Si:H cell, whereas for the HR cell, a rapid decrease of V_{OC} in the first 10 minutes was seen. Opposite dynamics are observed for the FF in both cases – for the LR cell, the time scales are on the order of 1000 minutes, whereas for the HR cell, most of the FF increase has taken place in the first 100 minutes.

Table II. Photovoltaic parameters for LR and HR cells. Values given are for as-deposited state (AD) and light-soaked state (LS). Cells were soaked in open-circuit conditions, but second column includes the current density measured for the light soaking intensity.

Cell	Light Soaking time and J_{SC}	FF_{AD}	FF_{LS}	$V_{OC,AD}$ [V]	$V_{OC,LS}$ [V]	$J_{SC,AD}$ [mA/cm^2]	$J_{SC,LS}$ [mA/cm^2]	η_{AD} [%]	η_{LS} [%]
LR C53a1	900mins 7.8mA/cm^2	0.63	0.59	0.894	0.908	9.7	8.9	5.5	4.8
HR C63a3	800mins 7.0mA/cm^2	0.51	0.55	0.863	0.853	8.6	8.6	3.8	4.0

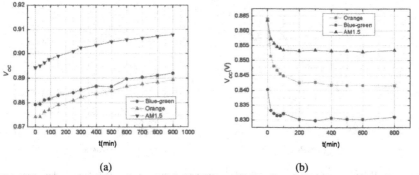

(a)   (b)

Figure 1. Dynamics of V_{OC} during LS for (a) LR-pm-Si:H cell, and (b) HR pm-Si:H cell. Curves are shown for AM1.5, blue-green light conditions (band-pass filter at 500nm), and orange light conditions (Kapton filter).

The graphs of Figures 1 and 2 display the dynamics of V_{OC} for three, constant light conditions, but the LS-VIM measurement acquires the JV curves over many orders of magnitude of illumination. The data collected in such curves are presented in Figure 3. In Figure 3a and 3b, one may observe that the light intensity-dependent V_{OC} can be roughly modelled over the entire range of intensities by two exponential sections (linear when $\log(J_{SC})$ is used as the x-axis). This suggests an equivalent circuit composed of two diodes in parallel, the changes in which can be used to observe the differences in the dynamics of these two devices. We term the diodes describing the low J_{SC} and high J_{SC} sections as the "low-light" and "high-light" diodes, respectively. Each diode current is determined in the classical way by a saturation current density J_0 and a diode ideality factor n.

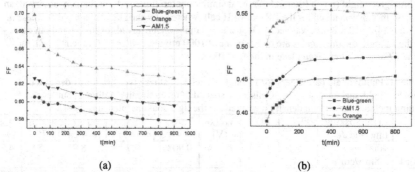

(a)   (b)

Figure 2. Dynamics of *FF* during LS for (a) LR-pm-Si:H cell, and (b) HR pm-Si:H cell. Curves are shown for AM1.5, blue-green light conditions (band-pass filter at 500nm), and orange light conditions (Kapton filter).

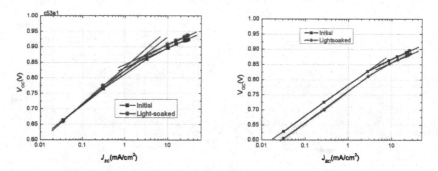

Figure 3. V_{OC} vs J_{SC} extracted from LS-VIM measurements on pm-Si:H PIN cells deposited at (a) LR (1.5 Å/s) and (b) HR (8 Å/s).

In the curves for the LR cell of figure 3(a), the low-light diode shows only a slight change in n (slope), whereas the "high-light" diode determining the high J_{SC} portion shows a **decreased** J_0 (horizontal shift) but an unchanged n (slope). The change in the "high-light" diode, therefore, is responsible for the increase in V_{OC}, whereas changes in the low-light diode occur independently of V_{OC}. Conversely, for the HR cell (Fig. 3b), both the low-light and high-light diode show an increase in J_0 but little change in n. This result underlines the fact that changes to the diode that can be measured under dark conditions do not necessarily predict if V_{OC} will increase with LS.

An interesting phenomenon in this series of cells is the increase in FF observed for the HR cells during light-soaking. This behaviour has been previously reported to occur in nanocrystalline cells, for example, when hydrogen profiling is used [8]. Although this is not a typical phenomenon in pm-Si:H cells, using LS-VIM to observe this process in-situ (Figure 4) is

126

an interesting venture. It can be observed in figure 4 that the gradual improvement in FF occurs over the entire range of light intensities, and that initially, the FF is relatively constant around AM1.5 (~10mA/cm^2), whereas after light-soaking, it decreases with increasing intensity.

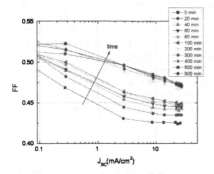

Figure 4. *FF* vs J_{SC} extracted from LS-VIM measurements on a pm-Si:H PIN cell deposited at 8 Å/s

The controlling physical process that determines the *FF* can be extracted by observing the light-dependent short circuit resistance, R_{SC}. The light intensity dependence of this parameter has been used (given certain assumptions) to extract an effective $\mu\tau$ product for carriers in the i-layer of PIN cells [7]. In Figure 5, it can be seen that during light-soaking, the $\mu\tau_{eff}$ is in fact *increasing* for the HR cells, as the R_{SC} vs J_{SC} curves are shifted up during LS. This observation, however, may better reflect the assumptions behind extracting $\mu\tau_{eff}$. For example, the assumption of a uniform field may not be accurate in such low quality cells, and light-soaking may redistribute defects in the gap (from band-edges to mid-gap) such that the field is more evenly distributed throughout the i-layer.

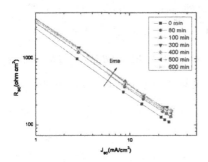

Figure 5. R_{SC} vs J_{SC} extracted from LS-VIM for HR-pm-Si:H PIN cell. The R_{SC} curves are shifted up with light-soaking time.

CONCLUSIONS

The dynamics of the changes in the light-dependent photovoltaic parameters during the light-soaking of pm-Si:H PIN cells reveals details about the internal workings of the cells. In particular, when observing changes in the V_{OC} of LR and HR-deposited cells, the light-dependent measurements suggest an equivalent circuit model of two parallel diodes. The changes in the "high-light" diode are responsible for the V_{OC} increase in the LR cell, whereas dramatic changes in both the "high-light" and "low-light" diodes are evident for the HR cell and result in a V_{OC} decrease. As well, for the HR cell, an increase in the FF during light soaking can be incorrectly interpreted as an improved $\mu\tau_{eff}$. However, cautious reconsideration suggests that during the light-soaking of these high-rate cells, a significant re-distribution of the field occurs. Energetic changes in the location of gap defects near the p-i and i-n interfaces is likely to be responsible for this homogenization of the field.

ACKNOWLEDGMENTS

This work was partially funded by European Project "SE Powerfoil" (Project number 038885 SES6)

REFERENCES

1. Y.M. Soro, A. Abramov, M.E. Gueunier-Farret, E.V. Johnson, C. Longeaud, P. Roca i Cabarrocas and J.P. Kleider, *Thin Solid Films* **516**, 6888 (2008).
2. A. S. Ferlauto, R. J. Koval, C.R. Wronski and R. W. Collins, *Appl. Phys. Lett.* **80**, 2666 (2002).
3. M. Meaudre, R. Meaudre, R. Butté, S. Vignioli, C. Longeaud, J. P. Kleider and P. Roca I Cabarrocas, *J. Appl. Phys.* **86**, 946 (1999).
4. E.V. Johnson, A. Abramov, Y.M. Soro, M.E. Gueunier-Farret, J. Méot, J.P. Kleider and P. Roca i Cabarrocas in *Thin Film Silicon: Amorphous and Microcrystalline Silicon* (23rd European Photovoltaic Solar Energy Conference, Valencia, Spain, 2008) pp. 2339-2342.
5. P. Roca i Cabarrocas, Y. Djeridane, Th. Nguyen Tran, E.V. Johnson, A. Abramov and Q. Zhang, *Plasma Phys. Control. Fusion* **50**, 124037 (2008).
6. P. Roca i Cabarrocas, J.B. Chévrier, J. Huc, A. Lloret, J.Y. Parey and J.P.M. Schmitt. *J. Vac. Sci. and Technol.* **A9**, 2331 (1991).
7. J. Merten, J.M. Asensi, C. Voz, A.V. Shah, R. Platz and J. Andreu, *IEEE Trans. Elec. Devices* **45**, 423 (1998).
8. G.Yue, B.Yan,G. Ganguly, J. Yang, S. Guha and C.W. Teplin, *Appl. Phys. Lett.* **88**, 263507 (2006).

Mater. Res. Soc. Symp. Proc. Vol. 1153 © 2009 Materials Research Society   1153-A07-12

Junction Capacitance Study of a-SiGe:H Solar Cells Grown at Varying RF and VHF Deposition Rates

Peter G. Hugger[1], Jinwoo Lee[1], J. David Cohen[1], Guozhen Yue[2], Xixiang Xu[2], Baojie Yan[2], Jeff Yang[2] and Subhendu Guha[2]
[1]Department of Physics, University of Oregon, Eugene, OR 97403, U.S.A.
[2]United Solar Ovonic LLC, 1100 West Maple Road, Troy, MI, 48084, U.S.A.

ABSTRACT

Significant advances have been made in increasing the deposition rate of hydrogenated silicon germanium alloys (a-SiGe:H) using a modified VHF glow discharge deposition method while also maintaining good electronic properties important for its application in photovoltaic devices. We examine the electrical and optical properties of these alloys deposited either by RF (13.56MHz) or the modified VHF methods over deposition rates varying from 1 to 10 Å/s. The electronic properties of a series of 1.4 eV optical gap a-SiGe:H i-layers, in many cases in solar cell device configurations, were characterized. Drive-level capacitance profiling was used to determine the deep defect densities, and transient photocapacitance measurements allowed us to determine the Urbach energies. Results were obtained for both the annealed and light-soaked degraded states and these results were correlated to the cell performance parameters. In general the a-SiGe:H layers deposited using the modified VHF excitation exhibited improved electronic properties at higher growth rates than the RF deposited samples.

INTRODUCTION

Hydrogenated amorphous silicon germanium alloys (a-SiGe:H) are technologically important for their incorporation as the low and mid-gap optical components in amorphous silicon (a-Si:H) based multijunction solar cells. Typically, RF (13.56MHz) glow discharge has been the predominant method of depositing these materials, albeit with the drawback that deposition rates have been limited to ≤ 3 Å/s. Alloys grown using RF methods at deposition rates higher than this have generally exhibited a significantly higher degree of light-induced degradation and lower stable efficiencies [1,2]. Depositing a-Si:H itself at rates higher than 3 Å/s while maintaining good electronic properties has been achieved at United Solar through its development of a "modified VHF" glow discharge deposition technique (hereafter also referred to as "VHF"). This technique utilizes significantly higher frequencies than for the RF glow discharge process and incorporates some further proprietary alterations to the standard VHF deposition method. Recently this method has been applied to deposition of a-SiGe:H films with some success of achieving better electronic properties at higher deposition rates [3].

In order to better understand the increased stability of a-SiGe:H films deposited using the modified VHF method, we have studied a series of low-gap ($E_{Tauc} \approx 1.4$ eV) a-SiGe:H devices grown using both RF and VHF techniques. Here we report the results of junction capacitance measurements such as drive-level capacitance profiling (DLCP) and transient photocapacitance and photocurrent (TPC and TPI). Using these methods, we characterize the deep state defect densities of these materials as well as their Urbach energies (E_U). We show that for most of these samples the deep defect densities are well correlated with these Urbach energies in accordance

Table I. Deposition conditions and structures of the a-SiGe:H samples examined in this study where n/i/p denotes the structure: $SS/n^+(a\text{-}Si\text{:}H)/i$ $(a\text{-}SiGe\text{:}H)/p^+(nc\text{-}Si\text{:}H)/ITO$ and n/i/Pd denotes the structure: $SS/n^+(a\text{-}Si\text{:}H)/i$ $(a\text{-}SiGe\text{:}H)/Pd$ Schottky diodes.

Sample No.	Deposition method	Structure	Deposition rate (Å/s)	Thickness (μm)	E_{Tauc} (eV)
17468	RF	n/i/p	0.9	1.3	1.37 ±0.03
17510	RF	n/i/p	4.4	1.3	1.37 ±0.03
17467B	RF	n/i/p	7.6	2.2	1.39 ±0.03
17469	RF	n/i/p	8.3	1.2	1.4 ±0.03
16647	RF	n/i/Pd	1	0.69	1.4 ±0.03
16648	RF	n/i/Pd	5	0.65	1.41 ±0.03
16646	RF	n/i/Pd	10	0.62	1.39 ±0.03
17473	VHF	n/i/p	5.5	1.3	1.41 ±0.03
17475	VHF	n/i/p	8.2	1.0	1.39 ±0.03
16660	VHF	n/i/Pd	5	0.68	1.37 ±0.03
16661	VHF	n/i/Pd	10	0.61	1.38 ±0.03

with the spontaneous weak bond breaking model [4,5] as has been demonstrated previously with good success for the a-SiGe:H alloys [6-8]. We will also discuss one particular RF sample (17469) which was deposited at 8.3 Å/s, but nonetheless exhibited very good electronic properties in the light degraded state.

SAMPLES AND SAMPLE PREPARATION

A total of eleven low-gap samples were deposited at United Solar Ovonic LLC for the purpose of this study. The samples were prepared using both RF and VHF glow discharge in the structures described in Table 1. The first structure: $SS/n^+(a\text{-}Si\text{:}H)/i(a\text{-}SiGe\text{:}H)/p^+(nc\text{-}Si\text{:}H)/ITO$, was a heterojunction configuration with extra thick (≈ 1.3 μm) i-layers in order to reduce the appearance of thin film interference fringes in the optical spectra. To ensure that experimental results were not a property of the p^+ layer, a second structure set with thinner i-layers (≈ 0.6 μm) in a SS/n^+ $(a\text{-}Si\text{:}H)/i$ $(a\text{-}SiGe\text{:}H)/Pd$ configuration were prepared. Semi-transparent Schottky Pd contacts were evaporated onto the intrinsic layers of these devices at the University of Oregon. All devices were classified as "low gap" with optical band gaps: $E_{Tauc} \approx 1.4$ eV and $E_{04} \approx 1.45$ eV. Here, E_{Tauc} was calculated using: $(\alpha h v)^{1/2} = B(h v - E_g)$ which is the original relation proposed by Tauc and Menth [12]. For our calculations TPI data was taken to be representative of the absorption coefficient, α in the above relation. The resulting bandgaps are listed in Table 1.

All samples were annealed for 1 hour at 460K before the initial series of measurements (State A). To investigate the effects of prolonged light exposure on the electrical properties of these devices, samples were exposed to 200 mW/cm^2 of 610 nm long-pass filtered white light (State B). During the lightsoaking process sample temperature was maintained below 315 K.

EXPERIMENTAL DETAILS

The densities of deep defects in these devices were investigated using both standard admittance measurements and drive-level capacitance profiling (DLCP). The DLCP method is now increasingly used to characterize deep defect distributions in disordered semiconductors and has previously been described in detail [9]. By investigating the quadratic response $(C=C_0+C_1\delta V+C_2(\delta V)^2+...)$ of a single-sided junction capacitance to an AC voltage perturbation δV, a defect density can be deduced which is directly related to the density of mid-gap states below the Fermi level. This density is then plotted against the spatial position variable $<x> = \varepsilon A/C_0$, where ε is the dielectric constant, and A is the area of the junction.

The transient photocapacitance (TPC) and photocurrent (TPI) methods have also been described elsewhere [10]. Both kinds of spectra closely resemble other more standard sub-bandgap absorption spectra with the important difference that the TPC measurement is able to resolve the proportion of each carrier *type* that is detected on the timescale of the measurement (≈ 1 s). Indeed, it can be shown that the TPC measurement is proportional to the number of majority carriers collected over the ≈ 1 s time window, minus the minority carriers collected $(n - p)$. Likewise, the TPI measurement is proportional to the sum of these carriers: $(n + p)$.

RESULTS AND DISCUSSION

Our initial experiments on the samples with Pd Schottky contacts showed a clear trend in the State B drive-level data, shown in Figure 1(a). Films deposited via RF exhibited the lowest deep defect densities of all the samples (≈ 2-$3 \cdot 10^{16}$ cm^{-3}) when the rate of deposition was low (1 Å/s). However, as the deposition rate increased to 5 Å/s and higher, deep defect densities were seen to increase by roughly a factor of 5 to levels near $1 \cdot 10^{17}$ cm^{-3}. Devices deposited using the modified

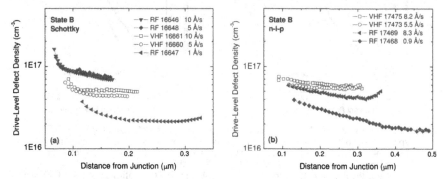

Figure 1. DLCP results in the light-degraded state (State B) for a series of (a) Schottky devices and (b) a subset of four *n-i-p* devices. These profiles were obtained at 390K using a 1.1kHz oscillatory voltage. Figure 1(b) shows a surprisingly low defect density for sample RF17469.

VHF deposition method, however showed maximum deep defect densities a factor of 2 lower than this, even at deposition rates of 10 Å/s.

In order to confirm this result for devices manufactured with a p^+-i heterojunction already in place, a second set of samples was prepared in an *n/i/p* geometry. As with the Schottky devices, materials having the lowest deep defect densities were the RF devices deposited near 1 Å/s. Also as before, the VHF films showed State B defect densities near 6-7 $\cdot 10^{16}$ cm^{-3} regardless of deposition rate. One RF sample, however (RF17469) showed a surprisingly low DLCP density of ≈ 5 x 10^{16} cm^{-3} even though it was deposited at the high rate of 8.3 Å/s. This was unexpected and an initial hypothesis to explain this result suggested that perhaps the DLCP measurements might have "missed" a portion of the deep defect band, or that by an effect such as area-spreading, was producing an incorrect result for the defect density. However, as discussed below, we ultimately concluded that the low defect density for this sample was indeed correct.

A full series of State A and State B TPC and TPI measurements were performed on a subset of four $n/i/p$ devices: a low-rate and high-rate RF device (RF17469 and RF17468) and the two VHF devices (VHF17475 and VHF17473). Representative TPC and TPI spectra are shown in Figure 2. The ratio of the TPI to TPC spectra in the bandtail energy region indicates the relative collection fraction of minority to majority carriers (roughly 80% for the two sets of spectra displayed) during the 1s measurement time window. In the sub-bandtail region, the evolution of the TPI to TPC ratio gives evidence for two gaussian defect bands roughly centered at optical energies of 0.8 eV and 1.0 eV, where the 1.0 eV defect contributes *negatively* to the TPC signal, as revealed by the large value of TPI/TPC in the neighborhood of 1.0 eV. This indicates that over the timescale of the measurements the 1.0 eV defect traps electrons optically excited from the

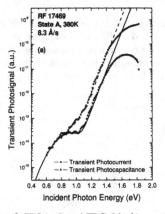 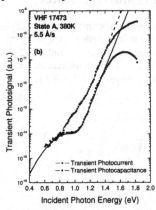

Figure 2. TPI (red) and TPC (black) spectra from (a) sample RF17469 and (b) VHF17473. These spectra have been deconvolved into three components: two sub-bandgap gaussian defects located near photon energies of 0.8 and 1.0 eV, and an exponential bandtail. Urbach energies were $E_U = 55$meV ± 2 and $E_U = 57$meV ± 2 for the RF and VHF devices, respectively. Comparisons of the deconvolved curves for TPI (dashed) and TPC (solid) show the 1.0 eV defect to contribute negatively to the TPC signal. This phenomenon is discussed fully in reference [6].

valence band. These general features are in good agreement with previous studies [6,11] of a-SiGe:H deposited by United Solar that contain germanium fractions above 20at. %. In the previous studies by C. Chen *et al* [6] it was proposed that the two optically active sub-bandgap defects in these materials indicate two charge states of the germanium dangling bonds. The positively charged D^+ (the 1.0 eV defect) and the negatively charged D^- (0.8eV).

In addition to agreeing with those previous results, we also noticed that the Urbach energies in these devices were well correlated to the deep defect densities measured using DLCP. That is, samples with lower Urbach energies also exhibited lower defect densities. We found this correlation to occur in accordance with the predictions of the spontaneous dangling-bond conversion model (SBC) originally proposed by Stutzmann [4,5]. In this model band tail states beyond a certain demarcation energy, E_{db}, above the valence band are considered to spontaneously convert to defect states that follow a gaussian distribution in energy. Moreover, the model predicts that if all deep defects in a material are created in this way the deep defect density can be expressed as:

$$N_D = N^* \cdot E_U exp[-(E_{db}-E^*)/E_U] \qquad (1)$$

where E^* is the energy along the bandtail density of states where the bandtail first becomes predominantly exponential (e.g. near 1.45 eV in Figure 2), N^* is the density of states at E^*, and E_U is the Urbach energy of the bandtail. Using this formalism, we have been able to accurately

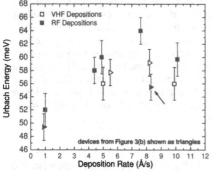
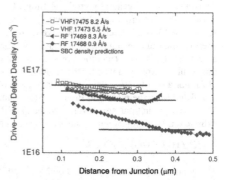

Figure 3(a). Urbach energies versus deposition rate for all RF and VHF deposited samples from Table 1. Although a clear trend is not seen here, in general, superior properties at high deposition rates are achieved by VHF. One notable exception to this finding is the high growth rate sample RF17469, indicated here by an arrow.

Figure 3(b). Here Figure 1(b) is replotted with solid lines indicating State B DLCP densities predicted using the spontaneous weak-bond conversion model (Eq. 1). Samples with small Urbach energies tended to have lower densities deep defects. Note the defect density of sample RF17469 and compare to Figure 3(a).

predict the DLCP densities (see Figure 3b) of the four samples mentioned above using Urbach energies determined from the optical spectra. In the calculation of these densities, the values of the variables N^*, E^*, and E_{db} were kept constant for all 4 samples (specifically: $N^* = 2 \cdot 10^{20}$ cm^{-3} eV^{-1}, $E^* = E_V + 0.44$ eV, and $E_{db} = E_V + 0.76$ eV). The deep defect densities predicted using this model (solid lines in Figure 3b) agree well with the spatially averaged DLCP data. Therefore, even though clear distinctions in Urbach energies between RF and VHF samples deposited at high rate was not seen (Figure 3a) this analysis indicates that samples with lower Urbach energies tend to have lower densities of deep defects. Note that scrutiny of Figure 3(a) shows this correlation between defect density and Urbach energy holds to within measurement uncertainty for the Schottky devices as well as the n/i/p devices. However, defect density predictions using the spontaneous bond breaking model have not yet been calculated for the Schottky devices.

Given a correct interpretation of these results, we return to the surprising case of the 8.3 Å/s sample RF17469 where it may now be understood that the low defect density is a likely consequence of a sharp Urbach energy (arrow, Figure 3a), which indicates relatively few weak-bond precursor sites. However, these surprisingly good material properties for sample RF17469 are not yet understood from a deposition standpoint. Further studies will address this issue.

SUMMARY

In an effort to better understand a-SiGe:H intrinsic layers deposited at high deposition rates using RF and modified VHF glow discharge methods, two series of devices were studied that were grown using these methods at various deposition rates between 1 Å/s and 10 Å/s. The first series of samples was grown in a Schottky geometry and the second series was grown in an n/i/p geometry. We have two primary conclusions regarding these devices. First, that the VHF deposition method generally produces films of lower deep defect density when high deposition rates (>5 Å/s) are used for film growth. One exception to this trend was the high deposition-rate RF sample "17469". Secondly, we have found an interesting correlation between the Urbach energies measured using transient photocapacitance and transient photocurrent and the deep defect density. This correlation was seen for both Schottky and n/i/p devices. We suggest that this result is explained well by the spontaneous bond conversion model proposed by Stutzmann [4,5]. For a set of 4 n/i/p devices, a comparison of true deep defect densities and densities calculated from the measured E_U using Stutzmann's bond conversion model is shown in Figure 3b). These predictions agree well with the DLCP data, even for the RF sample "17469", which exhibited a surprisingly small density of deep defects.

ACKNOWLEDGEMENTS

This work was partially supported by NREL under the Thin Film Partnership Program No. ZXL-5-44205-11 at the U. of Oregon, and by US DOE under the Solar America Initiative Program Contract No. DE-FC36-07 GO 17053 at both United Solar and the U. of Oregon.

REFERENCES

[1]   G. Yue, B. Yan, J. Yang and S. Guha, *Mat. Res. Soc. Symp. Proc.* **989**, 359 (2007).

[2]   S. Guha, J. Yang, S. Jones, Y. Chen and D. Williamson, *Appl. Phys. Lett.* **61**, 1444 (1992).

[3]   X. Xu, D. Beglau, S. Ehlert, Y. LI, T. Su, G. Yue, B. Yan, K. Lord, A. Banerjee, J. Yang, S. Guha, P. Hugger, and J. D. Cohen, Presented in 2009 *MRS Spring Meeting, Symp. A.* (2009).

[4]   M. Stutzmann, *Philo. Mag. B-Physics of Condensed Matter Statistical Mechanics Electronic Optical and Magnetic Properties* **56**, 63 (1987).

[5]   M. Stutzmann, *Philo. Mag. B-Physics of Condensed Matter Statistical Mechanics Electronic Optical and Magnetic Properties* **60**, 531 (1989).

[6]   C. C. Chen, F. Zhong, J. D. Cohen, J. C. Yang and S. Guha, *Physical Review B* **57**, R4210 (1998).

[7]   P. Wickboldt, D. Pang, W. Paul, J. H. Chen, F. Zhong, C.-C. Chen, J. D. Cohen and D. L. Williamson, *J. of App. Phys.* **81**, 6252 (1997).

[8]   C. C. Chen, F. Zhong and J. D. Cohen, *Mat. Res. Soc. Symp. Proc.* **420**, 581 (1996).

[9]   J. T. Heath, J. D. Cohen and W. N. Shafarman, *J. of App. Phys.* **95**, 1000 (2004).

[10]   J. D. Cohen, T. Unold, A. Gelatos and C. M. Fortmann, *J. of Non-Cryst. Sol.* **141**, 142 (1992).

[11]   F. Zhong, C.-C. Chen and J. D. Cohen, *J. of Non-Cryst. Sol.* **198-200**, 572 (1996).

[12]   J. Tauc and A. Menth, *J. Non-Cryst. Sol.* **8**, 569 (1972).

Mater. Res. Soc. Symp. Proc. Vol. 1153 © 2009 Materials Research Society   1153-A07-14

Nanosphere lithography of nanostructured silver films on thin-film silicon solar cells for light trapping

B. Ozturk[1], E. A. Schiff[1], Hui Zhao[1], S. Guha[2], Baojie Yan[2], and J. Yang[2]

[1]Department of Physics, Syracuse University, Syracuse, NY 13244-1130 U.S.A.
[2] United Solar Ovonic LLC, 1100 W. Maple Rd., Troy, MI 48084 U.S.A.

ABSTRACT

Sparse arrays of evaporated silver nanodisks were fabricated with nanosphere lithography (NSL) on glass substrates and on hydrogenated nanocrystalline silicon solar cells. The optical transmittance spectra for arrays on glass vary substantially with film thickness, and were reasonably consistent with previous work. The quantum efficiency spectra of hydrogenated nanocrystalline silicon solar cells show spectral shifts due to coupling of surface plasmons in the metal nanodisks to the planar waveguide modes of the cells, with overall photocurrent enhancement up to 10%.

INTRODUCTION

Nanostructured metal films prepared on top of thin silicon photodiodes can give surprisingly strong enhancements of the diode photocurrents at some wavelengths. The effect was discovered about ten years ago by Stuart and Hall [1] for silicon-on-insulator (SOI); it is due to the coupling of the surface plasmon excitations in the metal nanostructures to the planar waveguide modes in the photodiodes. The effect is now being assessed by several laboratories for its possible utility in thin-film solar cells [2,3,4]; it is an alternative to the use of textured substrates in solar cells, which produce enhanced photocurrents through stochastic light-trapping [5].

Surface plasmon resonance is the collective oscillation of the conduction electrons near metal surfaces. The resonant frequency of the sub-wavelength-size metallic particles depends on their size and shape as well as the dielectric into which they are embedded. At this frequency, particles interact with light and result in an extinction dip in the transmission spectrum reflecting both scattering of light and absorption (thermalization) within the nanoparticle. Absorption dominates if the particle size is very small compared to the wavelength [6] (i.e., $d \leq 10$ nm). The absorbed light is dissipated as heat to the system and this is not desirable for the solar cell application. Scattering dominates with the increasing size and can be utilized to couple certain wavelengths of the incident light into the waveguide modes of a thin-film solar cell. A schematic diagram of the thin film solar cell cross section is shown in Figure 1.

There are several technologies for creating nanostructured metal films on top of solar cells. Most previous research on plasmonic photocurrent enhancements in photodiode structures has used evaporated silver films that are subsequently annealed above 200°C to produce irregular nanostructured films. This method is the only one that has given substantial photocurrent enhancements (about tenfold) in photodiodes and silicon solar cells [1,3] but the high-temperatures of annealing are a difficulty for use with hydrogenated amorphous and nanocrystalline silicon (nc-Si:H) solar cells. Electron beam lithography is capable of fabricating highly uniform nanoparticle arrays [7, 8]. However,

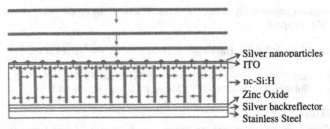

Figure 1. Cross section of the nc-Si:H solar cell design. Silver disks are deposited on the top ITO layer. The incident light is coupled into the waveguide modes of the active silicon layer by the silver nanoparticles. The thickness of the nc-Si:H layer was 1.75 micron.

this method is very slow for large area coverage. A third method is simple casting of metal nanoparticles from solution [2, 4].

In this paper we explore silver nanoparticle arrays with average size $d \sim 100$ nm fabricated using nanosphere lithography (NSL). In NSL, mono- and double- layers of submicron diameter spheres, typically latex, are formed on substrates by various methods; the voids between the spheres act as templates for metal evaporation. It is a simple method for depositing large area, uniform size metallic nanoparticle arrays on surfaces; it permits direct control of the lateral size and patterning of the silver films, as well as independent control of the film thickness.

Using NSL, we prepared silver nanoparticle arrays on top of glass substrates and on nc-Si:H solar cells. Consistent with the previous reports, we find that the optical properties of these arrays are sensitive to the thickness of the individual nanoparticles in the arrays; the shapes themselves can be modified by changing the metal evaporation conditions, surface conditions, and annealing [9]. For the present work we chose to use fairly sparse arrays of nanoparticles. These changed the quantum efficiency (QE) of the solar cells significantly in some cases. The largest photocurrent enhancement we found was about 10%. We describe next steps to achieving larger photocurrent enhancements in the discussion.

SAMPLE PREPARATION

Drop casting [9], spin coating [10] and convective self assembly (CSA) [11] are the most commonly used methods for the deposition of mono- and double-layers of microspheres. Here, the (CSA) method was used to deposit the spheres on various substrates. Latex spheres were purchased from Polysciences Inc. and their concentrations were 2-3% as received. The CSA method requires high concentrations of sphere suspensions. Hence, sphere suspension concentrations were increased to 20% by centrifuging and redispersing in de-ionized water.

Figure 2 depicts a schematic diagram of the CSA setup. As a test substrate, glass slides were placed on an inverted microscope stage. A 5 μl droplet of the concentrated microsphere suspension was deposited on the substrate. A fixed position cover slip spreader was used to trap the droplet as shown in Figure 2. The angle of the spreader was fixed at 25° as reported in the literature [12]. As the motorized stage moves, the trapped droplet gets dragged on the surface and convective self assembly of spheres occurs at the

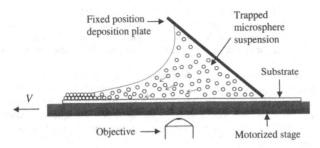

Fixed position → deposition plate

Trapped microsphere suspension

Substrate

V ←

Objective →

Motorized stage

Figure 2. Schematic diagram of the convective self-assembly setup.

tail of the dragged droplet. The motorized stage speed was varied to deposit mono-, double- or multi-layers of microspheres. After the formation of microsphere masks, silver (99.99% pellets) was deposited using a thermal evaporator. The mask was removed by sonication for 30 sec in absolute ethanol.

Initially, mono-layers of microspheres were deposited to fabricate triangular silver nanostructures as shown in Figure 3a and b. Fabrication of large-area, uniform triangular structures require defect-free, closed-packed monolayers of microspheres. Although, large patches of monolayers were deposited on glass substrates, larger gaps between monolayers were observed on the top ITO electrode of the solar cells. Larger gaps cause deposition of larger silver particles (Figure 3b), which block the incoming light and decrease the efficiency of the solar cells. This problem was eliminated by depositing double-layers of microspheres as NSL masks. Large gaps in the monolayers were filled by the spheres from the second layer, avoiding the formation of large silver particles (Figure 3c and d).

In order to explore the effect of the silver disk array thickness on extinction, 20 and

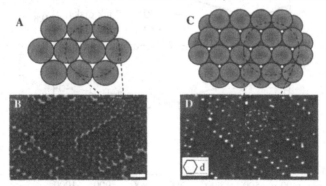

Figure 3. Schematic diagrams of a) mono- and b) double layer closed packed spheres and the shape and configuration of the voids between them. c) and d) Corresponding SEM images of silver particles obtained by thermal evaporation of silver through the voids. The sphere size was 450 nm. Scale bars represent 1 micron.

50 nm thick films of silver were deposited on the 750 nm double-layer NSL masks. As reported by Van Duyne and coworkers [13], the geometrical relation between the sphere diameter D and the size d of the hexagonally shaped (see Figure3d) deposited disks is given by $D = 0.155d$ for double-layers. Hence, the average size of the silver disks in this study was $d = 116$ nm.

We used NSL to fabricate silver disk arrays on indium-tin-oxide coated glass and onto special nc-Si:H silicon solar cells prepared at United Solar. The solar cells are prepared on stainless steel substrates coated with 500 nm of flat silver followed by a ZnO layer. The nc-Si:H solar cells were applied in the *nip* deposition sequence; more details are given in ref. [14]. 70 nm thick transparent conducting oxide contacts were deposited on top of the nc-Si:H cells.

OPTICAL PROPERTIES

Figure 4 shows the transmittance curves of silver disk arrays on glass substrates. The solid and the dotted curves represent the 20 and the 50 nm thick disk arrays, showing extinction dips at 500 and 920 nm wavelengths, respectively. These results are in accordance with previous findings [9]. The square-dash line was taken from reference [3] for comparison. It is the extinction curve of the 16 nm mass thickness annealed films that enhanced SOI detector photocurrents about tenfold. The extinction intensities of the disks in this study were low compared to the annealed silver films of [3]. This was most likely due to the lower density of silver disks on the surface. The average distance between the disks was approximately 750 nm, which is the diameter of the spheres (Figure 3(c)).

Silver disks with the same size and thickness were fabricated on the top ITO electrode of nc-Si:H solar cells. Before the silver deposition, a 30 nm thick lithium fluoride (LiF) layer was thermally evaporated through the voids of the NSL masks.

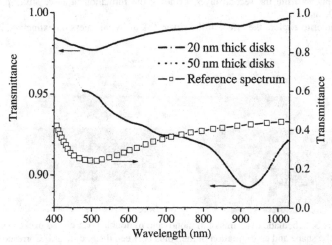

Figure 4. Transmittance curves for $d = 116$ nm silver particle arrays for different disk thicknesses.

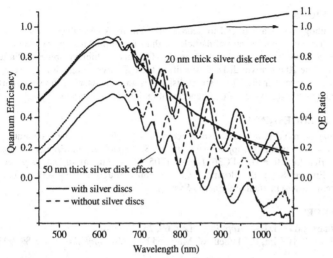

Figure 5. Quantum efficiency curves of nc-Si:H solar cell with (solid line) and without (dashed line) the silver disks. The y values of the two bottom QE curves were shifted by -0.3 for clarity. The solid line on the right-top corner represents the ratio of the fitted QE curves.

Figure 5 depicts the QE measurements of these solar cells with and without the silver disks. In the absence of the disks, the oscillations are due to the interference fringes in the reflectance of the 1.75 micron thick solar cells. The QE curve was blue-shifted in the long-wavelength range after the deposition of the 20 nm disks (solid line in the top QE curves in Figure 5). A slight overall enhancement with the 20 nm thick silver disks is also apparent in the long-wavelength range. In order to find the ratio of the non-spectral shift related enhancement, the curves were fitted to an exponential decay function. The ratio of the fitted curves shows enhancement up to 10% in the QE. Although this result is much below the reported average photocurrent enhancements of about tenfold in SOI photodetectors [1,3], it is comparable to other studies in thin-film solar cells [2,4,7].

The QE was red-shifted and decreased on average after the deposition of 50 nm thick silver disks (bottom curves in Figure 5). The QE shift and enhancement are thus correlated with the position of the extinction dip. Blue-shifting of the QE curves occurs when the dip is around 500 nm; red-shifting occurs when it's around 900 nm. Although not shown here, in the absence of the 30 nm LiF spacer layer, no spectral shift was observed, which is consistent with the results of [7,15].

We don't presently have a simple understanding of the relation between the position of the extinction dip and the plasmonic coupling, but both these and previous results suggest that the wavelength of the extinction dip of the silver particle arrays and its strength plays an important role in the enhancement of the QE of the solar cells.

CONCLUSIONS

NSL lithography was utilized to deposit different thickness silver disks on various substrates. Transmittance spectra of particles with different thickness were analyzed. The QE of a nc-Si:H solar cell increased 10% and the QE curve exhibited a blue-shift after the deposition of the 20 nm silver disks. Red-shift of the QE curve along with the decrease in the QE was observed for the 50 nm silver disks. Design parameters of the silver particles will be changed to improve the QE enhancement in thin-film solar cells.

ACKNOWLEDGEMENTS

We thank Arnold Honig and Gianfranco Vidali for the use of their apparatus. This research has been partially supported by the U. S. Department of Energy through the Solar America Initiative (DE-FC36-07 GO 17053). Additional support was received from the Empire State Development Corporation through the Syracuse Center of Excellence in Environmental and Energy Systems.

REFERENCES

1. H.R. Stuart and D.G. Hall, *App. Phys. Lett.* **69**, 2327 (1996)
2. D. Derkacs, S. H. Lim, P. Matheu, W. Mar, and E. T. Yu, *Appl. Phys. Lett.* **89**, 093103 (2006)
3. S. Pillai, K. R. Catchpole, T. Trupke, M. A. Green, *J. Appl. Phys.* **101**, 093105 (2007)
4. P. Matheu, S. H. Lim, D. Derkacs, C. McPheeters, and E. T. Yu, *Appl. Phys. Lett.* **93**, 113108 (2008)
5. Yablonovitch E, Cody GD., IEEE Transactions on Electron Devices, **ED-29**, no.2, 300 (1982)
6. C. F. Bohren and D. R. Huffman, *Absorption and Scattering of Light by Small Particles* (Wiley-Interscience, New York, 1983) p.309
7. C. Hägglund, M. Zäch, G. Petersson, B. Kasemo, Appl. Phys. Lett. **92**, 053110 (2008)
8. Z. N. Utegulov, J. M. Shaw, B. T. Draine, S. A. Kim, W. L. Johnson, Proc. SPIE 6641, 66411M (2007)
9. T. R. Jensen, M. Duval Malinsky, C. L. Haynes, R. P. Van Duyne, *J. Phys. Chem. B*, **104**, 10549 (2000)
10. J.C. Hulteen; R.P. Van Duyne, *J. Vac. Sci. Technol. A*, **13** , 1553 (1995)
11. B.G. Prevo and O.D. Velev, *Langmuir* **20**, 2099 (2004)
12. Y. Wang, L. Chen, H. Yang, Q. Guo, W. Zhou, M. Tao, *Sol. Energy Mater. Sol. Cells*, **93**, 85 (2008)
13. J. C. Hulteen, D. A. Treichel, M. T. Smith, M. L. Duval, T. R. Jensen, R. P. Van Duyne, *J. Phys. Chem. B* **103**, 3854 (1999)
14. B. Yan, G. Yue, J. Yang, S. Guha, D. L. Williamson, D. Han, *Appl. Phys. Lett.* **85**, 1955 (2004).
15. S. H. Lim, W. Mar, P. Matheu, D. Derkacs, and E. T. Yu, *J. Appl. Phys.* **101**, 104309 (2007).

Mater. Res. Soc. Symp. Proc. Vol. 1153 © 2009 Materials Research Society   1153-A07-15

Fabrication of Photonic Crystal Based Back-Reflectors for Light Management and Enhanced Absorption in Amorphous Silicon Solar Cells

Benjamin Curtin[1], Rana Biswas[1,2] and Vikram Dalal[1]
[1]Microelectronics Research Center; Dept. of Electrical and Computer Engineering, Iowa State University, Ames, Iowa 50011, U.S.A.
[2]Dept. of Physics & Astronomy; Ames Lab, Iowa State University, Ames, Iowa 50011, U.S.A.

ABSTRACT

Photonic crystal based back-reflectors are an attractive solution for light management and enhancing optical absorption in thin film solar cells, without undesirable losses. We have fabricated prototype photonic crystal back-reflectors using photolithographic methods and reactive-ion etching. The photonic crystal back-reflector has a triangular lattice symmetry, a thickness of 250 nm, and a pitch of 765 nm. Scanning electron microscopy images demonstrate high quality long range periodicity. An a-Si:H solar cell device was grown on this back-reflector using standard PECVD techniques. Measurements demonstrate strong diffraction of light and high diffuse reflectance by the photonic crystal back-reflector. The photonic crystal back-reflector increases the average photon collection by ~9% in terms of normalized external quantum efficiency, relative to a reference device on a stainless steel substrate with an Ag coated back surface.

INTRODUCTION

A critical need for all solar cells is to maximize the absorption of the solar spectrum. Optical enhancements and light trapping is a cross-cutting challenge applicable to all types of solar cells. Traditionally optical enhancements have involved use of anti reflecting coatings coupled with a metallic back-reflector. Solar cell efficiencies are improved by textured metallic back-reflectors which scatter incident light through oblique angles, thereby increasing the path length of photons within the absorber layer [1]. A completely random *loss-less* scatterer is predicted [2] to achieve an enhancement of $4n^2$ (n is the refractive index of the absorber layer), which has the value near 50 in a-Si:H. However the idealized limit of loss-less scattering is not possible to achieve in solar cells, and it is estimated that optical path length enhancements of ~10 are achieved in practice [3].

Although the analysis in this paper can be applied to any semiconductor absorber, we focus on a-Si:H, where the optical constants have been well-determined [4]. For a-Si:H with an energy gap of 1.75 eV typical of mid-gap cells, most photons with wavelengths below the band edge of 700 nm are absorbed. Short wavelength solar photons in the blue and green regions of the spectrum have absorption lengths less than 250 nm and are effectively absorbed within the thin absorber layer. However, the absorption length of photons grows rapidly for red light ($\lambda > 600$ nm) and even exceeds 6-7 μm for photons near the band edge. These red and near-IR photons are very difficult to absorb in thin a-Si:H layers and light-trapping schemes are critical to harvest these long-wavelength photons. Similar physical considerations apply to the band-edge photons in c-Si absorber layers [5] which are also difficult to harvest.

APPROACH

We develop a method for fabricating photonic crystal back-reflector structures that diffract the near-band edge photons. The back-reflector solar cell consists of a triangular lattice metallic photonic crystal, a-Si:H n-i-p solar cell device, and an indium tin oxide transparent top contact. The photonic crystal is etched into a patterned crystalline silicon wafer using reactive ion etching. Silver is then evaporated on the c-Si and used as both the back-reflector and back contact. Silver was chosen due to its high specular reflectance and low series contact resistance. A thin layer of zinc oxide is sputtered on the silver to prevent the diffusion of silver into the a-Si:H layer as well as silver agglomeration during high temperature a-Si:H processing. The a-Si:H n-i-p solar cell is deposited using standard plasma-enhanced chemical vapor deposition (PECVD) techniques. The thin ITO top contact is sputtered on the surface to complete the solar cell device.

The metallic photonic crystal structure was optimized using simulation methods presented in previous work [6], where Maxwell's equations are solved in Fourier space using a rigorous scattering matrix method. Through the diffraction of near-band edge photons, we found the optimal absorption enhancement occurs with the following dimensions: a transparent ITO top contact with a thickness (d1) of 100 nm, photonic crystal grating depth (d2) of 250 nm, pitch (a) of 0.74 μm, and radius R/a ~0.30. These dimensions were found for an a-Si:H n-i-p solar cell consisting of a p-layer thickness of 20 nm, intrinsic layer thickness of 500 nm, and an n-layer thickness of 200 nm. These dimensions are shown in Figure 1.

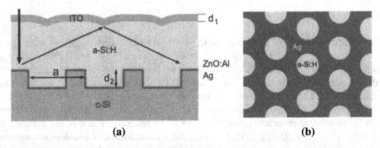

(a)   **(b)**

Figure 1. a) Schematic solar cell configuration with 2-d photonic crystal. b) Top view of 2-d photonic crystal layer.

PROCESS DEVELOPMENT

Crystalline silicon is used as the bulk photonic crystal structure for photolithography and etching purposes. The minimum feature size in the photonic crystal is approximately 300 nm, as defined by the spacing between etched c-Si cylinders. An ASML 193 nm step-and-repeat aligner is used to expose the photoresist with enough resolution to achieve the optimal dimensions. The photonic crystal is patterned into 480 nm of Rohm and Haas Epic 2135 photoresist and 80 nm of bottom antireflective coating. Since photoresist is the etch mask in this process, a sufficiently thick layer is required for reactive-ion etching.

We used a PlasmaTherm 700 series reactive-ion etching system to form the bulk c-Si photonic crystal in the patterned wafers. Dry etching is preferred over wet etching due to its greater amount of anisotropy and reproducibility between runs. Crystalline silicon etching is achieved with a 80:10 sccm CF_4:O_2 plasma, a chamber pressure of 50 mTorr, and RF power of 50 W. These parameters are the result of several experiments that investigated photoresist to c-Si etch selectivity and sidewall etch anisotropy. Once the etching is complete, an oxygen plasma is used to remove the remaining photoresist and bottom antireflective coating.

A thin silver layer is deposited on the surface using thermal evaporation. It is necessary to maximize the reflection of incident light that is not immediately absorbed within the intrinsic layer to determine the photonic crystal light-trapping enhancement. A 50nm Ag layer was found to be sufficiently thick, with a transmission of <2% for near-band edge photons. The zinc oxide film thickness and sputtering parameters are ideal for forming a thin layer to encapsulate the Ag. A low temperature deposition is necessary to prevent surface roughening due to silver agglomeration. Once the background chamber pressure reaches 1 μTorr, the ZnO:Al layer is sputtered at 150°C. The chamber pressure is held at 10 mTorr throughout this process with an argon ambient flow.

An a-Si:H n-i-p solar cell is deposited using PECVD process parameters that our group has developed. A typical device has an i-layer thickness of 250 nm and band-gap around 1.75 eV. Silicon carbide is used for the n-layer and is approximately 200-250 nm thick. DC sputtering is used to deposit the indium tin oxide top contacts. The sputtering parameters were developed to produce a 70 nm layer that is optimized for transparency, conductivity, and antireflective properties. Similar to the previous steps, the chamber background pressure is brought down to 1 μTorr before the substrate is heated to 225°C. During the deposition, the chamber pressure is held at 5 mTorr and a combination of argon and oxygen are introduced. Once the ITO sputtering is finished, the devices are annealed in atmosphere at a temperature of 200°C for 20 minutes.

RESULTS

A scanning-electron microscope was used to characterize the back-reflector structure between each processing step. This allowed us to measure changes in the cylinder diameter from Ag and ZnO:Al depositing on the sidewalls. As shown in Figure 2, R/a decreases from roughly 0.38 for bare c-Si, 0.36 after Ag evaporation, and 0.32 after sputtering ZnO:Al. The lattice spacing was measured to be 765 nm on average and had long range order across the 12 x 12 mm die. In addition to measuring the photonic crystal, surface roughness of the back-reflector was investigated. Figure 2b shows mild signs of silver agglomeration on the c-Si surface. The ZnO:Al sputtering appears to be slightly more uniform in Figure 2c. Figure 2d clearly shows conformal a-Si:H growth within the photonic crystal cavities and the resulting non-uniform surface.

External quantum efficiency was measured to determine wavelength dependent collection enhancement from the photonic crystal back-reflector. EQE was determined by taking the ratio of photo-generated current from the a-Si:H solar cell devices to a reference c-Si photodiode with known quantum efficiency. These measurements were taken from 400-800 nm in 20 nm increments and are normalized to a maximum EQE of 90%. As shown in Figure 3, the quantum efficiencies are similar for wavelengths below 550 nm where the wavelength-dependent photon absorption length is less than the i-layer thickness. The photonic crystal back-reflector device showed enhanced collection for wavelengths greater than 600 nm compared to the Ag and stainless steel reference devices.

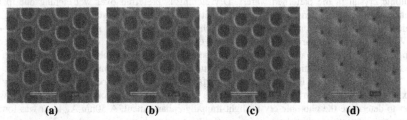

(a)   (b)   (c)   (d)

Figure 2. SEM images of the photonic crystal back-reflector taken after: a) RIE etching and plasma cleaning b) 50 nm Ag evaporation c) 70 nm ZnO:Al sputtering d) a-Si:H deposition and ITO sputtering. All images taken at 25,000x magnification (1 μm scale).

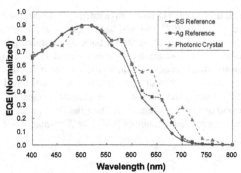

Figure 3. External quantum efficiency measurements for similar devices built on a stainless steel substrate, stainless steel substrate with 50 nm of Ag, and a c-Si wafer with a photonic crystal back-reflector. All measurements are normalized to 90% EQE.

An estimate of the solar cell short circuit current was also found by summing the product of device quantum efficiency and AMI 1.5 current at each measured wavelength. This is shown with:

$$J_{SC,EQE} = \sum_{\lambda=400nm}^{800nm} q\Phi(\lambda)EQE(\lambda) \tag{1}$$

Where q is the unit charge of an electron in Coulombs and Φ is the AMI 1.5 solar flux in photons/sec/cm^2. This quantity is not intended to be a substitute for the I-V measure for J_{SC}, but rather a means of comparing quantum efficiencies that are weighted against the solar spectrum. The photonic crystal back-reflector showed a 9% improvement in EQE J_{SC} over the silver reference device and 18% over the stainless steel reference. These values are shown in Table I.

Table I. Short circuit current from relative EQE for a-Si:H solar cells with different back substrates.

Substrate	EQE J_{SC} (mA/cm^2)
Stainless Steel	11.21
Stainless Steel with Silver	12.17
Photonic Crystal	13.26

DISCUSSION

We have used theoretical simulations to develop an a-Si:H n-i-p solar cell that utilizes a metallic photonic crystal back-reflector to increase the collection of near-band edge photons. As expected with photolithography, the PC lattice geometry was consistent across the 12 x 12 mm patterned die. Although the evaporated Ag was found to be relatively smooth, there was weak texturing across the entire sample that had features on the order of 100 nm. Reducing the roughness of the Ag film on c-Si will allow us to create an Ag reference immediately next to the photonic crystal device.

There is some variation in the structure parameters between our simulated and fabricated device. The R/a ratio was larger than expected after Ag evaporation, which can be attributed to isotropy during the RIE and possibly poor Ag sidewall coverage. R/a was closer to 0.3 after ZnO:Al sputtering but the structure was not simulated with this additional interface. A thinner absorption layer should result in greater EQE enhancement from the PC when compared to a smooth back-reflector.

The ratio of the quantum efficiency for the photonic crystal back-reflector to that of a reference device clearly shows considerable enhancement at near-infrared wavelengths and is shown in Figure 4. The most significant enhancement occurred near 720 nm, where the PC showed a factor of 8 improvement in collection over the Ag reference device. A secondary resonance was observed near 760 nm where the PC had an enhancement of ~6 over the Ag reference. Significant enhancement was also seen with the stainless steel reference. This is expected as stainless steel has poor reflectance compared to Ag.

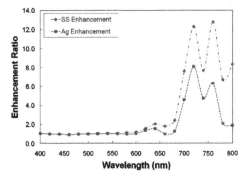

Figure 4. Relative EQE enhancement ratios for the photonic crystal back-reflector versus a stainless steel substrate with and without a 50 nm silver layer. The PC shows an enhancement of 8 near 720 nm compared to the Ag reference.

CONCLUSIONS

We develop a process and experimentally verify two dimensional metallic photonic crystal back-reflectors in a-Si:H solar cells. The photonic crystal pattern is etched into a c-Si wafer and then used as a back-reflector once Ag is evaporated and ZnO:Al is sputtered. This device shows a significant improvement in normalized EQE short-circuit current when compared to a stainless steel reference device that is half-coated with an Ag back-reflector. The EQE indicated that the PC back-reflector device enhanced near-band edge photon collection by a factor of 8 at 720 nm and 6 at 760 nm with respect to the Ag reference.

ACKNOWLEDGEMENTS

We thank K. Han, N. Chakravarty, and M. Noack for help with samples. We thank D. Vellenga and the North Carolina State University Nanofabrication Center for photolithography. We acknowledge support from the NSF under grant ECCS-0824091and the Iowa Powerfund. The Ames Laboratory is operated for the Department of Energy by Iowa State University under contract No. DE-AC0207CH11385.

REFERENCES

1. B. Yan, J. M. Owens, C. Jiang, S. Guha, Materials Res. Soc. Symp. Proc. **862**, A23.3 (2005).
2. E. Yablonovitch, J. Opt. Soc. Am. **72**, 899 (1982).
3. J. Nelson, The Physics of Solar Cells, (Imperial College Press, London, 2003), p. 279.
4. A.S. Ferlauto, G. M. Ferreira, J. M. Pearce, C. R. Wronski, R. W. Collins, X. Deng, G. Ganguly, J. Appl. Phys. **92**, 2424 (2002).
5. L. Zeng, Y. Yi, C. Hong, J. Liu, N. Feng, X. Duan, L.C. Kimmerling, B.A. Alamariu, Appl. Phys. Lett. **89**, 111111 (2006); Materials Res. Soc. Symp. **862**, A12.3 (2005). L Zeng et al, Appl. Phys. Lett. **93**, 221105 (2008).
6. R. Biswas, D. Zhou, J. Appl. Phys. **103**, 093102 (2008).

Mater. Res. Soc. Symp. Proc. Vol. 1153 © 2009 Materials Research Society   1153-A07-17

Passivation of Silicon Surfaces Using Atomic Layer Deposited Metal Oxides

Jun Wang[1], Mahdi Farrokh-Baroughi[1], Mariyappan Shanmugam[1], Roohollah Samadzadeh-Tarighat[2], Siva Sivoththaman[2], and Sanjoy Paul[1]
[1]Department of Electrical Engineering and Computer Science, South Dakota State University, Brookings, SD-57007, USA
[2]Electrical Engineering and Computer Science Department, University of Waterloo, Waterloo, N2L3G1, Canada

ABSTRACT

Surface passivation of silicon substrates using atomic layer deposited Al_2O_3 and HfO_2 thin films are assessed. Al_2O_3 and HfO_2 dielectric layers with various thicknesses were deposited on both sides of n-type (100) FZ-Si substrates (resistivity 4 – 6 Ω-cm) at 200°C by atomic layer deposition (ALD) system. The effective excess carrier lifetime of as-deposited oxide/Si/oxide structure was measured by microwave-photoconductivity-decay (MWPCD) measurement technique and it was observed that the thicker ALD dielectrics lead to higher effective excess carrier lifetime and better surface passivation. The measurements showed average excess carrier lifetime values of 302 μs and 347 μs for as-deposited Al_2O_3 and HfO_2 passivated Si substrates with 150 ALD cycles, respectively. MWPCD and capacitance-voltage (C-V) measurements suggest that as-deposited ALD HfO_2 layer leads to a better surface passivation compared to as-deposited ALD Al_2O_3 layer. Further, the results suggest that there exist fixed negative charges in the bulk of the ALD dielectrics and this contributes to the field effect passivation of the silicon surfaces.

INTRODUCTION

The need for lower cost silicon solar cells combined with rather high cost of pure Si material requires using thinner Si substrates (less than 150 μm). Surface passivation of advanced Si solar cells is becoming more important since the surface/volume ratio in thin Si substrates and the contribution of the surfaces in the overall performance of these solar cells is increasing. A well-passivated surface significantly reduces recombination of photogenerated carriers at front and back surfaces of Si substrate and enhance V_{OC}, I_{SC}, and even fill factor of solar cells. Recently, atomic layer deposited Al_2O_3 layer has been used to passivate surface of Si wafers [1]. ALD technique for surface passivation purpose is very appealing because it provides high level of control on the thickness of dielectrics, low temperature growth conditions, high quality films, and applicability over large areas.

Hoex et al. studied the passivation effect of the annealed ALD Al_2O_3 layer on p-type Si substrates [2-4]. They showed that as-deposited Al_2O_3 does not show passivation effect while a medium temperature post-deposition anneal at 425 °C on ALD Al_2O_3 films leads to outstanding passivation of Si surfaces. They analyzed the passivation mechanism of Al_2O_3 and concluded that the negative fixed charge present in Al_2O_3 films was especially beneficial for the surface passivation of p-type c-Si wafers [5]. J. Schmidt et al. have used ALD Al_2O_3 film to passivate p^+ emitter on n type Si solar cell and achieved a very high efficiency of 20.6% on n type Si substrates [6, 7, 8]. All of the previous works have focused only on passivation properties of ALD Al_2O_3 thin films. Furthermore, a medium temperature anneal was thought to be necessary in order to achieve high quality surface passivation. This may limit the application of these passivations to conventional Si solar cells which are made at medium to high temperatures and solar cells rather than those use amorphous silicon emitters such as heterojunction with intrinsic thin layer and hydrogenated silicon./crystalline silicon heterojunction solar cells [9,10].

In this article, we assess the passivation property of as grown ALD HfO_2 and Al_2O_3 on n-type Si substrates and compare the results of HfO_2 passivation with those of Al_2O_3. This work was conducted at low temperature, 200°C, without running any post-deposition anneal.

THEORY OF ATOMIC LAYER DEPOSITION FOR SURFACE PASSIVATION

ALD is a chemical vapor deposition (CVD) technique used to create extremely thin coatings. ALD is based on sequential self-terminating reactions of at least two reactants, or precursors [11]. The precursors react with a surface one-at-a-time in a sequential manner. The dielectric film is deposited by exposing the precursors to the growth surface repeatedly. This leads to growth of a very uniform, conformal, and reproducible thin film. The overall reactions in the deposition of Al_2O_3 and HfO_2 can be described by reaction 1 and 2 [11].

$$2Al(CH_3)_3 + 3H_2O \rightarrow Al_2O_3 + 6CH_4 \tag{1}$$

$$HfCl_4 + 2H_2O \rightarrow HfO_2 + 4HCl \tag{2}$$

In practice these reactions are the results of two half reactions. The half ractions for growth of Al2O3 and HfO2 are listed below.

$$2AlOH^* + 2Al(CH_3)_3 \rightarrow 2AlOAl(CH_3)_2^* + 2(CH_4) \tag{Eq. 1a}$$

$$AlCH_3^* + H_2O \rightarrow AlOH^* + CH_4 \tag{Eq. 1b}$$

$$yHfOH^* + HfCl_4 \rightarrow (HfO)_y HfCl_{(4-y)} + yHCl \tag{Eq. 2a}$$

$$HfCl^* + H_2O \rightarrow HfOH + HCl \tag{Eq. 2b}$$

where * are surface species. The density of dangling bonds at (100) surface of crystalline silicon is 1.36×10^{15} cm^{-2}. The chemical reactions at the surface of Si which growing ALD films, lead to termination of these bonds by strong covalent bonds and significant reduction in density of dangling bonds at the Si/ALD film interface. This significantly enhances the electronic quality of the Si surface and leads to significant reduction in recombination loss at this surface. In addition to the terminating of the dangling bonds of Si surfaces by ALD oxides, negative or positive charge at the Si/ALD film interface or the bulk of the ALD film would lead to a better passivation of Si surfaces.

Presence of a high density of fixed charges at the surface or bulk of the ALD film, which can be originated from defects in ALD layers [12, 13], would repel minority or majority carriers from the surface into the bulk of silicon wafer and hence reduce recombination rate at the surface. These two mechanisms play the primary roles in passivating surface of silicon wafers.

The recombination at surfaces of semiconductors is often represented by surface recombination velocity (cm/s), which combined with the density of minority carriers represents the flux of minority carriers which enter to the surface and recombine there. In a silicon wafer, recombination of minority carriers happens in the bulk and the surface of the wafer. The recombination in the bulk of wafer is characterized by bulk excess carrier lifetime (τ_B) and the recombination in the surface is characterized by surface recombination velocity (S_r). The effective excess carrier lifetime (τ_{eff}) in the whole structure relates to τ_B and S_r as below [14,15].

$$S_r = \sqrt{D\left(\frac{1}{\tau_{\mathit{eff}}} - \frac{1}{\tau_B}\right)} \times \tan\left(\frac{d}{2}\sqrt{\frac{1}{D} \times \left(\frac{1}{\tau_{\mathit{eff}}} - \frac{1}{\tau_B}\right)}\right) \tag{3}$$

where d is the wafer thickness and D is either the diffusion coefficient of minority carriers in moderately or highly doped Si wafers or the ambipolar diffusion coefficient of electron/holes in lowly doped Si wafers [15]. If the recombination of excess carriers is dominated by surface recombination (bulk lifetime is very high), equation (3) reduces to equation (4).

$$S_{r,MAX} = \sqrt{\frac{D}{\tau_{\mathit{eff}}}} \times \tan\left(\frac{d}{2\sqrt{D\tau_{\mathit{eff}}}}\right) \tag{4}$$

Equation (4) overestimates the value of surface recombination velocity by assuming that all the recombination (including surface and bulk) occurs at the surface. Therefore, we use $S_{r,Max}$ instead of S_r. By measuring the value of τ_{eff} and knowing the wafer thickness and the value of the diffusion coefficient, equation 3 or 4 can be used to calculate S_r and characterize the surface.

EXPERIMENT

Single-side polished n type FZ-Si substrates with resistivity of 4~6Ω-cm were cleaned by Prianha cleaning (H_2SO_4:H_2O_2=9:1) (v/v) for 10 min at 120°C. After 3 min oxygen plasma cleaning, using O_2 flow rate of 100 sccm at 300 mTorr and 150 W RF power, Si substrates were dipped in 1% HF solution for 20 sec to hydrogenate the dangling bonds at the Si wafer surfaces. The metal oxide layers, Al_2O_3 and HfO_2, were deposited on both sides of FZ-Si substrates in atomic layer deposition system. The substrates were positioned in the middle of the chamber so the gas precursors could access both sides of the substrates simultaneously. Tri methyl aluminum ($Al(CH_3)_3$) and Hafnium tetra chloride ($HfCl_4$) gases, as the Al and Hf precursors, and water vapor (H_2O), as the oxygen precursor, were applied sequentially into the deposition chamber to grow Al_2O_3 and HfO_2 in form of atomic layer by atomic layer at 200 °C. Deposition cycles were varied to get the different thickness of ALD layers.

Microwave photoconductivity decay (MWPCD) technique with wafer mapping capability was used to measure average effective excess carrier lifetime (τ_{eff}) of the oxide/n-Si/oxide structure. In this measurement, a laser with 904 nm wavelength and 1.2×10^{13} photons/pulse was used for creating excess carriers and a microwave signal with frequency of 10.352 GHz was used to measure the density and dynamics of the excess carriers in the structure. The measured lifetime values were used to calculate the interface (surface) recombination velocities.

After removing the ALD layers on the back side of Si substrates, Al layers were deposited on both sides oxide/Si structure by sputtering system. Photolithography was used to mask the front Al layer to form the metal/oxide/Si (MOS) structure in small areas. Capacitance-voltage characteristics and current-voltage characteristics of the MOS structures were studied by HP/Agilent 4192A impedance analyzer under 1MHz and Agilent 4155C semiconductor parameter analyzer, respectively.

RESULTS AND DISCUSSION

Figure 1(a) and (b) shows the map of τ_{eff}, measured by MWPCD, on as-deposited Al_2O_3 and HfO_2 passivated Si wafers, respectively. The MWPCD measurement resolution was 1mm^2, as shows in the figure by small pixels. The use of quarter of 4" Si wafers in the experiment guarantees that the contribution of recombination at the edge of the wafers in the measurements is only confined to a 5 mm wide strip around the edge of the wafer, as shows in low lifetime region (red) in the figures. To achieve reliable lifetime and surface recombination velocity readings, average values of τ_{eff}, $\tau_{eff,av}$, were obtained within a 2 cm X 2 cm squares at the middle of the sample, shown as black squares in the figures. The best $\tau_{eff,av}$ values for as-deposited Al_2O_3 and HfO_2 passivated wafers with 150 ALD cycles were 302 μs and 347 μs, respectively. These correspond to maximum surface recombination velocity values of 84.5 cm/s and 73.4 cm/s for Al_2O_3/Si and HfO_2/Si interfaces, respectively. It should be noted that the effective carrier lifetime reported by Boex et al. for as deposited Al_2O_3 passivation was 2 ~ 8 μs [2]. Comparing our results with those of Boex, we conclude that high quality surface passivation using as-deposited Al_2O_3 and HfO_2 is possible.

MWPCD measurement results also showed that HfO_2 thin film passivated Si substrate had lower surface recombination velocity and had better passivation effect on n-type Si wafer. This suggests that as-deposited ALD HfO_2 thin film can be used for high quality Si surface passivation.

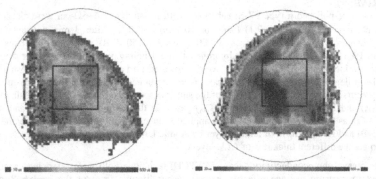

Figure 1: MWPCD carrier lifetime maps of (a) n-type Si wafer passivated by Al_2O_3 film, (b) n-type Si wafer passivated by HfO_2 film.

Figure 2 shows the dependence of the $\tau_{eff,av}$ and corresponding $S_{r,Max}$ values measured on oxide/Si/oxide structures on the thickness (number of ALD cycles) of the ALD oxide layer. Effective excess carrier lifetime experiences an almost linear increase as a function of the number of ALD cycles. Correspondingly, the $S_{r,Max}$ values experiences a hyperbolic decrease with increase in the thickness of the ALD oxide. This behavior is observed for both Al_2O_3/Si and HfO_2/Si samples. We believe this behavior points out to presence of fixed charges in the bulk of oxides. A thicker ALD oxide contains more fixed charges and hence it repels more of excess carriers. These results suggest the presence of field effect passivation where the charges are in the bulk of ALD oxide, as opposed to previously suggested negative charge at the oxide/Si interface [4]. The reduced surface recombination velocity may not be due to in-situ annealing during the subsequent cycles because 1) ALD is a fast deposition process and 2) the deposition temperature is only 200°C, which is much lower than the annealing temperature Hoex et al used [2]. Further, the experiments of Figure 2 show that the ALD HfO_2 layer consistently performs better in passivation Si surfaces than that of the ALD Al_2O_3 layer. The lowest $S_{r,Max}$, less than

100 cm/s,was observed for 150 cycles of ALD in both cases. It is also observed that to achieve a surface recombination velocity of less than 300 cm/s, a typical target surface recombination velocity for solar cell application, at least 40 ALD cycles in both cases are required.

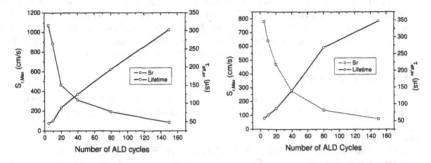

Figure 2: (a) surface recombination velocity (|) and effective carrier lifetime (O) of Al_2O_3 passivated Si substrate as a function of ALD cycles, (b) surface recombination velocity (|) and effective carrier lifetime (O) of HfO_2 passivated Si substrate as a function of ALD cycles

Figure 3 shows the high frequency capacitance-voltage (C-V) characteristics of the fabricated metal-ALD oxide-semiconductor structures. The results suggest that the gradient of C-V curve for HfO_2/Si structure is slightly greater than that of the Al_2O_3/Si structure, excluding the effect of dielectric constant. The smaller gradient in C-V curve of Al/Al_2O_3/Si structure mean that more traps at the Al_2O_3/Si interface charged or discharged when the applied voltages were changed [13]. C-V measurement results confirmed that the interface trap density at HfO_2/Si interface was less than that of the Al_2O_3/Si interface. The interface traps form recombination centers at the interface and increase the recombination rate of photogenerated carriers at the interface, hence increasing the surface recombination velocity, consistent with the results from MWPCD measurements. The C-V characteristics in the reverse sweep mode did not show a major shift, suggesting that there is no major trapping effect within the metal oxides.

Further, figure 3 showed that the threshold voltage of the Al-ALD oxide-Si structure shifts towards positive voltages by increasing thickness of the oxide. This observation suggests that the deposited metal oxides contain bulk negative charges, which is in a good agreement with the prediction of fixed bulk charges by MWPCD measurements.

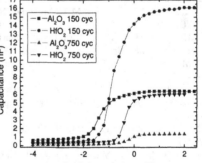

Figure 3: Capacitance-voltage measurement result of Al_2O_3/Si (■) and HfO_2/Si (●) structure under 150 and 750 ALD cycles

CONCLUSIONS

In summary, the measured τ_{eff} values for as-deposited Al_2O_3 and HfO_2 passivated wafers with 150 ALD cycles were 302 μs and 347 μs, respectively. Using the measured τ_{eff} values, maximum surface recombination velocity values of 84.5 cm/s and 73.4 cm/s were obtained for Al_2O_3/Si and HfO_2/Si interfaces. It was observed that the increasing thickness of the oxides lead to lower surface recombination velocities at the oxide/Si interface and also shift in C-V characteristics of the structure towards positive voltages. Both of these observations suggest that fixed negative charges are present in the bulk of ALD grown Al_2O_3 and HfO_2 thin films.

ACKNOWLEDGEMENTS

The authors would like to thank State of South Dakota for financial support and Nanofabrication Center of the University of Minnesota for sharing the ALD facility.

REFERENCES

1. G. Agostinelli, A. Delabie, P. Vitanoz, Z. Alexieva, H. F. W. Dekkers, S.De Wolf, and G. Beaucarne, Sol. Energy. Mater. Sol. Cells 90, 3438 (2006).
2. B.Hoex, S.B.S. Heil, M. C. M. van de Sanden, and W. M. M. Kessels, Appl. Phys. Lett. **89**, 042112 (2006).
3. B. Hoex, J. Schmidt, R. Bock, P. P. Altermatt, M. C. M. van de Sanden, and W. M. M. Kessels, Appl. Phys. Lett. **91**, 112107 (2007)
4. B. Hoex, J. J. H. Gielis, M. C. M. van de Sanden, and W. M. M. Kesselsb, J. Appl. Phys. 104, 113703 (2008).
5. B. Hoex, J.Schmidt, P. Pohl, M.C.M. van de Sanden, and W.M. M. Kessels, J. Appl. Phys. 104, 044903 (2008)
6. J. Schmidt, A.Merkle, R. Brendel, B. Hoex. M.C.M. van de Sanden, W.M.M. Kessels, Prog. Photovolt: Res. Appl. 2008; 16:461-466.
7. J. J. H. Gielis, B. Hoex, M. C. M. van de Sanden, and W. M. M. Kessels, Appl. Phys. Lett. 92, 253504 (2008).
8. J. J. H. Gielis, B. Hoex, M. C. M. van de Sanden, and W. M. M. Kessels, J. Appl. Phys. 104, 073701 (2008)
9. M.Taguchi, H.Sakata, Y. Yoshimine, E.Maruyama, A.Terakawa, M.Tanaka, S.Kiyama, 31st IEEE Photovoltaic Specialists Conference, 2005. p.p. 866- 871.
10. M. Farrokh-Baroughi and S. Sivoththaman, IEEE Trans. Electron Devices, Vol. 28, p.p. 575-577, 2007.
11. A. Londergan, O. Van der Straten, S. De Gendt, J. Elam, S. Bent, S. Kang , Atomic Layer Deposition 3, Proceedings: Atomic Layer Deposition Application Symposium (Washington, 2007).
12. K. Matsunaga, T. Tanaka, T. Yamamoto, and Y. Ikuhara, Phys. Rev. B 68, 085110 (2003).
13. W. J. Zhu, T. P. Ma, S. Zafar, and T. Tamagawa, IEEE Trans. Electron Devices, vol. 23, pp597-599.
14. A. G. Aberle, S. Glunz, and W. Warta, J. Appl. Phys. 71 (9), May 1992.
15. Dieter K. Schroeder, Semiconductor Material and Device Characterization (Wiley, New York, 2006).
16. E. H. Nicollian, and J. R. Brews, MOS Physics and Technology (Wiley, New Jersey, 2003).

Mater. Res. Soc. Symp. Proc. Vol. 1153 © 2009 Materials Research Society   1153-A07-19

Highly Transparent and High Haze ZnO:Al Film for Front TCO of a-Si:H and μc-Si:H Solar Cells by Controlling Oxygen Flow

Dong-Won Kang[1], Seung-Hee Kuk[1], Kwang-Sun Ji[2], Seh-Won Ahn[2], and Min-Koo Han[1]

[1]School of Electrical Engineering, Seoul National University, Seoul 151-742, Republic of Korea.
[2]Solar Energy group, LG Electronics Advanced Research, Seoul 151-742, Republic of Korea.

ABSTRACT

We fabricated highly transparent and high haze ZnO:Al film for front TCO of amorphous and microcrystalline silicon solar cells. We have sputtered ZnO:Al film of 1.3 μm on the thin seed layer of about 80nm which was previously sputtered on the glass substrate by using 4% dilution of oxygen to argon gas. The ZnO:Al film grown on the seed layer had much higher crystalline phase than one without any seed layer. Our bi-layer ZnO:Al film showed low resistivity of 2.66×10^{-4} Ω·cm and sheet resistance of 2.08 Ω/□ while conventional ZnO:Al film showed resistivity of 3.24×10^{-4} Ω·cm and sheet resistance of 2.46 Ω/□. After surface texturing by 0.5% HCl wet-chemical etching, the transmittance of ZnO:Al film was increased from 83.7% to 88.1% at wavelength of 550nm through the seed layer. Also the transmittance at 800nm was increased from 82.3% to 88.9%. Especially, haze values of the ZnO:Al film were drastically increased from 58.7% to 90.6% at wavelength of 550nm by employing the seed layer. Also haze values at 800nm were increased from 22.1% to 68.1%. It is expected that the seed layer method to improve the quality of ZnO:Al film will contribute to an increase of solar cell efficiency due to the high capability of light trapping and low electrical resistivity.

INTRODUCTION

Recently, thin film silicon solar cell has received considerable attention because of its low fabrication cost. To improve the efficiency of silicon thin-film solar cell, various topics related with transparent conductive oxide (TCO) may be important to increase short circuit current. As a front electrode of thin film solar cell, TCO requires high transmittance and low sheet resistance [1]. TCOs with optical transmittance more than 80% in the visible region and resistance less than 10^{-3} Ωcm are required to be used in solar cells [2]. Furthermore, high haze characteristic is also important to enhance an optical absorption in terms of light scattering [3]. ZnO:Al films have advantages over Asahi-U type TCOs in terms of low cost, low deposition temperature, high stability against hydrogen plasma, high haze values at long wavelength [4]. The ZnO:Al is the polycrystalline phase material, so the phase of substrate material affects the growth of the ZnO:Al film. Previous studies reported that ITO [5], ZnO[6] as a seed layer was introduced to improve the quality of ZnO:Al film before the deposition of ZnO:Al film. Many other buffer layers such as SiC [7], MgO [8], ZnS [9], and CaF$_2$ [10] have been studied to improve ZnO film growth. However, inserting those seed layer requires additional sputtering process of ITO or ZnO target and it could lead to increase the process cost. Also those substrates [7-10] could not be applied in thin film silicon solar cell deposited low cost soda lime glass substrate. ITO seed layer might have disadvantages in terms of free carrier absorption which could induce the loss of optical transmittance at the near infrared (NIR) wavelength region.

The purpose of this work is to investigate the ZnO:Al seed layer deposition controlling oxygen flow for improving the performance of the ZnO:Al film without increasing the process cost and steps. It should be pointed out that introducing the seed layer which controls oxygen flow using only ZnO:Al target sputtering has not been reported. We also aimed to avoid the optical transmittance loss of the NIR wavelength region. We investigated the electrical and optical characteristics of direct current magnetron sputtered ZnO:Al films with seed layer by controlling oxygen flow.

EXPERIMENT

ZnO:Al films were sputtered on soda lime glass(200 mm * 200 mm * 1.1mm) substrates by using ZnO:Al target consisting of ZnO:Al$_2$O$_3$ with 2 wt % Al$_2$O$_3$. The soda lime glasses were ultrasonically cleaned by using 20 % diluted tetra methyl ammomium hydroxide (TMAH). We applied a magnetron sputtering of dynamic mode, utilizing magnet scan for uniform deposition of ZnO:Al films due to large substrates. The base pressure was 1.0× 10^{-7} Torr. At first, the conventional ZnO:Al film which was not applied to oxygen control method was deposited on glass substrate. Only Ar gas was used to sputter the ZnO:Al target. Total magnet scan time and count during sputtering were 3500 seconds and 70, respectively. The substrate temperature of 300°C, working pressure of 3mT, target to substrate distance of 60 mm, and DC power of 450 W were fixed during the sputtering of all samples. The improved film which was introduced seed layer was deposited through mixed gas flow of 4% dilution of oxygen with argon gas. After thin oxygen-diluted seed layer deposition, bulk ZnO:Al layer was deposited without any oxygen flow. In other words, the scan time and count for oxygen dilution were 150 seconds and 3, respectively. Then, subsequent magnet scan time and count for bulk ZnO:Al deposition without oxygen feeding was 3000 seconds and 60, respectively. The dilution ratio of oxygen with Ar gas was about 4%. The thickness of deposited ZnO:Al films was measured by spectroscopic ellipsometry (J.A. Woollam, M2000-U) using Tauc-Lorentz optical model. And four-point probe method was used to measure the sheet resistance of ZnO:Al films. X-ray Diffractometer (XRD, M18XHF-SRA) was used to characterize structural properties of samples.

To investigate light scattering characteristics, surface texturing was performed. Wet-chemical etching of ZnO:Al films was carried out using diluted hydrochloric (HCl) acid of 0.5 % concentration. We used spin type wet etching system to improve the uniformity of texturing. We etched the films during 20 seconds (conv. ZnO:Al film) and 60 seconds (bi-layer ZnO:Al film) at 300 rpm (revolution speed of the substrate). It was difficult to measure final film thicknesses of surface textured samples using spectroscopic ellipsometry due to light scattering on the film surfaces which leads to the detection problem of reflected light. Thus, they were estimated on the base of the sheet resistance increase, assuming constant resistivity after etching process. Textured film thicknesses of conventional and bi-layer samples were about 751 nm and 717 nm, respectively.

Optical transmittance and specular component was measured by UV-Visible-NIR spectrophotometer (VARIAN CARY 5000). Total diffusive component was calculated by the difference between total transmittance and specular component. So haze values (=total diffusion/total transmittance) were calculated for characterizing the light trapping properties.

RESULTS & DISCUSSION

The thicknesses of the conventional and bi-layer ZnO:Al film with seed layer measured by spectroscopic ellipsometry were about 1.32 μm and 1.28 μm, respectively. The ZnO:Al film with oxygen-controlled seed layer showed lower resistivity and low sheet resistance than those of conventional ZnO:Al film. The sheet resistances after the deposition were 2.46 Ω/□ (conventional ZnO:Al) and 2.08 Ω/□ (bi-layer ZnO:Al). Thus the resistivity of them was 3.24×10^{-4} Ωcm (conv. ZnO:Al) and 2.66×10^{-4} Ωcm (bi-layer ZnO:Al). In terms of surface texturing, the etching rate of bi-layer film (9.3 nm/sec) from HCl etchant was lower than conventional film (28.7 nm/sec), implying that the ZnO:Al film with proposed seed layer method grew denser than conventional film. After surface texturing, the sheet resistance of the films was measured again to estimate the final thickness of the films. The sheet resistance after surface texturing was 4.31 Ω/□ (conv. ZnO:Al) and 3.71 Ω/□ (bi-layer ZnO:Al), respectively.

Figure 1 shows the results of XRD measurement. We applied K_α line of Cu (λ = 1.5406 Å) to measure ZnO:Al samples. (002) peak was strongly detected compared to other ones such as (100), (101), (103), (112), and (004) peak for conventional ZnO:Al film in Figure 1-(a). This is consistent to the results that ZnO has a preferential growth on the c-axis [11]. In Figure 1-(b), however, several weak peaks from Figure 1-(a) almost disappeared. Furthermore, (002) and (004) peak are drastically increased. It can be confirmed from Figure 1-(c). The bi-layer ZnO:Al film revealed much stronger (002) peak than conventional one. And preferred orientation of (004) peak was high at XRD data of the bi-layer film. These phenomena might be resulted from the seed layer which can affect the growth of ZnO:Al film. The bi-layer ZnO:Al film with oxygen-controlled seed layer showed higher crystallinity than conventional ZnO:Al film without seed layer. With measured FWHM values(θ=17.21° for (002) orientation) of 0.2755(conv. ZnO:Al) and 0.0984(bi-layer ZnO:Al), the grain sizes of the samples estimated from the Scherrer equation [12] were 30.2nm (conv. ZnO:Al) and 84.9nm(bi-layer ZnO:Al).

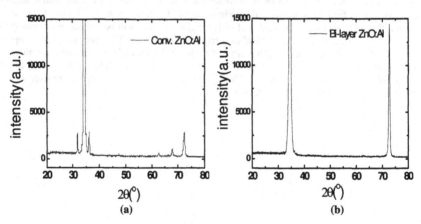

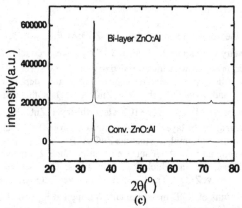

(c)

Figure 1. XRD data of surface textured (a) conventional and (b) bi-layer ZnO:Al film.

Optical transmittance and diffusive component of surface textured ZnO:Al films were measured by UV-Visible-NIR spectrophotometer. Figure 2-(a) shows that the transmittance of bi-layer ZnO:Al film is 88.1% and 88.9% at wavelength of 550nm and 800nm, whereas that of conventional film is 83.7% and 82.3%, respectively. The bi-layer film revealed high transmittance at the critical absorption region for amorphous and microcrystalline silicon tandem solar cells. The transmittance of ZnO film is increased when it is sputtered in oxygen atmosphere [13]. The thin ZnO:Al seed layer sputtering using oxygen-mixed argon gas improved the transmittance through the micro-structural changes of ZnO:Al film. Figure 2-(b) shows that the haze values of the bi-layer ZnO:Al film are about 90.6% and 68.1% at wavelength of 550nm and 800nm, whereas that of conventional film are 58.7% and 22.1% respectively. The bi-layer film showed a drastic increase of haze value in whole wavelength (UV-Vis-NIR) compared to that of the conventional film as it is showed in Figure 2-(b). From an increase of haze value it is expected that light scattering capability of bi-layer ZnO:Al film will be enhanced compared to that of the conventional film. The optical property of Asahi-U type TCO was added to compare with our samples. In some wavelength regions, Asahi-U TCO showed some higher transmittance than bi-layer ZnO:Al film. However, sheet resistance of bi-layer ZnO:Al (3.71 Ω/□) is lower than that of Asahi-U TCO (10~12 Ω/□). Also bi-layer ZnO:Al revealed much higher haze values than Asahi-U TCO in all wavelength regions.

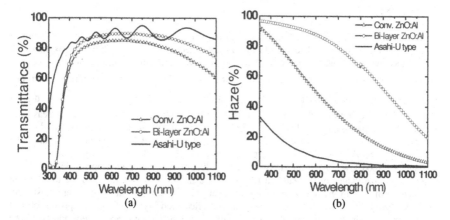

(a)   (b)

Figure 2. (a) The transmittance of the sputtered ZnO:Al films and (b) the haze values of the sputtered ZnO:Al films.

Figure 3 shows the FE-SEM(Carl Zeiss, SUPRA 55VP) images the samples. Figure 3-(a) and Figure 3-(b) shows the surface morphology of the conventional and bi-layer ZnO:Al films. The textured ZnO:Al film with seed layer had larger craters than conventional film. As shown in XRD data, ZnO:Al with seed layer had a higher crystalline phase than ZnO:Al without seed layer and when the surface texturing is performed, large craters and increased surface roughness which improve haze characteristics could be formed due to large grain size in the ZnO:Al film with seed layer.

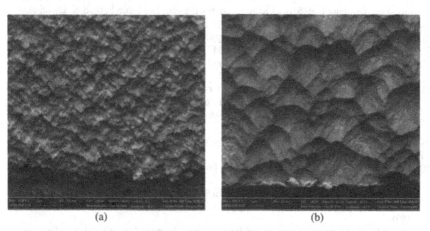

(a)   (b)

Figure 3. FE-SEM images (20000x resolution) of the surface morphology of textured (a) conventional ZnO:Al film and (b) bi-layer ZnO:Al film.

CONCLUSIONS

High quality bi-layer ZnO:Al film with high transmittance and haze properties was obtained by applying the seed layer. The seed layer was deposited by using 4% dilution of oxygen with argon gas when it was sputtered. The bi-layer ZnO:Al film showed lower electrical resistivity(ρ=2.66×10^{-4} Ωcm) and sheet resistance(R$_s$=3.71 Ω/$\square$, after texturing) than one without seed layer(ρ=3.24×10^{-4} Ωcm, R$_s$=4.31 Ω/$\square$). The ZnO:Al film grown on the seed layer showed higher crystallinity than one grown on the glass. Optical transmittance was increased by introducing oxygen-mixed seed layer. The haze characteristics were drastically increased due to an increase of grain size. After surface texturing by 0.5% HCl wet-chemical etching, the transmittance of ZnO:Al film was increased from 83.7% to 88.1% at wavelength of 550nm through the seed layer. Also the transmittance at 800nm was increased from 82.3% to 88.9%. Especially, haze values of the ZnO:Al film were drastically increased from 58.7% to 90.6% at wavelength of 550nm by employing the seed layer. Also haze values at 800nm were drastically increased from 22.1% to 68.1%. The oxygen-controlled seed layer technique improved optical and electronic characteristics simultaneously without increasing the process steps and cost. The bi-layer ZnO:Al film with the seed layer will be a promising candidate for front TCO of multi-junction silicon solar cells.

ACKNOWLEDGMENTS

This work is outcome of Development of Double Junction Si Thin Film Solar Module on Glasses supported financially by the Ministry of Knowledge Economy.

REFERENCES

[1]. Jinsu Yoo, Jeonghul Lee, Seokki Kim, Kyunghoon Yoon, and I. Jun Park, S.K. Dhungel, B. Karunagaran, D. Mangalaraj, and Junsin Yi : Thin Solid Films (2005) 480–481

[2]. Nakada, T.,Ohkubo, Y. and KunioKa : A., Jpn, J. Appl. Phys., 1991 30 3344

[3]. J. Krc˘, M. Zeman, O. Kluth, F. Smole, and M. Topic: Thin Solid Films 426 (2003)

[4]. Joachim Muller, Bernd Rech, Jiri Springer, and Milan Vanecek, Solar Energy 77 (2004)

[5]. X.L. Chen, X.H. Geng, J.M. Xue, and L.N. Li, Journal of Crystal Growth 299 (2007)

[6]. C.Y. Hsu, T.F.Ko, and Y.M.Huang: J. Euro. Ceramic Soc. 28 (2008) 3065-3070

[7]. Zhanga, Y., Zhenga, H., Sua, J., Lina, B. and Fu, Z.: J. Lumin. 124 (2007), 252.

[8]. Fujita, M., Kawamoto, N., Sasajima, M. and Horikoshi, Y.: J. Vac. Sci.Technol. B, 22 (2004) 1484.

[9]. Onuma, T., Chichibu, S. F., Uedono, A., Yoo, Y. Z., Chikyow, T., Sota, T. et al.: Appl. Phys. Lett., 85 (2004) 5586.

[10]. Koike, K., Komuro, T., Ogata, K., Sasa, S., Inoue, M. and Yano, M., Phys. E, 21 (2004) 679.

[11]. X.W. Sun, L.D. Wang, and H.S. Kwok: Thin Solid Films 360 (2000)

[12]. B.D. Cullity, p. 102, Elements of X-ray Diffraction, 2nd Edition,Addison Wesley, Reading, MA, 1978.

[13]. Oliver Kluth, Gunnar Scho¨pe, Bernd Rech, Richard Menner, Mike Oertel, Kay Orgassa, and Hans Werner Schock: Thin Solid Films, 502, (2006)

Mater. Res. Soc. Symp. Proc. Vol. 1153 © 2009 Materials Research Society   1153-A07-20

Optics in Thin-Film Silicon Solar Cells With Integrated Lamellar Gratings

Rahul Dewan, Darin Madzharov, Andrey Raykov and Dietmar Knipp
School of Engineering and Science, Jacobs University Bremen, 28759 Bremen, Germany

ABSTRACT

Light trapping in microcrystalline silicon thin-film solar cells with integrated lamellar gratings was investigated. The influence of the grating dimensions on the short circuit current and quantum efficiency was investigated by numerical simulation of Maxwell's equations by a Finite Difference Time Domain approach. For the red and infrared part of the optical spectrum, the grating structure leads to scattering and higher order diffraction resulting in an increased absorption of the incident light in the silicon thin-film solar cell. By studying the diffracted waves arising from lamellar gratings, simple design rules for optimal grating dimensions were derived.

INTRODUCTION

Efficient light management concepts are needed to increase the short circuit current and quantum efficiency of thin film solar cells. High efficiencies have been achieved by introducing randomly textured interfaces in the solar cell [1-3]. Introducing nano textured interfaces leads to reduced reflection losses and enhanced scattering and diffraction of light in the device. The optical path length is increased, which leads to a distinctly enhanced short circuit current and quantum efficiency in the red and infrared part (wavelength 600 – 1100 nm) of the optical spectrum [4-5]. For shorter wavelengths (300 – 600 nm) the short circuit current and quantum efficiency remain almost constant, since the absorption length is significantly smaller than the thickness of the solar cell. Subsequently the blue and green light will be absorbed even before reaching the back reflector. In order to understand the wave propagation within the nano textured solar cell and to optimize the nano texturing process, near field optics has to be considered when modeling the devices. Therefore, to describe the wave propagation in such devices simple geometric or wave optics is not sufficient, rather Maxwell's equations have to be solved rigorously [6]. Different approaches of using Maxwell's solvers have been used to analyze the wave propagation in solar cells [6-8]. In this work, a Finite Difference Time Domain (FDTD) simulation tool (OptiFDTD®) was used to investigate the wave propagation for nano textured microcrystalline silicon solar cells. So far FDTD has only been used to investigate the optical wave propagation within amorphous solar cells [7].

The analysis of the wave propagation within a randomly textured solar cell is complex; hence a simple model system was selected which approximates the randomly textured solar cell. The texturing was modeled by lamellar gratings. The results on smooth substrates are used as a reference to investigate the influence of the grating parameters on the solar cell parameters. Based on grating designs that were derived to maximize the short circuit current and the efficiency of the microcrystalline silicon solar cells, design rules towards optimal grating designs were formulated.

MODEL FOR OPTICAL SIMULATIONS

The optical model used in this study is based on the periodic arrangement of lamellar gratings. It is assumed that the solar cell on a textured substrate can be described by a unit cell, which provides all information on the behavior of the entire solar cell. The schematic cross section of a unit cell of a microcrystalline silicon solar cell on a smooth substrate and one with an integrated grating are shown in Fig. 1(a) and 1(b) respectively. The microcrystalline silicon solar cell structure consists of a 500 nm thick aluminum doped zinc-oxide (ZnO:Al) front contact, followed by a hydrogenated microcrystalline silicon diode (μc-Si:H) with a total thickness of 1000 nm and a back reflector consisting of an 80 nm thick ZnO:Al layer and a silver reflector. For the unit cells with lamellar grating, key parameters are the period, the groove height and the width of the groove. The groove height in case of Fig. 1(b) is just the height of the one-dimensional line grating. In the optical simulations, the period of the unit cell was varied from 500 nm to 3000 nm and the groove height of the structure was simulated for grooves from 0 nm to 500 nm. The width of the groove was always kept constant at 50 % of the period. The device structure is consistent with the standard microcrystalline silicon solar process developed by the Research Center Jülich, [9] which exhibits maximal conversion efficiencies of 10.3 % [10]. The optical constants were provided by the Research Center Jülich. [11]

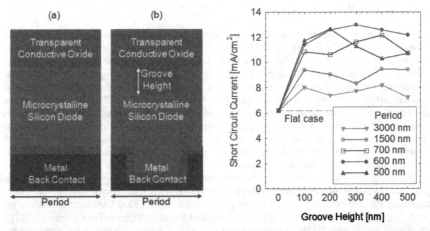

Figure 1. Schematic cross section of unit cell of a microcrystalline silicon thin-film solar cell (a) on a smooth substrate and (b) with an integrated lamellar grating.

Figure 2. Short circuit current for different period size of the lamellar grating under red illumination (wavelength 600 – 1100 nm) as a function of groove height of the unit cell.

TRANSMISSION AND REFLECTION GRATINGS

In order to compare different grating designs the quantum efficiency is utilized. The quantum efficiency is defined as the ratio of the power absorbed in the silicon layer with respect to the total power incident on the unit cell. The quantum efficiency is calculated by:

160

$$QE = \frac{1}{P_{opt}} \int Q(x,y)dxdy \qquad (1)$$

where Q(x,y) is the time averaged power loss. The power loss was determined by

$$Q(x,y) = \frac{1}{2}c\varepsilon_0 n\alpha \cdot |E(x,y)|^2 \qquad (2)$$

where c is the speed of light in free space, ε_0 the permittivity of free space, α is the absorption coefficient, with n being the real part of the complex refractive index and E is the electric field. The collection efficiency, taking the electronic properties of the material into account, is assumed to be 100%. In other words, the internal quantum efficiency is assumed to be 100 %. Therefore, the determined quantum efficiency defines an upper limit of the achievable external quantum efficiency. By considering the wavelength dependent AM 1.5 spectral irradiance, the short circuit current was calculated from the quantum efficiency. The influence of the grating dimensions on the short circuit current in the red and infrared part of the optical spectrum is shown in Fig. 2. The dashed line in Fig. 2 represents the short circuit current of a solar cell on a smooth substrate. For small grating periods in the range of 500 – 700 nm, the red and infrared short circuit current (600 nm – 1100 nm) is highly enhanced. For larger periods, the gain of the short circuit current is lowered and it almost converges towards that for solar cell on a smooth substrate. The highest short circuit current is observed for a period of 600 nm and a groove height of 300 nm. In comparison to the smooth substrate the short circuit current is increased by a factor of 2.1 resulting in a short circuit current of 13.0 mA/cm².

To gain a better understanding of the wave propagation in such thin film solar cells with integrated grating structures further investigations were carried out. The influence of the front and the back grating on the enhancement of the short circuit current structure was studied by dividing the grating structure into two segments. By inspecting both gratings individually, the effect of the front and the back grating on the wave propagation should be investigated step by step.

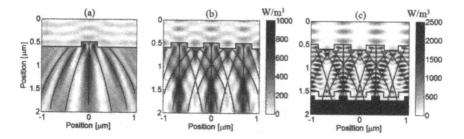

Figure 3. Power loss profile for infinitely thick silicon absorber layer with (a) one groove (analogous to single slit) and (b) lamellar gratings (acts as a transmission grating) at the front contact of the unit cell. The propagating modes are overlaid on (c) the investigated double grating structure.

The diffraction of the incoming wave at the front grating is shown in Fig. 3. In these simulations (Fig. 3(a) and 3(b)), an infinite silicon slab was assumed without any back reflector, thus allowing us to observe the contribution of the front grating. The period of the unit cell was kept fixed at 700 nm with a groove height of 100 nm. The wavelength of the incoming light was 700 nm. In the sequence of figures, Fig. 3(a) depicts the case with only one groove (analogous to a single slit) - the different diffracted orders can be distinctly observed for such a transmission grating. The lines indicate the different diffracted orders (m = 0, ±1, ± 2 and ±3). Fig. 3(b) shows the effect on the absorption pattern for a periodic transmission grating. Again the different diffracted orders were marked by colored lines. The high power loss at the center of the unit cell is determined by the constructive interference of higher diffraction orders of two neighboring unit cells. The maximum of the absorption is caused by the interference of the second diffraction orders of the neighboring cells. Fig. 3(c) shows the power loss profile of a structure with a double grating. The colored lines mark the different diffraction orders of the front grating. The different diffraction orders from the front grating are overlaid with the reflection of the back reflector and the diffraction from the back grating. Again the maximum of the absorption is determined by the interference of the 2nd order diffractions of the front grating.

The influence of the back grating on the power loss profile is depicted in Fig. 4. The input wavelength and dimensions of the grating were kept the same as described for the case in Fig. 3. Fig. 4(a) and 4(b) exhibit the power loss profile for a single groove and a periodic grating. Again the different diffracted orders were marked by colored lines. The power loss profile is dominated by the standing wave formed in front of the back reflector which is overlaid by the diffraction pattern. The power loss profile for the solar cell including the diffraction orders of the back contact is shown in Fig. 4(c). A comparison of Fig. 4(c) with Figure 3(c) reveals that the maximum of the absorption close to the back reflector is caused by the interference of the second order diffraction from the front rather than by the interference of the first order diffraction of the back reflector. Therefore, a comparison of the front and the back grating reveals that the power loss profile is dominated by the diffraction grating in the front of the solar cells. The grating in the back has only a minor influence on the power loss profile. This behavior is also observed for structures with different groove heights.

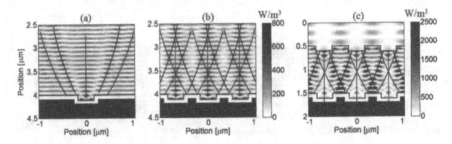

Figure 4. Power loss profile for a reflection grating with (a) a single groove and (b) lamellar gratings at the back contact of the unit cell. The propagating modes are overlaid on (c) the investigated double grating structure.

162

DESIGN RULES FOR OPTIMAL GRATINGS

To maximize the absorption within the absorber layer in a thin-film microcrystalline silicon cell, constructive interference of the different diffracted orders plays a key role. In order for the different orders to interfere, the propagating angles of the mode are an important parameter. These angles are governed by the grating equation (assuming normal incidence)

$$P \cdot n \cdot \sin(\theta_m) = m \cdot \lambda,\qquad(3)$$

where P is the grating period, n denotes the refractive index of the propagating media after diffraction, m specifies the diffraction order and θ_m being the diffraction angle. To have interference with the zeroth and first or second diffracted order of two neighboring unit cells, the period of the lamellar gratings should be small as the diffraction angles are higher for such cases. On the other hand, if the period is large, the angles of the diffracted orders are small and only higher diffracted orders (m>2) interfere with the zeroth diffracted order. Since the intensity of the diffracted waves scales by $1/m^2$, the influence of interference of higher orders on the power loss profile is reduced. The influence of the period on the absorption in the solar cell is shown in Fig. 5, where the short circuit current as a function of the period for different groove heights is plotted. Irrespective of the groove height, a drop in the short circuit current can be observed with increasing period. With an increasing period, the diffraction angles for integrated lamellar gratings are reduced. As a consequence the effective thickness of the solar cell is reduced. For periods approaching infinity, equation (3) dictates that the effective thickness will converge towards the real thickness of the solar cell, which implies that the optical path length of the cell is reduced for larger periods compared to shorter periods of the unit cell.

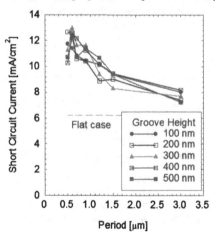

Figure 5. Short circuit current for different groove heights of the lamellar grating under red illumination (wavelength 600 – 1100 nm) as a function of groove height of the unit cell.

With an increasing period, the short circuit current converges against the short circuit current of a microcrystalline silicon thin film solar cell on a smooth substrate. In terms of design of a microcrystalline solar cell with integrated lamellar gratings the following design rules can be derived: The period of the diffraction grating should be small, so that the light is diffracted at large diffraction angles. However, if the period is too small higher order diffraction orders cannot propagate in the solar cells structure anymore. In the case of the investigated structure such behavior is observed for grating periods smaller or equal to 450 nm. The maximum of the short circuit current is observed for a grating period of 600 nm.

The results of the numerical simulations also indicate a trend that the short circuit current is maximized when the grooves height is approximately equal to half of the grating period. The investigated microcrystalline solar cell exhibits a maximum of the short circuit for a groove height of 300 nm. A detailed analysis of the wave propagation within the solar cell reveals that the increase of the short circuit current is mainly determined by the front rather than the back grating.

SUMMARY

The wave propagation in 1 μm thick microcrystalline silicon thin-film solar cells with integrated lamellar gratings was investigated using a Finite Difference Time Domain method. The simulation reveal that the short circuit current can be increased by more than 100 % in the red and the infrared spectrum by introducing a grating structure with a period of 600 nm and a height of 300 nm. If the period is smaller than 450 nm, higher diffracted orders cannot propagate in the cell. With an increasing grating period (> 900 nm), the short circuit current drops since only higher diffractions orders interfere, which carry less energy. In terms of the optimal grating height, it was found that the short circuit current is maximized if the grating height is approximately half the period size. A simple approach by looking at the diffraction patterns from individual gratings of the double grating was suggested and it was concluded that the increase of the short circuit current is mainly determined by the front rather than the back grating.

ACKNOWLEDGMENTS

The authors would like to acknowledge Christian Haase from the Research Center Jülich and Helmut Stiebig from Malibu Solar for helpful discussions. Furthermore, the authors like to acknowledge financial support from Embedded Microsystems Bremen.

REFERENCES

1. M. A. Green, K. Emery, Y. Hishikawa and W. Warta, Prog. Photovolt: Res. Appl., 17, 85 (2009)
2. K. Yamamoto, M. Yoshimi, Y. Tawada, Y. Okamoto, A. Nakajima and S. Igari, Appl. Phys. A 69, 179 (1999)
3. J. Müller, B. Rech, J. Springer and M.Vanecek, Sol. Energy 77, 917 (2004)
4. W.Beyer, J. Hüpkes and H. Stiebig, Thin Solid Films 516, 147 (2007)
5. C. Rockstuhl, S. Fahr, F. Lederer, K. Bittkau, T. Beckers, R. Carius, Appl. Phys. Lett. 93, 061105 (2008)
6. C. Haase and H. Stiebig, Appl. Phys. Lett. 91, 061116 (2007).
7. S. Lo, C. Chen, F. Garwe and T. Pertch, J. Phys. D: Appl. Phys. 40, 754 (2007)
8. Y. Lee, C. Huang, J. Chang and M. Wu, Opt. Express , 16, 7969 (2008)
9. N. Senoussaoui, M. Krause, J. Müller, E. Bunte, T. Brammer and H. Stiebig, Thin Solid Films 451- 452, 397 (2004)
10. Y. Mai, S. Klein, R. Carius, H. Stiebig, X. Geng and F. Finger, Appl. Phys. Lett. 87, 073503 (2005)
11. H. Stiebig, C. Haase, (personal communication)

Mater. Res. Soc. Symp. Proc. Vol. 1153 © 2009 Materials Research Society   1153-A07-21

Controlling Structural Evolution by VHF Power Profiling Technique for High-Efficiency Microcrystalline Silicon Solar Cells at High Deposition Rate

Guofu Hou[1], Xiaoyan Han, Changchun Wei, Xiaodan Zhang, Guijun Li, Zhihua Dai, Xinliang Chen, Jianjun Zhang, Ying Zhao, and Xinhua Geng
Institute of Photoelectronics, Nankai University, Tianjin 300071, P.R. China

ABSTRACT

High rate deposition of hydrogenated microcrystalline silicon (μc-Si:H) films and solar cells were prepared by very high frequency plasma enhanced chemical vapor deposition (VHF-PECVD) process in a high power and high pressure regime. The experiment results demonstrate that in high-rate deposited μc-Si:H films, the structural evolution is much more dramatic than that in low-rate deposited μc-Si:H films. A novel VHF power profiling technique, which was designed by dynamically decreasing the VHF power step by step during the deposition of μc-Si:H intrinsic layers, has been developed to control the structural evolution along the growth direction. Another advantage of this VHF power profiling technique is the reduced ion bombardments on growth surface because of decreasing the VHF power. Using this method, a significant improvement in the solar cell performance has been achieved. A high conversion efficiency of 9.36% (V_{oc}=542mV, J_{sc}=25.4mA/cm^2, FF=68%) was obtained for a single junction μc-Si:H *p-i-n* solar cell with *i*-layer deposited at deposition rate over 10 Å/s.

INTRODUCTION

Because of its improved stability and superior long wavelength response, μc-Si:H has been widely used as a stable and narrow bandgap absorber layers in both single-junction and multi-junction solar cells.[1~3] Since μc-Si:H is an indirect band-gap material, a thicker layer (~3μm) is necessary to make full use of solar spectrum, especially the infrared light(>800nm). Thus, high-rate deposition of μc-Si:H is a critical issue for low-cost production.[1] As well known that the high efficiency μc-Si:H solar cell are usually prepared with an intrinsic layer deposited near the a-Si:H/μc-Si:H transition region.[1~3] Such transition film is strongly dependent on the substrate and deposition condition, and its microstructure varies dramatically along the growth direction: an amorphous incubation layer regularly forms in the initial growth stage of layers, until fully microcrystalline growth sets in.[2] Then, the crystallinity increases gradually along the increase of thickness. This will cause serious deterioration for the device performance.[4, 5]

In order to control the microstructure evolution of μc-Si:H, B. Yan et al. developed a hydrogen dilution profiling technique.[6] Since then, this method has been adopted by many groups.[3, 7~8] Besides improving microstructure evolution by hydrogen dilution profiling technique, J. Gu reported that the μc-Si:H films by HWCVD also demonstrate higher compactness and thus high stability against the oxygen diffusion[7], while C. Niikura found improved carrier transport properties along the growth direction.[8] Optimized hydrogen dilution profiling can significantly improve both initial and stable μc-Si:H solar cell performance.[9]

---

[1] Corresponding author: Guofu Hou, E-mail: guofu_hou@yahoo.com.cn

In this paper, we firstly present thickness-dependent microstructure evolution, especially crystalline volume fraction and grain size, and solar cell performance. Then a VHF power profiling technique that the VHF power decreases step by step during i-layer deposition has been developed for the first time. Our aim is to control the microstructure evolution along the growth direction by this VHF power profiling technique, just as the hydrogen dilution profiling technique. With optimized VHF power profile, a significant improvement of the solar cell performance has been achieved.

EXPERIMENTAL DETAILS

All µc-Si:H films and p-i-n single-junction solar cells were deposited in a cluster-tool system with background pressure 5×10^{-6} Pa. The µc-Si:H p-layer and a-Si:H n-layer were deposited by PECVD in separate chambers with frequency of 60MHz and 13.56MHz, respectively. All i-layers were deposited with frequency of 70MHz, combining with gas pressure of 2Torr, substrate temperature of 175°C. The usual constant power is 60W, while it is decreased step by step with different power intervals (ΔP) for VHF power profiling process. All these above conditions lead to a relative high deposition rate around 12Å/s. Eagle2000 glass substrate are used for optoelectronic properties characterization of µc-Si:H films. SnO$_2$/ZnO and texture-etched ZnO are used as front electrode for p-i-n single-junction solar cell with configuration glass/TCO/p-µc-Si:H/i-µc-Si:H/n-a-Si:H/back reflector. A silver or aluminum back contact defines an active area of 0.253cm^2.

Raman scattering measurements were performed by Renishaw RM2000 microscope with 488nm laser with typical Raman Collection Depth (RCD) ~100nm for µc-Si:H. The Raman spectra were fitted with three Gaussian peaks and Raman crystallinity (X_c) were calculated by $X_c = (I_{520} + I_{500})/(I_{520} + I_{500} + I_{480})$. The x-ray diffraction (XRD) was performed by Rigaku D/max2500 with CuKα (0.153nm) x-ray to analyze the crystal orientations and grain size by Scherer formula. Light J-V characteristics were measured under AM1.5, 100mW/cm^2 at 25°C by a class A solar simulator.

RESULTS AND DICUSSIONS
Thickness Dependence of Microstructure Evolution and Solar Cell Performance

Firstly a series of intrinsic µc-Si:H films with various thickness were deposited on eagle2000 glass substrate at fixed VHF power and other deposition conditions at deposition rate of 12Å/s. Figure 1 shows the X_c and grain size of these µc-Si:H films as a function of film thickness. It can be seen that both X_c and grain size increase sharply with the increase of thickness for films with thickness less than 1µm, and then saturate for films with thickness more than 1µm. The X_c evolution along the thickness isn't as serious as that has

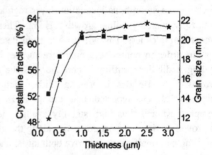

Figure 1 X_c (■) and (220) grain size (★) as a function of film thickness. Lines are guides to the eyes.

supposed for high-rate deposited μc-Si:H films. With the increase of i-layer thickness from 1μm to 3μm, the Xc only increases from 52.3% to 61.5%, however, the (220) grain size increase sharply from 12nm to 21nm.

Then the above intrinsic μc-Si:H films were applied as absorber layers in single-junction solar cells with SnO_2/ZnO front electrode and Al metal back contact. All solar cells were prepared with the same deposition conditions for p, i, n layers and only different i-layer thickness by varying the deposition time. The J-V characteristic parameters as a function of i-layer thickness were shown in Figure 2. It is observed that both FF and V_{oc} decreases with the increase of i-layer thickness, while J_{sc} increases firstly and then saturates. These results are in good agreement with previous work.[10, 11]

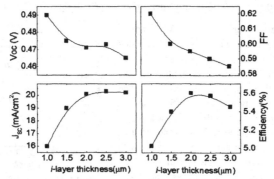

Figure 2 J-V characteristic parameters as a function of i-layer thickness.
Lines are guides to the eyes.

The decrease of FF suggests reduced carrier extraction efficiency in the thick devices, probably due to the reduced build-in electric field in i-layer. Higher J_{sc} results from more optical absorption and the saturation can attribute to a balance between increased optical absorption and reduced build-in electric field. One reason for V_{oc} decrease one with i-layer thickness is the reduced build-in electric field in thick i-layer, the other reason is supposed an increasing defect density caused by larger grain size, which will lead to much more grain boundary. In order to prove the second reason, the dark J-V characteristics of these solar cells were measured and analyzed to get the information about the charge carrier transport in the solar cells. The diode factor n and saturation current density J_0, which are calculated from the fitting of exponential part of the dark J-V

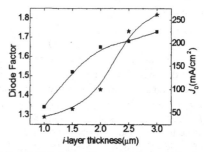

Figure 3 Diode factor n (■) and saturation current density J_0 (★) plotted as a function of i-layer thickness. Lines are guides to the eyes.

167

curves to the diode equation $J_{dark}=J_0$ [exp(eV/nkT)-1], are plotted in Figure 3. With the increase of i-layer thickness from 1μm to 3μm, the diode factor n monotonously increases from 1.34 to 1.73, while the J_0 increases from 41μA/cm^2 to 262μA/cm^2. The higher n values above 1.5 in a p-i-n diode suggest a dominated bulk recombination. The higher J_0 values indicate increased defect density and thus the decrease of V_{oc}. Besides above-mentioned two reasons, we suppose another non-neglectable factor. In order to get high deposition rate, high power density is necessary. However, the high-energy ion bombardment on the growing surface during i-layer deposition is foreseeable, which will further increase defect density in the bulk i-layer.[13, 14] This is why high pressure depletion technique is necessary to obtain high quality μc-Si:H films at high deposition rate.[14]

VHF Power Profiling Technique and Microstructure Evolution

The schematic diagram of the VHF power profiling technique is shown in figure 4. Compared with the popular constant power deposition, a step by step decreased VHF power is applied for the VHF power profiling technique. The power interval between neighboring two steps is defined as ΔP. The deposition time for every step is defined as T_{step}. In Figure 5 the X_c of μc-Si:H film, is plotted as a function of thickness. The former part of films with 500nm thickness were deposited with the same VHF power, the other part of films were deposited with different ΔP. It can be seen that by varying the steps, ΔP and T_{step}, the microstructure evolution along the growth direction can be purposefully controlled.

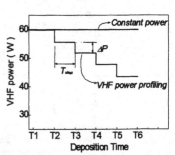

Figure 4 Schematic diagrams of the VHF power profiling technique.

VHF Power Profiling Technique and Solar Cell Performance

Then a series of single-junction μc-Si:H solar cells were prepared with SnO$_2$/ZnO front electrode and simple Al metal back contact. The same p and n layer deposition conditions were used, while i-layers with similar thickness ~2μm were deposited by VHF power profiling technique with different ΔP, just as the films shown in figure 5. In order to

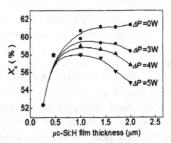

Figure 5 X_c plotted as a function of μc-Si:H film thickness deposited under different ΔP. ΔP=0 means constant power. Lines are guides to the eyes.

reduce the amorphous incubation layer at p/i interface, a lower-rate-deposited high-quality p/i buffer layer was used.[12] When ΔP ranges from 1W to 4W, V_{oc} and FF monotonously increase and J_{sc} behaves a slight increase. From figure 5 we can see that when the VHF power profiling technique is applied, the X_c for the latter part of the film (>1μm) will decrease. A decreased average X_c for the whole i-layer certainly will result in an increase of V_{oc}. Because of the decrease of VHF power, the high-energy ion bombardment on the growing surface also will reduce and defect density in the bulk i-layer will be lowered. Thus the transport property of i-layer is improved and results in increase of V_{oc} and FF. When ΔP is higher than 4W, V_{oc}, FF and J_{sc} begin to decrease. This probably has been caused by the too lower X_c for the latter part i-layer, which will deteriorate the charge carrier transport as an amorphous incubation layer does at p/i interface in p-i-n solar cell.[12, 15] In this case, a significant improvement of the solar cell performance has been achieved with ΔP of 4W.

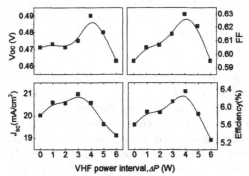

Figure 6 *J-V* characteristic parameters as a function of ΔP. $\Delta P=0$ means constant VHF power.
Lines are guides to the eyes.

At last we have transferred the μc-Si:H solar cell prepared by the primary optimized VHF power profiling technique onto texture-etched ZnO substrate. A ZnO/Ag back contact was used to enhance the light absorption with ZnO prepared by MOCVD. An initial active-area efficiency of 9.36% (V_{oc}=542mV, J_{sc}=25.4mA/cm^2, FF=68%) has been obtained for a single-junction μc-Si:H p-i-n solar cell at an average deposition rate over 10 Å/s. Using this recipe for depositing the bottom cell, an initial active-area efficiency of 11.14% for a-Si:H/μc-Si:H

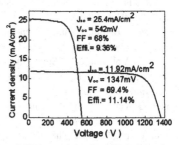

Figure 7 *J-V* characteristics of μc-Si:H single-junction and a-Si:H/μc-Si:H tandem solar cells with μc-Si:H *i*-layer deposited with an primary optimized VHF power profile.

tandem solar cell was achieved. By optimizing the current match and n/p tunnel recombination junction between the top and bottom cell,[16] the solar cell performance will be further improved.

SUMMARY

A novel VHF power profiling technique, which is designed by dynamically decreasing the VHF power step by step during depositing μc-Si:H *i*-layers, is proved to be an effective way to control the structural evolution along the growth direction. Another advantage of this VHF power profiling technique is the reduced ion bombardments on growth surface and thus lower defect density in the bulk *i*-layer. A primary optimized VHF power profiling has resulted in a high conversion efficiency of 9.36% for a single-junction µc-Si:H *p-i-n* solar cell at an average deposition rate over 10 Å/s.

ACKNOWLEDGEMENT

The work in this paper has been supported by the State Key Development Program for Basic Research of China (2006CB202602, 2006CB202603), the Tianjin assistant Foundation for the National Basic Research Program of China (07QTPTJC29500) and the Natural Science Foundation of Tianjin (07JCYBJC04000).

REFERENCES

1. Y. Mai, S. Klein, R. Carius, J. Wolff, A. Lambertz, F. Finger, and X. Geng, *Journal of Applied Physics*, 97, 2005, 114913
2. Corinne Droz, Ph.D Thesis, Université de Neuchâtel, Swiss, 2003
3. Aad Gordijn, PhD Thesis, Utrecht University, The Netherland, 2005
4. T. Brammer, and H. Stiebig, *Journal of Applied Physics*, 94, 1035 (2003)
5. T. Roschek, T. Repmann, J. Müller, B. Rech, H. Wagner, *Proc. of 28th IEEE Photovoltaic Specialists Conference*, Anchorage, AK, Sep. 15-22, 150 (2000)
6. B. Yan, G. Yue, J. Yang, J. Yang, S. Guha, D.L. Williamson, and C. Jiang, *Appl. Phys. Lett.* 85, 1955(2004)
7. J. Gu, M. Zhu, L. Wang, F. Liu, B. Zhou, Y. Zhou, K. Ding and G. Li, *Journal of Applied Physics*.98, 093505(2005)
8. C. Niikura, R. Brenot, J. Guillet, Jean-Eric Bourée, *Thin Solid Films* 516, 568 (2008)
9. B. Yan, G. Yue, Y. Yan, C. Jiang, C. W. Teplin, J. Yang, and S. Guha, *Mater. Res. Soc. Symp. Proc.* Vol.1066, 2008 Spring Meeting, 1066-A03-03
10. O. Vetterl, A. Lambertz, A. Dasgupta, F. Finger, B. Rech, O. Kluth, H. Wagner, *Solar Energy Materials & Solar Cells* 66, 345 (2001)
11. B. Yan, G. Yue, J. Yang, A. Banerjee, and S. Guha, *Mat. Res. Soc. Symp. Proc.* Vol.762, 2003 Spring Meeting, A4.1.1
12. G. Hou, X. Han, G. Li, X. Zhang, N. Cai, C. Wei, Y. Zhao, X. Geng, *Technical Digest of the 17th International Photovoltaic Science and Engineering Conference*, Dec. 2007, Fukuoka Japan, 1112-1113
13. B. Kalache, A. I.Kosarev, R.Vanderhaghen, P. Roca. I. Cabarrocas, *Journal of Applied Physics*, 93(2), 1262 (2003)
14. M. Kondo, M. Fukawa, L. Guo, A. Matsuda, *J. Non-Cryst. Solids* 266-269, 84 (2000)
15. U. K. Das, E. Centurioni, S. Morrison, A. Madan, *Proceedings of 3rd World Conference Photovoltaic Energy Conversion*, 12-16 May 2003, 1776-1779
16. G. Li, G. Hou, X. Han, Y. Yuan, C. Wei, J. Sun, Y. Zhao and X. Geng, *Chinese Phys.B* 18(4), 1674(2009)

Novel Device Applications

Mater. Res. Soc. Symp. Proc. Vol. 1153 © 2009 Materials Research Society   1153-A08-03

Voltage controlled amorphous Si/SiC photodiodes and phototransistors as wavelength selective devices: Theoretical and electrical approaches

M A Vieira[1,3], M. Vieira[1,2], P. Louro[1,2], M. Fernandes[1], A. Fantoni[1], M. Barata[1,2]
[1]Electronics Telecommunications and Computer Dept, ISEL, Lisbon, Portugal. [2]CTS-UNINOVA, Lisbon, Portugal. [3]CML-Traffic Department, Lisbon, Portugal

ABSTRACT

In this paper single and stacked structures that can be used as wavelength selective devices, in the visible range are analysed. Two terminal heterojunctions ranging from p-if-n to p-i-n-p-i'-n configurations are studied. Three terminal double staked junctions with transparent contacts in-between are also considered to increase wavelength discrimination. The color discrimination was achieved by ac photocurrent measurement under different externally applied bias. Experimental data on spectral response analysis and current –voltage characteristics are reported.
A theoretical analysis and an electrical simulation procedure are performed to support the wavelength selective behaviour. Good agreement between experimental and simulated data was achieved. Results show that in the single p-i-n configuration the device acts mainly as an optical switch while in the double ones, due to the self bias effect, the input channels are selectively tuned by shifting between positive and negative bias. If the internal terminal is used, the inter-wavelength cross talk is reduced and the signal-to-noise ratio increased.

INTRODUCTION

Application fields of polymer optical fibers (POF) technology is increasing mainly driven by the computing power incidence allied with everywhere multimedia home appliances [1]. Only the visible spectrum can be applied when using POF for communication. So, the demand of new optical processing devices is a request.Wavelength multiplexing devices have to accomplish the transient colour recognition of two or more input channels in addition to their capacity of combining them onto one output signal without losing any specificity (wavelength and transmission speed) [2].
Light wavelength discrimination depends on the structure of the sensor, thickness of each p-i-n cell, and on the selected sequence of the cells in the multilayer structure. In this paper we present results on the optimization of different multilayered a-SiC:H heterostructures for wavelength-division multiplexing applications in the visible spectrum. A theoretical analysis and an electrical simulation procedure are performed to support the wavelength selective behaviour.

DEVICE CONFIGURATION

Voltage controlled devices, with front and back indium tin oxide transparent contacts were produced by PECVD at 13.56 MHz radio frequency in three different architectures and tested for a proper fine tuning of the visible spectrum. In the first configuration (NC11), the device is a p-i-n photodiode where the intrinsic layer is a multilayered a-SiC:H/a-Si:H thin film. In the others (NC10 and NC12), the devices have an a-SiC:H (p-i-n)/ a-SiC:H(-p) /Si:H(-i)/SiC:H (-n) configuration. To test the efficiency of the internal n-p junction, in one of the architectures (NC12), a third transparent contact was deposited in-between. In all the devices the thickness (200nm) and the optical gap (2.1 eV) of the a-SiC:H intrinsic layer (i'-) are optimized for blue collection and red transmittance and the thickness (1000 nm) of a-Si:H one (i-) adjusted to achieve full absorption in the green and high collection in the red spectral range. As a result, both front and back diodes act as optical filters confining, respectively, the blue and the red optical carriers, while the green ones are absorbed across both [3].

OPTOELECTRONIC CHARACTERIZATION

Spectral photocurrent & ac I-V characteristics

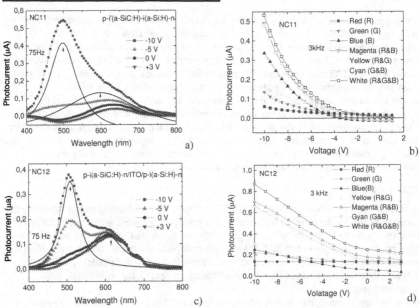

Figure.1-Spectral response (a, c,) and photocurrent-voltage characteristics (b, d) in different architectures. In a) and c) the solid lines show the multi peak curve fit at -10V.

All the devices were characterized through spectral response (400-800nm) and *ac* photocurrent-voltage (-10V <V <+3V) measurements. In this last measurement three modulated (3kHz) monochromatic lights: R (λ_R=650 nm); G (λ_G=550 nm) and B (λ_B=450 nm), and their polychromatic combinations; R&G (Yellow); R&B (Magenta); G&B (Cyan) and R&G&B (White) illuminated separately the device and the photocurrent was measured. In Fig. 1 the characteristics curves are depicted for NC11 and NC12 samples. Data show that in the graded cell (NC11) the individual contributions from the front and the back parts of the i'/i –layer are difficult to identify (solid lines). Here, the spectral response has the trend of a p-i-n structure with carbon in the active layer; it decreases as the negative bias decreases and is higher in the green region than in the red one. As expected, for a single cell, a reverse on the photocurrent regime is observed mainly in the blue/green spectral regions. If the tandem cells are analyzed

174

(NC 10 and NC12) the contribution of both front and back diodes are evident (dot curves). As the applied voltage changes from negative to positive, the response around 500 nm decreases sharply while the one around 600 nm remains constant.

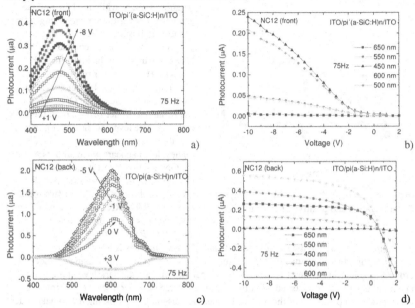

Figure 2 –Spectral photocurrent under different applied bias (a, c) and its trend with the applied voltage, at different wavelengths (b, d), for the front, p-i' (a-SiC:H)-n, and back, p-i (a-Si:H)-n.

In Fig. 2 (a, c) the spectral photocurrent under different electrical bias and its trend (b, d) with the applied voltage, under specific wavelengths, are displayed separately for the front, p-i' (a-SiC:H)-n, and back p-i (a-Si:H)-n, photodiodes. Here, the internal contact of the NC12 device was used. As expected from Fig. 1 the front and back photodiodes acts as optical filters. respectively in the blue and red spectral regions. The front diode, based on a-SiC:H, cuts the red component of the spectrum while the back one, based on a-Si:H, cuts the blue component. Each diode separately presents the typical responses of single p-i-n structures while the stacked configuration shows the influence of both front and back diodes modulated by its interconnection through the internal n-p junction.

Bias sensitive wavelength division multiplexing

A chromatic time dependent wavelength combination of red (λ_R=650 nm), green (λ_G=550 nm) and blue (λ_B=450 nm) input channels with different transmission rates, was shone on the device. The generated photocurrent was measured under negative and positive bias to readout the combined spectra. In Fig. 3 is displayed the input channels (lines) and the transient multiplexed signals (symbols) under negative (-10 V) and positive (+3V) applied voltages. The highest frequency (red

175

channel) was 3 KHz and the ratios between all of them was always one half. The reference level was assumed to be the signal when all the input channels were OFF (dark level).

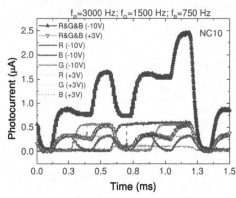

Figure 3 –Input channels and multiplexed signals under negative and positive bias (symbols)

As expected from Fig. 1 the red signal remains constant while the blue and the green decrease as the voltage changes from negative to positive. The lower decrease in the green channel when compared with the blue one can be ascribed to its red-like behavior (Fig. 1) under positive bias. The multiplexed signal depends on the applied voltage and on the ON-OFF state of each channel. Under reverse bias, there are eight separate levels while under positive bias they were reduced to one half. The highest level appears when all the channels are ON and the lowest if they are OFF. Furthermore, the levels ascribed to the mixture of three (R&G&B) or two input channels (R&B, R&G, G&B) are higher than the ones due to the presence of only one (R, G, B).

It is interesting to notice that the sum of the R, G and B input channels is lower than the multiplexed signal showing capacitive effects due to the time-varying input channels. Under forward bias, the blue component of the combined spectra falls into the dark level, tuning the red/blue input channels.As expected from Fig. 1 as the reverse bias increases the multiplexed signal exhibits a sharp increase if the blue component is present. By comparing the signals under positive and negative bias and using a simple algorithm that takes into account the different sub-level behaviors under reverse and forward bias (Figure 2b) it is possible to split the red from the green component and to decode their RGB transmitted information.

SPICE MODEL VALIDATION AND PHYSICS

Taking into account the experimental results (Fig. 1, 2 and 3) and the device configuration, an electrical model was developed and supported by an electrical simulation. The simple model considers that both front and back photodiodes are optically and electrically in series. The incident light traverses through the p-i'-n-p-i-n sequence and is absorbed according to its wavelength. As both are electrically in series, they must ensure that each one gives the same current, I, and the applied voltage, V, is shared between both ($V=V_1+ V_2$). Any diode whose current is over the other would have to reduce its current and consequently to self-bias. If the photocurrent, is higher than the net current the diode self forward bias ($V_{1,2}>0$) if not itself reverse biased ($V_{1,2}<0$). The flow of carriers across the internal p-n junction is proportional to the difference between V1 and V2 [4]. So, if the device is biased negatively the n-p internal junction is forward-biased and, the external voltage drops across the external junctions. The current is limited by the leakage current of the less excited diode, the front under red and the back under blue irradiations. Under green light the limiting factor is the photocurrent of the less irradiated diode. Under positive bias the internal junction becomes always reverse-biased.

176

To better understand the transient effects due to the time-varying irradiation, the linear state equation [5] was analyzed and used to design the equivalent electric circuit as displayed in Fig. 4a. The device will consist of two phototransistors connected back to back, modeling respectively the a-SiC:H p-i-n-p and a-Si:H n-p-i-n sequences. In order to simulate the n-p internal junction, the collector and the base of both transistors are shared. Two pulsed current sources with different frequencies, I1 and I2, are used to simulate the input blue and red channels, respectively. The green channel is simulated by two sources with the same frequencies, I3 and I4, since the green absorption occurs across both front and back intrinsic layers (Fig. 2). A *dc* voltage source, *V*, was applied giving rise to an output signal, *i(t)*. Time-varying versions of the linear circuit elements are the capacitors *C1* and *C2* used to simulate the transient capacitance due to the minority carriers trapped in both p-i-n junctions. Those capacitors exhibit time-varying charge/voltage characteristics being the current across them the instantaneous rate of change of charge ($i_{C1,2}(t) = C_{1,2} \, dv_{1,2}(t)/dt$) [6].

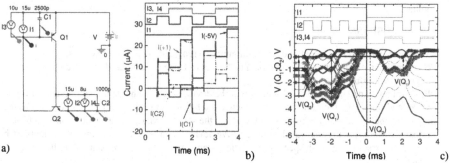

a)   b)   c)

Figure 4   a) Equivalent electric circuit used for simulation proposes. SPICE simulation under red (I2=15uA), blue (I1=15uA) and green (I3=10uA, I4=8uA) pulsed lights and negative and positive applied voltages: b) current; c) voltage drop across Q1 (symbols) and Q2 (lines).

In Fig. 4b is shown the transient input channels; the current across the capacitors, I(C1), I(C2), under negative bias; and the multiplexed signals, I, at -5V and +1 V. In Fig. 4c is displayed the base-emitter voltage, across the Q1 (symbols) and Q2 (lines), between -5V and 0V, during the first (t<0) and the second cycles (t>0). A good agreement between experimental (Fig. 3) and simulated (Fig. 4b) data was observed. The pi´npin device is a two-input cascode circuit with a common source amplifier as input stage. This input stage drives a common base amplifier as output stage. The upper transistor (Q1) acts as load of the input lower transistor (Q2) and also uses it sources (I1, I3) as input node. So, Q1 exhibits a low input resistance to Q2, making the voltage gain of Q2 very small.

The device is a transmission system able to stores and to transport the minority carriers generated by the current pulses, through capacitors C1 and C2. It acts as a charge integrator, keeping memory of the input channels (color and frequency). The control strategy of this transmission system, using the two-input cascode configuration, improves input-output isolation (or reverse transmission) as there is no direct coupling from the output to input. This eliminates the Miller effect and thus contributes to a higher bandwidth. Under negative bias, once the blue channel is *on*, the emitter-base of Q1 becomes optically forward biased and C2 is rapidly charged in inverse polarity of C1 ($i_{C_1}(t)C_2 = -i_{C_2}(t)C_1$; Fig. 4b) with an input voltage in which a threshold value is inserted for clamping (arrows in Fig. 4c) resulting in a reinforcement of the reverse bias at Q2. The current source keeps filling the capacitors for the duration of the pulse, Δt, of the input channel and the transferred charge between C1 and C2 will reach the output terminal as a capacitive charging current. The presence of the red channel changes, in the opposite way, the charge of both capacitors. If the green channel is *on* the current is the balance between the blue- and the red-like contributions (Figure 1, Figure 2). With several channels *on*, the packets of charge stored at C1 (I1, I3) are sequentially transferred to C2 and together with the minority carriers generated at the base of Q2 (I2, I4) flow across the circuit. When all the channels are *off*, the current is limited by the leakage current of both active junctions (dark level). If a small positive voltage is applied the junction capacitance across the internal n-p junction is charged and the only carriers collected come from the red-like channels (I2, I4) enabling the demultiplexing of the previous multiplexed signal (Figure 3) by switching between positive and negative voltages.

CONCLUSIONS

Voltage controlled multiplexing devices, in multilayered a-SiC:H pin architectures, were compared. An electrical model based on a two-input cascode circuit was developed to support the device functioning.
Experimental and simulated results show that the device acts as a charge transfer system. It filters, stores and transports the minority carriers generated by current pulses, keeping the memory of the input channels (color and transmission speed). In the stacked configuration, both front and back transistors act separately as wavelength selective devices, and are turned *on* and *off* sequentially by applying current pulses with speed transmissions dependent off the *on-off* state of all the input channels.
To enhance the bandwidth of the optical transmission system more work has to be done in order to enlarge the number of input channels and to improve the frequency response.

ACKNOWLEDGEMENTS
This work was supported by POCTI/FIS/70843/2006.

1. Mark G. Kuzyk, Polimer Fiber Optics, Materials Physics and Applications, Taylor and Francis Group, LLC; 2007.
2. Michael Bas, Fiber Optics Handbook, Fiber, Devices and Systems for Optical Communication, Chap, 13, Mc Graw-Hill, Inc. 2002.
3. P. Louro, M. Vieira, Yu. Vygranenko, A. Fantoni, M. Fernandes, G. Lavareda, N. Carvalho, Mat. Res. Soc. Symp. Proc., 989 (2007) A12.04.
4. M. Vieira, A. Fantoni, M. Fernandes, P. Louro, G. Lavareda, C. N. Carvalho, Journal of Nanoscience and Nanotechnology, Vol 9, Corrected Prof, Available online, February, 2009.
5. Wilson J. Rugh, "Linear System Theory" (2nd Edition) Prentice-Hall Information and Systems Science Series, Serie E (1995).
6. M. Vieira, P. Louro, M. Fernandes, M. A. Vieira, A. Fantoni, M. Barata Thin Solid Films, In Press, http://dx.doi.org/10.1016/j.tsf.2009.02.096.

Mater. Res. Soc. Symp. Proc. Vol. 1153 © 2009 Materials Research Society   1153-A08-04

Enzymatic Biosensors with Integrated Thin Film a-Si:H Photodiodes

A. T. Pereira[1,2], V. Chu[1], D. M. F. Prazeres[2,3] and J.P. Conde[1,3]
[1]INESC Microsistemas e Nanotecnologias and IN- Institute of Nanoscience and
Nanotechnology, Rua Alves Redol 9, Lisbon, Portugal
[2]Centro de Engenharia Biológica e Química, IBB – Institute of Biotechnology and
Bioengineering, Instituto Superior Técnico, Av. Rovisco Pais, Lisbon, Portugal
[3]Dept. of Chemical and Biological Engineering, Instituto Superior Técnico, Av. Rovisco Pais,
Lisbon, Portugal

ABSTRACT

A microfabricated amorphous silicon photodiode is used to detect chemiluminescent and colorimetric horseradish peroxidase (HRP) enzymatic reactions. Detections limits of 1 nM and 1 pM of HRP are obtained for chemiluminescent and colorimetric measurements, respectively, with the reactions carried out in solution volume of 50 μL in polystyrene microwells. Surface-adsorbed HRP can be detected with a limit of 1 fmol.cm^{-2} by both detection methods. Immunoassays were performed using HRP-labeled antibodies and the detection of specific antibody-antigen molecular recognition is demonstrated both in the plastic well and inside a microfluidic channel. The application of the a-Si:H/HRP system is extended by coupling HRP with oxidase enzyme systems for glucose detection and a sensitivity of 0.1 mmol/L was achieved.

INTRODUCTION

Horseradish peroxidase (HRP) is one of the most widely used enzymes in analytical applications and as an enzymatic label in medical diagnostics [1]. HRP can be covalently attached to biomolecules such as DNA or other proteins such as antibodies [1]. Being capable of reducing H_2O_2 and also some organic peroxides, HRP-based biosensors are used to monitor peroxides in several industries (e.g., pharmaceutical, dairy) [1]. Coupling HRP with H_2O_2-producing oxidases results in a system sensitive to the oxidase substrate, enabling the monitoring of a wide range of analytes such as glucose, ethanol, cholesterol, lactate, uric acid, pyruvate, and amino acids [1].

Thin-film amorphous silicon (a-Si:H) photodiodes present several characteristics which make them suitable for biological detection applications: low dark conductivity, high internal quantum efficiency in the visible, well established microfabrication processes, and device-quality material can be achieved at low temperatures in a variety of substrates [2].

In this paper, an a-Si:H photodiode was used to detect the chemiluminescent and colorimetric products of the enzymatic reactions of a biolabel (HRP). Measurements were performed in polystyrene microtiter plate wells using 50 μL of solution and replicated in a miniaturized disposable polydimethylsiloxane (PDMS) microfluidic format which was aligned to the photodiode array chip allowing faster reactions and the use of smaller quantities of reagents. Integration of the microfluidic channel with the a-Si:H integrated photodiode permits miniaturization and multiplexing and a portable device design. Glucose detection is performed by combining glucose oxidase and HRP opening the way for glucose miniaturized quantification.

MATERIALS AND METHODS

Photodiode fabrication

200 μm x 200 μm a-Si:H p-i-n photodiodes were microfabricated on glass substrates [3]. Briefly, an Al bottom electrode is deposited and patterned on a glass substrate followed by p-i-n a-Si:H deposition by rf plasma enhanced chemical vapor deposition (PECVD) and patterning by photolithography and reactive ion etching. An insulating layer of SiN_x deposited by rf PECVD is used as a sidewall passivation layer and vias are opened to allow electrical contact between the top transparent conductive oxide (ITO) and the a-Si:H p-layer. A TiW/Al line is deposited to connect the ITO contact to the contact pads to reduce resistive losses. Finally, a double layer passivation of SiN_x and SiO_2 is deposited both by PECVD. Vias for pad electric contact are opened by RIE and wire-bonding is performed to a PCB plate. After wire-bonding, epoxy glue is used to protect the metallic pads from contact with liquids during the measurements. A schematic vertical cut of the device is shown in Figure 1 *(a)*.

PDMS microfluidic channel fabrication

Microchannels 20 μm high and 200 μm wide were fabricated by soft lithography in polydimethylsiloxane (PDMS) using an SU-8 mold on a Si base substrate. An Al-on-quartz shadow mask was used for the UV lithography of the SU-8-2015 mold. PDMS was prepared by mixing curing agent and base in a 1:10 weight ratio followed by vacuum de-airing. In order to cast the microchannels together with access holes a PMMA mold was fabricated using a CNC milling machine. After assembling the SU-8 mold and the PMMA plates, PDMS was injected in the reservoirs and cured for 2 hours at 60 °C. The base of the channel was a 500 μm thick flat PDMS slab that was prepared by spinning the PDMS mixture at 250 rpm for 25 s. The channel and the base were sealed using a corona discharge.

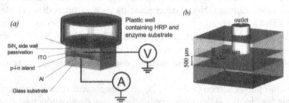

Figure 1 – *(a)* Side view of a-Si:H p-i-n photodiode. The plastic well for biochemical reactions is represented (not to scale). *(b)* Schematic side view of the microchannel, where reactions take place, aligned on top of the photodiode (not to scale).

Biosensing in the microtiter plate well

The colorimetric measurements [4] rely on the use of an enzyme, HRP (horseradish peroxidase) as the biolabel and the liquid substrate TMB Supersensitive (3,3',5,5' tetramethylbenzidine), both of which are transparent in the visible range. In the presence of HRP, TMB is converted to a blue colored product with maximum absorption at 655 nm. A clear polystyrene well (for volumes up to 200 μL) is placed at the top of the photodiode, as shown in

Figure 1 *(a)*. Before starting the assay, 5 µL of a solution containing the appropriate concentration of HRP is introduced in the well. The chromogenic reaction is initiated at t = 10 s by the addition of 45 µL of a solution containing the TMB substrate.

For chemiluminescence measurements the sensor is placed inside a light-tight metallic box. Before starting the assay, 10 µL of a solution containing a given concentration of HRP is placed inside the plastic well. At t=20 s, the chemiluminescent reaction is initiated by injecting 40 µL of a chemiluminescent substrate solution. The chemiluminescent substrate contains luminol that emits light at 425 nm during its oxidation by HRP.

Detailed description of surface immobilization and of immunoassay reactions of antibodies can be found in [4] for colorimetry and [5] for chemiluminescence. Briefly, HRP is adsorbed in the plastic well by incubation and then excess protein is washed. The measurement is initiated and the enzymatic substrate is inserted in the plastic well as described above. The first step of immunoassays was surface adsorption of the antigen (Mouse IgG or Goat IgG, both from SIGMA). Washing and surface blocking were performed using a solution containing bovine serum albumin, BSA. An antibody was used for specific molecular recognition of the adsorbed antigen (Anti-Goat IgG-HRP and Anti-Mouse IgG-HRP). The immunoassay was completed by washing with BSA and phosphate buffer 100mM pH 7.4, PB.

Glucose concentration was determined by mixing inside a plastic well 40 µL solution containing 12.5 units/µL of glucose oxidase 12.5 units/µL, 2.5 purpurogallin units of HRP per microliter (horseradish) and 40 µL of TMB. The plastic well is placed on top of the photodiode and aligned with the diode laser beam with wavelength of 658 nm. Electric current measurement is initiated at t=0 and at t = 20 s 80 µL of of D-glucose solution are added into the plastic well. Concentrations from 20 to 80 µg/mL of D-glucose standard solution were used. Absorbance was calculated by Beer-Lambert law and the values achieved 30 s after reaction is initiated were used for the calculations of the glucose concentration.

Biosensing in microfluidics

Anti-Goat IgG HRP labeled (SIGMA), 0.1 mg/mL in phosphate buffer saline (PBS) pH 7.4 was then injected into the channel with a constant flow rate (Q) of 0.5 µL/min for 10 minutes using a syringe pump. The channel is then washed by flowing PBS with Q = 5 µL/min for 3 minutes. For detection the PDMS channel was manually aligned to the top of the photodiode, as shown in Figure 1 *(b)*. A diode laser with wavelength of 658 nm is used as light source. The laser is positioned perpendicular to the photodiode and the light is reflected by a semi-transparent mirror, through the channel, and captured by the photodiode. TMB substrate was injected with Q = 2.5 µL/min for ~6 min. With the TMB formulation used the reaction product precipitates in the microchannels and accumulates as long as there is fresh substrate flow. The accumulated product absorbs the incident light, causing a decrease of the photon flux that reaches the photodiode (Figure 2*(d)*).

For immunoassays in the microfluidic format, the antigen, Goat IgG 0.1 mg/ml in CB, was injected in the microchannel with Q = 5 µL/min for 1 min and then incubated for 2 h at room temperature (RT) (Figure 2 *(a)*). In the second step the washing solution , composed of 0.1% (w/v) BSA and 0.05 % Tween 20 in PBS, was flowed with Q = 5 µL/min for 3 minutes to wash the channel and then flowed with Q = 0.5 µL/min for 10 minutes to block the surface, decreasing non-specific interactions (Figure 2 *(b)*). The secondary antibody, Anti-Mouse IgG HRP labeled or Anti-Goat IgG HRP labeled, 0.1 mg/mL in PBS was then injected in the channel

181

with Q = 0.5 µL/min for 10 minutes (Figure 2 (c)). Finally, the channel was washed by flowing PB with Q = 5 µL/min for 3 minutes. Detection was performed as described above (Figure 2 (d)).

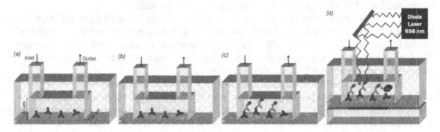

Figure 2 - (a) PDMS microchannel scheme with inlet and outlet highlighted. The first antibody is immobilized by adsorption on the channel walls. (b) In the second step the channel surface is blocked with BSA. (c) The secondary antibody is inserted in the microchannel reacting specifically with the first antibody. (d) The microchannel is aligned with the photosensor and with the light source. A photocurrent measurement is initiated and then the TMB substrate is inserted into the microchannel. In the presence of HRP TMB is converted into a blue-colored molecule that absorbs light in the red region of the spectrum.

RESULTS AND DISCUSSION

Photodiode characteristics

Square shape a-Si:H p-i-n photodiodes with sides of 100 µm and 200 µm were microfabricated. Dark I-V measurements are used to obtain the shunt resistance, 0.13 TΩ, the saturation current density, 8.65 pA/cm^2, and the ideality factor, 1.73. A photo-to-dark current ratio of ~10^4 is obtained using λ = 425 nm incident light at a photon flux of 3.13 × 10^{13} cm^{-2}s^{-1} under zero external bias. The low dark current measured allows detection of low intensity light such as that emitted by chemiluminescent reactions (minimum photon flux detected ~10^8 cm^{-2}s^{-1}). The responsivity of the photodiode is 0.2 A/W, which corresponds to an external quantum efficiency (electrons collected per incident photon) of 0.6, mostly limited by reflection losses.

Biosensing in the microtiter plate well

Figure 3 (a) shows colorimetric and chemiluminescent measurements performed with HRP in solution inside the plastic well. The colorimetric measurement is more sensitive than the chemiluminescence measurement. The two methods can be used complementarily using the same experimental set-up allowing a detection range from 10^{-6} to 10^{-12} M of HRP (Figure 3 (a)). The higher sensitivity of colorimetry over chemiluminescence arises from the accumulation of the blue-colored product inside the plastic well, allowing for signal integration. In the case of chemiluminescence the light emitted is transient.

HRP molecules adsorbed in the plastic well were detected with limits similar for both detection methods (around 10^{-9} mol of HRP (Figure 3 (b)) which corresponds to a minimum detected protein density of the order of 1 fmol cm^{-2}.

The possibility of quantifying HRP adsorbed in the plastic well with a low detection limit allows the use of this system for antibody-antigen specific recognition reaction, using HRP as an antibody label. A simple antigen-antibody direct-ELISA measurement was designed in which Goat IgG and Mouse IgG were used as antigens. The use of HRP-labeled Anti-Goat IgG and HRP-labeled Anti-Mouse IgG makes possible the distinction between specific and non-specific recognition reactions. Both detection methods were studied (Figure 3 (c)).

The combination of the enzymes glucose oxidase and HRP allows the detection of glucose. When oxidizing the glucose, glucose oxidase produces H_2O_2 which is necessary for the TMB oxidation by HRP [1]. The minimum concentration of glucose detected using this system was ~ 1 mmol/L (20 μg/mL). This value is lower than the blood glucose concentration (5-7 mmol/L), opening the possibility for the use of a-Si:H photodiode for blood glucose quantification.

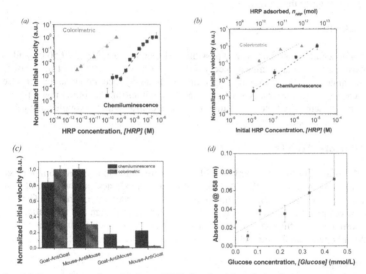

Figure 3 – (a) Comparison of sensitivity to HRP dissolved in solution between colorimetric and chemiluminescent measurements. The initial velocity of the enzymatic reaction was normalized. (b) Comparison of sensitivity between colorimetric and chemiluminescent measurements for HRP adsorbed in the plastic well walls. (c) Comparison of results obtained for immunoassay performed inside the plastic well and measured by chemiluminescence and colorimetry. (d) Absorbance levels measured for glucose detection inside the plastic well.

Biosensing in microfluidics

HRP-labeled Anti-Goat-IgG was adsorbed to the microchannel walls. Figure 4 shows that clear distinction between specific and non-specific antigen-antibody reaction was possible in the miniaturized assay.

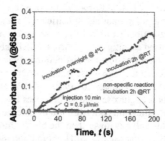

Figure 4 - Absorbance measured inside the microchannels for three different conditions of adsorption of the antigen. Non-specific reaction was tested using an antibody that does not specifically recognize the antigen.

CONCLUSIONS

Using a-Si:H photodiode it was possible to detect, in plastic wells of 50 μL and using both colorimetry and chemiluminescence: HRP in solution (range 10 fM to 1 μM); HRP adsorbed (range 10^9 to 10^{13} mol HRP); specific immunoassay recognition; and glucose sensing using bi-enzymatic system (range 0.1-0.45 mmol/L). In microfluidic format it was possible to detect by colorimetry HRP specific immunoassays reconition with the coupled microfluidic-a-Si:H photodiode system.

ACKNOWLEDGMENTS

The authors thank J. Bernardo, F. Silva, V. Soares for help in clean room processing and device packaging. Authors thank to A. Pimentel for photodiode design development to A. Gouvea for help in colorimetric measurements, and to F. Cardoso, J. Loureiro, J. Germano for help in PMMA plates design and fabrication and in PCB plate design and fabrication. This work was supported by FCT through research projects and Ph.D. grants. V. Chu acknowledges the Gulbenkian Foundation for a travel grant.

REFERENCES

[1]   A. M. Azevedo, *et al.*, "Horseradish peroxidase: a valuable tool in biotechnology," *Biotechnology Annual Review, Vol 9*, vol. 9, pp. 199-247, 2003.

[2]   R. A. Street, *Hydrogenated amorphous silicon*: Cambridge University Press, 1991.

[3]   A. C. Pimentel, *et al.*, "Detection of chemiluminescence using an amorphous silicon photodiode," *Ieee Sensors Journal*, vol. 7, pp. 415-416, Mar-Apr 2007.

[4]   A. Gouvea, *et al.*, "Colorimetric detection of molecular recognition reactions with an enzyme biolabel using a thin-film amorphous silicon photodiode on a glass substrate," *Sensors and Actuators B-Chemical*, vol. 135, pp. 102-107, Dec 2008.

[5]   A. T. Pereira, *et al.*, "Chemiluminescent Detection of Horseradish Peroxidase Using an Integrated Amorphous Silicon Thin Film Photosensor," *IEEE Sensors Journal*, Accepted for publication.

Film Growth I

Mater. Res. Soc. Symp. Proc. Vol. 1153 © 2009 Materials Research Society   1153-A09-02

Low temperature Si homoepitaxy by a reactive CVD with a SiH$_4$/F$_2$ mixture

Akihisa Minowa[1] and Michio Kondo[1,2]

[1]Tokyo Institute of Technology, Department of Innovative and Engineered Materials,
4259 Nagatsuta-cho, Midori-ku, Yokohama-shi, Kanagawa-ken, 226-8502, Japan

[2]Advanced Industrial Science and Technology, Research Center for Photovoltaics, Central 2,
1-1-1 Umezono, Tsukuba-shi, Ibaraki-ken, 305-8568, Japan

ABSTRACT

The technique for the preparation of single crystalline Si thin films termed "a reactive chemical vapor deposition (CVD)" is proposed, in which SiH$_4$ decomposes spontaneously by gas phase reactions with F$_2$ at low temperature and reduced pressure. Thus this technique provided us a variety of the films from "amorphous" to "single crystalline" in a wide range of the preparation conditions by a choice of the external parameters for the reactions. The technique was successfully applied for the homoepitaxial growth of single crystalline Si thin films. The films exhibited rather high deposition rates over 1 nm/sec.

INTRODUCTION

Single crystalline Si thin films on insulating substrates (SOI) have a variety of potential applications such as high mobility thin film transistor (TFT) and high efficiency and low cost solar cells. As an insulator, boro-silicate glass is a promising material because of the low cost and almost identical CTE (Coefficient of thermal expansion) to silicon. The SOI, however, is limited to a thin layer of the order of 0.1 micron and therefore it is necessary to develop an epitaxial growth technology to form active layers thicker than several microns at temperatures lower than a glass softening temperature.

Si molecular beam epitaxy (Si MBE) technology has been widely used for low temperature epitaxy with good crystal quality. However, the conventional Si MBE has several disadvantages for large area devices and up scaling due to the limitation of volume of silicon solid source. Another disadvantage is a trade-off between deposition temperature and deposition rate. The lower the temperature is, the lower the maximum deposition rate for good crystallinity. [1, 2]

We have attempted to establish a Reactive CVD technique for single crystalline Si thin films using silane and fluorine gas mixture where the advantage is the high quality epitaxy at low temperatures and at high deposition rate because of the exothermic reaction between source gases. [3, 4]

The purpose of this study is to develop a deposition technique of single crystalline Si thin films by a Reactive CVD method at temperatures less than 600°C utilizing gas-phase reaction (SiH$_4$, F$_2$). We investigated how the deposition condition affects a variety of the Si films directly.

EXPERIMENTAL DETAILS

The Reactive CVD apparatus is schematically shown in Fig.1. It consists of two chambers, i.e., the exchange chamber and the growth chamber based on a gas source MBE system where the back ground pressure is less than 1.0×10^{-8} Torr. A substrate holder was heated by carbon heater and the maximum temperature is 1000°C. A gas nozzle for mixing SiH_4 with F_2, was mounted right beneath the substrate holder at a distance of 50 mm.

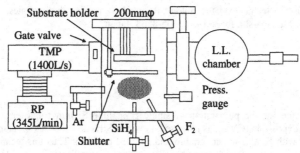
Fig.1 Schematic diagram of a Reactive CVD.

A c-Si (100) substrate was chemically cleaned with a solution (Ethanol and Acetone) and then chemically etched with a solution (10%-HF) to remove a thin oxide layer. In this study, the surface structure was monitored with reflection high-energy electron diffraction (RHEED). Even if the oxide is very thin, the oxide layer shows a halo diffraction pattern. In contrast, a clean Si surface shows a sharp diffraction pattern.

Pure SiH_4 and F_2 diluted by Ar ($F_2/Ar= 10\%$) were used as source gases, and were introduced through separate nozzles. The substrate temperature, Ts was monitored by a thermocouple attached to the other side of the substrate. Substrate-temperature was varied between 100 and 700°C, reaction-pressure 1 and 500mTorr, flow-rate between $SiH_4/F_2 = 1/1$ and 1/3, and the geometry of the substrate and the gas-outlet were optimized. SiH_4 reacted with F_2 spontaneously at the reduced pressure similar to the plasma CVD, affording radical species responsible fore the film growth. The reaction was exothermic and accompanied by an intense blue chemi-luminescence mainly from SiF* (420-480nm) [5, 6], as is shown in the photograph of Fig.2.

Fig.2 Photograph of the chemi-luminescence.

188

RESULTS AND DISCUSSIONS

Firstly, it was found that deposition rate was sensitive to the distance between the gas-outlet and the substrate and to the total pressure for four different combinations of pressures, 250 and 500 mTorr and distances, 50 and 150 mm. The deposition took place only for the combination of 500 mTorr and 50 mm, and otherwise the deposition rate was significantly lower or etching of Si wafer was observed. The deposition rate was observed as function of substrate temperature, in the range from 200 to 500°C for three different gas flow ratio. In a typical condition of $SiH_4/F_2=1/1$, d_{sub} = 50mm, and Press. = 500mTorr the deposition rate was 1.4 nm/s. A typical growth condition of the films is listed in Table.1.

Typical Condition

SiH_4/F_2	1/1
Ts (substrate)	350°C
Pressure	500mTorr
Substrate distance	50mm

Table.1 Typical preparation condition.

Secondly, as shown in Fig. 3 the deposition rate increases rapidly with substrate temperature for the flow rate $SiH_4/F_2=1/1$ up to 300°C and then slightly decreases. The maximum deposition rate was 1.7 nm/s at a substrate-temperature of 300°C, while for higher F_2 flow rate ratio, $SiH_4/F_2 = 1/2$ and 1/3, the deposition rates were 8.3×10^{-3} nm/s and etching, respectively.

The second result observed from Fig.3 is that the deposition rate vary more rapidly with substrate temperature for the flow rate $SiH_4/F_2=1/1$ case relative to that of the flow rate $SiH_4/F_2=1/2$, 1/3 case. SiH_4/F_2 of 1/1 was 1.7 nm/s at a substrate-temperature of 400°C, while for higher F_2 flow rate ratio, $SiH_4/F_2 = 1/2$ and 1/3, the deposition rates were 8.3×10^{-3} nm/s and etching, respectively.

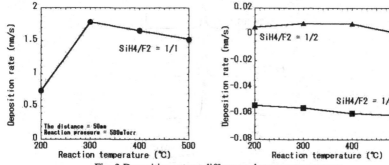

Fig. 3 Deposition rate at different substrate temperatures.

189

Thirdly, as observed in Fig. 4 the deposition rate varies more rapidly with substrate temperature for the reactive pressure 500mTorr case relative to that of the reactive pressure 250mTorr case at $SiH_4/F_2=1/1$ case.

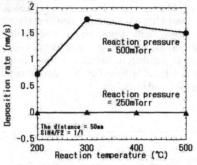

Fig. 4 Deposition rate at different pressure.

The crystallinity of the films deposited was characterized by Raman spectroscopy as shown in Fig.5. The crystallinity depends on the substrate-temperature; broad amorphous signal appears at 200°C, and microcrystalline signal at 300°C. At 400°C, a sharp signal comparable to single crystalline Si appears, while at a higher temperature of 500°C, the signal is broader because of the deterioration of crystallinity.

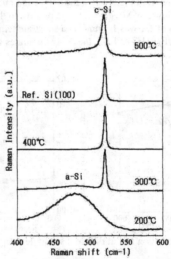

Fig. 5 Raman spectra of the films at variety substrate temperatures.

Fig.6 shows RHEED patterns for the films grown on (100) c-Si substrates under various conditions. [7] Correspondingly to the Raman results, the film deposited at 200°C is amorphous with a halo pattern. For 300°C and 400°C, a streak pattern with a 2×1 super structure was observed, showing high-quality epitaxial growth. The epitaxy quality is better for 400°C consistent with the Raman spectra. The deposition rate at 400°C was 1.4nm/s which is much higher than conventional solid source MBE. At the highest temperature of 500°C, the RHEED exhibits a spot pattern, indicating a single crystalline but a rough surface.

The presence of the optimum temperature of epitaxy around 400°C is not clear but surface passivation by hydrogen or fluorine could be responsible for the deterioration of the epitaxy at higher temperatures because of the desorption of hydrogen or fluorine at higher temperatures.

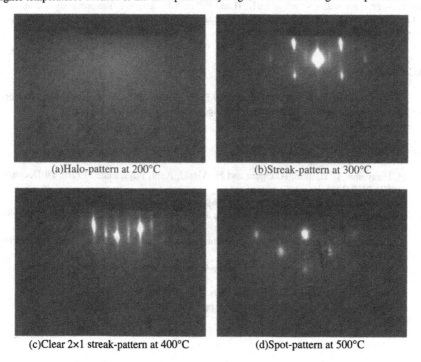

| (a)Halo-pattern at 200°C | (b)Streak-pattern at 300°C |
| (c)Clear 2×1 streak-pattern at 400°C | (d)Spot-pattern at 500°C |

Fig. 6 RHEED patterns from Si thin films on c-Si (100)

CONCLUSIONS

We have deposited homoepitaxial Si thin films by Reactive CVD based on the gas phase reactions of SiH_4 with F_2. The technique enables us to control the structure of the films intentionally with the external parameters in the chemical reaction and produces a variety of the films from "amorphous" to "microcrystalline". Experimental data suggest the direct contribution of F_2 to the chemical process on the surface.

It is noteworthy that the deposition rate of epitaxy obtained in this work is quite high, 1.7 nm/s even at 400°C. These observations are ascribed to the gas phase reaction between SiH_4 and F_2 and successive surface reactions. The SiH_4 and F_2 cause an exothermic reaction in the gaseous phases to generate radicals such as SiH_X, H and F. The SiH_X acts as a film precursor and others act as etchant. Under the conditions which radical density ratio SiH_X/F increases, therefore, the deposition rate decreases or etching occurs.

ACKNOWLEDGEMENTS

The authors would like to thank gratefully Prof. J.Hanna of Tokyo Institute of Technology, for his useful discussion on the reactive CVD mechanism with a SiH_4/F_2 mixture.

REFERENCES

1. H. Hirayama, T. Tatsumi, A. Ogura and N. Aizaki, Appl. Phys, Lett. 51 (26), 28 December (1987) 2213-2215
2. W. J. Varhue, J. L. Rogers and P.S. Andry, Appl. Phys. Lett. 68 (3) January 15 (1996)
3. J. Hanna , A. Kamo, T. Komiya, I. Shimizu and H. Kokado, J. Non-Cryst. Solids 114 (1989) 172-174
4. T. Komiya, A. Kamo, H. Kujirai, I. Shimizu and J. Hanna, Mat. Res. Soc. Symp. Proc. Vol.164. (1990)
5. C. R. Coner, G. W. Stewart, D. M. Lindsay and J. L. Gole, J. Amer. Chem. Soc., 99 (1977) 2540
6. H. U. Lee and J. P. Deneufville, Chem. Phys. Lett., 99 (1983) 394
7. T. Kitagawa, M. Kondo and A. Matsuda, Appl. Surf. Sci. 159-160 (2000) 30-34

Solar Cells

Mater. Res. Soc. Symp. Proc. Vol. 1153 © 2009 Materials Research Society 1153-A10-03

Photoelectron Spectroscopy Measurements of Valence Band Discontinuities for a-Si:H/c-Si Heterojunction Solar Cells

Tetsuya Kaneko[1] and Michio Kondo[1, 2]
[1] Innovative and Engineered Materials, Tokyo Institute of Technology, Nagatsuta-cho, Midori-ku, Yokohama, 226-8502, Japan
[2] RCPV, AIST, Umezono, Tsukuba, 305-8568, Japan

ABSTRACT

The valence band discontinuity (offset) between a-Si:H-based intrinsic thin layers and c-Si substrates was estimated using ultraviolet photoelectron spectroscopy (UPS) in combination with x-ray photoelectron spectroscopy (XPS). A core level shift measured by XPS was utilized to correct the shifts of UPS spectra after UV light illumination. Thin films of a-Si:H, a-SiO:H and a-SiC:H were prepared by plasma-enhanced chemical vapor deposition (PECVD) using SiH_4, CO_2 and CH_4 gases. The valence band offset of 0.11 eV was obtained from the a-Si:H/c-Si heterojunction, whereas 0.27 eV was obtained from the a-SiO:H/c-Si heterojunction. Moreover, the valence band offset between the c-Si and the a-SiC:H deposited with $[CH_4]$=10 SCCM and $[CH_4]$=20 SCCM were determined to be 0.25 eV and 0.36 eV, respectively. The c-Si-based heterojunction solar cells with estimated i layer in this study were fabricated, reduction of FF with increasing the valence band offset was observed. It is likely that increasing of the valence band offset contributes to the reduction of FF.

INTRODUCTION

Heterojunction solar cells consisting of thin a-Si:H layers on c-Si substrate such as "HIT" solar cells [1] have an advantage of higher potential efficiency over conventional c-Si solar cells as already proven by high open circuit voltage and its low temperature coefficient under operation. Although it has been suggested that the device characteristics are affected by the band alignment of a-Si:H and c-Si at the heterointerface, no consensus has been established for the value of the band-offset in spite of a variety of measurements using different methods [2-7]. For instance, Cuniot and Marfaing [3] measured the band discontinuities by internal photoemission; they reported the valence band discontinuities lower than 0.15 eV for a-Si:H/c-Si(p) heterojunctions. In contrast, Mimura and Hatanaka [7] estimated the valence band discontinuities (ΔE_V) of a-Si:H/ c-Si(n^+) heterojunctions using internal photoemission, they obtained ΔE_V=0.71 eV.

The purpose of the present study is to evaluate the valence band offset at heterojunctions between thin a-Si:H-based layers and c-Si substrates as employed in actual "HIT" devices using photoelectron spectroscopy.

EXPERIMENTAL DETAILS

Ultraviolet photoelectron spectroscopy (UPS) in combination with x-ray photoelectron spectroscopy (XPS) was used to determine the valence band offset. The UPS was performed with a He-I resonance line (21.22eV). Fermi level position was determined by measuring a clean Ag surface. The XPS was carried out with a non monochromatic Mg Kα radiation (400W 1253.6 eV). The base pressure of analysis chamber was 5×10^{-8} Pa. The valence band maximum (VBM) was determined by linear extrapolation of the leading edge of the UPS spectrum. The core level shift measured by XPS was utilized to correct the shifts of UPS spectra after UV light illumination. To estimate the valence band offset, the VBM of the samples after deposition was compared with the VBM of the c-Si substrate.

The samples in band offset study consisted of an undoped thin layer deposited on c-Si. To estimate the valence band offset between the undoped layer and c-Si of an actual HIT solar cell, the same deposition condition of intrinsic layer as HIT solar cell was employed. The interface quality has been confirmed by solar cell performance.

For film deposition, we used two types of plasma-enhanced chemical vapor deposition (PECVD) systems. The a-Si:H and a-SiO:H layers were deposited using the PECVD system operated at radio frequency [8]. The a-SiC:H layer was prepared using another PECVD system operated at 27.12 MHz. The deposition temperature of the a-Si:H layer was 130 °C, while that of the a-SiO:H and a-SiC:H layers were 180 °C. The undoped a-Si:H layer was deposited using a SiH_4 gas, the flow rate of which was 5 SCCM. The undoped a-SiO:H layer was also deposited using SiH_4 and CO_2 mixture gases. The gas flow ratio of the a-SiO:H layer was $[SiH_4]:[CO_2]=5:1$ SCCM. The a-SiC:H layers were prepared using SiH_4 and CH_4 mixture gases. For the a-SiC:H deposition, the flow rate of a SiH_4 was fixed to 20 SCCM, whereas the flow rate of a CH_4 was 10 and 20 SCCM. Thicknesses of all deposited layers were about 5nm estimated using SE.

Before film deposition, the c-Si substrates were rinsed by ammonium fluoride after modified RCA cleaning processes for surface passivation. In this study, the n-type c-Si(111) substrate was used. The resistivity of c-Si is 0.5 Ω cm for the a-Si:H and a-SiO:H deposition, while 1 Ω cm for the a-SiC:H deposition.

The structure of the heterojunction solar cells in this study is Ag grid/In_2O_3:Sn/p-i layers/c-Si(n)/Al. For p layer deposition, a B_2H_6 source gas was used. The solar cell characteristics were determined from solar cells having active area of 0.21 cm^2 under AM 1.5 illumination (100 mW/cm^2). The solar cell efficiencies were calculated using active area. Total area of solar cells in this study is 0.5×0.5 cm^2, where area of the Ag grid electrode is 0.04 cm^2.

RESULTS AND DISCUSSION

a-Si:H/c-Si and a-SiO:H/c-Si heterojunctions

To estimate the VBM of the c-Si substrate, the UPS measurement was performed at the H terminated c-Si surface. However, when measuring the core level spectrum of Si2s by XPS, shift of the spectrum was observed before and after the UPS measurement. Fig. 1 shows the shift of the XPS spectrum of Si2s. The shift of Si2s spectrum of -0.11 eV was observed. Assuming the binding energy of Si2s from vacuum level is constant, the shift of XPS spectrum may be attributed to the Fermi level shift due to the damage of UV light illumination.

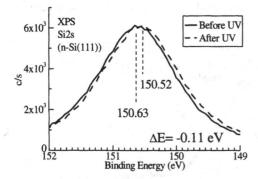

Figure 1. The shift of the XPS spectrum of Si2s before and after the UPS measurement.

Fig. 2 shows the energy diagram of estimation of the valence band offset in this study. Since a sample and an analyzer were connected to ground, an energy standard was Fermi level. Therefore, the shift of the VBM and the XPS spectrum of Si2s are the same. To obtain the unaffected VBM of samples, we corrected the energy position of UPS spectra using the shift value of Si2s XPS spectra.

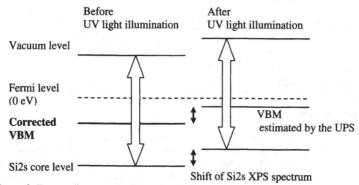

Figure 2. Energy diagram of valence band offset estimation

The UPS spectra of a-Si:H/c-Si and a-SiO:H/c-Si heterojunctions estimated by the method as described above are shown in Fig. 3. The VBM of the H terminated c-Si surface was estimated to be 1.09 eV. Similarly, the VBM of the a-Si:H and a-SiO:H were estimated to be 1.20 eV and 1.36 eV, respectively. The valence band offset of 0.11 eV was obtained from the a-Si:H/c-Si heterojunction, whereas ΔE_V=0.27 eV was obtained from the a-SiO:H/c-Si heterojunction. If the optical gap of a-Si:H (1.79 eV) estimated by SE and a band gap of a c-Si (1.12 eV) are used for evaluation, the conduction band offset (ΔE_C) of the a-Si:H/c-Si heterojunction is calculated to be 0.56 eV. In addition, if the optical gap of a-SiO:H (1.83 eV) is used, ΔE_C of the a-SiO:H/c-Si heterojunction is calculated to be 0.44 eV. These mean that the major band offset occurs in the conduction band.

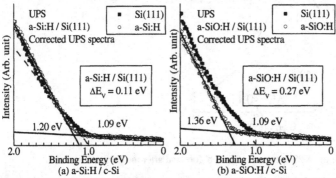

Figure 3. The valence band offset estimation of a-Si:H/c-Si and a-SiO:H/c-Si heterojunctions. The oxygen concentration in the a-SiO:H is 4 at. %.

Characteristics of a-Si:H/c-Si and a-SiO:H/c-Si heterojunction solar cells fabricated in this study, reported previously[8], was excerpted in Table I. Slightly degradation of FF and J_{SC} was observed when incorporating the a-SiO:H i layer.

Table I. Characteristics of a-Si:H/c-Si and a-SiO:H/c-Si heterojunction solar cells

Structure of p/i layers	V_{OC} (mV)	J_{SC} (mA/cm^2)	FF	Efficiency (%)
a-SiO:H(p)/a-Si:H(i)	624.7	33.55	0.771	16.16
a-SiO:H(p)/a-SiO:H(i)	624.7	33.35	0.768	16.00

a-SiC:H/c-Si heterojunctions

Figure 4 shows UPS spectra of a-SiC:H/c-Si heterojunctions in this study. For a-SiC:H deposition, another system was used instead of system used for a-Si:H and a-SiO:H deposition. In Fig.4, the shift of all spectra due to UV light illumination has been corrected. The VBM of the c-Si having resistivity of 1Ω cm was estimated to be 1.09 eV. The VBM of 1.34 eV was obtained from the a-SiC:H deposited by the gas flow rate of [CH$_4$]=10 SCCM, while 1.45 eV was obtained from the a-SiC:H deposited with [CH$_4$]=20 SCCM. Thus, the valence band offset of the a-SiC:H deposited with [CH$_4$]=10 SCCM and [CH$_4$]=20 SCCM were determined to be 0.25 eV and 0.36 eV, respectively. In addition, the carbon concentrations in the a-SiC:H layers were estimated to be 8 at. % ([CH$_4$]=10 SCCM) and 15 at. % ([CH$_4$]=20 SCCM) using XPS.

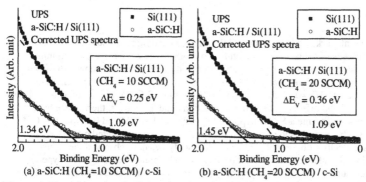

Figure 4. The valence band offset estimation of a-SiC:H/c-Si heterojunctions. For a-SiC:H deposition, we employed the CH₄ flow rate of 10 and 20 SCCM.

Figure 5 shows the characteristics of heterojunction solar cells, plotted as a function of CH₄ gas flow rate at the a-SiC:H i layer. For solar cell fabrication, a-Si:H p layer was employed. As evidenced from Fig. 5, V_{OC} and FF decrease with increasing the flow rate of CH₄, whereas J_{SC} slightly increases by incorporating the a-SiC:H i layer. Especially, large reduction of FF due to increasing series resistance is remarkable in spite of using the quite thin i layer of 5nm. As confirmed from Table I, the reduction of FF is also observed in the heterojunction solar cell incorporating the a-SiO:H i layer. From these results, it is likely that increasing of the valence band offset contributes to reduction of FF.

For the solar cell application using the n-type substrate, smaller valence band offset is preferable to minimize reduction of FF. Thus, the result indicates that a-Si:H is a good material for the fabrication of n-type c-Si-based heterojunction solar cells.

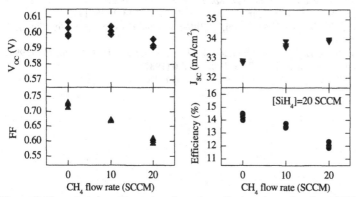

Figure 5. Characteristics of heterojunction solar cell, plotted as a function of CH₄ gas flow rate at the a-SiC:H i layer.

CONCLUSIONS

In the present work, we have evaluated the valence band offset at the heterojunction as employed in solar cells of actual "HIT" structure. To estimate the valence band maximum, ultraviolet photoelectron spectroscopy in combination with x-ray photoelectron spectroscopy was used. As a result, the valence band offset of 0.11 eV was obtained from the a-Si:H/c-Si heterojunction, whereas ΔE_V=0.27 eV was obtained from the a-SiO:H/c-Si heterojunction. These mean that the major band offset occurs in the conduction band. Moreover, the valence band offset between the c-Si and the a-SiC:H deposited by [CH_4]=10 SCCM and [CH_4]=20 SCCM were determined to be 0.25 eV and 0.36 eV, respectively. We have fabricated c-Si-based heterojunction solar cells with estimated i layer in this study, we have observed the reduction of FF with increasing the valence band offset. For the solar cell application using the n-type substrate, smaller valence band offset is probably preferable to minimize the reduction of FF. Based on above results, we expected that the conversion efficiency of n-type c-Si-based heterojunction solar cells can be improved by incorporating a material with smaller valence band offset compared with a-Si:H.

ACKNOWLEDGMENTS

The authors would like to thank Dr. H. Fujiwara and Dr. T. Matsui of AIST for support of sample preparation.

REFERENCES

1. Y. Tsunomura, Y. Yoshimine, M. Taguchi, T. Baba, T. Kinoshita, H. Kanno, H. Sakata, E. Maruyama and M. Tanaka, *Sol. Energy Mater. Sol. Cells* **93**, 670 (2009)
2. I. Sakata, M. Yamanaka and R Shimokawa, *Jpn. J. Appl. Phys.* **44**, 7332 (2005)
3. M. Cuniot and Y. Marfaing, *J. Non-Cryst. Solids* **77 & 78**, 987 (1985)
4. Chris G. Van de Walle and L. H. Yang, *J. Vac. Sci. Technol. B* **13**, 1635 (1995)
5. M. Sebastiani, L. Di Gaspare, G. Capellini, C. Bittencourt and F. Evangelisti, *Phys. Rev. Lett.* **75**, 3352 (1995)
6. H. Eschrich, J. Bruns, L. Elstner C. Swiatkowski, *J. Non-Cryst. Solids* **164 - 166**, 717 (1993)
7. H. Mimura and Y. Hatanaka, *Appl. Phys. Lett.* **50**, 326 (1987)
8. H. Fujiwara, T. Kaneko, and M. Kondo, *Appl. Phys. Lett.* **91**, 133508 (2007)

Mater. Res. Soc. Symp. Proc. Vol. 1153 © 2009 Materials Research Society   1153-A10-04

Device Physics of Heterojunction With Intrinsic Thin Layer (HIT) Solar Cells

Ana Kanevce[1,2] and Wyatt K. Metzger[1]
[1]National Renewable Energy Laboratory, Golden, CO 80401
[2]Colorado State University, Fort Collins, CO 80523

ABSTRACT

Heterojunction with intrinsic thin layer (HIT) solar cells have achieved conversion efficiencies higher than 22%. Yet, many questions concerning the device physics governing these cells remain unanswered. We use numerical modeling to analyze the role of a-Si:H layers and tunneling on cell performance. For cells with n-type c-Si (n-HIT cells), incorporating the indium-tin-oxide (ITO) as an n-type semiconductor creates an $n^+/p/n$ structure. Most device simulations do not work with this structure. Our modeling indicates that the $n^+/p/n$ device often produces irregular S-shaped current density–voltage (J-V) curves, which have been observed experimentally but were not previously understood. However, if tunneling is included, there are specific conditions where the $n^+/p/n$ structure performs as a robust solar cell with efficiencies exceeding 20%. Additional analysis examines voltage-dependent carrier collection in n-HIT cells, as well as material and interface properties that limit fill factor.

In p-HIT cells, modeling the ITO layer as a semiconductor, rather than as a metallic contact, significantly reduces the impact of a-Si:H layer parameters on device performance. In p-HIT cells, the a-Si:H layers adjacent to the ITO layer play the role of a buffer that reduces interface recombination at the a-Si:H/c-Si interface and prevents tunneling of electrons from the ITO layer to the c-Si absorber. Tunneling through the a-Si:H layers adjacent to the back contact is important to attain regular J-V curves.

I. INTRODUCTION

HIT solar cells present a promising alternative to c-Si cells made by diffusion. Their conversion efficiencies are only a couple of percent lower than that of traditional homojunction c-Si cells [1,2], and the deposition processes stay at temperatures below 200°C [3,4]. A schematic of HIT cells with n- and p-type absorbers is shown in Fig. 1. HIT cells are formed by depositing a thin hydrogenated amorphous silicon (a-Si:H) emitter on a crystalline silicon (c-Si) absorber. Inclusion of a thin intrinsic layer between them has been shown to improve the performance [5]. A similar c-Si/a-Si:H heterostructure is formed at the back to provide a back-surface field. A highly conductive transparent conductive oxide (TCO) layer is on the top of the cell. In general, transparent conductive oxides used in solar cells are n-type, and this can create a secondary n/p junction in solar cells with n-type absorbers. A HIT cell with an n-type absorber is one example. The TCO used in this cell is indium-tin-oxide, which has a large bandgap (3.7 eV) and is always highly n-doped. The ITO creates a junction with the adjacent p-doped a-Si:H layer, in addition to the main a-Si:H/c-Si heterojunction. We included ITO in the structure; this significantly varies the device physics and how material parameters impact device performance. In addition, we incorporated interband and intraband tunneling as possible transport mechanisms and investigated how they affect cell performance. We also analyze the importance of the a-Si:H layer in p-HIT cells once the ITO is included.

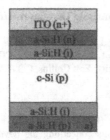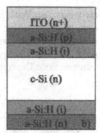

Figure 1. Schematics of HIT cells with a) *p*-type and b) *n*-type c-Si wafers. Layer thicknesses are not to scale.

II. MODEL

We use Sentaurus device simulation software [6]. The software solves Poisson and continuity equations as a function of position. The transport across the interfaces includes thermionic emission, tunneling, and recombination at the interface defects. The layer thicknesses, bandgaps, and carrier densities used in the model are shown in Table 1.

Table 1: Some Parameters Used in the Simulations

	c-Si	a-Si:H (doped)	a-Si:H (intrinsic)	ITO
Thickness	250 μm	5 nm	5 nm	70 nm
E_g [eV]	1.12	1.67	1.67	3.7
n, p [cm^{-3}]	10^{16}	$n = 10^{19}/$ $p = 3 \times 10^{19}$	10^{13}	10^{20}
Gaussian defect density [cm^{-3}]	10^{13}	10^{18}	10^{16}	10^{15}

The defects in c-Si and ITO are located in the middle of the bandgap. In addition to the Gaussian defects, the amorphous layers have exponential band tails located close to the band edges. Due to the crucial role of the hole density in the *p*-type emitter in *n*-HIT cells, it is doped slightly higher (3×10^{19} cm^{-3}) than the a-Si:H layer in *p*-HIT cells (10^{19} cm^{-3}). More details on electronic and optical parameters for the different layers used in the simulations are given in Ref. [7].

III. RESULTS AND DISCUSSION

A. HIT cells with *n*-type wafers

Figure 2a shows a simulated band diagram of a HIT cell with an *n*-type wafer, and Fig. 2b shows light and dark current density–voltage (J-V) curve using the baseline parameters. Neither the band diagram nor the light curve shows a typical diode behavior.

In addition to the *p/n* diode formed between the a-Si:H(p) and c-Si(n) layers, the ITO and a-Si:H in *n*-HIT cells form another n^+/p diode. The two diodes have opposite polarities (inset in

Fig. 2b). The second diode, as well as the large valence-band offset at the a-Si:H(p)/ITO interface, impede the transport of the light-generated holes. As a result, the light J-V curve is significantly distorted. The inclusion of ITO does not affect the forward current transport, and thus, the dark curve is well behaved (dashed curve in Fig. 2b).

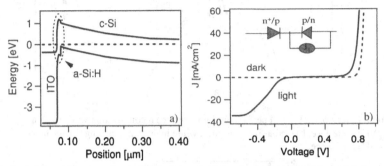

Figure 2. a) A band diagram of an n-HIT cell emphasizing the junction; b) light and dark current density–voltage curve of an n-HIT cell without an additional transport mechanism.

For this structure to work as a single diode, the J-V characteristic of the n^+/p diode has to have a slope at zero bias. We investigated two mechanisms that can introduce such a slope: recombination and band-to-band tunneling at the a-Si:H/ITO interface.

The experimentally observed distortions in light J-V curves for n-HIT cells usually have been explained by the valence-band offset (VBO) at a-Si:H/c-Si interface. To eliminate this possibility, in Figs. 3 and 4 we have reduced the VBO to 0.1 eV and concentrated on the a-Si:H/ITO interface transport.

Figure 3a shows simulated J-V curves for a HIT cell with an n-type wafer when recombination velocity at the a-Si:H(p)/ITO interface is varied for five orders of magnitude. The dark current is unaffected by the recombination at this interface. The fill factor (FF) increases for higher recombination velocities. Even at very high interface recombination velocities, the FF does not become higher than 57%, and the efficiency does not exceed 8%. This is obviously far from the excellent performances achieved experimentally. Therefore, a stronger leakage mechanism is needed. Tunneling of holes from the a-Si:H valence band to the ITO conduction band is a possible mechanism.

The solid curve in Fig. 3b shows a simulated J-V curve when the holes from the a-Si:H valence band are allowed to tunnel to the ITO conduction band. Once tunneling is allowed across the a-Si:H/ITO barrier, the calculated conversion efficiency for the device matches the high efficiencies obtained by Sanyo. Increased recombination at the a-Si:H/c-Si interface (dashed line in Fig. 3b) will increase the forward current, and thus, reduce the open-circuit voltage (V_{oc}) in the light curve and the turn-up voltage in the dark.

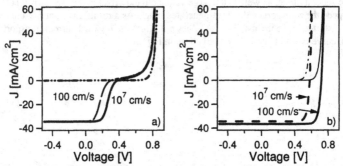

Figure 3. Impact of *n*-HIT cell performance on interface recombination velocity at: a) a-Si:H(p)/ITO interface, b) c-Si/a-Si:H interface, while including tunneling across the a-Si:H(p)/ITO interface.

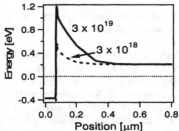

Figure 4. Changes in the conduction band for a hole density decrease from 3×10^{19} to 3×10^{18} cm^{-3} in the a-Si:H region.

In *n*-HIT cells, a-Si:H is the only source of holes between two *n*-doped materials. Therefore, a-Si:H parameters are important for the functioning of *n*-HIT cells. To ensure an exponential J-V curve, a minimum amount of holes needs to be provided. Figure 4 shows the impact of the carrier density in the a-Si:H emitter on the *n*-HIT cell band diagram. An order of magnitude variation of the hole density in a-Si:H lowers the conduction-band barrier by 0.7 eV. As the hole density is reduced, the V_{oc} plummets and the J-V characteristics transition to that of a resistor. If ITO is not included in the structure, the change in the band barrier for the same doping variation is less than 0.1 eV, and the V_{oc} changes by 30 mV.

B. HIT cells with *p*-type wafers

A simulated band diagram of a *p*-HIT cell is shown in Fig. 5a. Inclusion of the ITO in the structure decreases the importance of a-Si:H properties. The ITO creates a *p/n* junction with the *p*-doped wafer layer.

To determine the role of the a-Si:H layers on the junction formation, we varied the electron density and defect density in the a-Si:H layers. The results are plotted in Fig. 5b. A variation of five orders of magnitude in a-Si:H electron density increased the conversion efficiency by only 0.3%. The defect density in the intrinsic and doped a-Si:H layer did not affect the device performance, unless it is 5 x 10^{18} cm^{-3} or higher. In this severe case, the quality of the intrinsic layer has more impact on the device performance. The minor impact of a-Si:H properties on the device performance leads to a conclusion that the very thin a-Si:H layers are not major players in the junction formation; rather, they serve as a buffer and provide interface passivation.

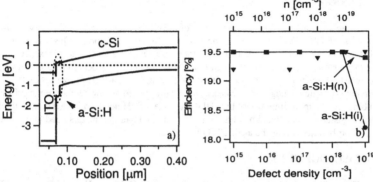

Figure 5. a) Band diagram of a *p*-HIT cell emphasizing the junction. b) Conversion efficiency of a *p*-HIT cell as a function of electron density in the *n*-doped a-Si:H (triangles) and defect density in the intrinsic a-Si:H layer (circles) and doped a-Si:H layer (squares).

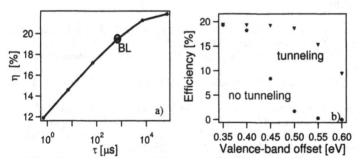

Figure 6. *p*-HIT cell dependence on a) c-Si wafer quality, and b) valence-band offset at the back contact.

Figure 6a shows the dependence of the *p*-HIT cell conversion efficiency on the minority-carrier lifetime in the c-Si wafer. Variation of lifetime from 1 μs to 10 ms changed the V_{oc} by 190 mV and the efficiency by 9%. Recombination at the a-Si:H/c-Si interface has a similar effect [7].

The valence-band offset between a-Si:H and c-Si at the back contact can impose an obstacle for the majority holes to reach the back contact. We have investigated the impact of the valence-band offset on the efficiency. The results are shown in Fig. 6b. If tunneling is not allowed through the c-Si/a-Si:H valence band spike (circles in Fig. 6b), then with our choice of parameters, the efficiency starts to decrease steeply for VBO higher than 0.4 eV. Thus, we conclude that a transport mechanism other than thermionic emission is necessary. If the holes are allowed to tunnel through (triangles in Fig. 6b), the significant decrease in efficiency does not appear until VBO > 0.5 eV.

CONCLUSIONS

There are major differences in the role of a-Si:H layers between HIT cells with n- and p-type wafers, and the ITO inclusion enhances those differences. If the c-Si layer is n- or p-type, the ITO/a-Si:H/c-Si layers in essence form either a $n^+/p/n$ or $n^+/n/p$ structure, respectively. Tunneling across the c-Si/a-Si:H and a-Si:H/ITO heterointerface is beneficial and can help avoid an irregularly shaped J-V curve. In p-HIT cells, the impact of a-Si:H properties on the conversion efficiency is negligible, whereas the quality of the c-Si wafer and the interface recombination are more important. In n-HIT cells, the a-Si:H needs to have a certain thickness and hole density to create band banding. In addition, the layer parameters need to provide high tunneling probability across the a-Si:H/ITO interface.

ACKNOWLEDGEMENTS

The authors would like to thank Richard Crandall, Qi Wang, Howard Branz, and David Young of the National Renewable Energy Laboratory for valuable discussion and input. Work was done under NREL Contract No. DE-AC36-08GO28308.

REFERENCES

1. S. Taira, Y. Yoshimine, T. Baba, M. Taguchi, T. Kinoshita, H. Sakata, E. Maruyama, and M. Tanaka, in 22nd EU PVSEC (Milan, 2007), p. 932–936.
2. M.A. Green, K. Emery, Y. Hisikawa, and W. Warta, *Prog. Photovoltaics* **15**, 425–430 (2007).
3. M. Taguchi, A. Terakawa, E. Maruyama, and M. Tanaka, *Prog. Photovoltaics* **13**, 481–488 (2005).
4. M. Taguchi, K. Kawamoto, S. Tsuge, T. Baba, H. Sakata, M. Morizane, K. Uchihashi, N. Nakamura, S. Kiyama, and O. Oota, *Prog. Photovoltaics* **8**, 503–513 (2000).
5. M. Tanaka, M. Taguchi, T. Matsuyama, T. Sawada, S. Tsuda, S. Nakano, H. Hanafusa, and Y. Kuwano, *Jpn. J. Appl. Phys., Part 1* **31**, 3518–3522 (1992).
6. Synopsys, Zurich, Switzerland, *TCAD DEVICE Manual* (2006), www.synopsys.com
7. A. Kanevce and W. K. Metzger, *J. Appl. Phys.* **105** 094507 (2009).

Mater. Res. Soc. Symp. Proc. Vol. 1153 © 2009 Materials Research Society   1153-A10-05

High efficiency amorphous and nanocrystalline silicon based multi-junction solar cells deposited at high rates on textured Ag/ZnO back reflectors

Guozhen Yue, Laura Sivec, Baojie Yan, Jeffrey Yang, and Subhendu Guha
United Solar Ovonic LLC, 1100 West Maple Road, Troy, MI 48084

ABSTRACT

We report our recent progress on nc-Si:H single-junction and a-Si:H/nc-Si:H/nc-Si:H triple-junction cells made by a modified very-high-frequency (MVHF) technique at deposition rates of 10-15 Å/s. First, we studied the effect of substrate texture on the nc-Si:H single-junction solar cell performance. We found that nc-Si:H single-junction cells made on bare stainless steel (SS) have a good fill factor (FF) of ~0.73, while it decreased to ~0.65 when the cells were deposited on textured Ag/ZnO back reflectors. The open-circuit voltage (V_{oc}) also decreased. We used dark current-voltage (J-V), Raman, and X-ray diffraction (XRD) measurements to characterize the material properties. The dark J-V measurement showed that the reverse saturated current was increased by a factor of ~30 when a textured Ag/ZnO back reflector was used. Raman results revealed that the nc-Si:H intrinsic layers in the two solar cells have similar crystallinity. However, they showed a different crystallographic orientation as indicated in XRD patterns. The material grown on Ag/ZnO has more random orientation than that on SS. These experimental results suggested that the deterioration of FF in nc-Si:H solar cells on textured Ag/ZnO was caused by poor nc-Si:H quality. Based on this study, we have improved our Ag/ZnO back reflector and the quality of nc-Si:H component cells and achieved an initial and stable active-area efficiencies of 13.4% and 12.1%, respectively, in an a-Si:H/nc-Si:H/nc-Si:H triple-junction cell.

INTRODUCTION

Hydrogenated nanocrystalline silicon (nc-Si:H), used as the intrinsic layer in the middle and bottom cells of multi-junction structures, has been intensively studied because of its low light-induced degradation and high long-wavelength response. However, the low absorption coefficient due to its indirect optical transition in crystalline phase requires a thick intrinsic layer to obtain high short-circuit current densities (J_{sc}). Therefore, high deposition rates are required to use nc-Si:H in large volume production. Moreover, an effective light trapping from textured back reflectors is also necessary to achieve high J_{sc}. It has been found that nc-Si:H materials deposited on textured surfaces usually have high defect densities due to crystallite collisions [1], which cause a reduction in fill factor (FF) of solar cells. An optimized surface morphology of back reflectors is needed to minimize this reduction. Bailat et al. [2] reported that by a surface treatment of ZnO coated glass substrates, which turned typical V-shaped into U-shaped valleys of the texture surface, they significantly improved the pin nc-Si:H cell efficiency. Li et al. [3] reported that strip-like structural defects are observed in nc-Si:H silicon layers by a cross-sectional transmission microscopy study, and a correlation between the opening angles of the valleys on the ZnO surface and the appearance of these strips was found. Further, they claimed that an opening angle larger than around $110°$ is desirable for the nc-Si:H growth.

In this paper, we report our findings of a systematic study on the cause of differences between *nip* solar cell performance on specular SS and Ag/ZnO coated SS substrates with different textures, using electrical, Raman, and XRD measurements. Based on this study, we have improved our Ag/ZnO back reflector and cell efficiency for a-Si:H/nc-Si:H/nc-Si:H triple-junction cells deposited at ~10 Å/s.

EXPERIMENTAL

A multi-chamber system consisting of three RF chambers and one MVHF chamber was used to deposit nc-Si:H single and multi-junction solar cells. The *nip* structure was deposited on specular and Ag/ZnO coated SS substrates. The a-Si:H and nc-Si:H intrinsic and buffer layers were prepared using the MVHF glow discharge at deposition rates of 10-15 Å/s, and the doped layers were deposited with RF glow discharge. The thickness of the nc-Si:H intrinsic layers is 1.5-2.0 μm. The solar cells were completed with indium-tin-oxide dots having an active-area of 0.25 cm^2 on the top *p* layer. Current density versus voltage, J-V, characteristics were measured under an AM1.5 solar simulator at 25 °C. Quantum efficiency (QE) was measured from 300 to 1200 nm at room temperature to calibrate the short circuit current (J_{sc}). Dark J-V characteristics of the cells were measured in a vacuum chamber using a programmable multi-meter at a controlled temperature. Raman and XRD measurements were conducted directly on the solar cells. The laser wavelength for Raman measurements is 633 nm, which ensures that the measurement detects the entire material along the growth direction. Atomic force microscopy (AFM) was used to characterize the morphology of the substrate surface.

RESULTS AND DISCUSSION

To compare the effect of the substrate texture on solar cell performance, we first made a series of nc-Si:H cells on SS and Ag/ZnO coated SS substrates with different textures. An identical recipe was used for all the nc-Si:H cell depositions. The texture of the Ag/ZnO was controlled. Root mean square (RMS) values from AFM measurements were used to characterize the surface roughness. Table I lists J-V characteristics of nc-Si:H single-junction cells made on SS and Ag/ZnO coated SS with different RMS values. From the table, a clear trend was observed that the V_{oc} and FF decrease while J_{sc} increases monotonically with increasing the substrate texture. A FF as high as 0.73 is achieved for the cell made on the bare SS with an RMS value of 6.0 nm, whereas it decreases to 0.54 when the RMS increases to 73 nm. In the meantime, the V_{oc} decreases from 0.56 to 0.48 V, and the J_{sc} increases from 15 to 26 mA/cm^2. These results are consistent with the reports from other researchers [2, 3].

Table I. J-V characteristics of nc-Si:H single-junction cells made on SS and Ag/ZnO coated SS with different textures. A 610 nm long wavelength cut-on filter was used to evaluate the red light response in quantum efficiency measurements.

Run No.	V_{oc} (V)	FF	Q (>610 nm) (mA/cm^2)	P_{max} (mW/cm^2)	RMS (nm)	Substrate
17108	0.561	0.729	14.85(4.09)	6.07	6	SS
17154	0.543	0.653	21.76(10.70)	7.72	20	ZnO/Ag/SS
17156	0.537	0.646	23.82(12.92)	8.26	43	ZnO/Ag/SS
17153	0.484	0.544	25.86(14.78)	6.81	74	ZnO/Ag/SS

Two major factors have been indentified as possible reasons for the reduced FF and V_{oc} of the cell on Ag/ZnO substrate. First, the light-intensity dependence could be a potential cause. Due to the light trapping effect, the light will undergo many paths, increasing the light absorption in nc-Si:H cells. This leads to a large quasi-Fermi level splitting, and increases the recombination probability of photo-excited carriers. In addition, the scattered light from the textured ZnO goes through the n layer first, which may potentially reduce FF due to the difference of electron and hole mobilities. Second, nc-Si:H materials deposited on textured surfaces may have a high defect density due to crystallite collisions [1]. To clarify the first issue, we made a solar cell on a Cr coated Ag/ZnO substrate. Table II lists the light J-V characteristics of the three solar cells made with the same recipe but on different substrates, SS, ZnO/Ag/SS, and Cr/ZnO/Ag/SS, respectively. The Cr is very thin, ~20 nm, and has a nearly conformal surface texture with ZnO. Because of the low light reflectivity of Cr, most of the light goes through only one path, similar to the cell on SS. As seen in the table, the J_{sc} is similar, however, the V_{oc} and FF are still much smaller than the cell made on SS. Therefore, the first possibility can be ruled out.

Table II. J-V characteristics of nc-Si:H single-junction cells made on SS, ZnO/Ag/SS, and Cr/ZnO/Ag/SS substrates. A 610 nm long wavelength cut-on filter was used to evaluate the red light response in quantum efficiency measurements.

Run No.	V_{oc} (V)	FF	Q (>610nm) (mA/cm^2)	P_{max} (mW/cm^2)	Substrate
17108	0.561	0.729	14.85(4.09)	6.07	SS
17110	0.541	0.600	23.18(12.54)	7.52	ZnO/Ag/SS
17109	0.514	0.664	14.78(4.07)	5.04	Cr/ZnO/Ag/SS

Dark J-V measurement is a very useful technique to characterize carrier recombination information in a-Si:H and nc-Si:H solar cells. The carriers recombine through defects, and the injected current is directly associated with the properties of the defect density [4, 5]. To check the effect of the substrate texture on the material properties, we carried out dark J-V measurement on these three cells. As shown in Fig. 1, the dark J-V curves show typical diode characteristics as described by $J=J_0[exp(qV/nkT)-1]$, where J_0 is the reverse saturation current density, q the unit charge, T the measurement temperature, k the Boltzmann constant, and n the diode quality factor. The dark J-V curves for the cells on Ag/ZnO and Cr coated Ag/ZnO substrates completely overlapped, indicating a similar material property, both having a larger dark current than the cell on SS. By fitting the data, we observed that n is around 1.25 for the cell on SS, and increases to 1.52 for cells deposited on the textured substrates. The reverse saturation current density J_0 increases 30 times from 3.58×10^{-8} to 9.39×10^{-7} A/cm^2, implying an enhanced recombination in the material on the textured substrates.

One may notice that the V_{oc} of the two cells deposited on ZnO/Ag/SS and Cr/ZnO/Ag/SS substrates are also different although they have similar dark J-V characteristics. To clarify this, we measured the light intensity dependence of the solar cell performance. The light intensity was varied by neutral filters with different transmissions. The result shows that J_{sc} has a very good linear relationship with the light intensity in the range of 3 to 100 mW/cm^2. In Figure 1 (b), we plotted V_{oc} as a function of J_{sc}. One can see that the data fit very well with typical logarithm relationship of $V_{oc} \sim kT \, ln(J_{sc}/J_0)/q$. The lines of V_{oc} versus J_{sc} from cells made on ZnO/Ag/SS and Cr/ZnO/Ag/SS substrates are almost identical and are 50-70 mV smaller than the cell on SS

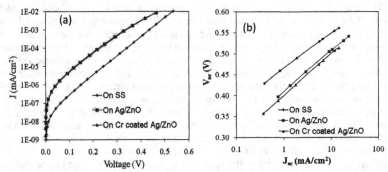

Figure 1. (a) Dark J-V characteristics of three nc-Si:H solar cells made on SS, Ag/ZnO, and Ag/ZnO/Cr coated SS, respectively. (b) V_{oc} of three solar cells as a function of their J_{sc} under different illuminated light intensities.

for the same J_{sc} values. This result is in agreement with the dark J-V measurement. The large difference of V_{oc} measured under AM1.5 illumination is mainly from the difference of photocurrent density. From Table II, we also find that the FF of the cell on the ZnO/Ag/SS substrate is smaller than that on the Cr/ZnO/Ag/SS substrate. Two factors could contribute to the observation. First, the series resistance is different, and second, the recombination centers could be different when the cells are subjected to different illuminations.

The low V_{oc} observed in the cells on the textured substrates could also be caused by the high volume fraction of the crystallinity due to the substrate dependence of the nc-Si:H growth [6]. To check the difference of the material structure, we carried out Raman measurements on two solar cells deposited on SS and Ag/ZnO coated SS substrates. A red laser with a wavelength of 633 nm was used for the measurements to ensure revealing information from the bulk of the intrinsic layer. Figure 2 (a) shows the Raman spectra of two solar cells. For a clear comparison, we normalized the two spectra based on the highest point of the c-Si peak. One can see that the two spectra are very similar except for a slightly higher amorphous component located at ~

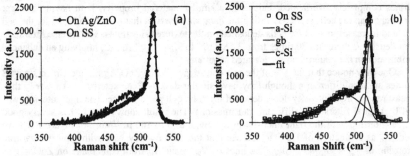

Figure 2 (a). Raman spectra of two nc-Si:H cells made on SS and Ag/ZnO substrates. (b) The deconvolution of Raman spectra of the cell on SS.

Table III. Deconvolution results of Raman spectra of two nc-Si:H solar cells.

Spectra	Components	Area	Peak (cm⁻¹)	FWHM (cm⁻¹)	Height	f_c (%)
on Ag/ZnO	a-Si	33883	483.9	70.7	449.6	46.5
	gb	11954	511.7	22.2	518.5	
	c-Si	17469	520.0	9.3	1772.3	
on SS	a-Si	44035	480.1	68.8	599.9	40.1
	gb	11954	511.7	18.5	593	
	c-Si	17547	520.0	9.3	1719	

480 cm⁻¹ for the cell on SS. We deconvoluted the Raman spectra using the Gaussian function into three components of amorphous (~480 cm⁻¹), grain boundary (gb) (~500 cm⁻¹), and crystalline (~520 cm⁻¹) modes as shown in Fig. 2 (b). The crystalline volume fraction (f_c) is determined using a formula of $f_c=(A_c+A_{gb})/[A_a+A_c+A_{gb}]$, where A_c, A_{gb}, and A_a represent the integrated areas of amorphous, grain boundary, and crystalline peaks, respectively. Table III lists the deconvolution results from Raman spectra of two nc-Si:H solar cells. Two important phenomena are observed. First, the cell on Ag/ZnO has crystalline volume fraction of 46%, compared to 40% on SS. Such difference could be due to the accuracy of the Raman measurements or slightly small difference in the material structure. From this result, we would conclude that the crystallinity for these two cells is similar. Second, the peak positions and the full width at half maximum (FWHM) of a-Si and c-Si components for two cells are similar. However, the grain boundary is different. The cell on Ag/ZnO has a broader grain-boundary peak as shown by the FWHM, indicating a more disordered gb structure than that on SS.

Although the two cells made on SS and Ag/ZnO substrates show similar crystallinity, we find a different crystallographic orientation in the crystalline phase. Figure 3 shows the XRD patterns for the two cells. Clearly, they all show three diffraction peaks of (111) at 28.3°, (220) at 47.2°, and (311) at 56.0°. However, the cell on SS has a strong (220) preferential orientation, while the cell on Ag/ZnO shows a much less preferential orientation. The intensity ratio of (220) to (111) is ~2 for the cell on SS, and ~1.2 for the cell on Ag/ZnO. We estimated the grain size d using the Scherrer equation $d = 0.9\ \lambda/(\beta\ cos\ \theta)$, where λ is X-ray wavelength (0.154 nm) and β is the FWHM of (220) the peak. The grain size is around 24 nm for the cell on SS, and 15 nm for the cell on Ag/ZnO. These results indicate that the textured surface deteriorates film crystallinity, including the random orientation and small grain size, which can be attributed to crystallite collisions as reported by Nasuno et al. [1].

Based on the above results, we have further optimized our Ag/ZnO back reflectors by considering the surface texture, ZnO total and diffusive reflections, and Ag plasmon absorption. The material quality and the short-circuit

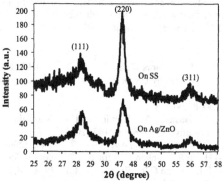

Figure 3. XRD patterns of nc-Si:H cells deposited on SS and the textured Ag/ZnO coated SS.

211

current are carefully monitored to obtain high efficiency a-Si:H/nc-Si:H/nc-Si:H triple junction cells. By combining the optimized back reflector and the improved component cells, we have achieved initial and stable small active-area efficiencies of 13.4% and 12.1%, respectively. The nc-Si:H intrinsic layers were made at a deposition rate of 10 Å/s. Figure 4 shows its initial J-V characteristics and quantum efficiency.

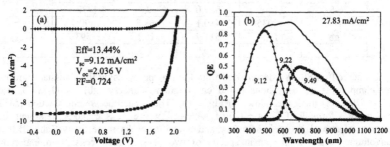

Figure 4. (a) J-V characteristics and (b) quantum efficiency of an a-Si:H/nc-Si:H/nc-Si:H triple-junction cell. The nc-Si:H intrinsic layers were made at 10 Å/s.

SUMMARY

We have studied the effect of the substrate texture on the cell performance in conjunction with dark J-V, Raman, and XRD measurements. We found that the cells deposited on the textured surface have low FF and V_{oc}. An enhanced carrier recombination from dark J-V measurements was observed on cells deposited on the textured substrates. Raman results showed that the substrate does not influence the nc-Si:H crystallinity, but does influence the grain boundary structure. Because the grain boundary could be the defective area, the cells on textured substrates show poorer FF than on flat substrates. XRD analysis demonstrated that the nc-Si:H on the textured substrate has a less preferential crystalline orientations and smaller grain size than that on SS. These features could also cause a high defect density in the material, thus lower FF and V_{oc} in solar cells. Finally, by combining the optimized back reflectors and the improved component cells, we have achieved initial and stable active-area efficiencies of 13.4% and 12.1%, respectively, for an a-Si:H/nc-Si:H/nc-Si:H triple-junction cell made at 10 Å/s.

The authors thank D. Bobela of NREL for Raman and XRD measurements. This work was supported by US DOE under SAI Program contract No. DE-FC36-07 GO 17053.

REFERENCES

[1] Y. Nasuno, M. Kondo, and A. Matsuda, *Proceedings of the 28th IEEE Photovoltaic Specialists Conference*, 142 (2000).
[2] J. Bailat, *et al.*, *Record of the 4th WCPEC* (2006. Hawaii, USA), p1533.
[3] H. Li, R. Franken, J. Rath, and R. Schropp, *Sol. Energy Mater. Sol. Cells* **93**, 338 (2009).
[4] J. Deng and C.R. Wronski, *J. Appl. Phys.* **98**, 24509 (2005).
[5] B. Yan, J. Yang, and S. Guha, *Mater. Res. Soc. Symp. Proc.* **910**, 713 (2006).
[6] S. Guha, J. Yang, D. Williamson, Y. Lubianiker, J. Cohen, and A.H. Mahan, *Appl. Phys. Lett.* **74**, 1860 (1999).

Mater. Res. Soc. Symp. Proc. Vol. 1153 © 2009 Materials Research Society   1153-A10-06

Nanocrystalline Silicon Superlattice Solar Cells

Atul Madhavan[1,2], Nayan Chakravarty[1,2], and Vikram L. Dalal[1,2]
[1]Dept. of Electrical and Computer Engineering , [2]Microelectronics Research Center,
Iowa State University, Ames, Iowa 50011, USA

ABSTRACT

We report on the growth and properties of nanocrystalline Si superlattice solar cells. The solar cells consisted of a stack of alternating layers of amorphous and nanocrystalline Si. The thickness of each of the two layers in the superlattice structure was varied independently. It was found that when the nanocrystalline layer thickness was low, increasing the thickness of the amorphous layer in the superlattice systematically reduced the <220> grain size, while the <111> grain size remained essentially invariant. This fact shows that by interposing an amorphous layer between two nanocrystalline layers forces the nano grains to renucleate and regrow. It was also found that when the amorphous Si layer was too thick, there were significant problems with hole transport through the device. Measurements of defect densities and effective diffusion lengths showed that there was an optimum thickness of the amorphous layer (about 10 nm) for which the defect density was the lowest and the diffusion length was the highest. We also show that the absorption coefficient in nano Si depends upon the grain size and can be increased significantly by increasing the grain size.

INTRODUCTION

Nanocrystalline Si is an attractive material system for solar cells [1-5]. It consists of small grains of Si, 10-20nm, whose grain boundaries are effectively passivated by H and potentially a thin layer of a-Si:H during growth[6]. As the grains grow, the smaller grains tend to agglomerate into larger conical grains, giving rise to a cauliflower-type structure which suffers from excessive recombination at the large grain boundaries [7]. To overcome this problem, many investigators use a hydrogen profiling technique where they reduce the hydrogen to silane ratio as the film grows, thereby forcing the film to remain close to the amorphous-crystalline boundary[8]. This arrangement usually results in an improvement inc ell characteristics. However, the precise dilution profile required to keep the sample just on the amorphous-nano phase change boundary is heavily dependent upon deposition parameters such as pressure, temperature, plasma power, precise reactor geometry, plasma potential etc [9]. Therefore, it would be useful to study alternative techniques whereby one can prevent the agglomeration of grains and the recombination at the large grain boundaries. We have previously shown that using a superlattice of alternating layers of amorphous and nanocrystalline Si may be such an approach [10]. In this paper, we systematically study the properties of such a superlattice cell. In particular, we systematically vary the thickness of the amorphous and nano Si layers individually, and show that the thickness of the amorphous layer plays a critical role in determining the properties of the superlattice layer.

We also study the influence of growth temperature on the properties of the solar cells, and show that as the temperature increases, one can obtain much higher currents due to higher optical absorption in the material, an unexpected result.

The devices were grown on two types of substrates-a planar steel coated with a thin layer of Ag, and a steel on which textured Ag/ZnO layers were applied so as to increase optical absorption by light trapping. The planar steel substrates were used for studying the enhancement of absorption in devices deposited at higher temperatures. The basic structure of the cell is shown in Fig.1.

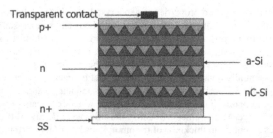

Fig. 1 Schematic diagram of the superlattice solar cell with alternating amorphous and nancorsytalline Si layers

It consists of a back n+ layer, usually n^+ a-Si:H, followed by a thin undoped a-Si:H layer. Next, a nano Si seed layer is deposited using a mixture of silane and hydrogen with a high dilution ratio (silane/hydrogen of 1:25). This seed layer is followed by the stack of superlattice layers, each superlattice consisting of alternating layers of amorphous Si and nano Si. The silane/hydrogen dilution ratio used for the superlattice is ~1:16. All the layers are deposited using a diode VHF plasma at ~45 MHz frequency and pressures of 50-100 mTorr. Growth temperatures are varied between 250 C and 500 C. After the higher temperature devices are grown, a plasma anneal in hydrogen at 25 mT pressure is performed while the sample is rapidly cooled (in about 15 minutes) to 250 C from the deposition temperature. For alternating between the amorphous and nanocrystalline layers, only the power is varied, a higher power leading to nano phase, and lower power leading to amorphous phase. Typical power levels are ~3W for the amorphous phase and 26W for the nano phase. The layer thicknesses can be individually varied by changing the growth times for each of the two layers. The total i layer thickness of the p-i-n device is controlled by controlling the number of superlattice layers in the stack. The i layer is followed by a capping amorphous i layer, which serves to reduce carrier recombination at the p-I interface. This layer is followed by a two level p layer, consisting of a first nano Si p layer and finally, a capping amorphous p layer. The deposition of ITO dots completes the cell.

After growth, the crystallinity is measured using Raman spectroscopy [11] and grain size using x-ray diffraction. For Raman measurements, half the sample is shielded so that no p layer is deposited on that part. Usual I-V measurements are performed on the cell, followed by a measurements of defect densities using capacitance-voltage measurements at varying frequencies [12]. Effective hole diffusion length is measured using quantum efficiency vs. voltage measurements coupled with C-V measurements [12].

Fig. 2 shows how the grain size varies as a function of the time of deposition of the amorphous layer when the thickness of nano layer is fixed at ~15 nm. Very clearly, the figure shows that as the time of deposition of the amorphous layer, and therefore its thickness, increases, the size of <220>grain, which is the dominant grain in device type nano Si [3], decreases. The size of the <111> grain remains relatively unchanged, since that is the grain that results from random nucleation, whereas <220> grain is the thermodynamically preferred grain [13]. The corresponding Raman data are shown in Fig. 3, which shows that as one goes from the thinnest (6001) to the thickest (6004) amorphous layers, while keeping the thickness of nano Si layer in the superlattice invariant at ~13 nm, the Raman signal corresponding to the crystalline phase decreases.

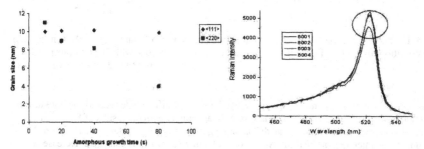

Fig. 2 Grain size of <111> and <220> nano Si in the superlattice as a function of growth time of intermediate amorphous Si layer

Fig. 3. Change in Raman crystallinity of superlattice solar cells as a function of the the thickness of the a-Si layer in the superlattice. 6001 had the thinnest a-Si layer and 6004, the thickest.

The corresponding I-V curves for the samples are shown in Fig. 4, which shows that as the thickness of amorphous layer increases significantly for this thin nano Si layer, the I-V curves become worse, implying a collection problem. We attribute this decrease in collection to the fact that there is a mismatch in the valence band edges of amorphous and crystalline Si, with the valence band of a-Si:H being as much as 0.5 eV below the corresponding valence band edge of crystalline Si. This mismatch would lead to a barrier being faced by holes generated in the nano Si in their travel towards the p layer. This reasoning is supported by the data on quantum efficiency vs. voltage curves for long wavelength (900nm) photons shown in Fig. 5, which shows that as the thickness of amorphous layer increases, there is a decrease in relative QE in the forward voltage direction. In the forward direction, there is less electric field available for tunneling and if the tunneling is inhibited because of a large thickness of the amorphous layer, the effect of a decreasing field will be felt more prominently in the forward direction of applied voltage. This is turn leads to a decrease in the fill factor of the cell, as shown in Fig. 4.

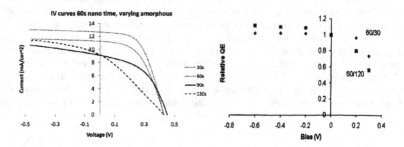

Fig. 4. I-V curves for superlattice cells as a function of the deposition times of a-Si layer in the superlattice

Fig. 5 Relative QE at 900 nm for two cells, one with 30s.a-Si layer in the superlattice and the other with 120 s.

Fig.6 shows the corresponding changes in defect density and diffusion lengths as the thickness of the amorphous layer in the superlattice is varied. In agreement with the results of Fig.4 and 5, it is clear that there is an optimum in the diffusion length and lowering of defect density as a function of the thickness of the a-Si layer, and that too high a thickness leads to a decrease in diffusion length.

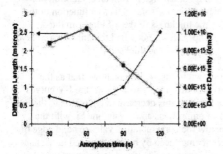

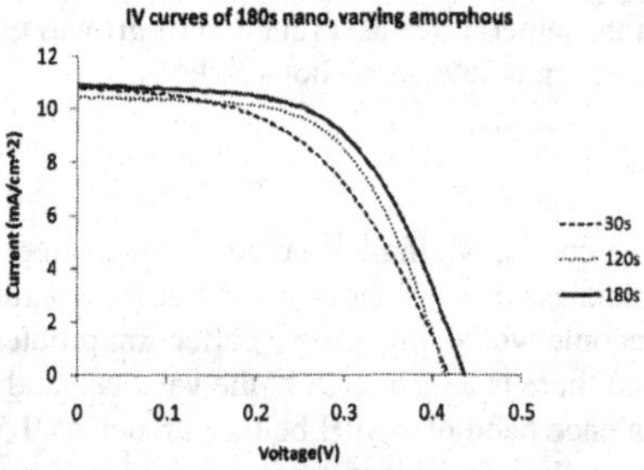

Fig. 6 Variation in the diffusion length and defect density of a superlattice as a function of the time of deposition (thickness) of the a-Si layer

Fig. 7 I-V curves of a superlattice cell with longer nano Si layer as a function of the growth time of the a-Si layer

Next, we show what happens when devices are made with longer nano Si growth times (180 s), and varying thickness of amorphous layer. See Fig. 7. In this case, a thicker nano layer, with larger grain size, needs a thicker a-Si layer to inhibit the formation of grain agglomeration which leads to fill factor losses.

Previously, we had shown that devices with H profiling grown at higher temperatures showed larger grain sizes and larger quantum efficiencies[10]. We did not know why. To understand this effect, we made such devices and then measured the absolute absorption coefficient from measuring QE under strong reverse bias. Under strong reverse bias, $QE = \exp(-\alpha t_1) [1 - \exp(-\alpha t_2)]$, where α is the absorption coefficient and t_1 and t_2 are the respective thicknesses of the p and i layers. t_1 is generally very small, of the order of 30 nm, and t_2 is of the order of 1 micrometer. t_2 is measured using capacitance measurements. In the above measurements, it is important to recognize that QE is the internal QE, i.e. we have to eliminate the influence of reflection. Therefore, we independently measured reflection at every wavelength, and then derived internal QE from the measured external QE. Back reflection, which could lead to multiple passes, was reduced by using a stainless substrate with low reflectivity and a n layer that absorbed most of the photons that were transmitted into it.

The results for grain size in devices vs. growth temperature are shown in Fig. 8. The results for α vs. photon energy for various grain sizes deduced from QE measurements under strong reverse bias are shown in Fig. 9. We see very clearly from Fig. 9 that there is an increase in absorption coefficient as the grain size increases. This is a remarkable result, at variance with the usual model of absorption in nano Si, which suggests that one way of increasing the efficiency of nano Si solar cells would be to make devices with larger grain sizes.

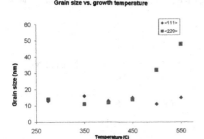

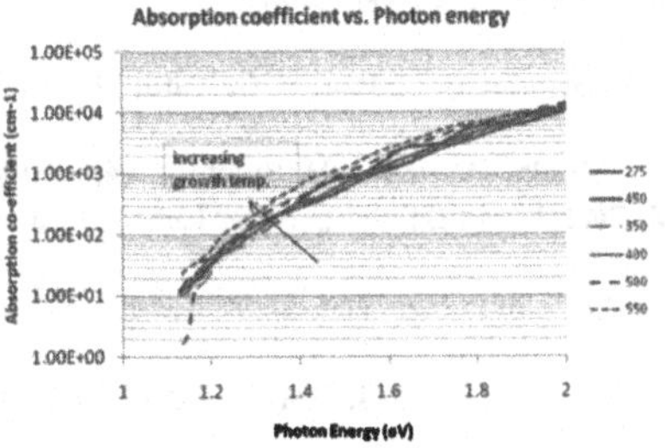

Fig. 8 Grain size vs. indicated growth temperature The real temperature is about 50 C lower.

Fig. 9 Absorption coefficient vs. growth temperature.

In summary, we have shown that superlattice devices can work to effectively suppress large grain boundary formation and consequent excessive recombination. One needs to be careful about the proper choice of individual layer thicknesses in a superlattice device so as to obtain low defect density, reasonable diffusion lengths and good fill factors. We also show that the absorption coefficient of nano Si is a strong function of grain size, and can be increased

217

significantly by increasing the grain size from 15 nm to 50 nm. One still needs to discover the best method for passivating grain boundaries at such high temperatures.

We thank Keqin Han, Max Noack, Nayan Chakravarty and Sambit Pattnaik for their experimental help. The project was supported in part by grants from NSF, from Iowa Energy center and from Catron Foundation.

1. Y. Tawada, H. Yamagishi and K. Yamamoto, Solar Energy Materials and Solar Cells, 78, 647(2003)

2. B. Rech, O. Kluth, T. Repmann, T. Roschek, J. Springer, J. Müller, F. Finger, H. Stiebig and H. Wagner, Solar Energy Materials and Solar Cells, 74, 439 (2002)

3. A. V. Shah, J. Meier, E. Vallat-Sauvain, N. Wyrsch, U. Kroll, C. Droz and U. Graf, Solar Energy Materials and Solar Cells, 78, 469 (2003)

4. Vikram L. Dalal, U.S. Patent 4,253,882 "High efficiency thin-film multiple-gap photovoltaic device" (1981)

5. T. Repmann, B. Sehrbrock, C. Zahren, H. Siekmann and B. Rech, Solar Energy Materials and Solar Cells, 90, 3047, (2006)

6. R. Biswas, B. C. Pan, V. Selvaraj , Proc. Of MRS, Vol. 862, 133(2005)

7. J. Koěka, A. Fejfar, H. Stuchlíková, J. Stuchlík, P. Fojtík, T. Mates, B. Rezek, K. Luterová, V. Švrček and I. Pelant, Solar Energy Materials and Solar cells, 78, 493(2003)

8. B. J. Yan, G. Z. Yue, Yang J and S. Guha , Appl. Phys. Lett. 85,1955 (2004)

9. Vikram L. Dalal, J. Graves and J. Leib, Appl. Phys. Lett., 85, 1413(2004)

10. A. Madhavan and V. L. Dalal, J. Non-Cryst. Solids, 354, 2403-2406(2008)

11. C. Smit C, RACCM van Swaaij , H. Donker , AMHN Petit , WMM Kessels , MCM van de Sanden., J. Appl. Phys. 94 ,3582(2003)

12. Vikram Dalal and Puneet Sharma, Appl. Phys. Lett. 86, 103510 (2005)

13. Vikram Dalal, Kamal Muthukrishnan, Xuejun Niu and Daniel Stieler, J. Non-Cryst. Solids, 352, 892(2006)

Crystallization

Mater. Res. Soc. Symp. Proc. Vol. 1153 © 2009 Materials Research Society 1153-A12-01

High Performance n- and p-channel Strained Single Grain Silicon TFTs using Excimer Laser

Alessandro Baiano, Ryoichi Ishihara and Kees Beenakker.
Delft University of Technology, Delft Institute of Microsystems and Nanoelectronics (DIMES),
Laboratory of Electronic Components, Technology and Materials (ECTM)
Feldmannweg 17, P.O. Box 5053, 2600 GB Delft, the Netherlands.

ABSTRACT

In this paper we investigate the carriers mobility enhancement of the n- and p-channel single-grain silicon thin-film transistors (SG-TFTs) by μ-Czochralski process at low-temperature process (< 350 °C). The high laser energy density nearby the ablation phenomenon that completely melts the silicon layer during the crystallization is responsible for high tensile strain and good crystal quality of the silicon grains, which lead carriers mobility enhancement.

INTRODUCTION

Thin-film electronics, such as thin-film transistors (TFTs) using amorphous silicon (a-Si) and low temperature polysilicon (LTPS) have been extensively employed for active matrix liquid crystal displays (AMLCDs), image sensors etc. A major reason for the success of these TFT-centric solution is that solution based on commercial silicon IC microelectronics would not be economically viable [1], beside the fact that it is based on high-temperature process not acceptable for unconventional electronics. Nowadays, many new applications have been coming up such as smart wireless systems that operate in the range of radio frequency fabricated on glass or even flexible substrates. 3D integrated circuits has also become an important research area causing many common interests between IC microelectronics and low-temperature TFTs technologies. Whereas 3D integration in package level and wafer level [2] can be made by the CMOS technology, 3D integration at device level (monolithic integration), which results in the largest decrease in interconnect length and so highest density of interconnects between nodes may only be approached by low-temperature TFTs process [3].

Many new recrystallization techniques have been developed to improve electrical performance, such as sequential lateral solidification (SLS) [4], CW-laser lateral crystallization (CLC) [5], selectively enlarging laser crystallization (SELAX) [6], phase modulated excimer laser annealing (PMELA) [7]. We have developed the μ-Czochralski process [8], which has many fundamental advantages over the aforementioned techniques in term of: 1) 2D location-control of large (up to 8 micros) grains, 2) wide energy density windows in which the location control is obtained, 3) the high throughput owing to the one shot process, and 4) capability of crystallographic orientation control [9]. Single grain silicon TFTs (SG-TFTs) by the μ-Czochralski has obtained mobility as high as SOI counterpart [8]. Transit frequency f_T in the range of 5 to 6 GHz has been obtained with a gate length of 1.5 μm. As a result, RF amplifiers have been attained well over 10 dB gain below 1 GHz [10]. However, the major goal of large-area electronics is to create novel technologies with so high mobility that enables electronics solutions not feasible with today's existing IC microelectronics-based methods [1].

The IC microelectronics technologies have developed the strained silicon technology to enhance the n- and p-channel mobilities and relax CMOS scaling issue. The prevalent technique used for forming tensile strained silicon is obtained by growing epitaxially silicon layer on relaxed-SiGe [11]-[13]. However, this technique is performed at high thermal budget (> 750 °C), which limits the use for IC microelectronics only.

It is well known, on the other hand, that polysilicon crystallized by excimer laser has intrinsic tensile strain [14]-[16]. The microscopic origin of the strain is due to the thermal stress. It could be originated by the different thermal expansion coefficients between substrate and polysilicon film [15]. However, the mobility enhancement in the poly-Si TFTs induced by intrinsic tensile strain has never been reported. This paper intends to show the electron and hole mobility enhancements of single-grain silicon TFTs by the μ-Czochralski process due to the intrinsic tensile strain and the high crystal quality of single grains.

EXPERIMENTS

Firstly, 1 μm diameter holes are formed by anisotropic dry etching of a 750 nm thick layer (it may be SiO_2). An 875 nm-thick SiO_2 layer is then deposited by TEOS PECVD at a substrate temperature of 350 °C in order to reduce the hole diameter to approximately 100 nm. A 250 nm thick a-Si film is then deposited by LPCVD. After a threshold-adjust implant of boron (B) ions for both n- and p-channel transistors (2.5×10^{11} cm^{-2}), a single 20 ns long pulse of light from a XeCl (308 nm) excimer laser irradiates the Si surface with very high energy densities, very close to the ablation. Thus, the silicon layer is entirely melted and crystallized [17] as can be seen from the bottom part of grain filter in the cross-sectional TEM image shown in fig. 1, as opposite to the partial melting condition used in the previously works [18]. Silicon grains with diameter as large as 8 μm are thereby obtained in predetermined positions via a crystallization known as the μ-Czochralski process [8]. Subsequently, the silicon film is patterned into islands by reactive ion etching. A 80 nm inductively coupled PECVD (IC-PECVD) SiO_2 layer is then deposited as a gate insulator at 250 °C [19]. Aluminum sputtered at room temperature forms the gate electrode. The source and drain are implanted with either P or B ions (1×10^{16} cm^{-2}) depending on the device type, which are then activated by the excimer laser annealing. After SiO_2 passivation and contact hole formation, Al interconnect metal is sputtered and patterned. SG-TFTs are fabricated inside the location-controlled grains obtained by the μ-Czochralski process, where no random grain boundaries (GBs), but only coincident site lattice (CSL) grain boundaries, are present. We have varied the position of the SG-TFT channel inside the grain, as referred in [20]. The channel can be placed in central C position, where CSLs may eventually be in all directions. It can be placed in X and Y positions, where radial grown CSLs may be parallel or perpendicular to the carrier flow, respectively [20]. As reference, TFTs are also fabricated on Silicon-on-Insulator (SOI-TFTs) wafer.

MEASUREMENTS RESULTS

The field-effect mobilities are evaluated by transconductance characteristic of the TFTs for 0.05 V drain voltage [21]. Concerning the n-channel SG-TFTs, the field-effect mobility is measured to be about 600 cm^2/Vs for a wide range of energy densities [22]. At the complete-melt condition, we achieve a mean field-effect mobility value of 883 cm^2/Vs, with a peak of 1400 cm^2/Vs at X position. Mean field-effect mobilities of 814 cm^2/Vs and 773 cm^2/Vs are obtained for C and Y positions, respectively, as shown in fig. 2. Concerning the p-channel SG-TFTs, in the previous work [23], the field-effect mobility is measured to be about 250 cm^2/Vs for wide range of laser energy densities. At the complete-melt condition, we achieve a mean field-effect mobility of 320 cm^2/Vs, with a peak of 500 cm^2/Vs.

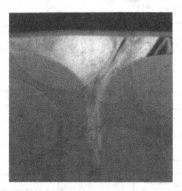

Figure 1. TEM image of single-grain silicon by the μ-Czochralski process cross section after laser crystallization at the complete-melt condition (laser energy density next to the ablation). The image shows the complete-crystallize silicon layer and basement of the grain filter.

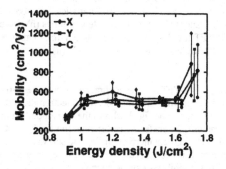

Figure 2. Electron field-effect mobility mean and standard deviation measured for n-channel SG-TFTs fabricated by different laser energy densities up to the complete-melt condition.

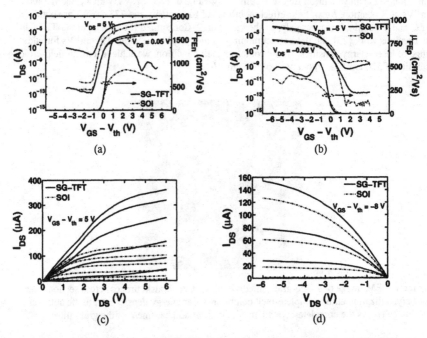

Figure 3. Transfer characteristics for 0.05 V and 5 V of n-channel SG-TFT and SOI-TFT and their field-effect electron mobilities (a). Transfer characteristics for -0.05 V and -5 V of p-channel SG-TFT and SOI-TFT and their field-effect hole mobilities (b). Output characteristics of n-channel SG-TFT and SOI-TFT for $V_{GS} - V_{th}$ from 0 V to 5 V (c). Output characteristics of p-channel SG-TFT and SOI-TFT for $V_{GS} - V_{th}$ from 0 V to -8 V (d).

Moreover, n- and p-channel SG-TFTs by μ-Czochralski process are also compared with SOI-TFTs. Fig. 3a shows transfer characteristics and field-effect electron mobilities of n-channel SG-TFT compared to the SOI counterpart and fig. 3c shows the corresponding output characteristics. Fig. 3b, on the other hand, shows the transfer characteristics and field-effect hole mobilities of p-channel SG-TFT compared to the SOI counterpart and fig. 3d shows the correlative output characteristics. Field-effect mobilities enhancements of about 1.6 are achieved for both electron and hole. Such electron and hole mobilities enhancements correspond to the ones yielded by 20% Germanium content in strained-Si on $Si_{80}Ge_{20}$ n- and p-channel MOSFETs, according to the S. Takagi *et al.* [24] and R. Oberhuber *et al.* [25] curves, respectively. Furthermore, the n-channel SG-TFT exhibits a subthreshold slope S as low as the one of the SOI-TFT (106 mV/dec). The p-channel SG-TFT shows a subthreshold slope S of 183 mV/dev, which is slightly higher than the one shown by SOI-TFT (156 mV/dec).

DISCUSSION

The polysilicon TFTs possess an intrinsic tensile strain. However, advantage with high mobility is not obtained because of 1) the presence of random grain boundaries in the channel, 2) high density of trap states at Si/SiO₂ interface, and 3) elevated roughness of the Si surface. Reasons for higher electron and hole mobilities of the single-grain Si TFT by μ-Czochralski process at complete-melt condition, nearby to the ablation, than the ones of the SOI-TFTs are suggested as follows. Considering a polysilicon TFT, the carrier mobility is strongly affected by density of trap states of random grain boundaries and Si/SiO₂ interface. Although thin polysilicon film possesses tensile strain [15], its benefit having mobility enhancement is to be inefficacious when high density of trap states at random grain boundaries (GBs), Si/SiO₂ interface and high silicon surface roughness occur.

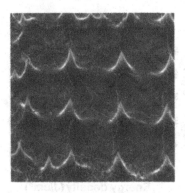

Figure 4. EBSD image of 6 μm location-controlled single-grains silicon by μ-Czochralski process: bright lines are random grain boundaries, which are only at the edge of single-grains, darker lines are $\Sigma 3$ and $\Sigma 9$ CSL grain boundaries.

Quality of single-grain silicon by μ-Czochralski process and IC-PECVD gate oxide

In the SG-TFTs, the defects are only CSL grain boundaries ($\Sigma 3$ and $\Sigma 9$) as shown in the EBSD image (fig. 4), whereas random GBs are arranged outside the channel area. The density of trap states compute for $\Sigma 9$ is much lower than the one of random GBs and the corresponding potential barrier is much smaller than the one of random GBs. The potential barrier of $\Sigma 3$ is even negligible. As a consequence, the carriers mobility are not affected by CSL grain boundaries [26]. Interface defects meanly due to the lack of Si to SiO₂ bonds lead density of trap states in the middle of the forbidden band gap. Those traps induce subthreshold characteristic worsening and mobility decrease since carriers are not free to flow through the channel but they are captured. As a consequence, decrease the interface trap states is an important issue to obtain beneficial performance. Indeed, the use of inductively coupled PECVD (IC-PECVD) as gate oxide substantially improves the mobilities due to the density of interface trap states as low as 2×10^{10} $eV^{-1}cm^{-2}$ [19]. The good quality of IC-PECVD gate oxide is also confirmed by the low subthreshold slope S shown by both n- and p-channel devices.

AFM analysis

The roughness of the silicon surface also plays an important rule. It leads additional scattering at interface with consequent mobility reduction when applied gate voltage becomes substantial. To analyze the roughness of silicon surface, atomic force microscopy (AFM) is performed (fig. 5). AFM reveals that the silicon surface roughness of single-grain silicon by μ-Czochralski process is substantially lower than the one measured for polysilicon. Furthermore for the polysilicon, the higher the laser energy density is, the higher the roughness is. That is due to taller hillocks forming during the grains collision. By the μ-Czochralski process, on the contrary, the hillocks are location-controlled out of the single-grain and, as a consequence, the roughness inside single-grains area is uniformly kept as low as 7 nm, independently from the used laser energy density. As a result, the increase of the laser energy density is not beneficial for the polysilicon roughness, whereas it is not influential for the roughness of single-grain by μ-Czochralski process.

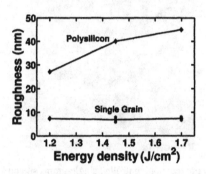

Figure 5. Mean roughness value of polysilicon surface and mean roughness and standard deviation values inside single-grains by μ-Czochralski process with grain filter distances of 8 μm. The roughness value is calculated by AFM.

Raman spectroscopy

In addition, we observe tensile strain by Raman spectroscopy as shown in fig. 6. The Raman shift measured for single grain is about -2.53 cm^{-1} for laser energy density of 1400 mJ/cm^2. Increasing the laser energy density up to the complete-melt condition (1700 mJ/cm^2), the Raman shift reaches -4.45 cm^{-1}. It is carefully checked that this shift is not due to heating of the sample by the Raman laser, since the Raman peak does not change with increasing of Raman laser energy. The FWHM is just slightly wider than the crystalline-silicon one, confirming the excellent crystal quality of the single-grain silicon by μ-Czochralski process. The thermal stress α_{th} can be estimated according to the following relation [15]:

$$(1) \qquad \alpha_{th} = \left(\alpha_f - \alpha_s \right) \Delta T \frac{E_f}{1 - y_p}$$

where α_f and α_s are the average thermal expansion coefficients of the film and substrate, respectively. ΔT is the difference between the crystallization and the initial temperature of the film. E_f is the elastic (Youngs) modulus of the film and y_p is the Poisson ration. The thermal stress incorporated in the film is tensile if $\alpha_f > \alpha_s$, as occur for this case.

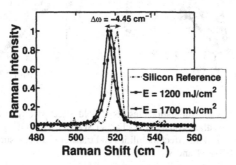

Figure 6. Raman spectroscopy of crystalline silicon reference and single-grain silicon by μ-Czochralski process obtained by two different laser energy densities.

The thermal tensile stress found out by this approximate analysis is about 1.22 GPa, which is in good agreement with the corresponding tensile stress calculated by the Raman shift using the following conversion rule [15]:

$$(2) \qquad \alpha_{th} = -0.27 \left(GPa/cm^{-1} \right) \Delta\omega \left(cm^{-1} \right)$$

where $\Delta\omega$ is the Raman shift. Eq. (2) yields a tensile stress of 1.20 GPa at the complete-melt condition.

However as shown in fig. 2, the carrier mobility variation of SG-TFT by μ-Czochralski process increases at complete-melt condition. That could be related to the additional dependence on surface orientations of electron and hole mobilities of silicon layer under tensile stress [27].

Device simulation

To further prove that the field-effect mobility enhancement is owing to induced tensile strain, we use the device simulator Sentaurus TCAD. Advanced strained silicon model that computes second-order stress-dependence isotropic enhancement factors for low-field mobility is taken into account [28]. First, parameters extraction are carried out by fitting n- and p-channel SOI-TFTs field-effect mobilities and transfer characteristics for 0.05 V and -0.05 V drain voltages, respectively [29]. Hence, strained model with tensile stress value of 1.2 GPa is added to the basic SOI-TFT extracted model. The result is a good fitting between simulation and measurement results of both n- and p-channel SG-TFTs, as shown in figures 7. That additionally proves the effect of mobility enhancement is due to the tensile strain induced by excimer laser, low density of trap states at Si/SiO2 interface, low surface roughness and high crystal quality (inactivity of CSL grain boundaries) of single-grains silicon by μ-Czochralski process.

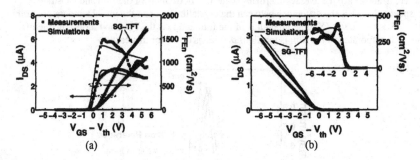

Figure 7. Measurements and simulations results of field-effect mobilities and transfer characteristics for 0.05 V drain voltage of n-channel SG-TFT and SOI-TFT (a). Measurements and simulations results of field-effect mobilities and transfer characteristics for -0.05 V drain voltage of p-channel SG-TFT and SOI-TFT (b).

CONCLUSIONS

High performance n- and p-channel strained single-grain silicon have been successfully obtained by μ-Czochralski process at low-temperature process. We have investigated that electron and hole field-effect mobility enhancements of SG-TFTs can reach peak of 1.6 times over SOI-TFTs at complete-melt condition. Reason for higher carrier mobilities of SG-TFTs than the ones of SOI-TFTs is related to the high crystal quality of the single-grains by μ-Czochralski process, low roughness of silicon surface and tensile strain induced by excimer laser at complete-melt condition. The tensile strain induces mobilities enhancements is also confirmed by TCAD device simulations.

ACKNOWLEDGMENTS

The authors are grateful to DIMES clean room and all process engineers. We express our special thanks to J. van der Cingel and B. Goudena of DIMES clean room.

REFERENCES

1. R.H. Reuss, B.R. Chalamala, A. Moussessian, M.G. Kane, A.Kumar, D.C. Zhang, J.A. Rogers, M. Hatalis, D. Temple, G. Moddel, B.J. Eliasson, M.J. Estes, J. K.unze, E.S. Handy, E.S. Harmon, D.B. Salzman, J.M. Woodall, M.A. Alam, J.Y. Murthy, S.C. Jacobsen, M. Oliver, D. Markus, P.M. Campbell, and E. Snow, "Macroelectronics:Perspectives on Technology and Applications", Proceeding of the IEEE, vol. 93, no. 7, July 2005.

2. T. Matsumoto, M. Satoh, K. Sakuma, H. Kurino, N. Miyakawa, H. Itani, and M. Koyanagi, "New three-dimensional wafer bonding technology using the adhesive injection method", Japanese Journal of Applied Physics Part 1-Regular Papers 37 (1998)

3. M.R.T. Mofrad, R. Ishihara, J. Derakhshandeh, A. Baiano, J. v.d. Cingel, C. Beenakker, "Monolithic 3D Integration of Single-Grain Si TFTs", Proceeding of MRS, vol. 1066, 2008.

4. R.S. Sposili and J.S. Im, "Sequential lateral solidification of thin silicon films on SiO_2", Appl. Phys. Lett., vol. 69, no. 19, pp. 2864, 1996.

5. A. Hara, F. Takeuchi, M. Takei, K. Suga, K. Yoshino, M. Chida, Y. Sano, and and N. Sasaki, "High-Performance Polycrystalline Silicon Thin Film Transistors on Non-Alkali Glass Produced Using Continuous Wave Laser Lateral Crystallization", Jpn. Journal of Appl. Phys. vol. 41, pp. L311, 2002.

6. M. Hatano, T. Shiba, and M.Okumura, "Selectively enlarging laser crystallization technology for high and uniform performance poly-Si TFTs", SID Int. Symp. Dig. Tech. Papers, vol. 33, pp. 158, 2002.

7. C.H. Oh, M. Ozawa, and M. Matsumura, "A Novel Phase-Modulated Excimer-Laser Crystallization Method of Silicon Thin Films", Jpn. Journal Appl. Phys, vol. 37, pp. L492, 1998.

8. R. Ishihara, P. v.d. Wilt, B. D. van Dijk, A. Burtsev, F. C. Voogt, G. J. Bertens, J. W. Metselaar, and C. I. M. Beenakker, "Advanced Excimer-Laser Crystallization Techniques of Si Thin-Film for Location Control of Large Grain on Glass", Flat Panel Display Technology and Display Metrology II (T. V. Edward F. Kelley, ed.), vol. 4295 of Proc. SPIE, pp. 14-23, 2001.

9. T. Chen, R. Ishihara, A. Baiano, C.I.M Beenakker, "Highly Uniform Single-Grain Si TFTs Inside (110) Orientated Large Si Grains", the 15th International Display Workshop 2008, AMD5-4L.

10. A. Baiano, M. Danesh, N. Saputra, R. Ishihara, J. Long, W. Metselaar, C.I.M. Beenakker, N. Karaki, Y. Hiroshima, and S. Inoue, "Single-grain Si thin-film transistors SPICE model, analog and RF circuit applications", Solid-State Electronics, vol. 52, no. 9, pp. 1345-1352, 2008

11. J. Welser, J. L. Hoyt, J. F. Gibbons, "Electron Mobility Enhancement in Strained-Si N-Type Metal-Oxide-Semiconductor Field-Effect Transistors", IEEE Electron Device Letters, vol. 15, March 1994.

12. J. Welser, J. L. Hoyt, S. Takagi, J. F. Gibbons, "Strain Dependence of the Performance Enhancement in Strained-Si n-MOSFETs", Inter. Electron Devices Meeting 1994.

13. T. Mizuno, S. Takagi, N. Sugiyama, H. Satake, A. Kurobe, A. Toriumi, "Electron and Hole Mobility Enhancement in Strained-Si MOSFET's on SiGe-on-Insulator Substrates Fabricated by SIMOX Technology", IEEE Electron Device Letters, vol. 21, May 2000.

14. I. Tsunoda, R. Matsuura, M. Tanaka, H. Watakabe, T. Sameshima, M. Miyao, "Direct Formation of Strained Si on Insulator by Laser Annealing", Thin Solid Films, vol 508, pp. 96-98, 2006.

15. P. Lengsfeld, N.H. Nickel, Ch. Genzel, W. Fuhs, "Stress in Undoped and Doped Laser Crystallized Poly-Si", Journal of Applied Physics, vol. 91, no. 11, June 2002.

16. H. S. Cho, W. Xianyu, X. Zhang, H. Yin, T. Noguchi, "Tensile-Strained Single-Crystal Si Film on Insulator by Epitaxially Seed and Laser-Induced Lateral Crystallization", Japanese Journal of Applied Physics, vol. 45, no. 10A, 2006, pp. 7682-7684.

17. R. Ishihara, D. Danciu, F. Tichelaar, M. He, Y. Hiroshima, S. Inoue, T. Shimoda, J.W. Metselaar, C.I.M. Beenakker, "Microstructure characterization of location-controlled Si-islands by excimer laser in the μ-Czochralski (grain filter) process", Journal of Crystal Growth, vol. 299, 2007, pp. 316-321.

18. R. Vikas, R. Ishihara, Y. Hiroshima, D. Abe, S. Inoue, T. Shimoda, J.W. Metselaar, and C.I.M. Beenakker, "High Performance Single Grain Si TFTs Inside a Location-Controlled Grain by μ-Czochralski Process with Capping Layer", Inter. Electron Devices Meeting 2005.

19. R. Ishihara, C. Tao, M. He, P. Deosarran, Y. van Andel, J. W. Metselaar and C. I. M. Beenakker, "Low-Temperature Deposition of High Quality SiO_2 by Inductively Coupled Plasma Enhanced CVD" to be published in Thin Solid Films.

20. V. Rana, R. Ishihara, Y. Hiroshima, D. Abe, S. Inoue, T. Shimoda, J.W.Metselaar and C. I.M. Beenakker, "Dependence of Single-Crystalline Si Thin-Film Transistor Characteristics on the Channel Position inside a Location-Controlled Grain", IEEE Transactions on Electron Devices, vol.52, no.12, Dec. 2005.

21. D. K. Scheoder, "Semiconductor Material and Device Characterization third edition", A Wiley-Interscience Publication.

22. V. Rana, "Single Grain Si TFT and Circuits based on the μ-Czochralski process", Ph.D. Thesis.

23. V. Rana, R. Ishihara, Y. Hiroshima, D. Abe, S. Higashi, S. Inoue, T. Shimada, J. W. Metselaar, and C.I.M. Beenakker, "High performance p-channel single-crystalline Si TFTs fabricated inside a location-controlled grain by μ-Czochralski process", IEICE Trans. Electron., vol.E87-C, no.11, pp. 1943-1947, Nov. 2004.

24. S. Takagi, J.L. Hoyt, J.J. Welser, and J.F. Gibbons, "Comparative study of phonon-limited mobility of two-dimensional electrons in strained and unstrained Si metal-oxide-semiconductor field effect transistors", J. Appl. Phys. 80, 1567 (1996).

25. R. Oberhuber, G. Zandler, and P. Vogl, "Subband structure and mobility of two-dimensional holes in strained Si/SiGe MOSFET's", Physical review B, vol. 58, no. 15, Oct. 1998.

26. N. Matzuki, R. Ishihara, A. Baiano, Y. Hiroshima, S. Inoue, and C.I.M Beenakker, "Local Electrical Property of Coincidence Site Lattice Boundary in Location-Controlled Silicon Islands by Scanning Spread Resistance Microscope", Proceeding of The 14th International Display Workshops, vol.2, pp. 489-492, 2007.

27. M.V. Fischetti, and S.E. Laux, "Band Stucture, Deformation Potential, and Carrier Mobility in Strained Si, Ge, and SiGe alloys", Journal of Appl. Phys 80, pp. 2234-2252, Aug. 1996.

28. Sentaurus TCAD manual.

29. A. Baiano, J. Tan, R. Ishihara, and K. Beenakker, "Reliability Analysis of Single Grain Si TFT using 2D Simulation", Electrochemical Society Transactions, vol. 16, pp. 109-114, 2008.

231

Light Trapping in Solar Cells II

Mater. Res. Soc. Symp. Proc. Vol. 1153 © 2009 Materials Research Society   1153-A13-01

Light Trapping Effects in Thin Film Silicon Solar Cells

F.-J. Haug, T. Söderström, D. Dominé, C. Ballif
École Polytechnique Fédérale de Lausanne (EPFL), Institute of Microengineering (IMT),
Photovoltaics and Thin Film Electronics Laboratory (PVLab),
Rue A.-L. Breguet 2, 2000 Neuchâtel, Switzerland

ABSTRACT

We present advanced light trapping concepts for thin film silicon solar cells. When an amorphous and a microcrystalline absorber layers are combined into a micromorph tandem cell, light trapping becomes a challenge because it should combine the spectral region from 600 to 750 nm for the amorphous top cell and from 800 to 1100 for the microcrystalline bottom cell. Because light trapping is typically achieved by growing on textured substrates, the effect of interface textures on the material and electric properties has to be taken into account, and importantly, how the surface textures evolve with the thickness of the overgrowing layers. We present different scenarios for the n-i-p configuration on flexible polymer substrates and p-i-n cells on glass substrate, and we present our latest stabilized efficiencies of 9.8% and 11.1%, respectively.

INTRODUCTION

Light scattering at textured interfaces has become a decisive feature for high efficiency thin film silicon solar cells. It allows using thinner absorber layers because the scattering enhances the effective light path in the absorbing film. While this is certainly important for production throughput, light trapping is also mandatory because of inherent material properties; in case of amorphous silicon, the impact of light induced degradation can be reduced in thinner films, in microcrystalline silicon it can, to some extent, compensate the low absorption of the indirect band gap. Light scattering is typically achieved at surface textures of the substrate or of the electric contact layer that precedes the silicon deposition [1]. For superstrate (p-i-n) devices, the transparent front contacts are either directly grown under conditions that favour preferential growth and faceting [2, 3], or they are structured by etching after growth [4]. Similar concepts are followed in substrate (n-i-p) devices for achieving textured metallic back contacts, e.g. the well known surface roughening of silver when it is grown on heated substrates [5]. Typical textures for amorphous silicon solar cells should have a root mean square (rms) surface roughness in the range from 50 to 90 nm, and a lateral feature size which varies between 300 and 500 nm.

For microcrystalline cells, the light trapping range lies between 800 and 1100 nm because of its lower band gap compared to amorphous cells. Empirical data suggest that the lateral feature size should be in the range from 1000 to 1400 nm, while higher rms roughness than in the amorphous case is not necessarily beneficial for the solar cell efficiency [6]. The latter observation might be related to the growth mechanism of microcrystalline silicon which often results in defective material above steep, V-shaped depressions [7, 8].

In micromorph tandem cells, where a microcrystalline and an amorphous absorber are combined in the same device, it becomes a challenging task to devise light scattering strategies that can effectively serve the different spectral ranges of the two individual cells. The first step to achieve this goal is the introduction of an intermediate reflector layer between the two cells, because light trapping in the amorphous top cell is quite simply impossible without reflection of light at its back surface; after its first realization with a thin film of ex-situ ZnO between the cells [9], in-situ solutions with P-doped SiOx have been realized [10], and the intermediate reflector is now established in production of micromorph tandem modules [11].

In this paper, we discuss the possibility of implementing different length scales for light trapping, either in the same interface by overcoating a large structure with smaller features, or by varying the texture between the individual interfaces. This approach must also take into account how a given surface structure evolves during the growth of amorphous and microcrystalline absorber layers, respectively.

EXPERIMENTAL

n-i-p cells

The n-i-p cells presented here are grown on a flexible polyethylene substrate. The surface of the substrate is textured with a periodic sinusoidal structure which is embossed into the surface by a roll-to-roll process [12]. The substrate is covered conformally by sputtering of a bilayer of silver and zinc oxide with thickness of 80 and 60 nm, respectively. The silicon layers are deposited at 180°C by very high frequency plasma enhanced CVD (PE-CVD) from a mixture of silane (SiH_4) and hydrogen (H_2), phosphine (PH_3) and tri-methyl-boron ($B(CH_3)_3$, TMB) are used as doping gases. The front contact consists of a 3.8 μm thick ZnO layer deposited by low pressure CVD (LP-CVD) which results in naturally textured growth [3]. The layer is lightly boron doped in order to suppress free carrier absorption [13]. In the n-i-p tandem cells shown here we use the same LP-CVD process for the deposition of the intermediate ZnO reflector with a thickness of 1.6 μm [14].

p-i-n cells

The p-i-n cells are grown on AF45 borosilicate glass substrates from Schott. First, the substrates are covered with a transparent front contact of LPCVD-ZnO; two different conditions are used, strongly doped films with a thickness of 1.9 μm (type A), and lightly doped films with a thickness of 4.8 μm. The sheet resistance of both substrates is 10 Ω/sq, their surface roughnesses are 66 and 180 nm, respectively. The thicker ZnO layer is subjected to a plasma treatment which changes the initial V-shaped morphology to U-shaped morphology which is better suited for the growth of microcrystalline silicon (type C) [13]. Depending on the duration of the treatment, the roughness can be reduced as much as down to 120 nm [15]. The amorphous and microcrystalline layers are deposited by VHF-PECVD under similar conditions to those of the previous section, but deposition temperatures up 220°C are tested. The intermediate reflector in the presented p-i-n cells is made from P-doped SiO$_x$ (SOIR) by in-situ processing [10]. The back contact of p-i-n cells consists again of LPCVD-ZnO, covered with a white reflector.

Characterization

The illuminated current voltage characteristics of all cells are measured in standard test conditions (25°C, AM1.5g spectrum, 100 mW/cm^2) with a dual source solar simulator (Wacom). The current density is determined independently by a measurement of the external quantum efficiency (EQE). Red and blue light bias is applied for measuring top- and bottom cells, respectively, and the photocurrent is determined by integration of the EQE weighted by the spectral photon density of the AM1.5g spectrum.

RESULTS AND DISCUSSION

n-i-p cells

As starting consideration for tandem development, we test the realistically attainable total current of the device by studying the dependence of the photocurrent on the bottom cell thickness. Figure 1 presents the EQEs of a series of microcrystalline cells on the 2D grating substrate. When the i-layer thickness increases from 1.1 μm to 2 μm, we observe a large increase in photocurrent from 23 mA/cm^2 to 24.5 mA/cm^2, but for thicker absorber layers the photocurrent tends to saturate at about for i-layers thicker than 2.5 μm. When the thickness is increased further, the electric performance degrades significantly which leads us to the conclusion the saturation reflects a problem with the collection of carriers rather than the limit of light trapping; with improved processing of the microcrystalline i-layer still higher current densities should be possible with our substrate texture.

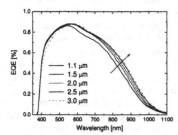 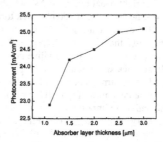

Figure 1: External quantum efficiency of microcrystalline solar cells on the periodic texture (left) and the dependence of photocurrent on the i-layer thickness (right).

Taking a limiting value of the total current of 25 mA/cm^2 for the time being, we can hope to get a matched tandem cell with a current of 12.5 mA/cm^2. However, even this current density is an ambitious goal for the amorphous top cell. If we consider a case where the amorphous top cell is stacked onto the microcrystalline bottom cell without any further enhancement, we can assume no more than one single pass through the amorphous absorber layer. After passing through the top cell, the remaining light has ample chance to be absorbed in the bottom cell where it can make one passes through a thick absorber layer, undergo diffusion at the back contact and make another pass before it could reach again the amorphous cell. Thus, the top cell in such a tandem does not benefit from the back reflector. The thickness series of top n-i-p amorphous cells in Figure 2 shows that we require an i-layer thickness of more than 600 nm to obtain a photocurrent of more than 12 mA/cm^2. Clearly, such a thickness is not desirable in terms of light induced degradation.

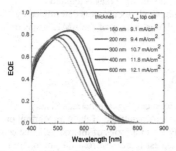 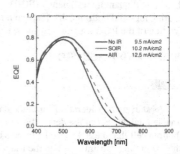

Figure 2: EQEs of the top cell in n-i-p/n-i-p tandem cells on the flexible substrate. The left panel illustrates the variation of i-layer thickness, the right panel illustrates the importance of a textured intermediate reflector layer (i-layer thickness: 200 nm).

By introducing a layer of lower refractive index between the bottom cell and the top cell, we can establish an interference condition where poorly absorbed light in the range between 600 and 750 nm is selectively reflected back into the amorphous absorber layer. The right panel in Figure 2 shows that a 100 nm thick, nominally flat SOIR improves the current in a 200 nm thick top cell from 9.5 mA/cm^2 to 10.2 mA/cm^2.

However, we are still far from the targeted photocurrent of 12.5 mA/cm^2, for two reasons. First, the periodic substrate with its period of 1200 nm is well matched to the requirements of the microcrystalline bottom cells, but not necessarily to amorphous cells. In 270 nm thick single junction amorphous cells deposited directly on this reflector, we obtained photocurrents between 12.8 and 14.4 mA/cm^2 [12], but the reflection at the SOIR in the tandem is very likely to be much inferior to the ZnO/Ag back contact. Second and more importantly, the texture of the back contact is changed by the growth of the microcrystalline bottom cell. Figure 3 shows a cross section image through a tandem cell on the periodic reflector. We observe that the amplitude of the substrate texture is reduced towards the top of the microcrystalline layer, and the sinusoidal shape is changed towards a mostly flat interface with small depressions located along the minima of the initial structure. Thus, the reflection at the SOIR can yield a second pass of light through the top cell, but we expect only little scattering of light at the flattened interface. Correspondingly, the top cell compares well to flat reference cells on a good reflector (e.g. 11 mA/cm^2 for a 270 nm thick cell [16]).

In order to achieve a real light scattering in the top cell, we introduce an asymmetric intermediate reflector (AIR) grown by LPCVD-ZnO [17]. Inherent to the growth process, LPCVD ZnO develops a textured surface regardless whether the substrate is flat or mildly textured, and it is well documented that this surface texture is well adapted to amorphous solar cells [18]. Figure 3 shows that a 1.6 μm thick LPCVD-ZnO layer completely fills the small depressions in the surface of the microcrystalline layer, and develops its own typical pyramidal texture independent of the original periodicity.

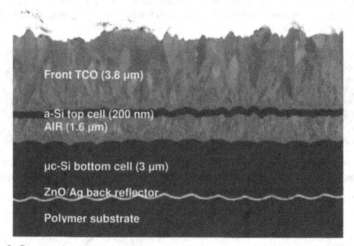

Figure 3: Cross section image through a micromorph tandem cell with LPCVD-ZnO AIR on the periodically textured polymer substrate.

Figure 2 shows that the asymmetric intermediate reflector results in a massive improvement of the top cell current, indeed, we achieve the goal of 12.5 mA/cm^2 in a 200 nm thick amorphous cell. The best cell of the AIR development showed an initial efficiency of 11.2%. After 1000 h of light soak, the efficiency stabilized at 9.8%.

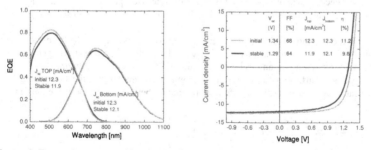

Figure 4: External quantum efficiency (left) and current voltage characteristic (right) of micromorph tandem cells with AIR on flexible polymer substrate in initial and stabilized state (1000 h light soak at 50°C and an illumination density of 100 mW/cm^2)

p-i-n cells

The p-i-n cell development is carried out on glass substrates covered with an LPCVD-ZnO front contact. Figure 5 shows typical surface morphologies for type A and type C substrates. Type A substrates consist of randomly distributed pyramids with a typical lateral feature size of 300 nm and clearly defined facets, whereas the surface treated type C substrates consist of large features (about 800 to 1000 nm) with rounded out bottoms of U-shape.

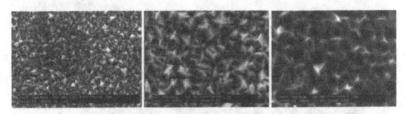

Figure 5: Surface morphology of type A substrate (left). Thicker films show similar shape, but bigger features (middle). A surface treatment yields type C substrate (right).

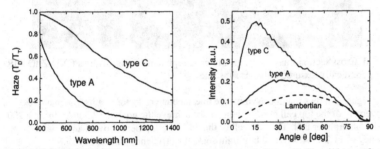

Figure 6: Spectral haze of type A and type C substrates (left panel). The right panel shows the sin-weighted ARS, the dotted line illustrates ideal Lambertian scattering.

Figure 6 compares the optical characteristics of the two layers; the haze is defined as ratio of diffuse to total transmission $H=T_D/T_T$, the angle resolved scattering (ARS) is the intensity scattered into angles between 0 and 90° for a fixed wavelength. For randomly rough surfaces with a correlation length much smaller than the wavelength, scalar scattering theory predicts the following relation between haze and wavelength [19, 20].

$$H = 1 - \exp\left\{ -\left(\frac{2\pi\sigma}{\lambda} |n_1 \cos\theta_1 - n_2 \cos\theta_2| \right)^{\beta} \right\}$$

The angles θ_1 and θ_2 represent the incident beam and the direction of the scattered beam, respectively; n_1 and n_2 are the corresponding refractive indices. The exponent β

240

should be equal to 2, but different values have been reported experimentally [20, 21]. The haze data are shown in the left panel of Figure 6; we find β values of 2.8 and 2 for the type A substrate and the type C substrate, respectively [22]. The ARS data in the right panel are plotted after weighting with the sinus of the scattered angle which provides the shape of the probability density function associated to the angular distribution. The type A substrate shows a maximum a 40° which is very close to the behaviour of an ideal Lambertian diffuser (maximum at 45°), indicating that it scatters effectively into high angles. The type C substrate scatters into a narrow angular distribution, the most probable scattering angle being only 20°. Assuming rotational symmetry with respect to surface normal (ARS not dependent on the polar angle φ), the integrated areas under the curves are proportional to the diffuse part of the transmittance T_D. Their variation reflects the different haze values for the wavelength of the ARS measurement, in this case at 543 nm.

$$T_D \sim \iint ARS \cdot \sin\theta \cdot \mathrm{d}\varphi \mathrm{d}\theta = 2\pi \cdot \int ARS \cdot \sin\theta \cdot \mathrm{d}\theta$$

Figure 7 compares EQEs of tandem cells on the two different types of LPCVD ZnO front contacts; the top and bottom cell thicknesses are 290 nm and 3.0 μm, respectively. The cells without intermediate reflector in the left panel show identical top cell currents of 10.9 mA/cm^2, but the moderately doped type C substrate yields better bottom currents because of the lower free carrier absorption in the TCO and because the large grained type C structure is better suited for light trapping in the bottom cell. The behaviour of the EQEs in the top cells suggests that the current is essentially produced in one single pass through the amorphous absorber. The right panel shows the situation after the introduction of a SOIR with 150 nm thickness. Both top amorphous cells gain in current because of reflection at the SOIR, but we observe a larger gain in the device with the type A front contact (2.6 mA/cm^2 compared to 2.1 mA/cm^2 on the type C substrate). We can understand the observations in terms of the optical measurements shown in Figure 6 when we assume that, when multiple reflections can occur within the top cell, the broad ARS of the type A substrate can offset its lower haze values. We have to keep in mind though, that the optical measurements in air are different from the situation in the cell where the actual scattering interface is between ZnO and silicon, not between ZnO and air.

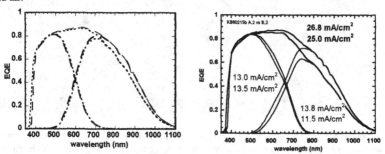

Figure 7: External quantum efficiencies of tandem solar cells on type A and type C substrates; left: no SOIR, right: with 150 nm thick SOIR.

We tried to combine the beneficial scattering effects of both types of substrates in a dedicated experiment by growing ZnO of type A on top of a substrate similar to type C. With this approach we intended to supply small features of type A for light scattering into the top cell. Furthermore, we anticipate that the flattening of the small features during the growth of the amorphous absorber should still maintain the larger features for scattering into the bottom cell. We compare a set of four different substrates including a type C substrate as reference (sample D3). The A3 sample is similar to the type C, but thinner; consequently it shows smaller features. The details of the substrate fabrication sequence are given in Table 1, further details can be found in [23].

Table 1: Design of the substrates for the double layer test

	A3	B3	C3	D3
Thickness 1st layer (µm)	3.6	3.6	3.6	4.5
SiOx layer		yes	no	
Thickness 2nd layer (µm)	-	1.1	1.1	
σ_{rms} (nm)	100	102	152	159
Comment:	"thin" type C	Figure 8 right	Figure 8 left	type C

The samples B3 and C3 are double structures, but care must be taken because the two constituent layers are made from the same material, ZnO. Thus, it turned out that simple stacking with surface treatment in-between just resumes the growth of the large grained ZnO in a form of local epitaxy, resulting in the formation of preferred surface facets very similar to the type C substrate before the surface treatment (c.f. left panel of Figure 8 compared to middle panel of Figure 5). The insertion of a thin SiOx layer using the same conditions as the SOIR layer, but only 5 nm thick, can effectively break the local epitaxy and force the ZnO to nucleate new grains. The right panel of Figure 8 shows that the result is a double structure that resembles to some extent the recently developed Asahi W structure [24].

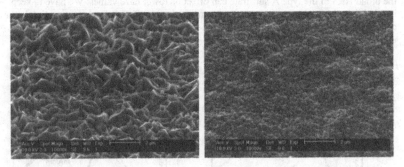

Figure 8: Stacking of type A ZnO on type C ZnO. In the left panel no interface treatment was used resulting in resumed grain growth and large features (sample C3), in the right panel a thin layer of SiOx was inserted, breaking the grain growth (sample B3).

Figure 9 compares the ARS of the four different substrates listed in Table 1. As expected, the single layer samples A3 and D3 are similar; their behaviour in the shallow angle range is identical, but we observe that D3 shows more scattering into large angles. The double structure without the SiOx treatment (sample C3) shows the best scattering into high angles, resembling in fact the type A structure of Figure 6. Surprisingly, the sample B3 with its clearly distinguishable double structure shows very poor scattering into high angles, but strong contributions into small angles around ~15°.

The EQEs of tandem cells with SOIR on the four different samples are shown in the right panel of Figure 9. We observe that the sample B3 shows poor light trapping in the bottom cell for wavelengths greater than 750nm, while the EQEs of the other samples are identical in this range. In the visible range between 550 and 700 nm, the inset shows that the top cell EQEs are higher in the double structures (samples B3 and C3) than in the single structures (samples A3 and D3). In the same range, the bottom EQEs of samples B3 and C3 are lower, indicating an efficient redistribution of light into the top cell. At this stage of development, it appears that among the double structures the configuration of sample C3 is preferable to B3, even though on that sample the character of a double structure is not obvious from Figure 8.

The poor performance of the B3 sample could possibly be explained by the poor ARS of this sample, but for the other substrates in this test the correspondence is much less evident. We should keep in mind that the measurement in air does not necessarily reflect the real light scattering properties in the cell. Recently, we proposed an optical model based on a Fourier approach [25] which is capable of predicting the behaviour of the ARS on the basis of AFM surface profile data [23]. We are confident that this tool will be useful in our further development of double structures because the calculations can be benchmarked against measured ARS data, and they can more reliably predict the behaviour in solar cells using realistic refractive indices.

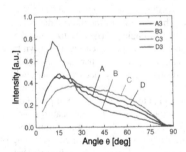 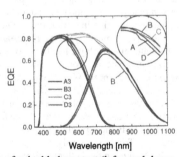

Figure 9: Comparison sin-weighted ARS data of a double layer test (left panel, layer types explained in text). The right panel shows the EQEs of tandem solar cells with SOIR on the different structures, the inset gives the details of the top cell EQEs.

After this brief outlook on new, but not yet fully conclusive double structures, we conclude this section on p-i-n tandem solar cells by reverting to the type C substrate; Figure 10 shows data on the electric performance of a configuration combining a top cell thickness of 300 nm, a bottom cell thickness of 3μm, and a SiOx intermediate reflector layer of 150 nm thickness. In this configuration, and without anti-reflection coating, we obtain an initial efficiency of 12.6% which shows a relative degradation of 12% after 1000 h of light soaking, reaching a stabilized efficiency of at 11.1%. The area of this cell is 1.2 cm^2.

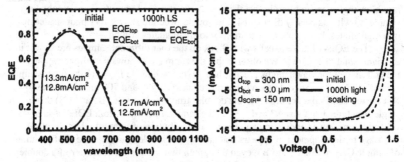

Figure 10: EQE and current voltage characteristics of a micromorph tandem solar cell with SOIR on the type C substrate in its initial and stabilized state (cell size 1.2 cm^2, no anti-reflection coating).

CONCLUSIONS

We presented thin film silicon tandem cells in n-i-p and p-i-n configuration. High efficiencies necessitate matched current densities higher than 12.5 mA/cm^2, but at the same time their thickness should not exceed 300 nm in order to avoid light induced degradation. Currently, an intermediate reflector between the top and bottom cells is the most successful route towards high top cell current densities. In n-i-p tandems, we observed that growth of the thick bottom cell results in adverse changes of the surface morphology, resulting in an almost flat interface. The potential of the intermediate reflector is very limited in these circumstances. By introducing surface texture into the intermediate reflector layer itself, we were able to supply to the top cell its very own light scattering interface, and we are at liberty to use a structure that is experimentally well proven for this purpose. We are able, on plastic substrate to apply that scheme to reach close to 10% stable efficiency micromorph cell, with an initial matched current of 12.3 mA/cm2 . We tried to apply the same line of arguments to the related case of p-i-n cells; a structure that combines small and large features should provide the fine features for light diffusion in the top cell while the flattening effect of growing the amorphous layer should not affect too seriously large features which could then diffuse longer wavelengths for light trapping in the bottom cell. We could successfully fabricate ZnO surfaces that show textures on two different length scales, but so far their application into tandem solar cells did not show a clear improvement. Single structure large grain TCO still leads to the best stable efficiencies over 11% with 12.5 mA/cm2 current.

ACKNOWLEDGEMENTS

Funding by the EU projects Flexcellence (contract No. 019948) and Athlet (contract No. 019670) as well as support from the Swiss Federal Office for Energy (OFEN) under project No. 101191 are thankfully acknowledged.

REFERENCES

1. H. W. Deckman, C. R. Wronski, H. Witzke, and E. Yablonovitch, *Optically Enhanced Amorphous-Silicon Solar-Cells.* Applied Physics Letters, 1983. **42**(11): p. 968-970.
2. M. Kambe, M. Fukawa, N. Taneda, Y. Yoshikawa, K. Sato, K. Ohki, S. Hiza, A. Yamada, and M. Konagai. *Improvement of light-trapping effect on microcrystalline silicon solar cells by using high haze transparent conductive oxide films.* in *Proc. 3rd World PVSEC.* 2003. Osaka. p. 1812-1815
3. S. Fay, J. Steinhauser, N. Oliveira, E. Vallat-Sauvain, and C. Ballif, *Opto-electronic properties of rough LP-CVD ZnO:B for use as TCO in thin-film silicon solar cells.* Thin Solid Films, 2007. **515**(24): p. 8558-8561.
4. O. Kluth, B. Rech, L. Houben, S. Wieder, G. Schöpe, C. Beneking, H. Wagner, A. Löffl, and H. W. Schock, *Texture etched ZnO: Al coated glass substrates for silicon based thin film solar cells.* Thin Solid Films, 1999. **351**(1-2): p. 247-253.
5. A. Banerjee and S. Guha, *Study of Back Reflectors for Amorphous-Silicon Alloy Solar-Cell Application.* Journal of Applied Physics, 1991. **69**(2): p. 1030-1035.
6. R. Franken, R. Stolk, H. Li, C. van der Werf, J. Rath, and R. Schropp, *Understanding light trapping by light scattering textured back electrodes in thin film n-i-p-type silicon solar cells.* Journal of Applied Physics, 2007. **102**: p. 014503.
7. M. Python, E. Vallat-Sauvain, J. Bailat, D. Dominé, L. Fesquet, A. Shah, and C. Ballif, *Relation between substrate surface morphology and microcrystalline silicon solar cell performance.* Journal of Non-Crystalline Solids, 2008. **354**(19-25): p. 2258-2262.
8. H. Li, R. Franken, J. Rath, and R. Schropp, *Structural defects caused by a rough substrate and their influence on the performance of hydrogenated nano-crystalline silicon n–i–p solar cells.* Solar Energy Materials and Solar Cells, 2009.
9. D. Fischer, S. Dubail, J. D. Anna Selvan, N. Pellaton Vaucher, R. Platz, C. Hof, U. Kroll, J. Meier, P. Torres, H. Keppner, N. Wyrsch, M. Goetz, A. Shah, and K.-D. Ufert. *The micromorph solar cell: extending a a-Si:H technology twoards thin film crystallline silicon.* in *Proc. 25th IEEE PVSC.* 1996. Washington D. C. p. 1053-1056
10. P. Buehlmann, J. Bailat, D. Domine, A. Billet, F. Meillaud, A. Feltrin, and C. Ballif, *In situ silicon oxide based intermediate reflector for thin-film silicon micromorph solar cells.* Applied Physics Letters, 2007. **91**(14): p. 143505.
11. K. Yamamoto, A. Nakajima, M. Yoshimi, T. Sawada, S. Fukuda, T. Suezaki, M. Ichikawa, Y. Koi, M. Goto, and T. Meguro, *A high efficiency thin film silicon solar cell and module.* Solar Energy, 2004. **77**(6): p. 939-949.

12.   F.-J. Haug, T. Söderström, M. Python, V. Terrazzoni-Daudrix, X. Niquille, and C.
       Ballif, *Development of micromorph tandem solar cells on flexible low cost plastic
       substrates.* To be published in Sol. En. Mat., 2009.
13.   J. Bailat, D. Dominé, R. Schlüchter, J. Steinhauser, S. Faÿ, F. Freitas, C. Bücher,
       L. Feitknecht, X. Niquille, R. Tscharner, A. Shah, and C. Ballif. *High efficiency
       pin microcrystalline and micromorph thin film silicon solar cells deposited on
       LPCVD ZnO coated glass substrates.* in *Proc. 4th World PVSEC.* 2006. Hawaii.
       p. 1533-1536
14.   T. Söderström, F. J. Haug, X. Niquille, V. Terrazoni-Daudrix, and C. Ballif,
       *Asymmetrid intermediate reflector for tandem micromorph thin film silicon solar
       cells.* Applied Physics Letters, 2009. **94**: p. 063501.
15.   D. Dominé, P. Buehlmann, J. Bailat, A. Billet, A. Feltrin, and C. Ballif. *High-
       efficiency micromorph silicon solar cells with in-situ intermediate reflector
       depositd on various rough LPCVD-ZnO.* in *Proc. 23rd European PVSEC* 2008.
       Valencia. p. 2091-2095
16.   T. Söderström, F. J. Haug, V. Terrazzoni-Daudrix, and C. Ballif, *Optimization of
       amorphous silicon thin film solar cells for flexible photovoltaics.* Journal of
       Applied Physics, 2008. **103**(11): p. 114509-114509.
17.   T. Söderström, *Single and multi-junction thin film silicon solar cells for flexible
       photovoltaics,* PhD Thesis, University of Neuchatel, 2009
18.   J. Meier, J. Spitznagel, U. Kroll, C. Bucher, S. Faÿ, T. Moriarty, and A. Shah,
       Potential of amorphous and microcrystalline silicon solar cells. Thin Solid Films,
       2004. **451**: p. 518-524.
19.   C. K. Carniglia, *Scalar scattering theory for multilayer optical coatings.* Optical
       Engineering, 1979. **18**(2): p. 104–115.
20.   M. Zeman, R. Van Swaaij, J. W. Metselaar, and R. E. I. Schropp, *Optical
       modeling of a-Si: H solar cells with rough interfaces: Effect of back contact and
       interface roughness.* Journal of Applied Physics, 2000. **88**: p. 6436.
21.   H. Stiebig, T. Brammer, T. Repmann, O. Kluth, N. Senoussaoui, A. Lambertz,
       and H. Wagner. *Light Scattering in Microcrystalline Silicon Thin Film Solar
       Cells.* in *Proc. 16th EU-PVSEC.* 2000. Glasgow. p. 549-552
22.   D. Domine, P. Buehlmann, J. Bailat, A. Billet, A. Feltrin, and C. Ballif, *Optical
       management in high-efficiency thin-film silicon micromorph solar cells with a
       silicon oxide based intermediate reflector.* Physica Status Solidi (RRL)-Rapid
       Research Letters, 2008. **2**(4).
23.   D. Domine, *The role of front electrodes and intermediate reflectors in the
       optoelectronic properties of high-efficiency micromorph solar cells,* PhD Thesis,
       University of Neuchatel, 2009
24.   T. Oyama, M. Kambe, N. Taneda, and K. Masumo. *Requirements for TCO
       substrate in Si-based thin film solar cells - toward tandem.* in *MRS Spring
       Meeting.* 2008. San Francisco. p. KK02-01
25.   J. E. Harvey and A. Krywonos. *A Global View of Diffraction: Revisited.* in *Proc.
       SPIE AM100-26.* 2004. Denver. p.

Mater. Res. Soc. Symp. Proc. Vol. 1153 © 2009 Materials Research Society   1153-A13-02

Light trapping in hydrogenated amorphous and nano-crystalline silicon thin film solar cells

Jeffrey Yang, Baojie Yan, Guozhen Yue, and Subhendu Guha
United Solar Ovonic LLC, 1100 West Maple Road, Troy, Michigan 48084

ABSTRACT

Light trapping effect in hydrogenated amorphous silicon-germanium alloy (a-SiGe:H) and nano-crystalline silicon (nc-Si:H) thin film solar cells deposited on stainless steel substrates with various back reflectors is reviewed. Structural and optical properties of the Ag/ZnO back reflectors are systematically characterized and correlated to solar cell performance, especially the enhancement in photocurrent. The light trapping method used in our current production lines employing an a-Si:H/a-SiGe:H/a-SiGe:H triple-junction structure consists of a bi-layer of Al/ZnO back reflector with relatively thin Al and ZnO layers. Such Al/ZnO back reflectors enhance the short-circuit current density, J_{sc}, by ~20% compared to the bare stainless steel. In the laboratory, we use Ag/ZnO back reflector to achieve higher J_{sc} and efficiency. The gain in J_{sc} is about ~30% for an a-SiGe:H single-junction cell used as the bottom cell of a multi-junction structure. In recent years, we have also worked on optimizations of the Ag/ZnO back reflector for nano-crystalline silicon (nc-Si:H) solar cells. We have carried out a systematic study on the effect of texture for Ag and ZnO. We found that for a thin ZnO layer, a textured Ag layer is necessary to increase J_{sc}, even though the parasitic loss is higher at the Ag and ZnO interface due to the textured Ag. However, a flat Ag can be used for a thick ZnO to reduce the parasitic loss, while the light scattering is provided by the textured ZnO. The gain in J_{sc} for nc-Si:H solar cells on Ag/ZnO back reflectors is in the range of ~60-75% compared to cells deposited on bare stainless steel, which is much larger than the light-trapping enhancement observed for a-SiGe:H cells. The highest total current density achieved in an a-Si:H/a-SiGe:H/nc-Si:H triple-junction structure on Ag/ZnO back reflector is 28.6 mA/cm^2, while it is 26.9 mA/cm^2 for a high efficiency a-Si:H/a-SiGe:H/a-SiGe:H triple-junction cell.

INTRODUCTION

Hydrogenated amorphous silicon (a-Si:H), silicon-germanium alloy (a-SiGe:H), and nano-crystalline silicon (nc-Si:H) are the three intrinsic materials commonly used in multi-junction solar cells [1,2]. Because of the nature of the disorder, the amorphous materials normally have a low mobility-lifetime product, which sets the fundamental limit of the solar cell thickness. For a-Si:H and a-SiGe:H cells, the intrinsic layers are in the range of a few thousand angstroms. For nc-Si:H cells, the intrinsic layer can only extend to one to two micrometers for maintaining a reasonable fill factor (FF). However, a thin intrinsic layer cannot produce enough short-circuit current density (J_{sc}). To resolve this issue, light trapping with textured substrates becomes an important technique for improving the J_{sc} in a-Si:H based solar cells without significant losses in the FF [3-7]. In addition to the conventional random textured substrate or superstrate light trapping approaches, advanced light managements have been proposed in recent years such as plasmon enhanced light trapping by metal nano-particles [8], two or three dimensional photonic structures [9], and periodic gratings [10,11]. However, random textured light trapping is still favored and practiced in large-volume productions. This paper focuses on light trappings from randomized scattering.

For a-Si:H based solar cells on transparent conductive oxide (TCO) coated glass superstrates, light trapping mainly arises from the texture of a thick TCO layer, which is usually a few thousand angstroms thick. For cells on flexible substrates, such as stainless steel (SS) or polymer, light trapping is mainly from the texture of substrates because the top transparent contact, normally indium-tin-oxide (ITO), is very thin for the anti-reflection effect and cannot be easily textured. In the substrate cell structure, a textured back reflector with a thin layer of highly reflective metal (Al or Ag) and a dielectric layer, such as ZnO, is coated on substrates to provide the light-trapping effect. The metal layer is used to reflect the light reaching the substrate, and the dielectric layer to enhance the light trapping and reduce the probability of potential metal diffusion into the semiconductor layer.

We have extensively studied Al/ZnO and Ag/ZnO back reflectors (BR) for high efficiency a-Si:H/a-SiGe:H/a-SiGe:H triple-junctions solar cells [3,4]. From the cost point of view, Al/ZnO BR are used in our current production lines. For high efficiency solar cells, Ag/ZnO BR are used in the laboratory. Recently, we have worked on the re-optimization of Ag/ZnO BR for a-Si:H and nc-Si:H multi-junction solar cells [12,13]. In this paper, we discuss various aspects of Ag/ZnO BR for high efficiency solar cells. The main emphasis is on the light trapping in nc-Si:H solar cells.

EXPERIMENTAL

Al/ZnO and Ag/ZnO back reflectors are deposited on SS substrates in a batch or a roll-to-roll system. Batch machines are used in the laboratory and roll-to-roll machines are used in both the R&D laboratory and production lines. Both the metal and ZnO layers can be textured by adjusting the deposition parameters and/or the thickness of the layers.

The surface and interface structures on the BR are measured with atomic-force-microcopy (AFM) and cross-sectional transmission electron microscopy (X-TEM). The crystalline structures are characterized by X-ray scattering spectroscopy. The optical properties of BR are characterized using light scattering with a He-Ne laser and reflection spectrum measurements. In the light scattering measurements, the scattered light intensities are measured as a function of angles with respective to the sample surface, where the incident light is perpendicular to the sample surface. In the reflection measurements, total and diffusive reflection spectra are measured in the wavelength range from 200 nm to 2500 nm.

Various thin film n-i-p solar cells with a-Si:H, a-SiGe:H, and nc-Si:H as intrinsic layers are studied. The intrinsic layers are deposited with radio frequency (RF) or very high frequency (VHF) glow discharge techniques at different deposition rates. Details of the solar cell optimization have been published elsewhere [14-16]. Current-density versus voltage (J-V) curves of solar cells are measured under an AM1.5 solar simulator at 25 °C. External quantum efficiency (QE) spectra are measured in the wavelength range between 300 nm and 1200 nm. Short-circuit current density (J_{sc}) values are obtained by convoluting the AM1.5 solar spectrum with the measured QE spectra.

ANALYSIS OF Ag/ZnO BACK REFLECTORS

The texture of Ag and ZnO layers can be controlled by the process parameters such as substrate temperature and film thickness. Normally, high substrate temperature and thick film result in high texture on the sample surface. The ZnO texture can be modified by plasma or chemical etching [6,7]. Except for texturing the ZnO layers by controlling the deposition process, we also use chemical etching with 0.5% HCl as reported by the Jülich group [7].

Figure 1 shows four AFM pictures of Ag/ZnO back reflectors made under various conditions and chemical treatments, where (a) is a conventional back reflector previously developed for a-SiGe:H solar cells [3,4], (b) is a back reflector with large micro-features created by adjusting the deposition parameters, (c) is a back reflector with a thick ZnO before chemical etching, and (d) is the same sample as in (c) but after a 30-second chemical etching in 0.5% HCl. The micro-feature size on the conventional back reflector is on the order of 0.1-0.2 μm with a root mean square (RMS) roughness of 38.2 nm. By adjusting the deposition parameters, we can increase the feature size to as large as 0.5-1.0 μm with RMS of 80.5 nm (Fig.1 (b)). In addition, the chemical etching modifies the micro-features on the BR surface. It seems that the small features are etched away and the feature size effectively increases. Correspondingly, the RMS increases from 39.6 nm to 63.2 nm after a 30-second etching in 0.5% HCl as shown by comparing Figs. 1 (c) and (d).

The surface texture is mainly from the crystalline structure. Normally, both Ag and ZnO have polycrystalline structures when the layers were deposited at an elevated substrate temperature. Figure 2 (upper) shows an example of XRD spectrum of a Ag/ZnO BR. It reveals that the ZnO layer has a strong (002) preferential orientation and the Ag layer a (111) preferential orientation. We estimated the grain size from the line width of XRD peaks and found that the calculated grain size is in the range of 17-25 nm without a clear trend in film thickness. The grain sizes are much smaller than the feature sizes on the surface of the BR.

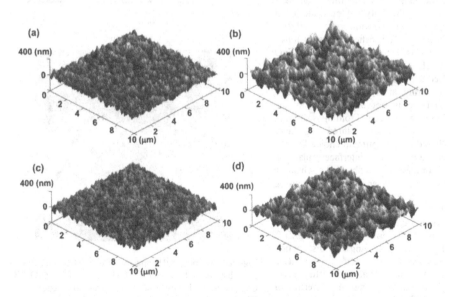

Figure 1. AFM pictures of four Ag/ZnO BRs made with various deposition and chemical etching parameters. Details of the sample preparation are given in the text.

We took X-TEM images of Ag/ZnO BR to analyze the crystalline structure and compare the images with the XRD results. Figure 2 (lower) shows an X-TEM image of Ag/ZnO layers deposited on c-Si. The image shows that both the Ag and ZnO layers have textured surfaces. The ZnO grain sizes are much larger than those estimated from XRD. The ZnO crystals have a columnar structure with the vertical grain size similar to the film thickness and the lateral size up to 0.3 μm. The lateral grain size determines the texture of the ZnO surface as seen by the Pt cap layer. There are two possible explanations for the different grain sizes measured by XRD and X-TEM. First, it could be that the XRD peak width is not determined by the grain size but by other factors such as the stress in the film. The average ZnO (002) peak position is at $2\theta=34.37°$, which is slightly lower than the peak position in ZnO powders, which could be an indication of stress in the film. Second, the large grains in the X-TEM images may contain many small grains as measured by XRD.

OPTICAL CHARACTERIZATION

The major factors affecting the light trapping in thin film solar cells are the reflectivity and the scattering at the semiconductor/back reflector interface. The scattering occurs at two locations: the semiconductor/ZnO interface and the ZnO/Ag interface. Because ZnO is transparent, the scattering at the semiconductor/ZnO interface is not as effective as the ZnO/Ag interface. From the scattering point of view, textured Ag layer should be the optimum choice for BR. However, the interface plasmon absorption is enhanced when a textured Ag layer is used [17], causing extra absorption (parasitic absorption) and reducing the reflectivity. Therefore, one has to consider these two effects for optimizing Ag/ZnO BR. We use angular distribution measurement and reflection spectroscopy (total and diffusive) to characterize the optical properties of Ag/ZnO BR.

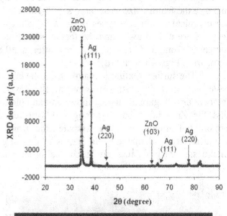

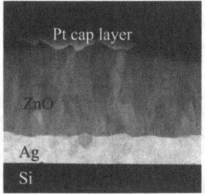

Figure 2. (upper) XRD spectrum of Ag/ZnO BR coated SS substrate and (lower) X-TEM image of Ag/ZnO BR. Using c-Si as the substrate is for the convenience of X-TEM sample preparations.

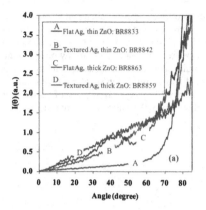 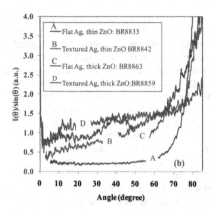

Figure 3. (a) angular distribution of scattering light intensity I(θ) and (b) the scattering light intensity with correction of viewing area.

Figure 3 (a) shows the measured scattering light intensity versus the angle, where the incident light is perpendicular to the substrate surface (90°). The sample with a flat Ag layer and a thin ZnO layer results in a large reflection with angles close to the direct reflection. The textured Ag with a thin ZnO layer produces both large scattering close to the direct reflection and wide angle scattering. Even though the ZnO layer is thin, the scattering is very pronounced, which means that one may obtain very good BRs with a textures Ag and a thin layer of ZnO. In order to increase the scattering further, we increase the ZnO thickness to cause more scattering at the ZnO surface. From the figure, one can see that the thick ZnO on textured Ag definitely enhances light-scattering for small angles, but reduces the scattering for angles close to the direct reflection. The small angle scattering is also significantly enhanced by the thick ZnO on flat Ag. An additional feature appears at around 40 degree, which could be related to the feature size and shape distributions of the ZnO layer. Theoretically, the perfect random surface for light scattering should produce a scattering light intensity proportion to cos(α), where α is the angle between the scattered light and the normal to the surface (in Fig. 3, α=90°-θ). The factor of cosine takes into account the effective area of the light spot on the surface to the viewer at a given angle. Such BR is referred to as Lambertian BR [18], which produces scattering light intensity following the cosine function. Figure 3 (b) plots the measured scattering light intensity divided by cos(90°-θ). It shows that the sample with textured Ag and thick ZnO layer is very close to a Lambertian BR.

Because the light scattering measurements are carried out with a single-wavelength light, it may not provide a complete picture of light trapping in solar cells. A complementary measurement is the total and diffusive reflection spectra. Figure 4 shows a comparison of two Ag/ZnO back reflectors with a thin ZnO layer (0.13 μm), where the upper figure plots the total reflection and the lower one plots the diffusive reflection. From the total reflection, one notices that the sample with a flat Ag layer shows high reflectivity in the long wavelength region, and a broad valley in the wavelength around 500 nm. The valley could be related to the surface plasmon absorption and the antireflection effect of the thin ZnO layer. Even though we attempted to make flat Ag, there are inevitably some textures on the Ag surface. For the sample with a textured Ag layer, the absorption valley moves to short wavelengths, but the long wavelength reflectivity is significantly reduced. Two possible reasons are given for the long wavelength loss in the samples with textured Ag. First, there could be some steep valleys on the Ag surface such that the incident light cannot escape from the valleys (from ray optics point of view); second, the highly textured Ag surface could also result in different surface plasmon modes and absorb the long wavelength light. The diffusive reflection spectra are very different between the two samples. The sample with a textured Ag layer shows significantly higher diffusive reflections than the one with a flat Ag layer over the entire wavelength range. This result is consistent with the light scattering measurements. Figure 5 shows the same measurement results from two samples with a thick ZnO layer (2.0 μm). Two observations can be made. First, the total reflections are reduced by the thick ZnO; the clear features of plasmon absorption observed for the thin ZnO layer disappear, although the decrease of reflectivity with the decrease of wavelength could still be caused by the surface plasmon absorptions. The difference between the two samples in total reflection becomes smaller than the case of the thin ZnO. Second, the long wavelength diffusive reflection increased significantly compared to the samples with the thin ZnO layer. Although the sample with textured Ag still shows high diffusive reflection, the difference becomes smaller.

From the optical measurements, we learn that for a thin ZnO layer, a textured Ag is necessary for high light scattering. With a thick ZnO layer, the scattering could be enhanced by the textures at the ZnO surface. Although the scattering from the samples with thick ZnO and textured Ag is still the highest, the reduction of surface absorption with a flat Ag could still be important for photocurrent enhancement.

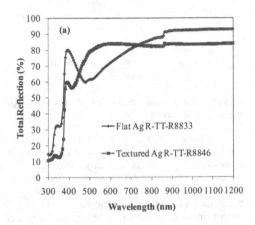

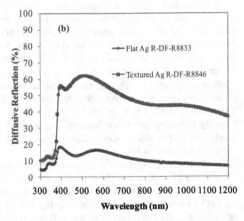

Figure 4. Total (a) and diffusive (b) reflection spectra of two SS/Ag/ZnO back reflectors. R8833 was made with a flat Ag and 0.13-μm thick ZnO layer and R8846 with a textured Ag and 0.13-μm thick ZnO.

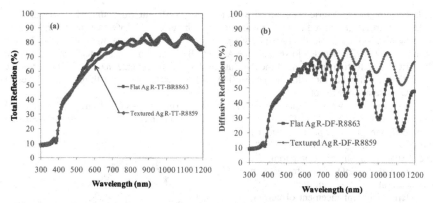

Figure 5. Total (a) and diffusive (b) reflection spectra of two SS/Ag/ZnO BR. R8863 was made with a flat Ag and 2.0-μm thick ZnO layer and R8859 with a textured Ag and 2.0-μm thick ZnO.

COMPARSION OF Al/ZnO AND Ag/ZnO BACK REFLECTORS

United Solar has been manufacturing a-Si:H/a-SiGe:H/a-SiGe:H triple-junction solar cells on Al/ZnO coated SS substrates. The advantage of using Al/ZnO back reflectors is mainly for the low cost. First, Al is less expensive than Ag. Second Al is much easier to be textured than Ag. Therefore, one can use a thinner Al to reach a reasonable texture for light trapping. The disadvantage of Al/ZnO is the lower reflectivity than Ag/ZnO. Figure 6 (a) shows the total reflection spectra of Al/ZnO and Ag/ZnO back reflectors from a roll-to-roll manufacturing machine. Comparing the total reflection curves, one notices a broad dip around 850 nm in the spectrum of Al/ZnO reflector, which could be partially caused by inter-band transition of Al as indicated by the arrow in Fig. 6 (b). The antireflection effect of the thin ZnO layer also reduces the total reflections, but it does not affect the light trapping effect when the BR is used in solar

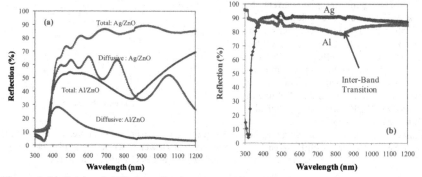

Figure 6. (a) Total and diffusive reflection spectra of an Al/ZnO back reflector and a Ag/ZnO back reflector, (b) total reflection of Al and Ag coated glass substrates.

cells because the dielectric constant of silicon is very different from air. However the low reflectivity of Al indeed reduces solar cell current. Therefore, Al/ZnO back reflectors have an intrinsic disadvantage from the reflectivity's point of view. In addition, the diffusive reflection from the Al/ZnO BR is much lower than the Ag/ZnO BR. The significant interference fringes in the diffusive reflection spectrum of the Ag/ZnO back reflector indicate that the ZnO layer in the Ag/ZnO BR is thicker than that in the Al/ZnO back reflector. Using a thinner ZnO layer in Al/ZnO is mainly due to cost.

Figure 7 shows a comparison of QE curves for (upper) a-SiGe:H and (lower) nc-Si:H solar cells made on SS, Al/ZnO back reflector, and Ag/ZnO back reflector. For the a-SiGe:H solar cell, the gains in photocurrent by the Al/ZnO and Ag/ZnO BR are 19.5% and 30.0%, respectively. Comparing the Al/ZnO and Ag/ZnO, the extra gain in photocurrent is around 10%. Transferring the enhancement of current to solar cell efficiency, we usually see an increase of a-Si:H/a-SiGe:H/a-SiGe:H triple-junction solar cell efficiency around 8-10% when Al/ZnO is replaced by Ag/ZnO back reflectors. Of course, one needs to adjust the solar cell design when the back reflector is changed, especially for the current mismatch in multi-junction structures. With the optimization of Ag/ZnO back reflectors and solar cell deposition parameters, we achieved initial and stable active-area cell efficiencies of 14.6% and 13.0%, respectively, using an a-Si:H/a-SiGe:H/a-SiGe:H configuration [14].

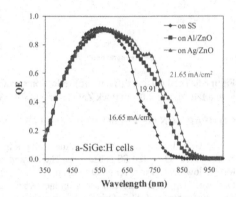

The gains in photocurrent by back reflectors are much larger for nc-Si:H than for a-SiGe:H solar cells. The example in Fig. 7 (lower plot) shows that the Al/ZnO back reflector leads to a 31.6% gain in photocurrent, while the gain by Ag/ZnO, when extending beyond 1000nm, is around 60%. By optimizing the Ag/ZnO BR, we have obtained a photocurrent gain of 75% [19]. The results clearly show that the advantage of using Ag/ZnO BR in nc-Si:H based solar cells.

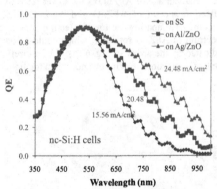

Figure 7. Comparison of QE curves of (upper plot) a-SiGe:H and (lower plot) nc-Si:H solar cells on SS, ZnO/Al//SS, and ZnO/Ag//SS substrates.

OPTIMIZATION OF Ag/ZnO BACK REFLECTORS FOR nc-Si:H SOLAR CELLS

Because of the dependence of nc-Si:H material structures on the substrate roughness, nc-Si:H solar cell performance depends on the Ag/ZnO back reflectors not only in terms of the photocurrent but also on the open-circuit voltage (V_{oc}) and FF. Normally, a flat substrate results in a lower J_{sc}, but better FF and V_{oc} [19]. It has been reported that the decrease of FF in nc-Si:H solar cells with the increase of substrate roughness arises from nano-crystallite collisions [20], because the nanocrystallites have the tendency of growing perpendicular to the local surface. Therefore the surface structure is a critical factor for nc-Si:H solar cells. In order to achieve good nc-Si:H performance, optimized Ag/ZnO BR is needed. Here, we compare the influence of the substrate structure on the nc-Si:H solar cell performance. Figure 8 shows the QE curves of nc-Si:H solar cells on the four Ag/ZnO BR characterized in the previous sections. The nc-Si:H cell on the Ag/ZnO with flat Ag and the thin ZnO layers has significant interference fringes, indicating a large portion of direct reflections. The nc-Si:H cell on the Ag/ZnO BR with textured Ag and thin ZnO layers has an improved spectral response and reduced interference fringes, indicating an enhanced scattering. The two curves for nc-Si:H cells on the Ag/ZnO BR with thick ZnO layers are very similar, which means that the texture of the Ag does not play an important role when the ZnO layer is sufficiently thick. As shown in the previous sections, for a thick ZnO, the texture on the ZnO layer is significantly increased, evidenced by more light scattering. In this case, the light scattering at the semiconductor/ZnO interface is enhanced. Comparing the flat and textured Ag layers, the texture at the ZnO/Ag interface could produce more scatterings, but the parasitic loss at the ZnO/Ag interface may offset the gain from the scattering.

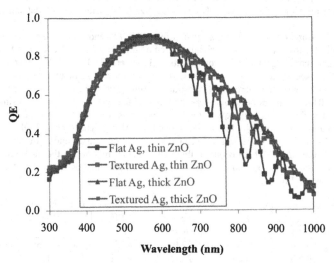

Figure 8. Quantum efficiency spectra of four nc-Si:H solar cells deposited on different Ag/ZnO back reflectors.

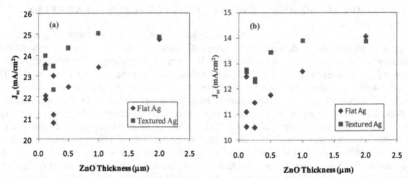

Figure 9. J_{sc} versus ZnO thickness for two sets of nc-Si:H solar cell, where (a) and (b) plot the values from 300 nm to 1200 nm, and from 610 nm to 1000 nm, respectively.

Figure 9 (a) plots the J_{sc} versus the thickness of ZnO layers, where the J_{sc} values are obtained from the convolution of the QE curves with the AM1.5 solar spectrum. Figure 9 (b) shows the long wavelength region from 610 nm to 1000 nm. For the set of nc-Si:H solar cells on BR with a flat Ag layer, the J_{sc} continuously increases with the ZnO thickness, especially for the long wavelength region. For the Ag/ZnO back reflectors with a textured Ag layer, the J_{sc} is much higher than the cells on Ag/ZnO with a flat Ag when the ZnO is thin. It also increases slightly with the ZnO thickness, but saturates around 1 μm. When the ZnO layer is around 2 μm, the J_{sc} values are similar for the nc-Si:H cells on Ag/ZnO BR with both the flat and textured Ag layers.

The texture of back reflector also affects the FF and V_{oc}. Table I lists the J-V characteristics of four nc-Si:H solar cells deposited on different Ag/ZnO BR. The J_{sc} follows the trend of light scattering and reflection results. However, the FF and V_{oc} decrease with the increase of surface texture on the substrates. Detailed analyses show that the nc-Si:H material quality decreases with the increase of substrate texture [19]. Therefore, we need to not only increase the RMS values but also optimize the shape of the surface features. By using improved Ag/ZnO back reflectors and high quality nc-Si:H materials, we have improved the nc-Si:H solar cell performance significantly. Figure 10 shows the J-V characteristics and QE curves of our best nc-Si:H solar cell deposited on a Ag/ZnO back reflector with around 2 μm thick ZnO. The J_{sc} reaches 27.6 mA/cm², which is one of the highest values reported in the literature. Using the high quality nc-Si:H as a bottom cell, we have achieved an initial active-area cell efficiency of 15.4% as shown in Fig. 11.

Table I: J-V characteristics of nc-Si cells deposited on different Ag/ZnO back reflectors.

Run No.	V_{oc} (V)	FF	J_{sc} (mA/cm²)	Eff (%)	Back Reflector
17363	0.540	0.637	22.07	7.59	Flat Ag, ZnO=0.13 μm
17389	0.556	0.629	23.40	8.18	Textured Ag, ZnO=0.13 μm
17367	0.510	0.559	24.75	7.06	Flat Ag, ZnO=2.0 μm
17372	0.511	0.562	24.85	7.14	Textured Ag, ZnO=2.0 μm

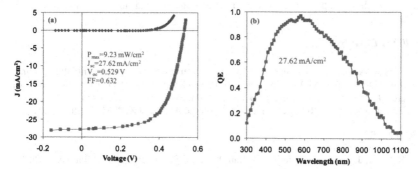

Figure 10. J-V characteristics and QE curves of a nc-Si:H solar cell on Ag/ZnO BR with a thick ZnO layer.

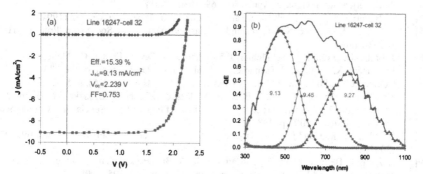

Figure 11. (a) J-V characteristics and (b) quantum efficiency curves of an a-Si:H/a-SiGe:H/nc-Si:H triple-junction cell with an initial active-area efficiency of 15.4%.

SUMMARY

This paper summarizes the current status of light trapping using Al/ZnO and Ag/ZnO BR. We use Al/ZnO BR in the current production because of the low cost. However, Al/ZnO BR has a low reflectivity. For high efficiency solar cells, Ag/ZnO BR is needed. Comparing the gains in photocurrent densities from a-SiGe:H cells and nc-Si:H cells, we found that the light trapping in nc-Si:H is more pronounced than in a-SiGe:H solar cells. For Ag/ZnO back reflectors with a thin ZnO layer, a textured Ag layer produces more light-scattering than a flat Ag layer. However, for Ag/ZnO BR with a thick ZnO layer, the light trapping originates from both the semiconductor/ZnO and ZnO/Ag interfaces. The one with textured Ag still produces high light scattering, but the parasitic loss at the ZnO/Ag interface reduces the reflectivity. With these two effects, the light enhancement in photocurrent density is similar for Ag/ZnO BR with flat and textured Ag layers. Combining the optimized Ag/ZnO back reflectors and high quality nc-Si:H materials, we have improved our nc-Si:H single-junction and multi-junction solar cells significantly.

ACKNOWLEDGEMENTS

This work was partially supported by DOE under the Solar America Initiative Program Contract No. DE-FC36-07 GO 17053. The authors thank the entire R&D group at United Solar for the great team work in the sample preparations and measurements. We also thank C.-S. Jiang and Y. Yan at the National Renewable Laboratory for the AFM and X-TEM measurements.

REFERENCES

[1]   S. Guha, J. Yang, A. Banerjee, B. Yan, K. Lord, Sol. Energy Mater. Sol. Cells **78**, 329 (2003).
[2]   A.V. Shah, J. Meier, E. Vallat-Sauvain, N. Wyrsch, U. Kroll, C. Droz, and U. Graf, Sol. Energy Mater. Sol. Cells **78**, 469 (2003).
[3]   A. Banerjee and S. Guha, J. Appl. Phys. **69**, 1030 (1991).
[4]   A. Banerjee, J. Yang, K. Hoffman, and S. Guha, Appl. Phys. Lett. **65**, 472 (1994).
[5]   T. Oyama, M. Kambe, N. Taneda, and K. Masumo, Mater. Res. Soc. Symp. Proc. **1101E**, KK02.1 (2008).
[6]   J. Bailat, D. Dominé, R. Schlüchter, J. Steinhauser, S. Faÿ, F. Freitas, C. Bücher, L. Feitknecht, X. Niquille, T. Tscharner, A. Shah, and C. Ballif, Record of the 4th World Conference on Photovoltaic Energy Conversion (2006. Hawaii, USA), p1533.
[7]   J. Müller, G. Schöpe, O. Kluth, B. Rech, M. Ruske, J. Trube, B. Szyszka, Th. Höing, X. Jiang, and G. Bräuer, Proc. of 28th IEEE Photovoltaic Specialists Conference (Anchorage, AK, 2000), p. 758.
[8]   S. Pillai, K. R. Catchpole, T. Trupke, and M. A. Green, J. Appl. Phys. **101**, 093105 (2007).
[9]   L. Zeng, P. Bermel, Y. Yi, B. A. Alamariu, K. A. Broderick, J. Liu, C. Hong, X. Duan, J. Joannopoulos, and L. C. Kimerling, Appl. Phys. Lett. **93**, 221105 (2008).
[10]   C. Eisele, C. E. Nebel, and M. Stutzmann, J. Appl. Phys. **89**, 7722 (2001).
[11]   H. Sai, H. Fujiwara, M. Kondo, and Y. Kanamori, Appl. Phys. Lett. **93**, 143501 (2008).
[12]   B. Yan, J. M. Owens, C.-S. Jiang, J. Yang, and S. Guha, Mater. Res. Soc. Symp. Proc. **862**, 603 (2005).
[13]   B. Yan, G. Yue, C.-S. Jiang, Y. Yan, J. M. Owens, J. Yang, and S. Guha, Mater. Res. Soc. Symp. Proc. **1101E**, KK13.2 (2008).
[14]   J. Yang, A. Banerjee, and S. Guha, Appl. Phys. Lett. **70**, 2975 (1997).
[15]   B. Yan, G. Yue, J. Yang, A. Banerjee, and S. Guha, Mater. Res. Soc. Symp. Proc. **762**, 309 (2003).
[16]   B. Yan, G. Yue, and S. Guha, Mater. Res. Soc. Symp. Proc. **989**, 335 (2007).
[17]   J. Springer, A. Poruba, L. Mullerova, M. Vanecek, O. Kluth, and B. Rech, J. Appl. Phys. **95**, 1427 (2004).
[18]   M. Y. Ghannam, A. A. Abouelsaood, and R. P. Mertens, J. Appl. Phys. **84**, 496 (1998).
[19]   G. Yue, L. Sivec, B. Yan, J. Yang, and S. Guha, Mater. Res. Soc. Symp. Proc., (2009), in press
[20]   Y. Nasuno, M. Kondo, and A. Matsuda, Proc. of 28th IEEE Photovoltaic Specialists Conference (Anchorage, Alaska, 2000), p. 142.

Thin-Film Transistors

Mater. Res. Soc. Symp. Proc. Vol. 1153 © 2009 Materials Research Society   1153-A14-02

Microcrystalline Silicon Thin-Film Transistors for Ambipolar and CMOS Inverters

Kah-Yoong Chan[1,2,3], Aad Gordijn[2], Helmut Stiebig[2,4], and Dietmar Knipp[1]

[1]School of Engineering and Science, Jacobs University Bremen, Bremen, 28759, Germany
[2]IEF5-Photovoltaics, Research Center Jülich, Jülich, 52425, Germany
[3]Faculty of Engineering, Multimedia University, Cyberjaya, 63100, Selangor, Malaysia
[4]Malibu GmbH & Co. KG, Bielefeld, 33609, Germany

ABSTRACT

Microcrystalline silicon (μc-Si:H) thin-film transistors (TFTs) have lately gained much attention due to their high charge carrier mobilities. We report on top-gate μc-Si:H TFTs fabricated by plasma-enhanced chemical vapor deposition at process temperatures below 180 °C with high electron and hole charge carrier mobilities exceeding 50 cm^2/Vs and 12 cm^2/Vs, respectively. Based on the μc-Si:H TFTs different thin-film inverters were realized including ambipolar and complimentary metal-oxide-semiconductor (CMOS) inverters. Microcrystalline CMOS inverters exhibit high voltage gains exceeding 22, whereas ambipolar inverters show reduced voltage gains of 10 at low operating voltages. The electrical characteristics of the μc-Si:H CMOS and ambipolar thin-film inverters will be discussed in terms of the voltage transfer curve, the voltage gain and the power dissipation.

INTRODUCTION

Thin-film transistors (TFTs) are key element for large area electronics. To date, TFTs based on amorphous silicon (a-Si:H) are commonly used as pixel switches for display backplanes [1]. However, the realization of more complex driver circuitry is not possible due to low charge carrier mobility and device instability of a-Si:H [2,3]. So far external drivers are needed or the circuitry has to be realized by polycrystalline silicon (poly-Si) TFTs with high charge carrier mobilities and stable threshold voltages [4]. However, the manufacturing cost of poly-Si TFTs is higher due to high processing temperatures or additional crystallization steps [4].

Nano or microcrystalline silicon (nc-Si:H or μc-Si:H) is a promising alternative to existing technologies due to its high electron and hole charge carrier mobilities [5-7]. The high charge carrier mobilities facilitate the realization of integrated thin-film circuits. However, the realization of complementary metal-oxide-semiconductor (CMOS) based integrated circuits requires complex processing of thin-film devices [8]. Ambipolar devices have been proposed as alternatives in realizing integrated circuits, since separate patterning steps for n- and p-doped layers can be eliminated [9]. In this paper, ambipolar and CMOS thin-film inverters based on μc-Si:H were realized. The electrical characteristics of the ambipolar and CMOS inverters are discussed in terms of the voltage transfer curve, the voltage gain and the power dissipation.

EXPERIMENT

The schematic cross-section of a microcrystalline silicon TFT is shown in Fig. 1a. The drain and source metal contacts were realized by evaporated chromium (Cr) with a thickness of 30 nm on glass substrates. Afterwards, an n- or p-type μc-Si:H film with a thickness of 25 nm was deposited by plasma-enhanced chemical vapor deposition (PECVD) at 180 °C to form ohmic contacts between the drain and source electrodes and the intrinsic channel material. In the case of ambipolar TFTs, the doped layers were not inserted. The channel material was formed by an intrinsic microcrystalline silicon (i-μc-Si:H) film with a thickness of 100 nm. The i-μc-Si:H layer was deposited by PECVD at a temperature of 160 °C and plasma excitation frequency of 13.56 MHz, in the high pressure (1330 Pa) and high power (0.3 W/cm²) regime, which facilitates the deposition of material at high deposition rates of up to 25 nm/min [10]. The i-μc-Si:H channel layer was prepared in the transition to amorphous growth regime to ensure high device performance [11]. Following the deposition of the i-μc-Si:H channel layer, a gate dielectric (silicon oxide, SiO_2) of 300 nm was prepared by PECVD at 150 °C. Finally, the gate metal electrode was formed by an evaporated aluminum (Al) film with a thickness of 100 nm.

The fabricated TFTs were integrated into thin-film inverters according to the schematic circuit given in Fig. 1b. The inverters are realized by integrating two TFTs, with the upper TFT operated in p-channel mode while the lower TFT operated in n-channel mode [12]. The input signal of the inverter, V_{IN}, is applied to the gate terminals of both TFTs, while the drain terminal of the lower TFT serves as output voltage, V_{OUT}, for the inverter. The source terminal of the upper TFT is connected to the supply voltage of the inverter, V_{DD}. If the input signal is equal to V_{DD}, which is considered as logical high, the lower TFT is conducting while the upper TFT (which has its gate-source voltage, V_{GS}, equal to 0 V) is cut off. Hence the output voltage of the inverter, V_{OUT}, is approximately equal to 0 V, representing a logical low. When the input signal is at ground or equal to 0 V which is considered as logical low, the lower TFT is cut off while the upper TFT is conducting, so that the V_{OUT} is very close to V_{DD}, representing a logical high.

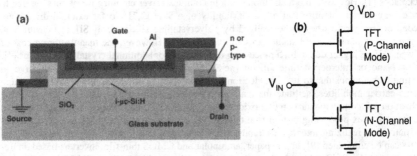

Figure 1: a) Schematic cross-section of a top-gate microcrystalline silicon TFT. b) Schematic circuit of a thin-film inverter, which is realized by integrating n- and p-channel TFTs.

RESULTS AND DISCUSSION

The n-channel and p-channel transfer characteristics of an ambipolar μc-Si:H TFT with a channel length, L, of 20 μm and a channel width, W, of 1000 μm are shown in Fig. 2. The n-channel transfer characteristics are plotted for positive gate voltages, whereas the p-channel transfer characteristics are shown for negative gate voltages. An electron charge carrier mobility of 37 cm^2/Vs and a threshold voltage of 2.7 V were extracted from the n-channel transfer characteristics of the ambipolar TFT in the linear region of operation. For p-channel operation, the ambipolar μc-Si:H transistor exhibits a hole charge carrier mobility of 10 cm^2/Vs. The achieved electron charge carrier mobility is about four times higher than the hole charge carrier mobility. The ratio of the electron to hole charge carrier mobility seems comparable to the electronic properties of crystalline silicon. The threshold voltage of the ambipolar μc-Si:H transistor in the p-channel operation mode is increased to 6 V in comparison to 2.7 V extracted for the ambipolar TFT in the n-channel operation mode.

TFTs with high electron and hole charge carrier mobilities can only be realized if electrons and holes are effectively injected in the μc-Si:H channel material. The formation of a distinct Schottky barrier in the drain and source contact region would hinder the effective injection of charges resulting in a high contact resistance. In order to qualitatively determine the contact properties of the ambipolar μc-Si:H TFTs, contact effects of the transistors in the n- and p-channel operation modes were investigated according to a model described in reference 7. A normalized drain and source contact resistance in the range of 1 kΩ·cm and 4 kΩ·cm was extracted for the ambipolar μc-Si:H TFTs in the n- and p-channel operation mode, respectively. The higher normalized contact resistance for the ambipolar TFTs in the p-channel operation mode in comparison to the n-channel operation mode can be understood as the charge carrier mobilities of the ambipolar transistors have an influence on the contact resistance.

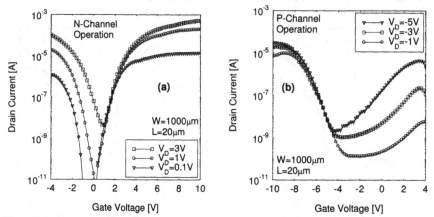

Figure 2: a) N-channel and b) p-channel transfer characteristics of the ambipolar microcrystalline silicon TFT with a channel width, W, of 1000 μm and a channel length, L, of 20 μm.

The same behavior was observed for µc-Si:H TFTs prepared at different conditions. Transistors with high charge carrier mobilities exhibit low contact resistances, whereas TFTs with low charge carrier mobilities show high contact resistances [11]. In general, the extracted contact resistances are low given the fact that the highly doped n- or p-layers were not inserted between the drain and source metal contacts and the i-µc-Si:H channel layer. It has been proposed that the low contact resistances are caused by the formation of highly conductive chromium silicide layer. Such layers are already formed at temperatures below 200 °C [13]. However, the high drain current observed in the nominal off-state of the ambipolar TFTs in n-channel (p-channel) mode is a result of an effective injection of holes (electrons) into the channel.

The n- and p-channel transfer characteristics of an unipolar µc-Si:H TFT with a channel width of 1000 µm and a channel length of 20 µm are shown in Fig. 3. TFT fabricated with p-doped layers as ohmic contacts exhibits high hole charge carrier mobilities exceeding 12 cm^2/Vs. The achieved hole charge carrier mobility is 4 times lower than the electron charge carrier mobility of the unipolar TFT with n-doped layers (>50 cm^2/Vs). The threshold voltage of the p-channel unipolar TFTs is increased to higher than 4 V in comparison to 2 V for the n-channel unipolar TFTs. In the nominal-off state with reverse gate voltage, the n- and p-channel unipolar TFT exhibit a considerably lower drain current in comparison to the ambipolar TFT operated in n- and p-channel mode. The contact resistance extracted for the unipolar µc-Si:H TFTs is in the range of 0.9 kΩ·cm.

For the µc-Si:H TFTs reported in this paper, a systematic study of the transistors behavior under bias stress has not been performed. However, subsequent measurements of the transistors lead to a shift of the device threshold voltage by 50-150 mV. In the following the ambipolar and unipolar µc-Si:H TFTs were employed to realize thin-film inverters. The experimental voltage transfer characteristics of the ambipolar and CMOS inverters measured for positive supply voltages are shown in Fig. 4. The ambipolar and CMOS inverters were realized using an 'n-channel' TFT with a channel length of 20 µm and a channel width of 200 µm, and a 'p-channel' TFT with a channel length of 20 µm and a channel width of 1000 µm.

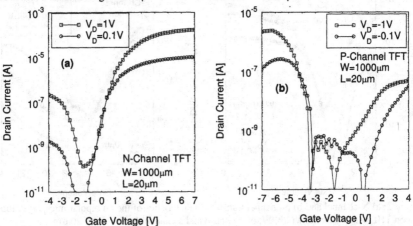

Figure 3: a) N-channel and b) p-channel transfer characteristics of the unipolar microcrystalline silicon TFT with a channel width, W, of 1000 µm and a channel length, L, of 20 µm.

The increased channel width of the transistor meant for p-channel operation compensates for the reduced hole charge charier mobility of the transistor. The voltage transfer curve of the ambipolar thin-film inverter exhibits the typical features of ambipolar inverters. The voltage gains of the ambipolar and CMOS inverters, v_{GAIN}, were extracted by taking the slope of the measured voltage transfer curves in the transition from logical high to logical low (v_{GAIN} = $\partial V_{OUT}/\partial V_{IN}$, where V_{OUT} and V_{IN} are the output and input voltage of the inverters, respectively). The voltage gains of both ambipolar and CMOS inverters are higher than the voltage gains of μc-Si:H based n-channel metal-oxide-semiconductor (NMOS) inverters fabricated under comparable processing conditions [14]. However, the ambipolar inverter exhibits a lower voltage gain comparing to CMOS inverters apparently. The experimentally extracted voltage gains for the ambipolar and CMOS inverters are 5 to 10 and 15 to 22, respectively. Furthermore, the output voltage of the ambipolar inverter in the logical low-state increases with increasing input voltage, which can be explained by the high off-current of the constituent 'p-channel' ambipolar μc-Si:H TFT at high drain voltages. As a consequence, the noise margin of the ambipolar inverter is reduced and the power consumption is increased in comparison to the CMOS inverter. These limit the realization of complex thin-film circuitry with ambipolar devices. The problem can only be reduced by selecting the right operating voltages of the ambipolar inverter. The CMOS inverter exhibits an increase of the noise margin with increasing operating voltages. However, this is not the case for ambipolar inverters. The increase of the output voltage of the ambipolar inverters with increasing operating voltage leads to a drop of the noise margin for high operating voltages. The given ambipolar inverter curves in Fig. 4a exhibit the highest noise margin for an operation voltage of 4 V, while maintaining a high voltage gain of the inverter.

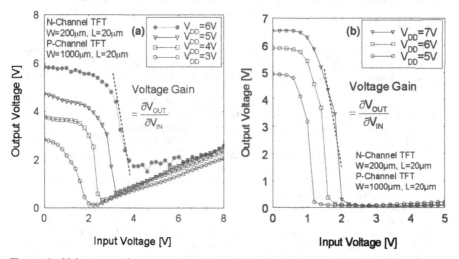

Figure 4: Voltage transfer curves for a) ambipolar and b) CMOS inverter based on microcrystalline silicon TFTs. The drive transistor (n-channel TFT) has a channel width and channel length of 200 μm and 20 μm, respectively, and the load transistor (p-channel TFT) has a channel width and channel length of 1000 μm and 20 μm, respectively.

It can be concluded that ambipolar inverters are an interesting addition to existing CMOS inverter technology. However, the specific disadvantages of the technology have to be taken into account when designing ambipolar inverter based circuitry.

CONCLUSIONS

In summary, microcrystalline silicon TFTs were realized at temperatures below 180 °C with distinctively high electron and hole charge carrier mobilities exceeding 50 cm^2/Vs and 12 cm^2/Vs, respectively. Based on the microcrystalline silicon TFTs, ambipolar and CMOS inverters were demonstrated which exhibit high voltage gains of 5 to 10 and 15 to 22, respectively. Though the processing of ambipolar inverters is simplified in comparison to CMOS inverters, the performance of the ambipolar inverter is limited by the high off-current of the ambipolar TFTs. When designing ambipolar inverter based circuitry, the specific disadvantages of the technology have to be taken into account. Microcrystalline silicon TFTs and inverters facilitate the realization of different low cost large area integrated circuit technologies.

ACKNOWLEDGMENTS

The authors are thankful to S. Bunte and Y. Mohr (IBN-PT) for preparation of the SiO$_2$, M. Hülsbeck, J. Kirchhoff, T. Melle, S. Michel, and R. Schmitz for technical assistances and E. Bunte, R. Carius, D. Hrunski, S. Reynolds and V. Smirnov for helpful discussions.

REFERENCES

1. T. Tsukada, Technology and Applications of Amorphous Silicon, Springer Series in Material Science, 37, edited by R. A. Street (Springer-Verlag, Berlin, Germany, 2000).
2. W. S. Wong, S. E. Ready, J. P. Lu, and R. A. Street, *IEEE Electron Device Lett.* **24**(9), 577-579 (2003).
3. H. Kavak and H. Shanks, *Solid-State Electron.* **49**, 578-584 (2005).
4. S. D. Brotherton, *Semicond. Sci. Technol.* **10**, 721-738 (1995).
5. I. C. Cheng and S. Wagner, *Appl. Phys. Lett.* **80**(3), 440-442 (2002).
6. C. H. Lee, A. Sazonov, and A. Nathan, *Appl. Phys. Lett.* **86**, 222106 (2005).
7. K.-Y. Chan, E. Bunte, H. Stiebig, and D. Knipp, *Appl. Phys. Lett.* **89**, 203509 (2006).
8. Y. Chen and S. Wagner, *Appl. Phys. Lett.* **75**(8), 1125-1127 (1999).
9. C.-H. Lee, A. Sazonov, M. R. .E. Rad, G. R. Chaji, and A. Nathan, *Mater. Res. Soc. Symp. Proc.* **910**, 0910-A22-05 (2006).
10. B. Rech, T. Roschek, T. Repmann, J. Müller, R. Schmitz, and W. Appenzeller, *Thin Solid Films* **427**, 157-165 (2003).
11. K.-Y. Chan, D. Knipp, A. Gordijn, and H. Stiebig, *J. Appl. Phys.* **104**, 054506 (2008).
12. D. A. Hodges and H. G. Jackson, Analysis and Design of Digital Integrated Circuits (McGraw-Hill Book Company, New York, 1983) p. 98.
13. T. E. Schlesinger, R. C. Cammarata, and S. M. Prokes, *Appl. Phys. Lett.* **59**(4), 449-451 (1991).
14. K.-Y. Chan, E. Bunte, D. Knipp, and H. Stiebig, *Solid-State Electron.* **52**, 914-918 (2008).

Transport

Mater. Res. Soc. Symp. Proc. Vol. 1153 © 2009 Materials Research Society   1153-A15-01

Carrier drift-mobilities and solar cell models for amorphous and nanocrystalline silicon

E. A. Schiff
Department of Physics, Syracuse University, Syracuse NY 13244-1130, U.S.A.

ABSTRACT

Hole drift mobilities in hydrogenated amorphous silicon (a-Si:H) and nanocrystalline silicon (nc-Si:H) are in the range of 10^{-3} to 1 cm^2/Vs at room-temperature. These low drift mobilities establish corresponding hole mobility limits to the power generation and useful thicknesses of the solar cells. The properties of as-deposited a-Si:H *nip* solar cells are quite close to their hole mobility limit, but the corresponding limit has not been examined for nc-Si:H solar cells. We explore the predictions for nc-Si:H solar cells based on parameters and values estimated from hole drift-mobility and related measurements. The indicate that the hole mobility limit for nc-Si:H cells corresponds to an optimum intrinsic-layer thickness of 2-3 μm, whereas the best nc-Si:H solar cells (10% conversion efficiency) have thicknesses around 2 μm.

INTRODUCTION

The mobility μ of a charge carrier describes its drift-velocity v in the presence of an electric field F: $v = \mu F$. Mobilities are significant in solar cells because they affect the useful thickness of the layer of material that absorbs the sunlight. For crystalline solar cells, this thickness is typically that of the "ambipolar diffusion length":

$$L_{amb} = \sqrt{2(kT/e)\mu\tau_R} \ , \qquad (1)$$

where μ is the mobility of the minority carrier (holes in n-type material), and τ_R is its recombination lifetime [1]. In crystalline silicon, the hole mobility of about 500 cm^2/Vs and a recombination lifetime of about 100 microseconds yield a diffusion length of 500 μm.

Many materials that are interesting for solar cells have much lower carrier mobilities, and much shorter recombination lifetimes, than are typical for crystalline silicon. Under the conditions present in solar cells, holes in hydrogenated amorphous silicon (a-Si:H) have drift mobilities more than 10^5 smaller than in c-Si, and recombination lifetimes about 10^2 smaller. The corresponding ambipolar diffusion lengths are around 0.1 μm.

The absorber layers in a-Si:H can be several times thicker than this. The reason that this extra thickness is useful is that, for low-mobility solar cells, the *space-charge* layer near the junction of the cell also makes a significant contribution. In c-Si the space-charge layer is the depletion region, whose width is determined by the dopant density. In a-Si:H and in other low-mobility, highly insulating materials, the width of the space-charge layer L_{SC} is determined directly by the carrier mobilities and the photocarrier generation rate G [2]. Holes drift so slowly in such materials that the space-charge from slowly drifting holes screens the built-in potential, thus limiting the width of the region from which holes can be collected. Denoting the limiting drift mobility as μ_p, the expression for the width of this region is [2]:

$$L_{SC} = (\Delta V)^{1/2} \left(\frac{4\mu_p \varepsilon\varepsilon_0}{eG} \right)^{1/4} \ , \qquad (2)$$

where ΔV is the electrostatic potential dropped across the space-charge layer and $\varepsilon\varepsilon_0$ is the dielectric constant of the layer. This expression is based on the minimal "5-parameter" model for a semiconductor layer described below.

In Fig. 1 we illustrate the ambipolar diffusion length and the space-charge width as a function of the hole mobility μ_p; we assume a uniform photogeneration rate $G = 10^{21}$ cm^{-3}s^{-1} (for calculating L_{SC}) and recombination lifetime $\tau_R = 10^{-6}$ s (for L_{amb}). The voltage dropped across the space-charge width is 1.0 V, which is a nominal value similar to the built-in potentials and open-circuit voltages for silicon-based solar cells.

The figure also indicates the corresponding lengths for c-Si, for which the diffusion length is very much larger than the depletion-width. The regime for c-Si, in which the diffusion length is dominant, is well known and is the basis for most semiconductor device physics texts. However, when the mobility less than 1 cm^2/Vs, the space-charge width is larger than the diffusion length. This is the regime applicable to amorphous silicon, for which the hole drift-mobility is less than 10^{-2} cm^2/Vs; we expect the same regime to apply with many organic, polymeric, and nanostructured hybrid materials.

In this paper we elaborate on the "mobility perspective" for hydrogenated amorphous silicon and nanocrystalline silicon (nc-Si:H) solar cells. In the next section we review a relatively simple device model that has been proposed for as-deposited amorphous silicon solar cells [3,4]; there are about seven essential electronic parameters, and with conventional estimates for their values, the model describes the solar cells well. Of the seven parameters, those determining the hole drift mobility have a significance exceeded only by the bandgap itself. We next propose a set of values for the same minimal parameter set that apply to nanocrystalline silicon (the material formerly known as microcrystalline silicon). The values for the parameters are not as well established as for a-Si:H; additionally, nc-Si:H lies at the boundary of space-charge and diffusion domination, so the mobility parameters are not as dominant as for a-Si:H. Nonetheless, the mobility perspective offers an interesting perspective on the cells. With the present choices of values, it predicts an optimum thickness (2-3 μm) and a voltage (0.60-0.64 V) that are somewhat larger than the values for optimized cells (2 μm, 0.60 V).

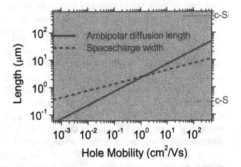

Fig. 1: Ambipolar diffusion length and space-charge width as a function of carrier mobility for the "5-parameter" model of a low-mobility semiconductor. The associated generation rate and recombination lifetime are $G = 10^{21}$ cm^{-3}s^{-1} and $\tau_r = 10^{-6}$ s; the voltage dropped across the space-charge width is 1.0 V. The markers on the right axis indicate corresponding lengths for c-Si assuming $\tau_r = 10^{-4}$ s and a doping level 10^{16} cm^{-3}. Based on ref. [2].

MODELING a-Si:H CELLS WITH A MINIMAL PARAMETER SET

Six electronic parameters seem to be the minimum required for modeling any semiconductor layer used in a solar cell; these parameters are listed in the unshaded lines in Table 1. We shall refer to this model as the "5-parameter" model because the larger of the carrier mobilities usually has no significant effect on its predictions. This model has been developed at some length elsewhere [2,5, 6], and may be directly useful in some materials [7]. For a-Si:H, we think that two additional parameters are required to give a good account of as-deposited cells (before light-soaking is significant). The first is the width of the exponential tail of the valence band ΔE_V, which is usually just called the valence bandtail width. The second parameter is the coefficient b_T describing trapping of a density of mobile holes p onto a density of bandtail traps N_t ($dp/dt = -b_T pN_T$). The recombination coefficient b_R describes recombination of a density of electrons n onto a density of holes N_T^+ trapped in the

Table 1: Parameter values for a-Si:H solar cell modeling (from ref. [3]).

Parameter	Value
Bandgap E_G	1.74 eV
Band mobilities μ_p^0 (holes) and μ_n (electrons)	0.3 cm^2/Vs 2.0
Band densities-of-states N_V and N_C	4x10^{20} cm^{-3} 4x10^{20}
Recombination coefficient b_R	10^{-9} cm^{3}s^{-1}
Valence bandtail width ΔE_V	0.04 eV
Hole trapping coefficient b_T	1.6x10^{-9} cm^{3}s^{-1}
Defect density	unused

valence bandtail ($dn/dt = -b_R nN_T^+$).Deep levels, usually identified as dangling bonds, are not included in the model; they do need to be incorporated to model the light-soaked state of a-Si:H cells.

We shall refer to this model as the "7-parameter" model for a-Si:H, as originally proposed by Zhu, *et al.* [3]; although the parameter count is apparently 8, the electron drift-mobility is so much larger than the hole drift-mobility that it is inconsequential in practice. Because we shall shortly extend this model to nanocrystalline silicon, we note its additional assumptions. First, the valence bandedge at E_V is assumed to lie within the exponential valence bandtail; this viewpoint is consistent with the interpretation of drift-mobility measurements, and most workers believe that E_V marks a "mobility edge" separating localized trap states and delocalized transport states. This assumption sets the value for the density of bandtail traps g_V^0 at E_V (see Appendix 1). Second, the conduction bandtail is neglected; at about 22 meV [8], the conduction bandtail in a-Si:H is sufficiently narrow that this neglect is justified near room-temperature. The assumption should be reassessed for work at low temperatures below about 225 K or for a-SiGe:H alloys, which have broader conduction bandtails [9]. Recombination of free holes with electrons is also neglected.

The 7-parameter model is motivated by general arguments [2, 5, 6] that the space-charge of slowly drifting holes is likely to

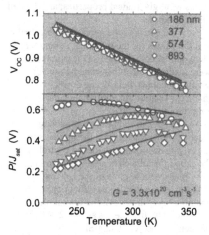

Fig. 2: a-Si:H solar cell parameters for four cells measured at varying temperatures. The symbols indicate measurements with near infrared laser illumination on cells prepared at United Solar Ovonic LLC. The curves are calculations of the "hole mobility limit" based on the parameters in Table 1. Based on ref. [4].

dominate a-Si:H solar cells. The values for the modeling parameters were chosen for consistency with hole drift-mobility measurements. We think that it is important for solar cell models of low-mobility semiconductors to be constrained this way. To facilitate such constraints, in Appendix 1 we give the expressions for calculating the drift-mobility based on the values of the bandtail parameters; the calculations can then be checked for consistency with the experimental results. Electron and hole drift-mobilities for a-Si:H and related materials have been reviewed fairly recently [8].

For a-Si:H, the 7-parameter model has been tested by comparing its predictions with the temperature-dependent properties of *nip* cells of varying thickness; the results are presented in Fig. 2 [4]. The cells were made at United Solar Ovonic, LLC, and they did not incorporate a back reflector or a textured substrate; the optoelectronic properties of the cells should be comparable to the optimized, high-efficiency cells made by United Solar. For the measurements, intense near-infrared laser illumination was used instead of solar simulator illumination; since the laser is nearly uniformly absorbed in the intrinsic layer, this procedure simplifies the photogeneration profile compared to that obtained with solar illumination while maintaining a similar photocurrent. We think that the comparison with uniformly absorbed illumination is more stringent than a comparison of the power vs. thickness relation using solar illumination; the solar spectrum contains a significant component that is absorbed close to the *p/i* interface, and hardly tests the electronic properties of the a-Si:H. The calculations were done using the the AMPS 1D program (Pennsylvania State University®) and the parameter values of Table 1. The *p* and *n* layers were given ideal properties; as intended, the details of these have no noticeable effect on the calculations.

We consider the agreement of calculation and measurement to be very good, noting also that the values for the modeling parameters were published before these measurements were made, and that this same parameter set also accounts well for the power-thickness relation for similar cells under solar illumination [10].

What can such device modeling teach us? One lesson from the model is that the hole drift-mobility is indeed central to the properties of a-Si:H solar cells. In Fig. 3 we present three plots of the calculated power from a solar cell as a function of the absorber layer thickness; the lowest plot used the same parameters as Fig. 2. As can be seen, the power from a-Si:H cells saturates for absorber layer thicknesses greater than about 300 nm, which is our estimate for the useful thickness of the as-deposited cell. If one could increase the hole band mobility tenfold, the calculations show that the power output would rise markedly for thicker cells, and that the useful thickness rises to about 600 nm.

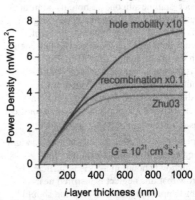

Fig. 3: Model calculations for the power density from *pin* solar cells with a uniform photogeneration rate in the intrinsic layer. The curve labeled Zhu03 uses the parameter valaues from Table 1. The other curves indicate the results after increasing μ_p^0 tenfold, or reducing b_R tenfold.

This behavior is due to the increase in the width L_{SC} of the space-charge region, and is qualitatively consistent with the behavior expected from eq. (2). It should be noted that texturing and back reflectors modify the power-thickness relation, which then shows a weak maximum instead of smooth saturation [11].

On the other hand, if we reduce the recombination coefficient tenfold, there is a smaller increase, presumably due to the increased ambipolar diffusion length. The fact that mobility has a larger effect on the cell's power than its recombination coefficient is the "recombination paradox" of low-mobility solar cells. On the one hand, the substantial fraction of photogenerated carriers that do not contribute to the cell's photocurrent are recombining somewhere in the cell; on the other hand, the parameters of this recombination have little effect on the power. In effect, recombination establishes an absorbing boundary to the space-charge region; the width of the space-charge region itself is mainly dictated by the hole mobility.

HOLE MOBILITY LIMIT FOR NANOCRYSTALLINE SILICON SOLAR CELLS

In this section we propose values for the parameter set of Table 1 that are useful as a starting point for understanding two aspects of nanocrystalline silicon solar cells: the best open-circuit voltages of about 0.6 V [12], and the typical thickness of the most efficient cells, which is about two microns [13,14]. A series of modeling studies on nc-Si:H solar cells has been published by a Kolkata-Palaiseau collaboration ([15] and further references therein), and there has been a recent paper from a Jülich-Delft collaboration [16]. These papers treat several important issues for nc-Si:H cells that we leave untouched here, where we are primarily trying to establish fundamental limits to the performance of these cells. They generally use more parameters and more intricate models than we have. More importantly, the values for the bandtail parameters vary noticeably. In Appendix 1 we present a procedure for checking these values for consistency with hole drift-mobility measurements. The set of parameter values in the present work was constrained to be consistent with these measurements, and we hope that the procedure in Appendix 1 will assist other modelers to do the same.

Table 2 presents our proposed parameters values. The three fundamental electronic parameters, E_G, N_V, and N_C are given the same values as for c-Si. In their recent study, Pieters, *et al.* [16], have made a similar proposal; their measurements on dark currents suggest a slightly increased value $E_G = 1.17$ eV. In Table 2, the valence bandtail width $\Delta E_V = 30$ meV and the hole mobility band parameter $\mu_p^0 = 0.7$ cm^2/Vs are set directly from hole drift-mobility experiments on cells from Forschungszentrum Jülich [17]. It is interesting that the value for the hole band mobility μ_p^0 is essentially the same as for a-Si:H, which suggests that transport in these nc-Si:H materials also occurs at a mobility-edge [8]. The reduced valence bandtail width in nc-Si:H compared to a-Si:H seems reasonable.

In Table 2, the bandtail trapping coefficient b_T is markedly lower in nc-Si:H than in a-Si:H. The associated

Table 2: Parameters for nc-Si:H solar cell modeling.

Parameter	Value
Bandgap E_G	1.12 eV
Band mobilities μ_p^0 (holes) and μ_n (electrons)	0.7 cm^2/Vs [17] 3.0 [18]
Band densities-of-states N_V and N_C	1x10^{19} cm^{-3} 2.8x10^{19}
Recombination coefficient b_R	10^{-10}-10^{-9} cm^{3}s^{-1} [20,21]
Valence bandtail width ΔE_v	0.03 eV [17]
Hole trapping coefficient b_T	1.0x10^{-10} cm^{3}s^{-1}
Defect density	unused

fitting parameter from hole drift-mobility measurements is the bandtail trap emission frequency prefactor v, which is connected to b_T through the detailed balance expression $v = N_V b_T$. v is about 10^9 s^{-1} in nc-Si:H, which is about 10^3 times smaller than in a-Si:H [17]. About a factor 40 of this reduction in v is expected from detailed balance and from the reduced value for N_V compared to a-Si:H; the reduction in b_T is the unexplained remainder. Schiff has proposed that this reduction is a manifestation of "Meyer-Neldel" behavior for bandtail trap emission in a-Si:H and nc-Si:H [19].

The bandtail recombination coefficient b_R has been estimated from transient measurements on photogenerated space-charge layers by Juška, et al. [20,21] using nc-Si:H diodes from Université de Neuchâtel and Prague. Their method seems well suited to determining the recombination parameter for solar cell modeling; the values of b_R varied from $10^{-9} - 10^{-10}$ cm^{3}s^{-1}.

Several groups have measured the ambipolar diffusion length L_{amb} in nc-Si:H films using the steady-state photocarrier grating (SSPG) method; the largest values are about 300 nm [22,23,24]. This value is much smaller than expected from the parameters of Table 2; the procedure for calculating L_{amb} is presented in Appendix 2. As the original authors suggested, the recombination traffic of electrons and holes is probably passing through defect levels instead of the valence bandtail.

As for a-Si:H, we neglect the conduction bandtails and equate the electron band mobility with the measured drift-mobility [18]. Unlike the situation for a-Si:H, the conduction bandtail width has not been extensively studied in nc-Si:H. The assumption that its breadth remains negligible near room-temperature does appear consistent with transient photocurrent measurements by Reynolds, et al. [25].

The 7-parameter model is simple enough that an analytical expression for the open-circuit voltage is available. Presuming ideal n and p layers and spatially uniform photogeneration G [2],

$$eV_{OC} = E_G + \frac{kT}{2}\left\{\ln\left(\frac{G}{b_R N_C^2}\right) + 2\ln\left(\frac{G}{b_T N_V^2}\right)\right\} - \frac{(kT)^2}{2\Delta E_V}\ln\left[\frac{b_R}{b_T}\left(\frac{G}{b_T N_V^2}\right)\right] \quad (3)$$

The results are presented as Fig. 4, which also incorporates the temperature-dependent bandgap [26]; we neglected the temperature-dependences of N_V and N_C. For a-Si:H, the calculation is consistent with the measurements in Fig. 2. The generation rate G corresponds to a short-circuit current from a 1 micron cell of 16 mA/cm^2, which is comparable to the values reported for solar illumination with nc-Si:H cells. The recombination lifetime due to bandtail recombination is $(b_R G)^{-1/2}$. With the range of values from Table 2, the lifetime evaluates as 1-3 μs, which is several times longer than some experimental estimates for recombination lifetimes in nc-Si:H diodes [27]. Fig. 4 shows a range of V_{OC} from 0.60-0.64 V at 298 K, depending on the value of b_R; the results agree fairly well with the largest open-circuit voltages of about 0.60 V for nc-Si:H solar cells [12]. To the best of our knowledge, the temperature dependence of V_{OC} for these high V_{OC} cells has not been reported; Fig. 4 corresponds to a predicted value dV_{OC}/dT ranging from -1.3 meV/K to -1.4 meV/K near 298 K.

In Fig. 5 we present the power as a function of the intrinsic layer thickness for the a-Si:H and the nc-Si:H models. The a-Si:H model, which suggests a useful thickness of about 300 nm for as-deposited cells, is in good agreement with measurements [4]. The nc-Si:H model indicates a useful thickness of 2-3 μm. This is somewhat larger than the thicknesses of high efficiency nc-Si:H cells (9.8%, 2.0 μm thickness, FF= 0.70 [13] and 9.9%, 1.8 μm thickness, FF=0.74 [14]). These best efficiencies are from nc-Si:H cells with lower values of V_{OC} of about 0.55 V

274

CONCLUSIONS

The best nc-Si:H solar cells are close to the limit due to fundamental hole transport and recombination processes involving the valence bandtail. We refer to this limit as "the hole mobility limit"; as valence bandtails are improved, most likely by narrowing, hole transport will improve (by diminished trapping). While b_R is not very consequential in a-Si:H modeling (cf. Fig. 3), it affects the model results for nc-Si:H noticeably, and additional measurements on materials more similar to those used for current, optimized cells would be valuable. It is worth noting that the fundamental physics that determines b_R is not well established. The fact that optimized cells are somewhat thinner than the present hole mobility limit calculation needs further research.

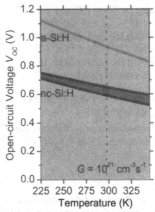

Fig. 4: Open-circuit voltage calculated as a function of temperature from eq. (3) using the parameter values of Table 1 (for a-Si:H) and Table 2 (for nc-Si:H, and including a range of values for b_R).

If the model and parameter choices prove to be correct in essence, one possibility for the difference between realized cells and the calculation would be that nc-Si:H layers become coarser grained and more defective for positions that are further from the substrate. However, contemporary deposition technology reduces this problem by modifying the growth conditions during the growth process to maintain a homogeneous material throughout [28]. Furthermore, the hole drift mobility measurements on which the present model was based were done on cells with intrinsic layers thicker than 3 μm [17]. At present, deterioration of the electronic quality of the nc-Si:H for thicker layers can't be established as the mechanism that determines the 2 μm thickness for optimized cells.

A second mechanism would involve defects, which may play a comparable role in recombination and transport to the bandtails. This possibility can't be excluded, but it isn't very satisfactory to rely on a coincidence that two supposedly independent physical features – the bandtail and the defect density – compete fairly evenly for hole trapping and recombination.

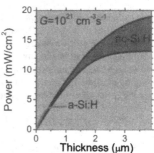

Fig. 5: Predictions of the power output of *pin* solar cells for varying intrinsic-layer thickness at room temperature. The photogeneration rate G in the intrinsic layer is uniform throughout the layer. The model is based on the parameter values of Table 1 (a-Si:H) and Table 2 (nc-Si:H, including a range of values for b_R).

ACKNOWLEDGMENTS

The author thanks Richard Crandall (National Renewable Energy Laboratory-NREL), Vikram Dalal (Iowa State Univ.), Gytis Juška (Vilnius Univ.), Jianjun Liang (Silicon Valley Solar, Inc.), Qi Wang (NREL), Chris Wronski (Pennsylvania State Univ.), Baojie Yan (United Solar Ovonic LLC), and Kai Zhu (NREL) for many enjoyable discussions related to this research. This research was supported by USDOE under the Solar American Initiative Program Contract No. DE-FC36-07 GO 17053 to United Solar Ovonic LLC. Additional support was received from the Empire State Development Corporation of New York State through the Syracuse Center of Excellence in Environmental and Energy Systems.

APPENDIX 1: CONNECTING SOLAR CELL MODELS TO DRIFT MOBILITY MEASUREMENTS

Drift-mobilities established using the time-of-flight technique are somewhat tricky to interpret in materials that exhibit "anomalously dispersive" transport. The drift-mobility is defined as:

$$\mu_D = \frac{L}{Ft_T}, \tag{4}$$

where L is the mean displacement of the photocarriers following their photogeneration [29], t_T is the "transit time" corresponding to this displacement, and F is the electric field used for the measurement. In practice, for samples exhibiting dispersive transport, μ_D depends noticeably on the thickness of sample used for the measurement, and there is also a dependence on the magnitude of the electric field that (misleadingly) suggests nonlinear transport. These effects reflect the fact that the mean displacement of a carrier $L(t)$ following its photogeneration at time $t = 0$ is sublinear in time:

$$L(t) \propto Ft^\alpha ,$$

where α is called the "dispersion parameter". In the presence of dispersion, it is necessary to compare the drift-mobilities of differing materials and differing experiments at some specific value of the ratio L/F [9]; we have generally adopted $L/F = 2\times10^{-9}$ cm^2/Vs for our work on a-Si:H, but this value can be too small for convenient work with higher mobility materials.

In amorphous and nanocrystalline silicon, dispersive behavior is a usually a consequence of trapping by a bandtail of states. For an exponential valence bandtail, the "multiple-trapping" model for the drift-mobility of holes yields (see Appendix 2):

$$\mu_D(T) = \left(N_V b_T \left(L/F \right) \right)^{1-1/\alpha} \left(\mu_p^0 \frac{N_V}{kTg_V^0} \frac{\sin(\alpha\pi)}{\alpha\pi} \right)^{1/\alpha} \left(\mu_D < \mu_V^0 \right) , \tag{5}$$

where (L/F) is the ratio of the hole displacement L to the electric field F in the drift-mobility experiment, kT is the thermal energy, $\alpha = kT/\Delta E_V$ is the dispersion parameter, and g_V^0 is the bandtail density-of-states evaluated at E_V. The expression applies for temperatures such that $\mu_D < \mu_V^0$; at higher temperatures $\mu_D = \mu_V^0$.

The parameter g_V^0 is the factor governing the density of valence bandtail traps $g_V(E) = g_V^0 \exp(-(E - E_V)/\Delta E_V)$. While it does not appear in Tables 1 and 2, it is commonly specified in solar cell modeling. The approach we have taken is to assume that the valence bandedge E_V lies within the exponential bandtail, in which case one can derive the equation relating it to the effective density-of-states N_V [2]:

276

$$\frac{N_V}{\Delta E_V g_V^0} = \frac{\alpha}{1-\alpha} . \tag{6}$$

If E_V does not lie in the bandtail, the band mobility parameter fitted to time-of-flight measurements at low temperatures ($kT < \Delta E_V$) would be smaller than the drift mobility measured at high temperatures; this possibility can be excluded for electrons in a-Si:H [8].

Equation (5) is unambiguous for calculating the drift-mobility at a specified temperature. To calculate the temperature-dependent drift-mobility, we presume that g_V^0 and b_T are temperature-independent, and that the temperature-dependence of N_V follows eq. (6):

$$\frac{N_V(T)}{N_V^0} = \frac{\alpha(1-\alpha_0)}{\alpha_0(1-\alpha)} ,$$

where N_V^0 is the value of N_V at a reference temperature T_0 (presumably room temperature), and $\alpha_0 = kT_0/\Delta E_V$. We obtain [30]:

$$\mu_D(T) = \left[\frac{\alpha(1-\alpha_0)}{\alpha_0(1-\alpha)}\right] \left(N_V^0 b_T \left(L/F\right)\right)^{1-1/\alpha} \left(\mu_p^0 \frac{N_V^0}{kTg_V^0} \frac{\sin(\alpha\pi)}{\alpha\pi}\right)^{1/\alpha} . \tag{7}$$

In Fig. 6 we compare one set of measurements on a-Si:H (Dinca, et al., [31]) with the corresponding mobilities calculated using the three different sets of parameter values of Table 1. The curve identified as Zhu03 corresponds to Table 1, for which the parameter values were intentionally matched to the hole drift-mobility measurements. Both of the other parameter sets have poor agreement with the hole drift-mobility, despite the fact that no single parameter value is unreasonable; the parameter values are typical of modeling papers that don't explicitly check for consistency with hole drift measurements. Efforts to use the set of values "B" for solar cell modeling lead to cell power-thickness functions that are inferior to actual a-Si:H cells. Efforts to use the set of values "A" lead to power-thickness relations that are superior to real cells, but the difficulty is masked in practice because it is usually compensated by defects.

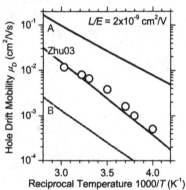

Fig. 6: The symbols indicate hole drift-mobility measurements on a BP Solar cell as reported by Dinca, et al. [31]. The smooth curves are the predictions using bandtail parameter values from Table 3.

Table 3: Bandtail parameter proposals for a-Si:H			
Parameter	Zhu03[3]	A	B
Band mobility μ_p^0 (cm²/Vs)	0.3	5.0	0.3
Band densities-of-states N_V (295 K)	4×10^{20}	1.0×10^{20}	2.5×10^{20}
Valence bandtail width ΔE_V (eV)	0.040	0.047	0.048
Trapping coefficient b_T (cm³s⁻¹)	1.6×10^{-9}	3×10^{-10}	7×10^{-9}
Trap density g_V^0 at E_V (cm⁻³eV⁻¹)	6×10^{21}	4×10^{21}	1.0×10^{22}

APPENDIX 2: CONNECTING THE 7-PARAMETER MODEL WITH STEADY-STATE PHOTOCARRIER GRATING MEASUREMENTS

For the model of Tables 1 and 2, which does not incorporate either defects or conduction bandtail trapping, the ambipolar diffusion length L_{amb} can be calculated as follows. For a field-free region, the model predicts a recombination time $\tau_R = 1/\sqrt{Gb_R}$. The bandtail multiple-trapping expression for the drift L_{drift} of a hole distribution following its photogeneration at time $t = 0$ is obtained by integrating eq. (6) in [32] with respect to time:

$$L_{drift}(t) = F\left(\mu_h^0/(N_V b_T)\right)\left(\frac{N_V}{kTg_V^0}\frac{\sin(\alpha\pi)}{\alpha\pi}\right)(N_V b_T t)^\alpha , \qquad (8)$$

where F is the electric field; we have also used the detailed balance relation $v = b_T N_V$. Eq. (8) is the fundamental equation from which eq. (5) is obtained; for a given value of L_{drift}/F, one solves for the corresponding "transit time" t_T, and then uses the fundamental definition for the drift mobility in eq. (4).

There is a "generalized Einstein relation" connecting this drift with diffusion of a carrier [33]; the root-mean-square width L_D of an initially narrow distribution is related to L_{drift} as:

$$(L_D(t))^2 = (kT/e)(L_{drift}(t)/F) , \qquad (8)$$

where e is the electronic charge.

Ambipolar diffusion involves the spreading of an initially narrow distribution with equal densities of both charges of photocarrier. If one carrier is much more mobile than the other, the details of its drift properties are unimportant, and the ambipolar diffusion length can be written as:

$$L_{amb} = \sqrt{2}L_D(\tau_R) = \left\{(2kT/e)(L_{drift}(\tau_R)/F)\right\}^{1/2} . \qquad (9)$$

Gu, et al. [33] reported that this prediction agreed fairly well with measurements for a-Si:H, which is thus an experimental confirmation of the generalized Einstein relation.

For $G = 10^{21}$ cm^{-3} and $b_R = 10^{-10}$ cm^{3}s^{-1}, the recombination time is 3.1 μs. Using the parameter values of Table 2, the ambipolar diffusion length is 2.5 μm at room temperature and is thus very close to the useful thickness of Fig. 5.

REFERENCES

1. M. A. Green, *Silicon Solar Cells: Advanced Principles & Practice* (University of New South Wales, Sydney, 1995).
2. E. A. Schiff, *Solar Energy Materials and Solar Cells* **78**, 567 (2003).
3. K. Zhu, J. Yang, W. Wang, E. A. Schiff, J. Liang, and S. Guha, in *Amorphous and Nanocrystalline Silicon Based Films - 2003*, edited by J.R. Abelson, G. Ganguly, H. Matsumura, J. Robertson, E. A. Schiff (Materials Research Society Symposium Proceedings Vol. 762, Pittsburgh, 2003), pp. 297--302.
4. Jianjun Liang, E. A. Schiff, S. Guha, Baojie Yan, and J. Yang, *Appl. Phys. Lett.* **88** 063512-063514 (2006).
5. A. M. Goodman and A. Rose, *J. Appl. Phys.* **42**, 2823 (1971).
6. R. S. Crandall, *J. Appl. Phys.* **55**, 4418 (1984).
7. V. D. Mihailetchi, J. Wildeman, and P. W. M. Blom, *Phys. Rev. Lett.* **94**, 126602 (2005).
8. E. A. Schiff, *J. Phys.: Condens. Matter* **16**, S5265-S275 (2004).
9. Qi Wang, Homer Antoniadis, E. A. Schiff, and S. Guha, *Phys. Rev. B* **47**, 9435 (1993).
10. E. A. Schiff, *J. Non-Cryst. Solids* **352**, 1087 (2006).
11. X. Deng and E. A. Schiff, in *Handbook of Photovoltaic Science and Engineering*, Antonio Luque and Steven Hegedus, editors (John Wiley & Sons, Chichester, 2003), pp. 505 - 565.
12. Y. Mai, S. Klein, X. Geng, M. Hulsbeck, R. Carius, and F. Finger, *Thin Solid Films* **501**, 272 (2006).
13. Y. Mai, S. Klein, R. Carius, J. Wolff, A. Lambertz, and F. Finger, *J. Appl. Phys.* **97**, 114913 (2005).
14. J. Bailat, D. Domine, R. Schluchter, J. Steinhauser, S. Fay, F. Freitas, C. Bucher, L. Feitknecht, X. Niquille, T. Tscharner, A. Shah, C. Ballif, in *Conference Record of the 2006 IEEE 4th World Conference on Photovoltaic Energy Conversion, Vol. 2* (IEEE, 2006), p. 1533.
15. M. Nath, P. Roci I Cabarrocas, E. V. Johnson, A. Abramov, P. Chatterjee, *Thin Solid Films* **516**, 6974—6978 (2008).
16. B. Pieters, H. Stiebig, M. Zeman, and R. A. C. M. M. van Swaaij, *J. Appl. Phys.* **105**, 044502 (2009).
17. T. Dylla, F. Finger, and E. A. Schiff, *Appl. Phys. Lett.* 87, 032103-032105 (2005).
18. T. Dylla, S. Reynolds, R. Carius, F. Finger, *J. Non-Cryst. Solids* **352**, 1093—1096 (2006). Note that these authors use the $L = d$ definition of the drift-mobility (see [29]).
19. E. A. Schiff, *Phil. Mag. B*, in press.
20. G. Juška, M. Viliūnas, K. Arlauskas, J. Stuchlik, and J. Kočka, *Phys. Stat. Sol. (a)* **171**, 539 (1999).
21. G. Juška, K. Arlauskas, J. Stuchlik, and J. Österbacka, *J. Non-Cryst. Solids* **352**, 1167 (2006).
22. C. Droz, M. Goerlitzer, N. Wyrsch, and A. Shah, *J. Non-Cryst. Solids* **266-269**, 319 (2000).
23. R. Schwarz, P. Sanguino, S. Klynov, M. Fernandes, F. Macarico, P. Louro, and M. Vieira, *Mat. Res. Soc. Symp. Proc. Vol. 609*, A32.4.1 (2000).
24. S. Okur, M. Gunes, F. Finger, and R. Carius, *Thin Solid Films* **501**, 137 (2006).
25. S. Reynolds, V. Smimov, C. Main, F. Finger, and R. Carius, in *Mat. Res. Soc. Symp. Proc. Vol. 808* (Materials Research Society, Pittsburgh, 2004), p. A.5.7.1.

26. The temperature-dependence of the bandgap for a-Si:H is -0.47 meV/K [4]. For nc-Si:H we've used the value for c-Si, which is -0.27 meV/K near room-temperature; see J. Weber, in *Properties of Crystalline Silicon*, R. Hull, ed., Institution of Engineering and Technology, Stevenage, 1999, pp. 391-393.

27. S. Saripalli, P. Sharma, P. Reusswig, V. Dalal, *J. Non-Cryst. Solids* **354**, 2426 (2008).

28. B. Yan, G. Yue, J. Yang, S. Guha, D. L. Williamson, D. Han, and C.-S. Jiang, *Appl. Phys. Lett.* **85**, 1955 (2004).

29. Most experimental papers cited here calculate the drift-mobility assuming that the mean displacement L at the transit-time is half the sample thickness d ($L = d/2$) [9]. Some experimenters use the older expression $L = d$ (cf. [18]), which yields mobilities that are twice as large.

30. This expression in square brackets differs slightly from eq. (4) of ref. [8] because that reference implicitly assumed that the product $N_V b_T$ is temperature-independent. This assumption requires that the temperature-dependence of b_T compensates that of N_V, which seems arbitrary. The fittings to drift-mobilities are not substantially affected; this can be seen in Fig. 6, where the fitting Zhu03 seems satisfactory with the original parameters.

31. S. Dinca, G. Ganguly, Z. Lu, E. A. Schiff, V. Vlahos, C. R. Wronski, Q. Yuan, in *Amorphous and Nanocrystalline Silicon Based Films - 2003*, edited by J.R. Abelson, G. Ganguly, H. Matsumura, J. Robertson, E. A. Schiff (Materials Research Society Symposium Proceedings Vol. 762, Pittsburgh, 2003), pp. 345--350.

32. E. A. Schiff, *Phys. Rev. B* **24**, pp. 6189 (1981).

33. Q. Gu, E. A. Schiff, S. Grebner, F. Wang, and R. Schwarz, *Phys. Rev. Lett.* **76**, 3196 (1996).

Mater. Res. Soc. Symp. Proc. Vol. 1153 © 2009 Materials Research Society   1153-A15-02

Bulk-heterojunction based on blending of red and blue luminescent silicon nanocrystals and P3HT polymer

Vladimir Švrček, Michio Kondo

Novel Si Material Team, Research Center for Photovoltaics, National Institute of Advanced Industrial Science and Technology (AIST), Central 2, Umezono 1-1-1, Tsukuba, 305-8568, JAPAN
Corresponding e-mail: vladimir.svrcek@aist.go.jp

ABSTRACT

Blending of red and blue photoluminescent silicon nanocrystals (Si-ncs) with poly(3-hexylthiophene (P3HT) conjugated polymer is demonstrated. The room temperature luminescent and ambient conditions stable Si-ncs prepared by electrochemical etching and laser ablation in water are used for the blend fabrication. Furthermore photo-electric properties in parallel configuration on platinum interdigitated contact are shown. Both types of Si-ncs results the bulk-heterojunction formation and photoconductivity is observed when the blends are irradiated AM1.5. The increase in photoconductivity is rather the same and ratio between photo- and dark-conductivity is about 1.7. The nanocrystal oxidation during laser ablation fabrication process in water hinders the transport properties of the blend.

INTRODUCTION

Nowadays technologies and materials for solar cells are near their potential peak for converting photons to electricity. Therefore, new energy material development is timely and higly needed. Different innovative approaches are reported in literature for both cell and material design [1-4]. For instance, recent mean involving quantum dots have been propossed to control and increase the absorption as a function of the dot size [1]. Quantum dots and polymer based blends might provide low cost, and perspective solution for fabrication of low-cost hybrid solar cells [1, 3]. In particular silicon nanocrystals (Si-ncs) due to the quantum confinement effects, carrier multiplication, low-toxicity and photovoltaic technologies compatibility emerged to be very promising photovoltaic material [4-6]. Formation of bulk-heterojunction by blending of freestanding Si-ncs with conjugated polymers can be achieved [6], which results photosensitive blend fabrication.

The blending of Si-ncs with different energy band gap, in principle, could allow so called a "tandem cell" fabrication. The multiple cells can be used with diferent gap controled by Si-ncs size, then each one converting a range of photons energies close to its bandgap. It has been shown that the electrochemical etching and laser ablation allow preparing the colloidal dispersible Si-ncs with well distinguished band gaps [7, 8]. On the other hand, it is expected that also new solar cell design might considerably improve the solar cell performance. It has been reported that the placing both negative and positive contact on the backside of the active layer has multiple advantages over the conventional sandwich structures [9]. An elimination of the front contact provides a potential to improve short circuit current and could lead to interconnection in an easier and simpler way with a higher packing density during the module fabrication [9]. In this frame using of parallel-interdigitated contacts for new design solar cell has been explored for investigation of bulk-heterojunction properties [6]. It is expected that advantages upcoming from parallel configuration eliminate shunting of Si-ncs/polymer based bulk heterojunction and might facilitate the tandem solar cells fabrication as well.

In this contribution a blending of Si-ncs with different band gaps and poly(3-hexylthiophene (P3HT) polymer is explored. The blending of room temperature luminescent and ambient conditions stable red and blue luminescent Si-ncs prepared by electrochemical etching and laser ablation in water is shown. The blends transport and photo-transport properties are examined in parallel electrode configuration when the blends are spun cast on interdigitated platinum contacts evaporated on glass substrate. An increase in photo-conductivity is observed when both types of the nanocrystals are introduced in the P3HT polymer. Results suggest that the exciton extraction and photo-conductivity response is rather the same however the transport is more efficient in the case of blending the red luminescent Si-ncs.

EXPERIMENT

Red room temperature photoluminescent Si-ncs were prepared by electrochemical etching of silicon wafer. Boron doped wafer with a resistivity of 0.5-0.75 Ω ·cm (p-type, <100>, B concentration of ~3 x 10^{16} cm^{-3}) was used. The wafers were electrochemically etched in a mixture of hydrofluoric acid with pure ethanol (HF:C$_2$H$_5$OH = 1:4). The Si-ncs were harvested by mechanical scratching [6, 7]. The synthesis of blue luminescent Si-ncs is based on a recently developed water-confined nanosecond laser plasma approach [8]. In this work the Si-ncs are prepared by nanosecond excimer pulsed laser (KrF, 245 nm 20 Hz, 10 ns, ~23.5 mJ/cm^2). The silicon wafer with same parameters as used for electrochemical etching is adhered to the bottom of a glass container and immersed in 10 ml water. The laser beam is focused onto a 1.5 mm diameter spot on the wafer surface by a lens. The container is rotated during the ablation process. The ablation process is maintained for 2 hours at room temperature and ambient pressure. In order to obtain enough amounts of Si-ncs the process was repeated for several times. Each time the water has been evaporated and harvested powder used for blend fabrications.

A commercially available (ALDRICH) polymer P3HT was dissolved in chlorobenzene. In our experiments the blends were prepared by dissolving 12 mg of polymer in 10 g of chlorobenzene. Polymer solution was used as a host matrix for Si-ncs. Both blue and red luminescent Si-ncs powders at ~ 55 wt. % were mixed to make layers on two types of substrates. Firstly, the 300 nm thin films on quartz substrates. Secondly, for a photoconductivity measurement the thin films at the same thickness were spun cast on a glass covered by interdigitated platinum contacts. The interdigitated contacts consisted of 20 fingers, with a length of 6 mm and a width of 200 μm, separated by 200 μm. At the same conditions another set of samples just with pure P3HT polymer was fabricated for comparison. For all cases the samples were dried at 140 ˚C for 30 min in vacuum.

The PL measurement of the colloids was performed at room temperature and ambient atmosphere. For PL measurements (Shimadzu, RF–5300PC) an excitation wavelength of 300 nm was used, obtained from the monochromated output of a Xe lamp. For the temperature dependence of the PL, a HeCd laser has been used. The samples were placed in the cryostat with a varying temperature from 4 to 300 K. The conductivity and photoconductivity measurements were conducted in ambient conditions. A voltage from a regulated DC power supply was applied and the resulting current was measured with a amperemeter (Sub Femtoamp, KEITHLEY 6430). For photoconductivity measurement a white light of 1.5AM was used.

RESULTS AND DISCUSSION

In our previous works, a theoretical description of Si-ncs formation was proposed to describe the formation of blue luminescent Si-ncs in liquid media by Nd:YAG nanosecond pulsed laser ablation [10]. A similar description is adopted to interpret the Si-ncs growth and features by excimer laser ablation in water used in this work. The basic physics of Si-ncs growth can be briefly described as follow. Immediately after the laser pulse hits target, a dense plasma plume of silicon atoms is created at the solid-liquid interface. The increased pressure (15 GPa) and temperature (5000 K) in water confined plasma [11] enables formation of the Si-ncs [10]. Silicon atoms form Si-ncs by the diffusion and collision of atoms within the liquid-confined plum [12] until these nanocrystals reach the water. Then self-limiting oxidation in water is believed to provide defects passivation and a natural stable surface resulting in room temperature stable blue PL [8]. Contrary to that the Si-ncs prepared by electrochemical etching are larger in diameter [7] and show smaller optical band gap. As a result visible red-orange photoluminescence can be observed at ambient conditions.

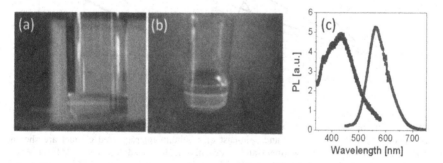

Figure 1. Photos (a) and (b) show typical photoluminescence (PL) of Si-ncs prepared by electrochemical etching and laser ablation dispersed in ethanol and water, respectively. The excitation in both cases is assured by He:Cd laser at 325 nm. (c) Corresponding room temperature PL spectra of Si-ncs in respective solutions are shown.

The results for our Si-ncs used for blend fabrication are summarized in Fig. 1. Image (a) and (b) show luminescent Si-ncs prepared by electrochemical etching and laser ablation in colloidal solution, respectively. Visible room temperature PL is observed when both colloids are excited by He:Cd laser at 325 nm. Corresponding PL spectra of colloidal solutions used for fabrication of the blends are shown in Fig. 1 (c). At ambient conditions stable red-band with maximum at 580 nm (Fig.1, red line) and blue-band with maximum at 420 nm (Fig.1, blue line) is recorded. Corresponding optical gaps of ~ 2 eV and ~2.9 eV could be evaluated, respectively. The origin of the bands is most likely due to the many possible electronic states in Si-ncs and the quantum confinement effects in nanocrystals with large size distribution. Electron–hole pair radiative recombination in Si-ncs with different sizes is responsible for the broadening of the emission spectra. [13]. It has to be pointed out that the broadening originates also from carrier trapping and recombination at surface silicon oxygen bonds that produce stable states in the band-gap [14]. Laser ablation produced nanocrystals in water offers peculiar oxidation conditions with unique surface chemistry especially for the small Si-ncs (< 2 nm) [8]. One can argue that the electron-hole recombination in the nanocrystals with high quality surface passivation in water promotes carriers trapping in oxygen-related localized states. The oxide shell provides a natural stable surface resulting in room temperature stable blue PL emission, which is more efficient in smaller sized Si-ncs. Those states are stabilized by the widening of the nanocrystals band gap induced by quantum confinement in blue emitting Si-ncs.

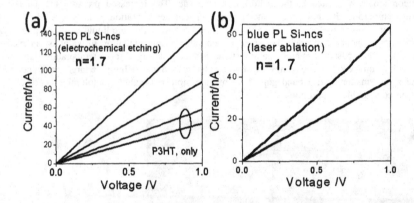

Figure 2. Current-voltage (I-V) characteristics for red (a) and blue (b) photoluminescent (PL) Si-ncs blended with P3HT polymer and spin-cast on platinum interdigiteted contact are shown, respectively. The dark (black lines) and under illumination (red lines) at AM1.5 I-V are presented after annealing at temperature 140 °C. The I-V characteristics of pure P3HT are shown for comparison (Fig. 2 (a), indicated by circle).

Furthermore, we investigate temperature dependent PL of the blends. An excitation transfer energy from a conjugated polymer to a nanocrystal results in a red-shift of nanocrystals/polymer blends as a function of the temperature. Contrary to blue shift of PL pure polymer films, a systematic temperature-dependent PL maxima shift is recorded. In the temperature range 4-300 K the red-shift is of about 40 nm in the case of red luminescent Si-ncs and the ~70 nm-shift for the blue luminescent ones. When the nanocrystals are capped by an organic layer (e. g. ligands) the excitation transfer is dominated by the Förster mechanism. This transfer is based on a dipole-dipole coupling process [15]. However, in our case during preparation process the Si-ncs surface is not covered by any surfactant (i.e. ligands), which limits potential dipole-dipole coupling. Contrary to that when the nanocrystal do not contains such groups and acceptor-donor are in close distance the excitation transfer is dominated by the charge exchange process (i.e. Dexter like energy transfer) [16]. In the Dexter mechanism, the spectral overlap is independent of the oscillator strength of the transitions and is efficient at very small distances (< 10 Å) [17]. Then a close proximity of the donor and acceptor overlaps of theirs wave functions [17]. We assume that in our case the donor and acceptor are in direct or at least in van der Waals contact. Therefore we expect that the Dexter energy transfer is most likely responsible mechanism in excitation energy transfer [18]. It is assumed that after the energy transfer the blend luminescence is attributed to zero-phonon electron–hole recombination in Si-ncs due to strong enhancement of the quantum confinement effect. However at this stage of the research the mechanism is not clear yet anf more studies are required.

At the same time, when both types of nanocrystals are introduced in the P3HT polymer, an improvement in dark- and photo-conductivity response is recorded. It has to be noted that at chosen concentration (~ 55 wt.%) and polymer properties the ratio (n) between photo- and dark-photoconductivity reached maximum. As the concentration is further increased the n decreases. This is most likely due to the perturbation of percolation trajectories of the polymer chains. A typical influence of the nanocrystal on blend dark- and photo-transport properties is shown in figure 2. Current-voltage (I-V) characteristics of the blends spin-cast on platinum interdigiteted contact are shown: (Fig. 2 (a)) red and (Fig. 2 (b)) blue luminescent Si-ncs blended with P3HT polymer, respectively. The dark lines measured in dark and red lines under illumination at AM1.5.The I-V characteristics are presented after annealing at temperature 140 °C. I-V curves of pure P3HT are shown for comparison (indicated by circle, Fig. 2 (a)). The ratio between photo- and dark-conductivity after introduction of Si-ncs increases and is rather the same for both types of Si-ncs ($n=1.7$). Resulting electronic interaction of Si-ncs with P3HT polymer leads the bulk-heterojunction formation [6]. The highest occupied molecular orbital (HOMO) is at (~5 eV) and lowest unoccupied molecular orbital (LUMO) level is at ~2.9eV [3, 6]. Since both types of Si-ncs having a large optical band gaps (~2 eV, ~2.9 eV) and work function being around −4.1 eV alignment with polymer is assured. Different electron affinity and ionization potential between the nanocrystal and polymer provide a driving force for dissociating the excitons. A electron is swept into the nanocrystal while the hole goes to the polymer. As a result an increase in photocurrent generation is observed when the blends are irradiated by AM1.5. It has to be noted that a more important increase in conductivity is noticed for the films containing the red luminescent nanocrystals. This is most likely due to the thinner oxide layer of nanocrystals prepared by electrochemical etching since the Si-ncs prepared by laser ablation in water might naturally contain thicker oxide layer that hinders the overall blend transport properties.

CONCLUSIONS

In conclusion, the blending of room temperature red/ blue luminescent and ambient conditions stable Si-ncs prepared by electrochemical etching and laser ablation in water has been successfully achieved. Contrary to blue shift of PL pure polymer films, an excitation transfer energy from a conjugated polymer to the nanocrystals results in a red-shift of both blends as a function of the temperature. The red-shift of PL spectra is more pronounced in the case of blue luminescent Si-ncs. Both types of Si-ncs blended with P3HT polymer spun on platinum interdigiteted contact showed photosensitivity and photoconductivity generation under AM1.5 irradiation. The ratio between photo- and dark-conductivity after introduction of Si-ncs is rather the same for both types of Si-ncs and reaching value of ~1.7. The self-limiting oxidation and oxide shell in the case of Si-ncs prepared by laser ablation hinders overall transport properties of the blend. The results indicate that conjugated polymers blended with Si-ncs owing a quantum confinement effects may provide a prospective for development of low-cost photovoltaic devices.

ACKNOWLEDGMENTS

This work was also partially supported by a NEDO project.

REFERENCES

1   B. Sun, E. Marx, and N. C. Greenham, *Nano Lett.* **3** 961 (2003).
2   C. J. Brabec and V. Dyakonov, in Organic Photovoltaics: Concepts and Realization, edited by C. J. Brabec, V. Dykonov, J. Parisi and N. S. Sariciftci (Springer-Verg, Berlin, 2003).
3   N. Greenham, X. Peng, and A. Alivisatos, *Phys. Rev. B* **54**, 17628 (1996).
4   A. Nozik, *Chem. Phys. Lett.*, **457** 3 (2008).
5   M. C. Beard, K. P. Knutsen, P. Yu, J. M. Luther, Q. Song, W. K. Metzger, R. J. Ellingson, A.J. Nozik, *Nano Lett.* **7**, 2506 (2007).
6   V. Švrček, H. Fujiwara and M. Kondo, *Appl. Phys. Lett.* **92**, 143301 (2008).
7   V. Švrček, A. Slaoui and J.-C. Muller, *J. Appl. Phys.*, **95**, 3158 (2004).
8   V. Švrček, D. Mariotti, and M. Kondo, *Optics Express*, **17**, 520 (2009).
9   Nakamura K, Isaka T, Funakoshi Y, Tonomura Y, Machida T and Okamoto K 20 th European Photovoltaic Solar Energy Conference, Barcelona, 2005.
10   V. Švrček, T. Sasaki, T. Shimizu, and N. Koshizaki, *Appl. Phys. Lett.* **89**, 213113 (2006).
11   R. Fabbro, J. Fournier, P. Ballard, D. Devaux, and J. Virmont, *J. Appl. Phys.* **68**, 775 (1990).
12   A. A. Oraevsky, V. S. Letoshkov, and R. O. Esenafiev, *Proceedings of the Workshop"laser ablation: Mechanism and Applications", Pulsed laser ablation of biotissue. review of ablation mechanisms* (Springer, Berlin, 1991).],
13   C. Garcia, B. Garrido, P. Pellegrino, R. Ferre, J. A. Moreno, J. R. Morante, L. Pavesi and M. Cazzanelli, *Appl. Phys.Lett.* **82**, 1595 (2003).
14   D. I. Kovalev, I. D. Yaroshetzkii, T. Muschik, T. V. Petrovakoch and F. Koch F *Appl. Phys. Lett.* **64**, 214 (1994).
15   J. H. Warner, A. A. R. Watt, E. Thomsen, N. Heckenberg, P. Meredith, and H. Rubinsztein-Dunlop, *J. Phys. Chem. B*, **109**, 9001 (2005).
16   D. L. Dexter, *J. Chem. Phys.* **21**, 836 (1953).
17   L. Nayak, M. K. Raval, B.Biswal and U. C. Biswal, *Photochem. Photobiol. Sci.*, **1**, 629 (2002).
18   A. Monguzzi, R. Tubino, and F. Meinardi, *Phys. Rev. B* **77**, 155122 (2008).

Mater. Res. Soc. Symp. Proc. Vol. 1153 © 2009 Materials Research Society   1153-A15-03

Imaging electron transport across grain boundaries in an integrated electron and atomic force microscopy platform: Application to polycrystalline silicon solar cells

M. J. Romero[1], F. Liu[1], O. Kunz[2], J. Wong[2], C.-S. Jiang[1], M. M. Al-Jassim[1], and A. G. Aberle[2]

[1]National Renewable Energy Laboratory (NREL), 1617 Cole Boulevard
Golden, CO 80401-3393
[2]ARC Photovoltaics Center of Excellence, The University of New South Wales (UNSW)
Sydney NSW 2052, Australia

ABSTRACT

We have investigated the local electron transport in polycrystalline silicon (pc-Si) thin-films by atomic force microscopy (AFM)-based measurements of the electron-beam-induced current (EBIC). EVA solar cells are produced at UNSW by *EVAporation* of *a*-Si and subsequent *solid-phase crystallization* (SPC)–a potentially cost-effective approach to the production of pc-Si photovoltaics. A fundamental understanding of the electron transport in these pc-Si thin films is of prime importance to address the factors limiting the efficiency of EVA solar cells. EBIC measurements performed in combination with an AFM integrated inside an electron microscope can resolve the electron transport across individual grain boundaries. AFM-EBIC reveals that most grain boundaries present a high energy barrier to the transport of electrons for both *p*-type and *n*-type EVA thin-films. Furthermore, for *p*-type EVA pc-Si, in contrast with *n*-type, charged grain boundaries are seen. Recombination at grain boundaries seems to be the dominant factor limiting the efficiency of these pc-Si solar cells.

INTRODUCTION

Silicon is the leading semiconductor in terrestrial solar energy applications, and is expected to dominate the photovoltaic industry for at least another decade. Although both expansion of the silicon production and advances in *solar-grade silicon* (such as innovations in *upgraded metallurgical* silicon) will drive down the price of the feedstock, the added costs from fabricating wafers will continue at current levels. A *wafer-replacement* proposal such that provided by polycrystalline silicon (pc-Si) thin films grown on inexpensive substrates is therefore of great interest for the large-scale deployment of silicon-based photovoltaics at low cost. If solar-grade silicon is obtained and appropriate light-trapping strategies are implemented, pc-Si thin films (5– 40 μm in thickness) on foreign substrates (not to be confused with silicon thin wafers) can realistically reach 15% solar conversion efficiency.

There are many different strategies for the fabrication of pc-Si thin films at temperatures compatible with borosilicate glass substrates (< 650 °C) [1,2]. One approach towards improved solar cells makes use of an ultrathin *seed layer* as *'template'* for the silicon epitaxy, with the purpose of producing high-quality epitaxial pc-Si. AIC (*Aluminum-Induced Crystallization*) is the most widely used *seeding method* [3]. As for the epitaxy, HWCVD (*Hot-Wire Chemical Vapor Deposition*)[4,5], IAD (*Ion-Assisted Deposition*) [6], and SPC (*Solid-Phase Crystallization*) [7,8] have all been explored on AIC seeds. We have confirmed that the resulting silicon epilayer is preferentially oriented along <100> and the orientation of the underlying *seed layer* is effectively imprinted on the *epilayer* [9]. Well-developed (up to 10-20 μm) <100>–oriented grains in the epitaxial film are of high quality, in contrast with grains of <111> orientation, which are ill-developed and have a high density of structural defects. The occurrence of *seed-to-epi* misorientation seems to be the Achilles' heel of this technology.

An alternative to epitaxial *pc*-Si has been developed at UNSW [7]. These researchers found that electron-beam evaporation of *a*-Si under non-ultra-high-vacuum conditions (doped *in-situ* using phosphorus and boron effusion cells) produces precursor diodes that are well suited for SPC. These solar cells of the structure borosilicate (BS) glass/SiN/p^+/n^-/n^+ or n^+/p^-/p^+ are named EVA for "*EVAporation* of *a*-Si and subsequent *solid-phase crystallization*". This approach is a *seedless* method and consequently results in a *pc*-Si film without preferential orientation. With a grain size in the micron range, the best *open-circuit voltage* V_{oc} reported to date for EVA solar cells is 517 mV [10].

In all *pc*-Si thin-film solar cells, one of the most significant contributions to efficiency losses is represented by recombination at grain boundaries. EVA solar cells are expected to be very sensitive to these losses because the distance between adjacent grain boundaries (~ *1-2 μm*) is well below the projected diffusion length (*L*) required for practical solar cells. Thus, a fundamental understanding of the electron transport across grain boundaries is of prime importance to address the factors limiting the efficiency of EVA solar cells.

EXPERIMENTAL

Measuring the electron transport across individual grain boundaries requires a *multiprobe* setup with high spatial resolution. Our approach is based on integrating in one platform conductive atomic force microscopy (cAFM) and electron-beam-induced current (EBIC) measurements inside a scanning electron microscope (SEM), which has been shown to improve the resolution of conventional EBIC measurements [11]. In our case, the conductive ultrasharp tip (sensing the current) and the highly localized electron beam (exciting the current) are independent *probes* that can be controlled simultaneously. Fig. 1 illustrates how these AFM-EBIC measurements are applied to EVA thin films of BS/SiN/n^-(or p^-) structure –phosphorous or boron, uniformly doped to 1-5×10^{16} cm^{-3} (nominal [12])– which are fabricated for contact resistance measurements with aluminum patterned contacts on the top. These films are preferred over complete solar cells to avoid the interference of the cell's *p–n* junction on the current sensed by the tip.

The AFM sensor consists of an ultrasharp metallic tip (*W*) attached to a self-sensing and self-actuating piezoelectric tuning fork (TF). The TF is excited at the resonant frequency (~ 2^{15} Hz) and the oscillation amplitude (measured by the TF piezoelectric current at the resonant frequency) is adjusted to a setpoint (force) under feedback control during the scanning of the tip. The tip (accessible to the electron beam) is an independent electrode in the AFM sensor, allowing the EBIC current to be measured without interference from the TF. In the measurements described here, (i) an AFM image is acquired; (ii) the tip establishes contact (monitored by the current at the picoamplifier) over preselected location(s) on the AFM image; (iii) an image of the current sensed by the tip (scanning the e-beam) is acquired.

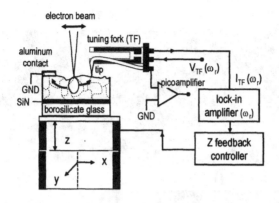

Figure 1. Schematics of the AFM-EBIC measurements.

RESULTS AND DISCUSSION

Figure 2 illustrates the application of AFM-EBIC measurements to uniformly doped n-type EVA thin films. Figs. 2(a) and 2(b) are SEM and AFM images revealing the microstructure of the pc-Si film and Figs. 2(c)–(e) the corresponding images of the current sensed by the tip over different locations $p1$, $p2$, and $p3$ of Fig. 2(b). Location $p1$ in Fig. 2(c) is a good example for the two sources of contrast in the current images. First, the tip can establish a local Schottky diode and the holes excited by the electron beam (minorities in n-type) can be collected at the tip; this collected current, of positive polarity in our setup, is laterally confined by the tip and the local (hole) diffusion length at the tip's location [see Fig. 2(c)]. Second, in this contact scheme with *remote* ground, the silicon film acts as a *current divider*. Thus, a fraction of the primary electrons from the electron beam (the specimen current) flows to the tip and the rest to the ground, depending on the relative resistance along these two different paths –this is referred to as *R*EBIC for *remote* EBIC [13]. This current is of negative polarity in our setup (i.e., of opposite polarity compared to the tip-induced Schottky collected current). Fig. 2(c) shows, originating from the tip, a *stepped* current contrast defined at the grain boundaries [this is more clearly seen in the linescan of Fig. 2(f)], which is due to the much higher resistance to the *electron flow* imposed by the grain boundaries when compared to the grain interiors (the REBIC depends on the local value of the resistivity). Although the *electron flow* is not impeded at all across certain grain boundaries [see, for example, *A–B* in Figs. 2(b) and (c)], most of the boundaries present a very high energy barrier to the electron transport (see *A–C–D* in the same figure) and the current towards the tip decreases rapidly when increasing the number of boundaries to be crossed by the electrons in order to reach the tip. The REBIC contrast present on most locations ($p1$ and $p2$ shown here) is absent on $p3$ [Fig. 2(e)] where the collected current at the local Schottky diode is also improved, indicating the action of a low-resistivity, boundary-assisted, electron *percolation-diffusion* to the remote ground. Thus, the grain boundary associated to the $p3$ grain interior presents a low energy barrier to the electron transport. Unfortunately, this is the exception and not the rule for these pc-Si films.

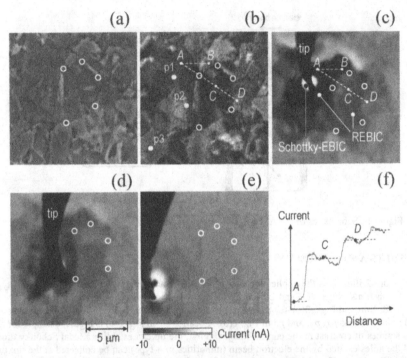

Figure 2. AFM-EBIC measurements on uniformly doped *n*-type EVA thin-films. Figs. 2(a) and 2(b) are the SEM and AFM images acquired over the same area of the *pc*-Si film. Figs. 2(c), (d), and (e) are the corresponding images of the current sensed by the tip (acquired during the scanning of the electron beam) over the locations *p1*, *p2*, and *p3* selected on the AFM image. The current linescan along *A–C–D* on Fig. 2(c) is shown in Fig. 2(f). Electron-beam energy E_b = 15 keV and current I_b = 500 pA. To better establish a correlation between these images, the contours of one grain are outlined for all images shown (hollow circles).

The results of AFM-EBIC measurements on *p*-type EVA films are comparable to those on *n*-type films, with an added feature: a considerable fraction of grain boundaries are *charged*. This is illustrated in Fig. 3, which shows an AFM image of a *p*-type *pc*-Si film and the current image on the location *p4*. On top of the stepped REBIC contrast, there is a reversal in the polarity of the current (– to +) when crossing the grain boundary [see linescan *E–F*, Fig. 3(d)]. The depletion region on either side of the *charged* boundary sets up collecting electric fields of opposite directions [Fig. 3(d)], giving a *peak-and-trough* contrast [13]. The *peak* occurs on the side where the collected current at the depletion region subtracts from the overall current and the *trough* where it adds to it. Recombination losses increase dramatically when *charge* is *stored* at grain boundaries and are thus very detrimental to the EVA cell efficiency. This result can explain that *n*-type absorber layers are preferred over *p*-type for EVA solar cells (glass/p^+/n over glass/n^+/p polarity).

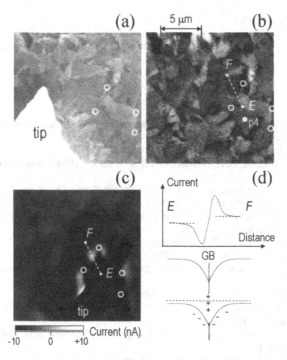

Figure 3. Results on uniformly doped *p*-type EVA *pc*-Si films are similar to those on *n*-type films, with an added feature: a considerable fraction of grain boundaries are *charged*. Figs. 3(a) and (b) are SEM and AFM images over the same area of the *p*-type EVA thin film. Fig. 3(c). EBIC image acquired over the location *p4* selected on the AFM image. Fig. 3(d) current linescan from point *E* to point *F* (top) shows the characteristic *peak-and-trough* contrast of *charged* grain boundaries. A model of the grain boundary is also shown. See reference [13] for a detail description of the *peak-and-trough* contrast. To better establish a correlation between these images, the contours of one grain are outlined for all images shown (hollow circles).

SUMMARY AND CONCLUSIONS

In summary, we have investigated the local electron transport in EVA *pc*-Si thin films and solar cells by combining conductive AFM with EBIC measurements in an integrated platform. AFM-EBIC reveals that most grain boundaries present a high energy barrier to the electron transport, for both *p*-type and *n*-type EVA thin films. Furthermore, for *p*-type EVA films, *charge* is stored at grain boundaries, which is very harmful to the cell efficiency. Mitigation of these adverse effects can significantly improve the efficiency of EVA *pc*-Si thin-film solar cells.

ACKNOWLEDGMENTS

This work was supported by the U.S. Department of Energy under Contract No. DE-AC36-08-GO28308.

REFERENCES

[1] K. R. Catchpole, M. J. McCann, K. J. Weber, and A. W. Blakers, Solar Energy Materials and Solar Cells **68**, 173 (2001).
[2] A. G. Aberle, Thin Solid Films **511-512**, 26 (2006).
[3] O. Nast, S. Brehme, D. H. Neuhaus, and S. R. Wenham, IEEE Trans. Electr. Dev. **46**, 2062 (1999).
[4] J. Stradal, G. Scholma, H. Li, C.H.M. van der Werf, J.K. Rath, P.I. Widenborg, P. Campbell, A.G. Aberle and R.E.I. Schropp, Thin Solid Films **501**, 335 (2006).
[5] C. W. Teplin, H. M. Branz, K. M. Jones, M. J. Romero, P. Stradins, and S. Gall, Mat. Res. Soc. Symp. Proc. **989**, 133 (2007).
[6] A. G. Aberle, A. Straub, P. I. Widenborg, A. B. Sproul, Y. Huang, and P. Campbell, Prog. in Photovolt: Res. and Appl. 13, 37 (2005).
[7] M. L. Terry, A. Straub, D. Inns, D. Song, and A. G. Aberle, Appl. Phys. Lett. 86, 172108 (2005).
[8] P. I. Widenborg and A. G. Aberle, J. Crystal Growth 306, 177 (2007).
[9] F. Liu, M. J. Romero, K. M. Jones, A. G. Norman, M. M. Al-Jassim, D. Inns, and A. G. Aberle, Thin Solid Films **516**, 6409 (2008).
[10] A. G. Aberle, Proceedings of the 21st European Photovoltaic Solar Energy Conference, Dresden, 2006, p. 738.
[11] M. Troyon and K. Smaali, Appl. Phys. Lett. **90**, 212110 (2007).
[12] B-doped *p*-type films can be partially compensated because of residual phosphorous contamination in the deposition chamber.
[13] D. B. Holt, B. Raza, and A. Wojcik, Materials Science and Engineering B**42**, 14 (1996).

Poster Session:
Characterization

Mater. Res. Soc. Symp. Proc. Vol. 1153 © 2009 Materials Research Society   1153-A16-01

Blue/White Emission From Hydrogenated Amorphous Silicon Carbide Films Prepared by PECVD

V.I. Ivashchenko[1], A.V. Vasin[2], L.A. Ivashchenko[1], P.L. Skrynskyy[1]
[1]Institute for Problems of Material Science, NAS of Ukraine, 03142 Kyiv, Ukraine
[2]Institute of Semiconductor Physics, NAS of Ukraine, 03026 Kyiv, Ukraine

ABSTRACT

Photoluminescence (PL) from hydrogenated silicon carbide (SiC:H) films is studied at room temperature. The films were deposited by plasma-enhanced chemical vapor (PECVD) technique with and without substrate bias using methyltrichlorosilane as a main precursor. After the deposition the samples were annealed at various temperatures in vacuum. The films were characterized by atomic force microscopy (AFM), Fourier transform infrared spectroscopy (FTIR), X-ray photoelectron spectroscopy (XPS) and X-ray diffraction (XRD). The samples deposited without substrate bias (series A) were amorphous, whereas the samples deposited with negative substrate bias -100V (series B) were nanocrystalline. The one-peak (470 nm) and double-peak (415 and 437 nm) PL structures of the as-deposited samples A and B were observed, respectively. Annealing strongly enhanced the intensity of the PL of the samples B and transformed the PL spectrum from a double-peak into broad featureless bands with intensity at about 470 nm. The blue PL in as-deposited films B is supposed to be assigned to the radiative recombination in the sites located at the nanocrystallite surface, whereas the photo excitation of carriers mostly occurs in nanocrystallite cores. A further increase in the annealing temperature causes hydrogen effusion, which leads to an increase of the concentration of non-raidative recombination centers associated with dangling-bonds and as a result, to the quenching of PL.

INTRODUCTION

Photoluminescence properties of silicon-based nanostructures have been widely studied in the last years, since the discovery of visible light emission from porous Si at room temperature [1]. It was reported earlier that some Si-based nanostructures exhibit the bright blue PL with very specific PL spectrum composed by two peaks at 415-420, 430-440 nm and a shoulder at approximately 460-480 nm. Such a "magic" spectrum was observed in: i) anodized microcrystalline Si thin films [2]; ii) Si^+-implanted SiO_2 films [3]; iii) strongly oxidized porous silicon (PS) [4]; iii) nc-Si/SiO_2 multi-layers [5]; iv) PS embedded in $Pb(Zr_xTi_{1-x})O_3$ [6]; v) nc-SiC films deposited on Si substrate using pulsed laser ablation of a polycrystalline SiC target [7]; vi) the C_{60}-coupled PS annealed in N_2 [7]; vii) anodized n-type Si wafers annealed in N_2 [7]; viii) Si^+-implanted n-type Si wafers annealed in N_2 [7]; ix) SiC:H films [8,9]. Some preliminary discussion of the Si-based nanostructures that exhibit bright blue emission was presented in Refs. 8-10. The thermal treatment of the Si nanostructures having the average grain size (D) in the range of 1-5 nm did not change PL peak position [2-7], which indicates that the observed emission cannot arise from band-to-band recombination in the quantum confined nano-crystallite core. Several models of this blue emission have been proposed. The most common models consider: oxide-related defects [6]; surface electronic states [3,8,9]; vacancies in silicon nanocrystallites [7].

In this work we examine the peculiarities of the blue/white PL of PECVD SiC films deposited using methyltrichlorosilane (MTCS, CH_3SiCl_3) under various deposition conditions. In particular, we focus on the comparative investigation of the effects of substrate bias and annealing temperature on the PL properties of SiC films.

EXPERIMENTAL

Silicon carbide films were deposited in a rf (40.68 MHz) PECVD reactor using methyltrichlorosilane (MTCS) as a main precursor. The deposition was carried without substrate bias (series A) and with 100 V negative bias (series B) using a 5.27 MHz generator. The MTCS vapor was delivered into the reaction chamber with hydrogen that passed through a thermostated bubbler with MTCS. The bubbler temperature was 32 ^{0}C. H_2 and H_2 + MTCS flow rates were 15 and 5 sccm, respectively. The substrate temperature (T_S) was 350 ^{0}C. The background and working pressures were 10^{-6} and 0.2 Torr, respectively. The average power density was estimated to be about 0.14 W/cm^2. The films were deposited for 40 min resulting in ~0.10-0.13 μm thickness. The film thickness has been determined by an Alpha Step profilometer. The substrates were phosphor-doped (111) silicon wafers (4.5 Ohm×cm) and glass. Before deposition, the wafers were etched in an HF solution and in pure H_2 plasma. After deposition, the samples were cut into several segments. One sample was left initially untreated, whereas others were annealed in vacuum (10^{-6} Torr) at 450 - 1050 ^{0}C for 30 min. The chemical composition of the high temperature film was determined with an Auger-electron spectrometer JUMP-10s. The content of Si, C and O in the films was estimated to be 35, 45 and 20 at %, respectively. The oxygen incorporation in the films occurs due to oxygen containing residuals (oxygen, water vapors, carbon oxides) adsorbed at the chamber wall. The photoluminescence (PL) was excited by an ILGI-503 nitrogen laser irradiation with a wavelength of 337.1 nm and was measured at room temperature using an MDR-23 monochromator and an FÉU-100 photomultiplier. Chemical bonding was analyzed using an FTIR spectrometer (Infralum FT-801). The band gap (E_g) was derived from optical absorption spectra, measured with a spectrometer SPECORD M40. XRD measurements were carried out with the help of a DRON-UM1 diffractometer (Cu K_α, λ=0.15406 nm). XPS analysis was performed with VGS ESCLAB MKII instrument using non-monochromatic Mg K_α x-radiation (hν=1253.6 eV). Surface morphology was analyzed by atomic force microscopy (Nanoscope IIIa).

RESULTS AND DISCUSSION

Figure 1 shows the AFM topography of the as deposited films of series A and B. It may be noted from the surface morphology of both series consists of semi-spherical features. An application of substrate bias leads to refining semi-spherical objects.

The effect of substrate bias on the Si 2p and C 1s core levels XPS spectra of the as-deposited films is demonstrated in Figure 2. One can see that application of substrate bias leads mainly to enhancement of the C-O and Si-O bonds and to weakening C-C and Si-Si bonds. One of a possible reason of oxidation under biasing condition can be an enhancement of nano-porous morphology that promotes oxygen absorption (cf. Figure 1).

The FTIR absorption spectra of the as-deposited and annealed films are shown in Figure 3 and 4. FTIR spectra of the films are represented by several main absorption bands at about 800 cm^{-1}, 900 cm^{-1}, 1040-1100 cm^{-1}, 1200 cm^{-1}, 1560-1650 cm^{-1} and 3300-3500 cm^{-1}. A feature at about 2300 cm^{-1} is an instrumental artifact. Unambiguous identification of these absorption bands is rather difficult as the several absorption bands are expected in the same spectral regions. For example, absorption at 800 cm^{-1} can be ascribed to Si-C stretching vibration mode [12] as well as to Si-O-Si symmetrical stretching vibrations. The band located around 900 cm^{-1} can be attributed to Si-OH or to OSi-H$_2$ [11, 13]. Some contribution to absorption in spectral range of 850-900 cm^{-1} can be also due to Si-H$_n$ bonds [14]. Strong absorption band at 1050 cm^{-1} can be originated from superposition of Si:C-H$_n$ bending modes and Si-O-Si asymmetrical in-phase stretching vibrations [11-13]. Si-O vibrations are also shown as a shoulder around 1100 cm^{-1} [14]. Absorption at 1200 cm^{-1} can be due to asymmetrical out-of-phase stretching vibration of Si-O-Si bridges as well as C-O vibration modes [11]. Two narrow bands at 1560 and 1650 cm^{-1} can be assigned to the sp^2 C-C bonds [12]. A comparison of the FTIR spectra of the as-deposited A and B films shows that an applying of substrate bias leads to an appearance of sp^2 C-C and Si-H bonds and to an enhancement of C-H, Si-O and O-H bonds. Such a redistribution of bonds does not contradict to the bonding picture implied the XPS measurement of the as-deposited samples (cf. Figure 2).

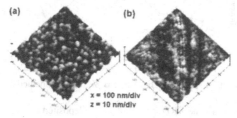

Figure 1. AFM images of the films A (U$_d$=0) (a) and B (U$_d$=-100V) (b).

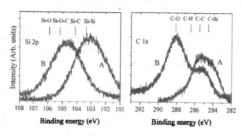

Figure 2. Si 2p and C 1s core level XPS spectra of as-deposited A and B films. The arrows denote the binding energies of corresponding signals, according to Refs. 11.

The main effects of annealing on the FTIR spectra of the samples A are the enhancement of Si:C-H$_n$ and Si-O bonds (Figure 4). For the samples B, the annealing at 650 ^{0}C causes a shift of the main absorption band at 1050 cm^{-1} towards low wavenumbers. A further increase in annealing temperature leads to disappearance of the band around 900 cm^{-1} and to ordering of Si-O bonds (enhancement of the shoulder at 1200 cm^{-1}).

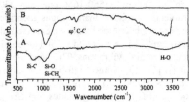

Figure 3. FTIR absorption spectra of the as-deposited A and B SiC films.

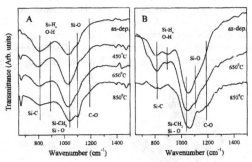

Figure 4. FTIR absorption spectra of the A and B SiC films annealed at various temperatures.

In Figure 5 we show the XRD spectra of the as-deposited and annealed B samples. In the spectrum of the as-deposited film one can see a small peak at ~ 42° that can be assigned to superposition of β-SiC (200) [073-1665] and 4H-SiC (103) [029-1127] diffractions. This indicates the formation of the nc-SiC:H structure. FWHM of this peak is 2.8° and the average size of the nanocrystallites evaluated according to the Scherrer formula is less than 3 nm. After annealing of this film at 650 °C we observe additional peaks at ~ 38, 43 and 49° that can be mainly assigned to reflection from 4H-SiC nanocrystallites [029-1127]. As-deposited and annealed samples of the series A were amorphous.

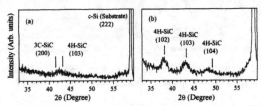

Figure 5. XRD spectra of the as-deposited (a) and annealed at 650 °C (b) films B.

Photoluminescence spectra of both of the sets of samples are shown in Figure 6. One can see that the as- deposited amorphous sample A has a broad emission spectrum centered around 470 nm. On the contrary, the as-deposited nanocrystalline sample B exhibits a bright blue PL

298

having a distinct double-peak spectrum with peaks at 415 and 437 nm. Annealing at 650 ^{0}C of the film B leads to strong increasing of the integral intensity of light emission and crucial change of the PL spectrum. Specific double-peaked spectrum was transformed after annealing into broad featureless white band with maximum intensity at about 470 nm. PL spectrum shape of the films A is unchanged under thermal treatment. High temperature annealing quenches the PL of both of the sets of the samples.

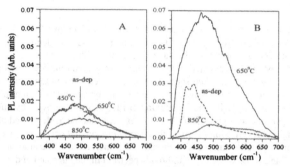

Figure 6. PL spectra of SiC:H films.

The possible mechanism of the observed blue PL of the nano-crystalline oxygen-rich SiC layers (series B) was suggested and discussed in our previous work [9]. Here we use the new experimental data on amorphous SiC:H films to add some new details to this mechanism. Basing on our XPS, FTIR and XRD data for the as-deposited sample B, we suppose that nanocrystallites are surrounded by the a-SiO$_x$:H environment, whereas the amorphous intergranular tissue is represented by the Si:C:O:H matrix. It is worth noting that the Tauc optical band gap of our amorphous SiC:H films is about 2.6-2.7 eV (477-460 nm), that is a typical value for the a-SiC:H films deposited from organosilanes [15]. We have supposed earlier that surface states of nanocrystallites are of critical importance in light-emitting mechanism: the recombination occurs in a radiative center (RC) located at the nanocrystallite surface, whereas the photo excitation of electron-hole pairs mainly occurs in nanocrystallite cores. The confining materials in this case could be the nanocrystallite cores and the surrounding a-SiO$_x$:H matrix that can have the band gap higher than the RCs. Surface Si=O groups could be considered as the RCs [9]. Such a mechanism predicts unchanged PL peak wavenumber under thermal treatment, which is often observed for un-hydrogenated Si-based nanostructures [2-7]. A moderate annealing of the film B leads to an increase in crystallite size and modification of the SiO$_x$ surrounding towards weakening of confinement effect, destroying of radiative recombination sites and enhancement of non-radiative recombination paths. This leads to quenching of specific blue PL related to SiC nano-crystallites. On the contrary, the contribution of the Si:C:O:H tissue in light-emission rises after annealing at 650^{0}C possibly due to passivation of carbon related paramagnetic defects in the amorphous tissue by the mechanism similar to that observed earlier for annealed a-SiC:H films deposited by reactive magnetron sputtering technique [16]. Further increase in annealing temperature leads to hydrogen effusion from films and to a strong increase of the concentration of paramagnetic centers that suppress radiative recombination in the amorphous tissue, which results in quenching of the white PL.

In the amorphous as-deposited sample A crystallites are absent, and the amorphous tissue plays the major role in photoemission. Therefore the shape of the PL spectra resembles that of the annealed samples B. Judging from the FTIR spectra of this sample (cf. Fig. 4), it is less hydrogenated than the sample B and moderate annealing leads only to an insignificant increase in PL intensity.

CONCLUSIONS

Hydrogenated SiC films were deposited by a PECVD technique from methyltrichlorosilane. The effects of negative substrate bias and post-deposition vacuum annealing were studied. It was shown that layers deposited without biasing were amorphous whereas -100V substrate bias assisted to formation of nanocrystalline SiC structure. Bright blue emission with specific double-peaked spectrum was observed at room temperature from the as-deposited nano-crystalline film. An increase of annealing temperature led first to an enhancement of integral intensity and broadening of PL spectrum followed by quenching of PL at annealing temperature as high as 850^0C. It is suggested that the as-deposited nanocrystalline film exhibits bright blue emission owing to the recombination that occurs in the radiative center located at the nanocrystallite surface. The radiative recombination in the amorphous Si:C:O:H tissue is responsible for the white PL with broad spectrum in the annealed and as-deposited amorphous film.

REFERENCES

1. L.I. Canham, *Appl. Phys. Lett.* **57**, 1046 (1990).
2. X. Zhao, O. Schoenfeld, Y. Aoyagi and T. Sugano, *Appl. Phys. Lett.* **65**, 1290 (1994).
3. W. Skorupa, R.A. Yankov, E. Tychenko, H. Fröb, T. Böhme and K. Leo, *Appl. Phys. Lett.* **68**, 2410 (1996).
4. V.V. Filippov, P.P. Pershukevich, V.V. Kuznetsova, V.S. Khomenko and L.N. Dolgii, *J. Appl. Spectrosc* **67**, 852 (2000).
5. Z. Ma, L. Wang, K. Chen, W. Li, L. Zhang, Y. Bao, X. Wang, J. Xu, X. Huang and D. Feng, *J. Non-Cryst. Solids* **299-302** (2002).
6. Q.W. Chen, D.L. Zhu, C. Zhu, J. Wang and Y.G. Zhang, *Appl. Phys. Lett.* **82**, 1018 (2003).
7. X.L. Wu, S.J. Xiong, G.G. Siu, G.S. Huang, Y.F. Mei, Z.Y. Zhang, S.S. Deng and C. Tan, *Phys. Rev. Lett.* **91**, 157402 (2003).
8. L.A. Ivashchenko, V.A. Vasin, V. Ivashchenko, M. Ushakov and A.V. Rysavsky, *Mater. Res. Soc. Symp. Proc.* **910**, 279 (2006).
9. V. Ivashchenko, V.A. Vasin, L.A. Ivashchenko, M. Ushakov, *Proc. SPIE*, **7041X**, 1 (2008).
10. J.Y. Fan, X.L. Wu and P.K. Chu, *Progr. Mater. Sci.* **51**, 983 (2006).
11. E. Ech-chamikh,. E.L. Ameziane, A. Bennouna, M. Azizan, T.A. Nguyen and T. Lopez-Rios, *Thin Solid Films* **259**, 18-24 (1995).
12. J. Bullot, M.P. Schmidt, *Phys. Stat. Sol.* (b) **143**, 345-418 (1987).
13. D. Seyferth, C. Prud'homme, G.H. Wiseman, *Inorg. Chem.* **22**, 2163 (1983).
14. F. Demichelis, G. Grovini, C.F. Pirri and E. Tresso, *Phyl. Mag.* **68**, 329 (1993).
15. J. Seekamp and W. Bauhofer, *J. Non-Cryst. Solids* **227**, 474 (1998).
16. A.V. Vasin, S.P. Kolesnik, A.A. Konchits, A.V. Rusavsky, V.S. Lysenko, A.N. Nazarov, Y. Ishikawa, Y.Koshka, *J. Appl. Phys.* **103**, 123710 (2008).

Mater. Res. Soc. Symp. Proc. Vol. 1153 © 2009 Materials Research Society   1153-A16-02

Defect study of polycrystalline–silicon seed layers made by aluminum-induced crystallization

Srisaran Venkatachalam[1, 2], Dries Van Gestel[1], Ivan Gordon[1], Guy Beaucarne[1], Jef Poortmans[1,2]
1 IMEC, Kapeldreef 75, Leuven BE-3001 Belgium
2 ESAT, Katholieke Universiteit Leuven, BE-3001 Belgium

ABSTRACT

A polycrystalline silicon (pc-Si) thin film with large grains on a low-cost non-Si substrate is a promising material for thin-film solar cells. One possibility to grow such a pc-Si layer is by aluminum-induced crystallization (AIC) followed by epitaxial thickening. The best cell efficiency we have achieved so far with such an AIC approach is 8%. The main factor that limits the efficiency of our pc-Si solar cells at present is the presence of many intra-grain defects. These intra-grain defects originate within the AIC seed layer. The defect density of the layers can be determined by chemical defect etching. This technique is well suited for our epitaxial layers but relatively hard to execute directly on the seed layers. This paper presents a way to reveal the defects present in thin and highly-aluminum-doped AIC seed layers by using defect etching. We used diluted Schimmel and diluted Wright etching solutions. SEM pictures show the presence of intra-grain defects and grain boundaries in seed layers after defect etching, as verified by Electron Backscatter Diffraction (EBSD) analyses. The SEM images after diluted Wright etching of pc-Si seed layers show that grain boundaries much more clearly than with diluted Schimmel etch.

INTRODUCTION

A thin-film pc-si solar cell combines the benefits of crystalline Si with the high potential for cost reduction of a thin film technology [1]. Additionally, pc-Si thin films on non-Si substrates have become widely used as material for thin-film transistors integrated in liquid-crystal display panels [2]. The pc-Si layers with grain sizes ranging from 1 to 100 μm are useful for solar cell applications. The main advantages of such pc-Si used in solar cells are its stability under sunlight, unlike amorphous silicon which undergoes light-induced degradation, and its production cost that is much less than that of single-crystalline wafers. We prepare pc- Si layers in two steps, namely a seed layer formation step followed by a high-temperature epitaxy step. The seed layer can be prepared by several techniques, in which aluminum-induced crystallization (AIC) drew our attention since it allows us to obtain pc-Si seed layers with very large grains. The AIC seed layer itself cannot be used as absorber layer due to its high Al doping level of ~ 3×10^{18} cm^{-3} and due to the fact that it is only around 200 nm thick. Therefore, we grow epitaxial layers of 3 to 6 μm on these seed layers. The doping densities are controlled during epitaxy to obtain a p^+ - p structure in which the p^+ layer acts as Back Surface Field (BSF) and the p-type layer is the absorber layer. We have reached solar cell efficiencies of up to 8 % so far [1]. However, to compete with other solar cells in current market, it is necessary to improve the efficiency of thin-film solar cell even more.

The main factor that limits the efficiency of our pc-Si solar cells is the presence of many intra-grain defects in our epitaxial layers [3]. The crystallographic defects in the epitaxial layers can be revealed using a wet chemical etching. The estimated defect density in our layers is around 10^9 cm^{-2}. These defects are electrically active and responsible for recombination activity. They therefore limit the performance of our solar cells [3]. Our objective is to reduce the intra-grain defect density and hence to improve the efficiency of our solar cells. Most of these intra-grain defects originate from the seed layer. A simple technique like defect etching should be developed to reveal the defects in the seed layer. Standard defect etching is difficult to perform on the AIC seed layers since they are very thin and highly doped. This paper shows, however, that defect etching of the seed layer can be possible if the chemical solution is diluted enough.

EXPERIMENTAL DETAILS

Pc-Si seed layer preparation

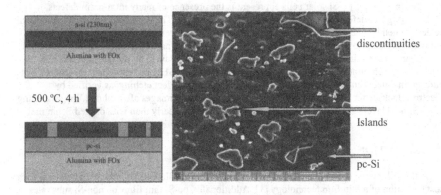

Figure 1 A Schematic sketch of aluminum induced crystallization process (left) and a typical SEM image of a pc-Si seed layer prepared by AIC (right)

We prepared pc-Si seed layers by AIC on ceramic alumina (Coorstek ADS996R) substrates. The substrates were subjected to RCA cleaning and spin-coating of FOx (Flowable oxide) before AIC. The FOx layers reduce the surface roughness of the substrates, and additionally it acts as barrier layers against diffusion of impurities from the substrates. The AIC process involves the deposition of a stack of aluminum (Al) and amorphous-silicon (a-Si) in a high vacuum e-beam evaporator. The nominal thickness of the deposited Al and a-Si layers was 200 and 230 nm respectively. In between two depositions, the Al layer was oxidized by exposure to ambient air for 2 minutes before the a-Si deposition.

The stack was then annealed at 500 °C for 4 hours in a N_2 atmosphere. During the annealing process, a layer exchange takes place while simultaneously the a-Si layer crystallizes into pc-Si. A schematic illustration of the process is shown in Figure 1. The Al layer is selectively etched using an Al etch solution (16 H_3PO_3: 1 CH_3COOH: 1 HNO_3: 2 H_2O) at 40 °C. The morphology of the resulting pc-si layer includes the following additional features: islands (secondary crystallites on top of the seed layer) and discontinuities as shown in a SEM image (Figure 1) [4].

Island removal

Silicon islands (secondary crystallites with straight sidewalls and thickness of ~ 230 nm) are present on the surface of the seed layers. These islands may interrupt the appearance of fine features after defect etching and also interfere with the Electron Backscatter Diffraction (EBSD) measurements. Therefore, these islands were selectively removed from part of the investigated seed layers by reactive ion etching (RIE) system using O_2 and SF_6. The Al layer was hereby used as a self-aligned etching mask [4].

Defect etch

We have used two different solutions to reveal defects in the pc-si seed layers:

i) *Schimmel solution*:

The composition of Schimmel solution used for the seed layers was CrO3 [0.075M]: HF [49%]: DI H_2O in the ratio of 1:2:2. During optimization, the amount of DI water is varied from 1.5 to 3 parts of the solution. The Schimmel solution can be used for silicon with resistivities below 0.2 Ohm cm if diluted and it is ideal for (100)-oriented grains [5]. A freshly prepared solution is used every time because the solution has a shelf-life of less than a day [5].

ii) Wright solution:

The composition of the Wright solution is as follows: 60 ml HF (49 %), 30ml HNO_3 (69%), 30ml of 5M CrO_3, 2 g Cu $(NO_3)_2.3H_2O$, 60 ml acetic acid (glacial) and 60 ml H_2O. The solution should be made by first dissolving the Cu $(NO_3)_2$ in the given amount of water followed by adding the other chemicals in any order [6]. For seed layers, additionally one part of DI water is mixed. In general, Wright etching is ideal for (100) and (111) grains and it has a shelf life of about six weeks [6].

RESULTS AND DISCUSSION

i) Schimmel etching

The concentration and the etching time should be optimized for seed layers to obtain clear results. The etching rate can be controlled by diluting the solution. Hence a diluted solution helps to execute the defect etching on the thin and highly-doped seed layers. During etching CrO_3 acts as an oxidizing agent and HF dissolves the oxide formed at the defect sites.

We use Schimmel solution with ratio of 1:2:1.5 for our epitaxial layers [3]. For the seed layers, the best results were obtained with the solution of compositional ratio 1:2:2 and etching time between 8 and 12 seconds (shown in Figure 2). We can observe the following features: grain boundaries (GB), intra-grain defects (IGD), deeply etched (DE) regions and etch-holes (EH). These IGDs look more or less similar to IGDs observed in our epitaxial layers [3].

The defect etching of seed layers with islands are also executed to observe the effect of islands and it is observed that the islands were etched anisotrophically.

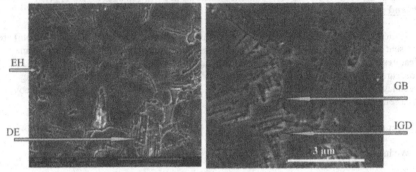

Figure 2 The SEM images of seed layer after Schimmel defect etching solution of 1:2:2 for 8 seconds

Figure 3 A comparison of an EBSD map with a SEM image of the same region of a

An EBSD analysis gives a 2D map of the crystallographic orientation of the grains, and the different colors correspond to different orientations. For instance, the red color regions are grains with a (100) orientation. The EBSD analysis was executed on seed layers after defect etching and it shows the grain boundaries visible in the SEM picture exactly match with the grain boundaries in the EBSD map (Shown in Figure 3).Low-resistivity material tends to etch faster than material with a higher resistivity for the same dopant type. Furthermore, high-index-oriented planes such as (111) tend to etch faster initially than planes of a lower index planes such as (100) [6]. We think these could be possible reasons for the appearance of deeply-etched regions and etch holes. Unfortunately these deeply etched regions did not give a beneficial signal in the EBSD measurement and hence we can not conclude which orientation they originally had.To rectify this problem, the Wright solution was investigated as an alternative etching solution. The Wright etch is interesting because it preferentially delineates defects on (100) and (111) orientations of p- and n- type silicon.

ii) Wright defect etch:

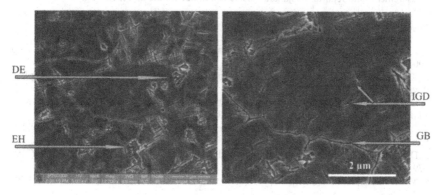

Figure.4 The SEM images of pc-Si seed layer after diluted Wright etch for 24 seconds.

The etching mechanism behind the Wright etch is similar to Schimmel but it involves additional chemicals. The function of the acetic acid is to maintain the smoothness of the etched surface and that of cupric nitrate is to make localized oxidation [6].

The SEM picture (Figure 4) of the Wright etched pc-si seed layer shows that grain boundaries become much better visible than with Schimmel etch. The optimal etching time for diluted Wright etching was found to be 24 seconds. The etching rate of the Wright solution is relatively low and the etching time can therefore be more easily controlled. The etch-holes and deeply etched regions were also observed here. Further studies on those regions are required. For instance, a comparative analysis of deeply-etched regions and etch-holes with the EBSD map before defect etching will provide beneficial information.

CONCLUSIONS:

Aluminum-induced crystallization can be used for preparing pc-si layers with large grains but the resulting layers contain numerous defects. These defects originate in the seed layers and limit the performance of the solar cells. To improve the seed layer quality even more, a simple way to characterize those seed layers must be developed initially. The first results of defect etching directly on the pc-si seed layer were obtained. Diluted versions of Schimmel and Wright solutions can be used for defect etching of the thin and highly-doped pc-Si Seed layer. The diluted Wright etch reveals grain boundaries more precisely when compared to diluted Schimmel etch. Further investigations are necessary to understand the origin of unknown features like deeply etched regions on the etch surface. The intra-grain defect density of AIC seed layers will be estimated if the technique is optimized further.

ACKNOWLEDGEMENTS:

This work is partly funded by the European Commission under contract number 019670-FP6-IST-IP ('ATHLET'). The authors would like to thank Jan D'Haen for the EBSD measurements and Jennifer Alejandra Amaya Rodriguez for the defect etching.

REFERENCES:

1. I. Gordon, L.Carnel, D.Van Gestel, G.Beaucarne & J. Poortmans, Prog. Photovolt: Res. Appl. 2007; 15:1575-586.
2. K. Kitahara, H.Ogasawara, J.Kambara, M.Kobata, & Y.Ohashi, Japanese Journal of Applied Physics 2008; 47(1): 54-58.
3. D.Van Gestel, M.J.Romero, I.Gordon, L.Carnel, J.D'Haen, G. Beaucarne, M.Al-Jassim & J.Poortmans, Applied Physics Letters 2007; 90:092103.
4. D.Van Gestel, I. Gordon, A. Verbist, L.Carnel, G.Beaucarne, and J.Poortmans, Thin Solid Films 516, 6907-6911 (2008).
5. D.G.Schimmel, J. Electrochemical, soc: solid -state science and technology. 1979; 126 (3): 479-483.
6. M.W Jenkins, J. Electrochemical, soc: solid -state science and technology, 1977; 124 (5): 757-762.
7. D.Van Gestel, I.Gordon, Y. Qiu, S.Venkatachalam, G.Beaucarne and J. Poortmans,this Conference, Symposium A: Oral presentation A10.2.

Mater. Res. Soc. Symp. Proc. Vol. 1153 © 2009 Materials Research Society 1153-A16-03

Infrared Ellipsometry Investigation of Hydrogenated Amorphous Silicon

Franco Gaspari[1]*, Anatoli Shkrebtii[1], Tom Tiwald[2], Andrea Fuchser[2], Shafiq Mohammed[1], Tome Kosteski[3], Keith Leong[3] and Nazir Kherani[3]
[1] Faculty of Science, University of Ontario Institute of Technology
2000 Simcoe Street North, Oshawa, ON, Canada L1H 7K4
[2] J.A. Woollam Co., Inc. 645 M Street, Suite 102, Lincoln, NE 68508, USA
[3] Dept. of Electrical & Computer Engineering, University of Toronto,
Toronto, ON, Canada M5S 1A4

ABSTRACT

Hydrogenated amorphous silicon (a-Si:H) has been extensively investigated experimentally in the infrared spectral region via techniques such as Fourier Transform Infrared (FTIR) and Raman spectroscopy. Although spectroscopic ellipsometry has been proven to be an important tool for the determination of several parameters of a-Si:H films, including dielectric constant, surface roughness, doping concentration and layer thickness, the spectral range used in these studies has rarely covered the infrared region below 0.6 eV, and never over the complete spectral region of interest (0.04 – 0.3 eV), which contains atomic vibration frequencies.
We have measured for the first time the dielectric function of a-Si:H films grown by the saddle field glow discharge technique by spectroscopic ellipsometry in the energy range from 0.04 eV to 6.5 eV, thus extending the analysis into the far infrared region. The a-Si:H films were deposited on germanium substrates for the ellipsometry studies, and on crystalline silicon substrates for the comparative FTIR analysis. Preparation parameters were chosen to obtain films with different hydrogen content. In this paper, we present the results of the ellipsometry analysis, evaluate different fitting techniques, and compare the results with the corresponding FTIR spectra. The similarities and differences between the spectra are discussed in terms of the a-Si:H properties.

INTRODUCTION

Hydrogenated amorphous silicon (a-Si:H) has been the subject of numerous theoretical studies and experimental characterizations over the past 40 years; yet, quite a few issues still remain to be clarified [1-2]. The authors have initiated a systematic study of a-Si:H based on ab-initio molecular dynamics simulations [3,4]. The initial results have shown that vibrational spectroscopy data, in particular Fourier Transform Infrared Transmission Spectroscopy (FTIR), provide a good basis for validation of models. IR spectroscopic ellipsometry has also been used in the past by, among others, Drevillon [5] and Darwich [6]. However, these investigations have usually been limited to exploring the stretching modes (at about 2000 cm^{-1}). To better understand the nature of vibrational absorption in a-Si:H, we have investigated the results obtained from IR spectroscopic ellipsometry, extending the range of the investigation into all the relevant wavenumbers for amorphous silicon, i.e., from 400 cm^{-1} and up. This allows for comparison of the FTIR data with the ellipsometry data in the specific regions of the hydrogen stretching modes (~ 2000-2100 cm^{-1}), the bending/scissors modes (840-890 cm^{-1}), and the wagging mode (640 cm^{-1}). Understanding how the vibrating dipoles within the silicon matrix interact with different forms of excitation will provide useful information for the refinement and improvement of the models being used to analyze the properties of a-Si:H.

EXPERIMENT

Sample preparation

Three samples were prepared with the saddle field glow discharge technique [7] using preparation parameters conducive to different types of hydrogen bonding. In particular, sample 1 (S1) has a high mono-hydride content, sample 2 (S2) on the other hand has a large content of polyhydride (PH) bonds and/or monohydride (MH) clusters or chains, sample 3 (S3) was similar to sample 1 but with a lower total hydrogen concentration. The films were grown on crystalline silicon wafers for FTIR and on Germanium wafers for ellipsometry. Samples were approximately 0.4 μm, 0.6 μm, and 0.8 μm thick, respectively.

Ellipsometry

Ellipsometry measures the change in polarization of reflected light as compared with the polarization of the incident light. A detailed description of ellipsometry principles and IR spectroscopic ellipsometry can be found in reference [8]. One of the advantages of IR ellipsometry over FTIR is the insensitivity to the thickness of the film. Infra-red spectroscopic ellipsometry measurements were carried out using IR-VASE variable angle spectroscopic ellipsometer from J.A. Woollam [9]. The data were collected in the range 1.7-33 μm bases on a Fourier-transform spectrometer. The angle of incidence was computer controlled from 30° to 90°. The Tauc-Lorentz model, based on the Tauc joint density of states and the Lorentz oscillator was used to model the absorption in the 0.6-6.5 eV range. A thin native oxide layer was included in the analysis, and required for best fit of the data. The model also includes some thickness nonuniformity to fit the depolarization data. This model is particularly well suited for amorphous materials [10]. Five Gaussian oscillators were first used to model the absorption in the IR range. However, during preliminary fits to the data it was found that these oscillator model (and similar ones) tend to obscure subtle details in $n(\lambda)$ and $k(\lambda)$. In order to overcome this problem, a novel approach was used to analyze the data, the k-k layer technique [11]. We fit $k(\lambda)$ on a wavelength-by-wavelength basis, while using numerical integration of the Kramers-Kronig transform to determine $n(\lambda)$. This allows us to see all of the details of the a-Si film absorptions, while maintaining Kramers-Kronig consistency despite the large number of free parameters.

FTIR

Fourier Transform Infrared spectroscopy has been used extensively over the years to characterize hydrogenated amorphous silicon. A general review of FTIR and a-Si:H can be found in reference [1]; however, identification and nomenclature of some of the main features is still a matter of debate [2]. In this paper, we will adopt the convention from [2] when addressing the IR region related to stretching modes. The main features of FTIR applied to a-Si:H can be summarized as follows: there are three main absorption areas at 640 cm^{-1}, 840-890 cm^{-1}, and 2000-2100 cm^{-1}, although the latter region can be extended in both directions; these peaks are related to different vibrational modes. The signal at about 2000 cm^{-1} is usually a convolution of two different peaks, a high stretching mode (HSM) at 2070-2100 cm^{-1}, and a low stretching mode (LSM) at 1980-2010 cm^{-1}; other additional feature might be present, however, the assignment of all these modes is still a matter of debate [2]. The two sub-signals have different proportionality factors with respect to the hydrogen content, i.e. MH vs. PH, and therefore the signal cannot be easily used to determine the total concentration of hydrogen; the signal at 840-890 cm^{-1} is normally used to identify the bending and scissors modes and is associated with non-MH bonds [1]. Comparing this signal with the HSM can help distinguish between the PH's and the MH chains contributing to the latter; finally, the peak at 640 cm^{-1} should include all (Si—H$_n$)$_n$ wagging modes and it is used to calculate the hydrogen concentration by using a single proportionality constant.

FTIR measurements were carried out using a Perkin-Elmer system 2000 FTIR. The spectrophotometer sample compartment was purged with dry air. A bare c-Si substrate, from the same wafer used for the a-Si:H substrate, was used as reference.

RESULTS AND DISCUSSION

Figure 1a shows the FTIR results for three samples in the 400-2200 cm^{-1} wave number range. Each sample exhibits the distinct signature peaks of a-Si:H, however, the evident shift to higher wavenumbers in the 2000 cm^{-1} region and the larger signal in the 840-890 cm^{-1} range (shown with an expanded scale in the inset) confirms that the second sample, S2, has indeed a higher content of polyhydrides. In particular, the doublet features at ~ 840 cm^{-1} and at ~ 890 cm^{-1} indicate the presence of Si-H$_2$, Si-H$_3$ and (Si-H$_2$)$_n$ bonds.

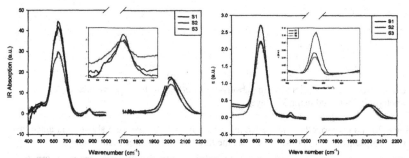

Figure 1. The left panel (1a) shows FTIR spectra for three samples under study. The inset shows the doublet feature. The right panel (1b) demonstrates infrared ellipsometry spectra for the same three samples derived from the Gaussian oscillator model.

The same three main features are present in the IR portion of the ellipsometry results, as shown in Figure 1b. However, the ellipsometry spectra indicate a stronger signal for the 640 cm^{-1} peak relative to the signal in the 2000 cm^{-1} region. Another main difference between the two spectra is related to the nature of the peaks, with the ellipsometry data showing a more symmetric signal. As indicated previously, a computationally intensive approach was used to verify whether the symmetry is a by-product of the modeling procedure. In this study, it is important to determine the $k(\lambda)$ values as accurately and with as much detail as the measurement allowed. Therefore, $n(\lambda)$ values were explicitly obtained from $k(\lambda)$ by numerically calculating of the Kramers-Kronig integral at each measured wavelength, using the following equation:

$$n(\omega) = \sqrt{\varepsilon_\infty} + \frac{A}{\omega_o^2 - \omega^2} + \frac{2}{\pi} P \int_{\omega_{min}}^{\omega_{max}} \frac{\omega' k(\omega')}{\omega'^2 - \omega^2} d\omega'$$ [1]

In the equation, ω and ω' have units of wavenumber, with $\omega_{min} = 400$ cm^{-1} and $\omega_{max} = 5000$ cm^{-1}. The quantity ε_∞ in combination the pole parameters A and ω_o model the index dispersion that occurs because of electronic transitions above the a-Si bandgap.

During analysis, $k(\lambda)$ is allowed to vary at every wavelength, and $n(\lambda)$ is calculated from $k(\lambda)$ with A, ω_o and ε_∞ providing the dispersion described by the above equation. The film thickness is also varied. By tying $n(\lambda)$ and $k(\lambda)$ together via the Kramers-Kronig integral, we were able to guarantee that the optical constants remain physical, despite the large number of free parameters.

The data for the extinction coefficient, k, and the absorption coefficient, α, are shown in Figures 3a and 3b.

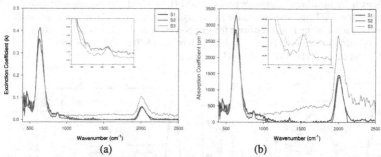

(a)   (b)

Figure 3 – Extinction coefficient (a) and absorption coefficient (b) vs. wavenumber for ellipsometry data obtained using k-k layer fit.

By tying $n(\lambda)$ and $k(\lambda)$ together via the Kramers-Kronig integral, we were able to guarantee that the optical constants remain physical, despite the large number of free parameters.

Figure 4 shows the signal in the stretching mode region for S2 obtained from FTIR. The signal has been fitted using two Gaussians. A similar procedure was implemented for the other samples. The ratio of the 2000 cm^{-1} signal over the 640 cm^{-1} signal obtained from ellipsometry is plotted in Figure 5a *vs.* the structure factor R obtained from FTIR, defined as the ratio of the HSM over the LSM. It is interesting to note that the ratio decreases as the structure factor increases. An increase in the structure factor is generally attributed to a greater content of di-hydride bonds [1], suggesting that ellipsometry has a lower sensitivity to the stretching mode of Si-H$_2$ bonds. In Figure 5b we show the ratio of the stretching mode signal over the wagging mode signal at 640 cm^{-1} for both techniques. The negative trend again suggests a lower sensitivity to the di-hydride stretching modes for ellipsometry, while it appears that this is not the case for the other vibrational modes. The presence of a strong doublet peak for the high R-factor sample in the ellipsometry spectra (see inset in Figure 3b) also confirms a good sensitivity to rocking-scissors modes.

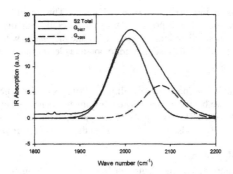

Figure 4 - Double Gaussian fit of the 2000 cm^{-1} signal for sample S2. The subscript numbers for the G symbols indicate the center position of the Gaussian. The black line (S2 Total), representing the experimental data, has been shifted to help the viewer.

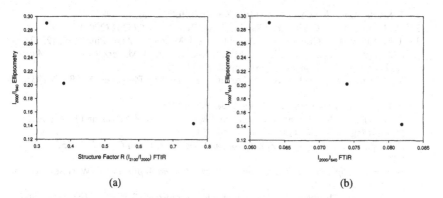

(a)   (b)

Figure 5 – (a) Relative ratio of the 2000 cm^{-1} signal over 640 cm^{-1} signal from ellipsometry vs. structure factor; (b) comparison of I_{2000}/I_{640} for FTIR and Ellipsometry.

The structure factor R shows that sample S2 has a much larger content of PH's and/or MH chains. An isolated monohydride has been modeled as a vibrating dipole in a cavity, producing a depolarizing field [12, 13], which could affect the ellipsometry measurements. On the other hand, Smets argues that this interpretation is faulty, and defines a nanostructure parameter K, which gives a measure of the averaged number of MH's per unit volume of a Si atom in the a-Si:H network [2]. The ellipsometry signal in fact appears to be less sensitive to HSM rather than LSM, indicating that other factors must be considered, including the impact of hydride modes on hydrogenated silicon surfaces, and the correlation between the MH configurations with the nanostructure parameter K. Another possibility is that the different directions of the polarization due to the two silicon-hydrogen stretching vibrations effect the

311

interaction between the input signal and the Si-H_2 bond. The authors have developed a theoretical model to simulate this process. This analysis could help distinguish between Si-H_2 bonds with different bond directions, or could correlate the results with different K values. Further experimental results are also needed to obtain a clearer picture and validate the model.

CONCLUSIONS

The FTIR and the Infrared Spectroscopic Ellipsometry spectra for 3 a-Si:H sample with different hydrogen bonding have been compared. The ellipsometry technique appears to be less sensitive to the overall HSM bonds. This loss in sensitivity could be the results of different net polarization of the Si-H_2 bonds, or to a variation in the nanostructure parameter K.

ACKNOWLEDGMENTS

The research was supported by the Centre for Materials and Manufacturing/Ontario Centres of Excellence (OCE/CMM) "Sonus/PV Photovoltaic Highway Traffic Noise Barrier" project, Discovery Grants from the Natural Sciences and Engineering Research Council of Canada (NSERC).

REFERENCES

1. T. Searle, *Amorphous Silicon and its Alloys* (INSPEC, London, 1998).
2. A.H Smets and M.C.M. van de Sanden, Phys. Rev. B. **76**, 073202 (2007).
3. I. M. Kupchak, F. Gaspari, A. I. Shkrebtii, and J. M. Perz, J. Appl. Phys. **104**, 123525-1 (2008) and F. Gaspari, I. M. Kupchak, A. I. Shkrebtii and J. M. Perz. Condensed Matter Archive, Nov. 17, 2008. http://arxiv.org/abs/0811.2282.
4. F. Gaspari, I. M. Kupchak, A. I. Shkrebtii, and J. M. Perz, Phys. Rev. B. **79**, 224203 (2009)
5. B. Drevillon, Microelectronics Journal, **24**, 347 (1993).
6. R. Darwich, P. Roca Cabarrocas, S. Vallon, R. Ossikovski, P. Morin, and K. Zellama, Philos. Mag. B. **72**, 363 (1995).
7. F. Gaspari, S.K. O'Leary, S. Zukotynski, and J.M. Perz, J. non-Cryst. Sol., **155**, 149 (1993).
8. H. Fujiwara, *Spectroscopic Ellipsometry Principles and Applications*, (West Sussex, John Wiley & Sons, 2007).
9. J.A. Woollam, "Ellipsometry, Variable Angle Spectroscopic", in *Wiley Encyclopedia of Electrical and Electronic Engineering*, New York: Wiley (2000) 109-117.
10. G. E. Jellison, jr. and F.A. Modine, Appl. Phys. Lett. **69**, 2137 (1996).
11. J. A. Dobrowolski, Yanen Guo, Tom Tiwald, Penghui Ma, and Daniel Poitras, APPLIED OPTICS, Vol. 45, No. 7, (2006).
12. A.A. Langforf, M.L. Fleet, B.P. Nelson, W.A. Lanford, and N. Maley, Phys. Rev. B **45**, 13 367 (1992).
13. M. Cardona, Phys. Status Solidi B **118**, 463 (1983)

Mater. Res. Soc. Symp. Proc. Vol. 1153 © 2009 Materials Research Society   1153-A16-04

Raman Characterization of Protocrystalline Silicon Films

A. J. Syllaios[1], S. K. Ajmera[1], G. S. Tyber[1], C. Littler[2], R. E. Hollingsworth[3]

[1]L-3 Communications Infrared Products, 13532 N.Central Expressway, MS37, Dallas, TX 75243
[2]Physics Dept., University of North Texas, 1155 Union Circle #311427, Denton, TX 76203
[3]ITN Energy Systems, 8130 Shaffer Parkway, Littleton, CO 80127

ABSTRACT

An increasingly important application of thin film hydrogenated amorphous silicon (α-Si:H) is in infrared detection for microbolometer thermal imaging arrays. Such arrays consist of thin α-Si:H films that are integrated into a floating thermally isolated membrane structure. Among the α-Si:H material properties affecting the design and performance of microbolometers is the microstructure. In this work, Raman spectroscopy is used to study changes in the microstructure of protocrystalline p-type α-Si:H films grown by PECVD as substrate temperature, dopant concentration, and hydrogen dilution are varied. The films exhibit the four Raman spectral peaks corresponding to the TO, LO, LA, and TA modes. It is found that the TO Raman peak becomes increasingly well defined (decreasing line width and increasing intensity), and shifts towards the crystalline TO energy as substrate temperature is increased, H dilution of the reactants is increased, or as dopant concentration is decreased.

INTRODUCTION

Thin amorphous silicon and silicon alloy films such as silicon germanium ($Si_{1-x}Ge_x$) are widely used in several applications including solar cells and thin film transistors (TFTs) [1] as well as uncooled infrared detector arrays [2,3,4]. Such films exhibit a wide range of electrical and structural properties [5]. One of the key material properties that affect an array of electrical parameters is the degree of structural order in the film. This structural order manifests physically in terms of crystallinity or embedded crystallinity in the amorphous matrix. In addition to crystallinity, strained bonds, dangling bonds, bond energy distribution, voids, and grain structure can all influence the effective bandgap and hence the electrical transport properties of the material. Deposited film morphology can be controlled by the selection of specific deposition conditions such as temperature, reactant and dopant concentrations, the use of nucleation or seed layers, and film thickness. Hydrogen dilution in the deposition plasma affects the nucleation of crystallites and results in a phase transition from amorphous to protocrystalline. In some applications such as solar cells, enhanced performance is observed for films that are deposited near the amorphous to crystalline phase transition [6].

For uncooled infrared detectors, film morphology directly influences key properties such as resistivity, temperature coefficient of resistance (TCR) and noise characteristics which impact detector performance. Microbolometer arrays consist of a focal plane of pixels that are suspended above an integrated circuit and thermally isolated through the use of extremely narrow legs. The pixels consist of suspended membrane structures with an electrically active layer that changes in resistance as the pixel changes temperature. Infrared flux from a scene is focused on the array through the use of optics and causes the pixels to heat up proportional to the IR flux incident upon the pixels. The integrated circuit below the pixels measures the change in resistance and generates a composite thermal image of the entire scene generated by the array. Therefore, materials which can be deposited as a thin film with high TCR values and low electrical noise characteristics are ideal for microbolometer applications. Amorphous silicon has some advantages for use in microbolometers. The resistivity and TCR of the material fall into the useful range for high performance uncooled microbolometers. Because it is a predominantly silicon and hydrogen based film, a wide range of parameters can be optimized for improved performance and standard techniques such as doping can be used to tune the bandgap to give desired electrical properties. In addition, amorphous silicon is directly integratable with standard CMOS fabrication processes such as depositions, etches, photolithography, and other processes that make amorphous silicon microbolometers highly manufacturable. In this study, we review the structural characterization of p-type doped amorphous silicon by Raman spectroscopy for films grown by plasma enhanced chemical vapor deposition (PECVD) at various substrate temperatures, hydrogen dilutions of silane reactant, and dopant concentrations.

EXPERIMENTAL DETAILS

Amorphous silicon thin films were grown in load locked PECVD systems using capacitively coupled 13.56 MHz plasmas with the substrate on the ground electrode. Source gases for α-Si:H growth were silane (SiH$_4$) and hydrogen in argon, and boron trichloride (BCl$_3$) at growth temperatures ranging from 285 °C to 390 °C. The α-Si:H films were approximately 200 nm thick. Raman spectroscopy was used to characterize the differences in film microstructure due to various deposition conditions, i.e., temperature, H dilution, and doping. Hydrogen dilution experiments were performed at a substrate temperature of 350 °C. Doping level experiments were conducted at [BCl$_3$] / [SiH$_4$] flow rate ratio ranging from 0.05 to 0.30. A Thermo Electron Almega XR Raman Spectrometer with a 532 nm laser and a 2.5 micron spot size was used to characterize the films grown over Si$_x$N$_y$:H coated Si <100>, or Si$_x$N$_y$:H coated borosilicate glass. The spectrometer provides a spectral resolution of 2 cm^{-1} full width at half maximum amplitude (FWHM), a diffraction limited spatial resolution down to 1 μm, and a confocal depth profiling resolution down to 2 μm.

RESULTS AND DISCUSSION

Raman Spectroscopy of Amorphous Silicon

Thin silicon films deposited by PECVD exhibit a wide range of structural configurations. Depending on the deposition conditions chosen, various phases can be grown that range from purely amorphous to nanocrystalline, microcrystalline, and polycrystalline. Mixed phases are also common such as amorphous-crystalline where crystallites are embedded in an amorphous matrix or grow in a columnar or conical fashion within the amorphous matrix. Deposition conditions that affect the structure and properties of these films include the dilution of precursor reactants with hydrogen, varying the substrate temperature, and dopant levels. In general, as the hydrogen dilution ratio increases, the deposited silicon film undergoes a transition from amorphous to microcrystalline [6]. Figure 1 shows a representative Raman spectra of p-type amorphous silicon and single crystal silicon measured on the same measurement system.

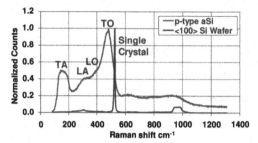

Figure 1. Representative Raman spectra for p-type amorphous silicon and single crystal silicon.

As seen above, the Raman spectrum for amorphous silicon is characterized by four bands [7,8], the most intense of which is the transverse optical (TO) mode located at 480 cm^{-1}. The other modes are located at ~ 380 cm^{-1} (longitudinal optical LO), 310 cm^{-1} (longitudinal acoustic LA), and 150 cm^{-1} (transverse acoustic TA). By contrast, single crystal silicon has a predominant peak representing tetra-coordinated Si-Si located at about 521 cm^{-1} which is noticeably absent in the amorphous silicon spectra. Shifts of the amorphous TO peak towards higher energy represent increased order in the film and can indicate crystallinity [9]. In addition, narrowing of the peak also indicates an increase in structural order which can be an indication that the film is moving towards crystallinity. These effects will be discussed below relative to various process conditions that influence the deposition of these films.

Deposition Condition Effects on Crystallinity

Figure 2 shows Raman spectra for p-type doped amorphous silicon films deposited under two different hydrogen dilutions as defined by the ratio of gas flow rate of H$_2$ to silane in the deposition plasma. The crystalline silicon peak centered ~521 cm^{-1} is markedly larger and well defined for the higher dilution sample consistent with crystallinity embedded in the film. In addition, Gaussian peak deconvolution shows the broad TO peak shifting towards higher energy, indicative of higher structural order for higher dilution as is shown in Figure 3.

Figure 3 examines the impact of H dilution for p-type amorphous silicon on the TO peak energy and peak width for a range of H dilutions. As can be seen, the TO peak shifts towards higher energies with increasing H dilution consistent with prior observations that higher H dilution enhances crystallinity in these films [6]. Further, TO FWHM decreases with increasing H dilution indicating an increase in structural order of the film consistent with the onset of crystallinity. This shows that Raman spectroscopy can discern silicon crystallinity embedded in a largely amorphous matrix.

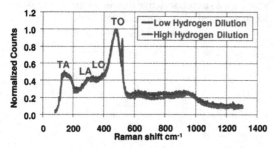

Figure 2. Raman shift of the amorphous TO mode for two p-type doped amorphous silicon thin films deposited under different H dilutions.

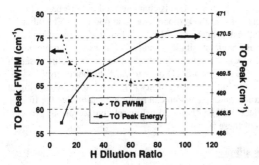

Figure 3. H dilution dependence of the TO mode energy and FWHM for p-type α-Si:H.

Another key parameter in the deposition of amorphous silicon is doping. Varying the dopant level by modulating the flow or composition of the dopant gas can effect both the incorporation of dopant into the silicon matrix as well as the reaction and transport mechanisms that govern the deposition. Two sets of p-type amorphous silicon were examined with different dopant gas flow rates. Figure 4 shows the Raman spectra for a film deposited with dopant gas flow ratio 6 times higher than a sister sample where all other conditions were held constant. The FWHM of the TO mode is narrower for the lower doped sample indicating that the film microstructure is tending toward higher order (microcrystallinity) for less dopant atoms in the structure. Also the TA mode amplitude decreases significantly with the lower dopant concentration.

316

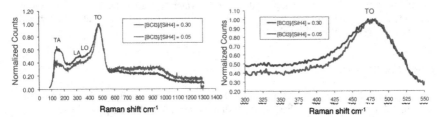

Figure 4. Raman shift spectra of p-type silicon films grown at two different doping levels. The full spectrum is shown on the left. The TO mode is shown in greater detail on the right.

The Raman spectra indicate that increasing doping by increasing the flow rate of dopant gas may retard the formation of microcrystals in these films and increase the level of structural disorder. This makes physical sense when considering that relative to an undoped film, adding dopant atoms adds a variety of new bonding states to the film that increases the number of energy states throughout the matrix. The increased number of local structural configurations manifest as an increase in aggregate disorder that can be observed in the Raman spectra.

In addition to doping, the effect of deposition temperature on the structural order of amorphous thin films was examined. Three different deposition temperatures were used ranging from 285 °C up to 390 °C. Gaussian peak deconvolution was performed and the TO peak energy and FWHM was determined and is reported in Figure 5. Higher substrate temperature leads to a narrower and higher energy TO peak. Lower substrate temperatures show more disorder or amorphous behavior in the film structure.

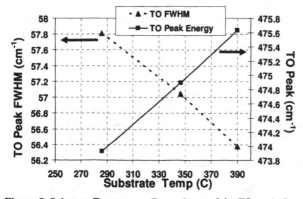

Figure 5. Substrate Temperature Dependence of the TO mode for amorphous silicon.

This behavior can be explained based on the nature of PECVD. Higher substrate temperature during deposition provides added surface mobility to absorbed reactant species which increases the likelihood of lower energy bond configurations. In addition, higher temperatures increase the desorption or dissociation of weakly bonded species, which typically lowers the net deposition rate and increases the likelihood that stronger bond configurations are prevalent in the depositing film. Raman spectra reflect this behavior.

CONCLUSIONS

Raman spectroscopy is able to discern differences in crystallinity or structural order in thin amorphous silicon films. These PECVD films were all broadly amorphous in nature but displayed varying levels of structural order due to differences in deposition conditions such as substrate temperature, hydrogen dilution of silane reactant, and dopant concentration. Using Raman spectra, it is observed that deposition at higher substrate temperatures and hydrogen dilution promotes higher atomic order in the films as indicated by peak width and peak energy of specific characteristic Raman modes, namely the TO mode. In addition, increasing dopant concentration shows increasing disorder in the film structure displayed in changes in the peak profiles of the TO and TA Raman modes. Future work envisioned involves the correlation of Raman data to optical data from spectroscopic ellipsometry where relative crystallinity of the films can be independently measured, and study the electrical properties of these films relative to the requirements for state-of-the-art microbolometers.

ACKNOWLEDGEMENTS
The technology development reported in this paper has been funded through the DARPA HOT MWIR Program (NBCH3060001, DARPA Program Manager, Stuart Horn; NVESD COTR, Dieter Lohrmann); and L-3 and ITN Internal Funds.

Cleared for Public Release by U. S. Army NVESD, dated April 15, 2009, reference number PR008585.

REFERENCES

1. R. A. Street (ed.), *Technology and Applications of Amorphous Silicon*, Springer, 2000.
2. J. Brady, T. Schimert, D. Ratcliff, R. Gooch, B. Ritchey, W. L. McCardel, K. Rachels, S. Ropson, M. Wand, M. Weinstein, and J. Wynn, *Proc. SPIE* **Vol. 3698**, 161 (1999).
3. J. L. Tissot, F. Rothan, C. Vedel, M. Vilain, and J. J. Yon, *Proc. SPIE* **Vol. 3436**, 605 (1998).
4. A. Kosarev, M. Moreno, A. Torres, and C. Zuniga, *J. Non-Cryst. Solids* **358** (2008) 2561.
5. K. Tanaka, E. Maruyama, T. Shimada, and H. Okamoto, *Amorphous Silicon*, John Wiley & Sons, 1993.
6. R. W. Collins, A. S. Ferlauto, G. M. Ferreira, J. Koh, C. Chen, R. J. Koval, J.M. Pearce, C. R. Wronski, M. M. Al-Jassim, and K. M. Jones, *Mat. Res. Soc. Symp. Proc.* **Vol. 762**, A10.1.1 (2003).
7. G. Viera, S. Huet, and L. Boufendi, *J. Appl. Phys.* **90**, 4175 (2001).
8. K. Wu, X. Q. Yan, and M. W. Chen, *Appl. Phys. Lett.* **91**, 101903 (2007).
9. P. Roura, J. Farjas, and P. Roca i Cabarrocas, *J. Appl. Phys.* **104**, 073521 (2008).

Mater. Res. Soc. Symp. Proc. Vol. 1153 © 2009 Materials Research Society   1153-A16-06

Erik.V. Johnson, Laurent Kroely, Mario Moreno, and Pere Roca i Cabarrocas
LPICM, CNRS, Ecole Polytechnique, 91128 Palaiseau, France

ABSTRACT

One of the primary challenges in the application of hydrogenated microcrystalline silicon (μc-Si:H) to photovoltaic cells is achieving high growth rates while maintaining good material quality over a wide process window. The rapid characterization of the material without generating a complete cell is thus a useful tool to determine said process window. Infrared absorption due to the various vibrational modes of the material has been used as a coarse tool towards this purpose, but the use of FTIR to perform this diagnosis limits the substrates upon which the analysis can be performed. We report on the use of high wave-number (1800-2200 cm^{-1}) Raman scattering to perform a similar role of telltale peak detection directly on solar cells and on substrates suitable for thin-film photovoltaics. We evaluate material grown from SiF$_4$ by RF-PECVD and from SiH$_4$ by Matrix Distributed Electron Cyclotron Resonance (MDECR-) PECVD.

INTRODUCTION

The high wavenumber (1800-2200cm^{-1}) infrared absorption in hydrogenated silicon thin-films has long been used to characterize their quality. Absorption peaks at ~2000cm^{-1} have been assigned to isolated stretching Si-H bonds, with higher energy peaks variously assigned to clustered Si-H, platelet-like configurations, and the higher hydrides. This characterization tool has been extended to μc-Si:H thin films destined for use in photovoltaics. Smets, Matsui and Kondo [1] have recently made the case that the appearance of narrow, twin peaks around 2100 cm^{-1} in the infrared absorption spectrum, as measured by Fourier Transform Infrared (FTIR) spectroscopy, coincides with porous μc-Si:H material that will perform poorly in solar cells. Narrow absorption peaks in the high wavenumber (2000-2200 cm^{-1}) FTIR absorption spectra of μc-Si:H thin films have been reported for many years [2,3,4], but have only recently been directly correlated with the oxygen-related degradation of solar cells.

The use of FTIR to examine μc-Si:H in this fashion is limited in that it requires the material to be deposited on lightly-doped crystalline silicon (c-Si) wafers. It is well-known that μc-Si:H film deposition is highly substrate dependent [5]. In addition, exposure of c-Si wafers to H environment results in the formation of hydrogen platelets within the first 200nm of the wafer surface [6]. Although in the case of thin-film deposition the exposure time of the bare substrate to the hydrogen-rich plasma (before the film begins to grow) is quite short, atomic hydrogen is quite mobile within the film during growth [7] and may continue to interact with the substrate throughout the film growth process. It is therefore possible that the FTIR peaks may be originating from the c-Si wafer rather than the film itself.

As opposed to the stringent requirements of FTIR absorption spectroscopy, Raman spectroscopy can be performed on any type of substrate, and therefore eliminates the necessity of IR-transparent, low-impurity, intrinsic crystalline Si substrates. The appearance of a given vibrational peak in the Raman spectrum will be determined by the changes in polarizability of the bond during the vibration. However, due to the lack of order, the selection rules are not strict for bonds in amorphous and μc-Si:H films, therefore the presence of a given peak must be checked experimentally. As well, although Raman spectroscopy in the backscattering configuration has the advantage of the sampling beam not passing through the substrate, it is difficult to use this method to get absolute density values (by assigning oscillator strengths to individual peaks). However, if simply the *presence* of tell-tale peaks is sufficient, Raman scattering may be a useful tool.

In this work, we apply high-wavenumber Raman scattering to analyze materials for applicability to photovoltaic devices. We present results showing the appearance of the twin Si-H stretching peaks in the Raman scattering spectrum of μc-Si:H grown by MD-ECR, as well as using this tool to evaluate μc-Si:F:H material directly in solar cells.

EXPERIMENT

This study concerns microcrystalline silicon grown from two different precursor gases and by two different deposition methods. Firstly, we examine the use of Matrix-Distributed Electron Cyclotron Resonance (MD-ECR) Plasma Enhanced Chemical Vapour Deposition (PECVD). This deposition technology has been used to perform the deposition of μc-Si:H at rates up to 28Å/s [8], although the films for this study were deposited at lower rates (~5 Å/s). The films were deposited from a gas mixture consisting of 8 sccm of SiH_4, 5 sccm of Ar and 75 sccm of H_2, and at a pressure of 5 mTorr. The substrate temperature during deposition was ~230°C, as determined by pyrometry, and a total microwave power of 1.5 kW was fed into the seven microwave antennas. A relatively low RF power of 11 W was applied to the substrate-holder during growth to establish a DC self-bias voltage.

The second set of films was deposited by standard radio-frequency (RF)-PECVD at 13.56 MHz, but by using SiF_4 as the gas pre-cursor. This set of films was deposited from varying mixtures of SiF_4, Ar, and H_2. As SiF_4 is more difficult to dissociate than SiH_4, Ar was used during the deposition of these films to increase gas dissociation and therefore the growth rate. These films were deposited at 200°C as the i-layer in PIN solar cells, and measurements were acquired on the devices themselves (between evaporated back-contacts) without removing the doped n-layer. The Raman spectra for both sets of films were acquired using a backscattering configuration and a HeNe laser to provide optical excitation at 632 nm.

DISCUSSION

Characterization of Material Grown from SiH₄ by MD-ECR

In Figure 1, FTIR spectra are presented that were acquired on a μc-Si:H film grown by MDECR PECVD on a 10-20kΩcm, float-zone crystalline silicon wafer. The presence of the twin peaks around 2100cm^{-1} in the spectra is clear, and indicates that this material is porous and that the grain boundaries will oxidize upon exposure to air. With the subsequent passage of time, these twin peaks decrease in magnitude and another peak at 2250cm^{-1} appears, presumably due to O-Si-H$_x$ bonds. These results are consistent with the results of Ref [1], and indicate that this material is unsuitable for photovoltaic applications.

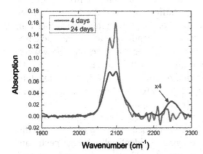

Figure 1. FTIR absorption spectra obtained for a μc-Si:H thin film (a80929b) deposited by MDECR-PECVD on a c-Si wafer. Spectra are shown for four and 24 days after deposition, during which sample was kept in air.

To test the detection of the same peaks on a non-IR transparent substrate suitable for PV applications, the Raman spectrum was acquired for a film co-deposited on a Corning Glass borosilicate substrate. This spectrum is shown in Figure 2, along with subsequent measurements taken up to 210 days after the deposition. During this time, the sample was kept in air, and a similar disappearance of the peaks was observed as in the case of the FTIR measurements. In comparison with the FTIR results, however, the strength of the O-Si-H_x peak is much less in the case of Raman. Indeed, whereas in the spectra of Fig. 1 a strong peak at 2250 cm^{-1} appears, only a shoulder is visible in the Raman spectra of Fig. 2 for the films measured after 10,17, and 210 days (for the other scans, the acquisition window did not include 2250 cm^{-1}).

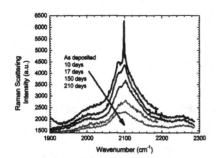

Figure 2. Time dependence of Raman scattering spectra obtained for a μc-Si:H thin film (a80929b) deposited by MDECR-PECVD on a borosilicate substrate. Sample was kept in air in between measurements.

Characterization of Material Grown by RF-PECVD from SiF₄

The use of a pre-cursor gas mixture of H_2, Ar and SiF_4, a chamber pressure above 1Torr, and areal power densities > 50mW/cm^2 can result in growth conditions in which a significant portion of the chemistry takes place in the gas phase. Such conditions have been recently shown to result in photovoltaic devices with both high Raman crystallinity values (80%), as well as an elevated open-circuit voltage, V_{OC} (523mV) [9]. We therefore applied high wavenumber Raman scattering characterization to films grown under these general conditions, and which have been incorporated as the intrinsic layer of pin solar cell devices. The spectra presented below were measured directly on pin cells without removing the n-layer (which for these devices is ~180Å thick). Although these measurements were thus acquired through the n-layer, for the excitation wavelength used in this work (632nm, at which α^{-1} = 0.5-1 μm for these materials), the presence of the doped layer should have little influence on the spectra obtained.

In Figure 3, representative Raman scattering spectra from films obtained for three different sets of deposition conditions are presented to give a broad overview of the peaks that are observed in this material system. Four features are typically identifiable in the spectra and are represented here – a shoulder at 1983cm⁻¹, a broad peak at 2023cm⁻¹, and two narrow peaks at 2104 and 2134 cm⁻¹. It should be noted that when deconvoluting these spectra, a peak at 1983cm⁻¹ must be included, as the spectra cannot be fit with a 2023cm⁻¹ peak alone. The twin Raman peaks observed in the MDECR films are not observed in any of the films, although the deposition rate of the RF-PECVD μc-Si:H films are lower (1Å/s). In Figure 3, the spectrum with a single peak at 2023cm⁻¹ was measured directly on a cell with an efficiency of 8.3% [9] and as a general rule, only films showing the single peak at 2023cm⁻¹ give solar cell characteristics with efficiencies above 7%. This rough classification of material may therefore be useful in optimizing deposition conditions to obtain device quality material.

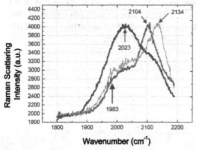

Figure 3. Raman scattering spectra acquired for three different sets of μc-Si:F:H thin film deposition conditions. Four features are identifiable – a shoulder at 1983 cm⁻¹, a broad peak at 2023 cm⁻¹, and two narrow peaks at 2104 and 2134 cm⁻¹.

The interpretation of the various peaks present when depositing μc-Si:F:H is subject to the complication of the possible presence of fluorine in the films. The presence of an electronegative element near the Si-H bond being detected is known to shift the vibrational energy to higher wavenumber [10]. The peak at 2134 cm⁻¹ may therefore be variously interpreted as SiH₃ bonds, or as SiH on a F-back bonded Si atom.

To further explore the deposition condition dependence of these spectral features, high-wavenumber Raman spectra were obtained for a set of films grown varying only the H₂ flow. The results from this series of films are present in Figure 4. The H₂ flow and RF-power are

322

listed beside each of the spectra, and the remaining conditions are as follows: Pressure=2.2Torr, SiF_4=3sccm, and Ar=80sccm.

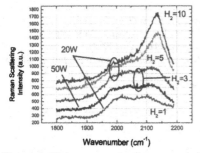

Figure 4. High-wavenumber Raman scattering spectra for series of μc-Si:F:H films deposited with varying hydrogen flux. The applied RF-power and H_2 flow in sccm are noted.

It can be clearly seen from Figure 4 that increasing the H_2 flow during deposition from SiF_4 increases the intensity of the peak at $2134cm^{-1}$. This is a logical result, as the flow of H_2 is the only source of H (due to the use of a fluorinated pre-cursor). As well, the addition of H_2 to the admixture dramatically increases the growth rate (from 1 to 2.5Å/s when going from 1 to 5 sccm). However, in the case of these cells, the devices with the highest H_2 flow showed the poorest device performance (~2%) whereas the devices with a lower peak at $2134cm^{-1}$ showed better performance (4.8-5.8%). None of the devices, however, showed the excellent performance of the cell with only a dominant 2000 cm^{-1} peak (8%).

CONCLUSIONS

The use of high-wavenumber Raman scattering to characterize μc-Si:H and μc-Si:F:H thin films presents distinct advantages as this analysis can be performed for films on any substrate. We have applied this technique to films and devices with an aim to (a) detect the tell-tale peaks in high-deposition rate μc-Si:H that will be prone to oxidation, and (b) determine equivalent tell-tale peaks in μc-Si:F:H that may help to rapidly evaluate material for suitability in PV applications. We have shown that the twin peaks visible in FTIR in oxidation-prone μc-Si:H are also visible in high wavenumber Raman scattering. As well, the best quality μc-Si:F:H films possess no dominant peaks at 2100 cm^{-1} and above.

REFERENCES

1. A. H. M. Smets, T. Matsui and M. Kondo, *J. Appl. Phys.* **104**, 034508 (2008).
2. S. Veprek, Z. Iqbal, H.R. Oswald and A.P. Webb, *J. Phys. C: Solid State Phys.* **14**, 295 (1981).
3. T.Imura, K. Mogi, A. Hiraki, S. Nakashima and A. Mitsuishi, *Solid State Commun.* **40**,161 (1981).
4. D.C. Marra, E.A. Edelberg, R.L. Naone and E.S. Aydil, *J. Vqc. Sci. Technol.* A **16**, 3199 (1998).

5. P. Roca i Cabarrocas, N. Layadi, T. Heitz, B. Drévillon and I. Solomon, *Appl. Phys. Lett.* **66**, 3609 (1995).
6. D. Stryahilev, F. Diehl, B. Schröder, M. Scheib and A.I. Belogorokhov, *Phil. Mag. N* **80**, 1799 (2000).
7. F. Kail, A. Hadjadj and P. Roca i Cabarrocas, *Thin Solid Films* **487**, 126 (2005).
8. P. Roca i Cabarrocas, P. Bulkin, D. Daineka, T. H. Dao, P. Leempoel, P. Descamps, T. Kervyn de Meerendre and J. Charliac, *Thin Solid Films* **516** 6834 (2008).
9. Q. Zhang, E. V. Johnson, Y. Djeridane, A. Abramov and P. Roca i Cabarrocas, *physica status solidi (RRL)* **2**,154 (2008).
10. G.Lucovsky, *Solid State Commun.* **29**, 571 (1979).

Mater. Res. Soc. Symp. Proc. Vol. 1153 © 2009 Materials Research Society 1153-A16-07

Hole Drift Mobility Measurements on a-Si:H Using Surface and Uniformly Absorbed Illumination

Steluta A. Dinca[1], Eric A. Schiff[1], Subhendu Guha[2], Baojie Yan[2], and Jeff Yang[2]
[1]Department of Physics, Syracuse University, Syracuse, NY 13244-1130
[2]United Solar Ovonic LLC, Troy, Michigan 48084

ABSTRACT: The standard, time-of-flight method for measuring drift mobilities in semiconductors uses strongly absorbed illumination to create a sheet of photocarriers near an electrode interface. This method is problematic for solar cells deposited onto opaque substrates, and in particular cannot be used for hole photocarriers in hydrogenated amorphous silicon (a-Si:H) solar cells using stainless steel substrates. In this paper we report on the extension of the time-of-flight method that uses weakly absorbed illumination. We measured hole drift-mobilities on seven a-Si:H *nip* solar cells using strongly and weakly absorbed illumination incident through the *n*-layer. For thinner devices from two laboratories, the drift-mobilities agreed with each other to within a random error of about 15%. For thicker devices from United Solar, the drift-mobilities were about twice as large when measured using strongly absorbed illumination. We propose that this effect is due to a mobility profile in the intrinsic absorber layer in which the mobility decreases for increasing distance from the substrate.

INTRODUCTION

Hole drift mobilities are crucial to understanding amorphous silicon (a-Si:H) solar cells [1]. Experimentally, electron and hole drift mobilities are generally measured using the time-of-flight technique in which a pulse of illumination is absorbed near an electrode interface. Depending on the direction of the electric field, electrons or holes are swept across the structure, and an average drift-mobility μ_d is calculated from their transit time t_T. To measure holes in a-Si:H solar cells, the standard method requires fairly strong illumination through the *n*-layer to create a sheet of carriers near the *n/i* interface. One typically uses a 500 nm wavelength and reverse electrical bias on the cell. Such illumination is not possible for a-Si:H *nip* solar cells deposited onto opaque substrates such as stainless steel, and an alternative would be very desirable [2].

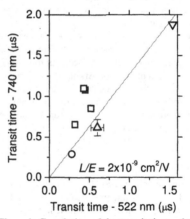

Figure 1: Correlation of the transit times obtained with 740 nm illumination with transit times obtained with 522nm for seven a-Si:H cells. The symbol shapes indicate the sample substrate: 2 US, ○ PSU, (▽△) BP1, BP2. The line represents the ratio predicted by Arkhipov, *et al.* (see text). The error bar was determined from multiple measurements on one device.

One alternative is to use uniformly absorbed illumination, which leads to two complications *vis a vis* conventional time-of-flight. The first is that both electron and hole motions contribute to the photocurrent. Because electrons are far more mobile than holes in a-Si:H, their half of the photocharge is swept out quickly, and the long-time photocurrents are dominated by holes. The second difficulty is that the initial positions of the holes are uniformly distributed; as we show later, a hole transit-time t_T can be measured even for uniform photoexcitation; the standard expression for the drift-mobility is:

$$\mu_D = L/(Et_T) \tag{1}$$

where L is the average displacement of the carriers at the transit-time and E is the electric field. We discuss the relation of L to the intrinsic layer thickness d later in this paper.

While this method should be suitable to measurements on samples with opaque substrates, it has never been carefully tested; we do this testing in the present paper. The method must accommodate the fact that holes in a-Si:H exhibit "anomalous dispersion" in their motions. When anomalous dispersion obtains, drift mobilities for different materials or different techniques must be compared for a specific value of the ratio L/E [3]. With this proviso, the work of Arkhipov, *et al.* [4,5] shows that one expects a non-unity ratio of the transit times for weakly absorbed light t_T^u to the conventional estimate based on strongly absorbed light t_T^s:

$$\frac{t_T^u}{t_T^s} = (4/3)^{1/(2\alpha)} \tag{2}$$

where α is the "dispersion parameter". For holes near room-temperature, $\alpha \approx 0.6$, so the predicted ratio is 1.3. The derivation assumes that hole transport properties are uniform throughout the material.

To our knowledge, there have been no experimental tests of this prediction. In Figure 1 we show a summary of room-temperature measurements of the transit time for seven a-Si:H devices at approximately the displacement to field (L/E) ratio of 2×10^{-9} cm^2/V. Several of the devices are consistent with the Arkhipov, *et al.* ratio of 1.27 (see the error bar for one point). However, some of the devices yielded transit times with ratios that are systematically larger than expected. We believe that this behavior is probably evidence for a hole mobility that declines for positions that are increasingly distant from the substrate. The possibility of inferring a mobility profile is an unexpected outcome of the present work. In addition, our results suggest that drift mobilities measured in special, thick samples may not be representative of thinner samples.

SPECIMENS

We studied seven a-Si:H *pin* devices on four substrates. Four of the devices were on one TCO-coated glass substrate prepared at United Solar Ovonic LLC (*nip* deposition sequence, VHF deposition, 2.0 μm a-Si:H intrinsic layer, no evidence for microcrystallinity). A second sample was prepared at Pennsylvania State University (*pin* deposition sequence, specular TCO as substrate, 0.59 μm a-Si:H intrinsic layer, semitransparent top Cr contact). Two additional samples were made in 2002 at BP Solar, Inc. (*pin* deposition sequence, DC plasma). BP1 was made with a hydrogen/silane dilution ratio of 10 (thickness 0.89 μm); BP2 was made with a hydrogen dilution of 20 (intrinsic layer thickness 1.13 μm). A semitransparent ZnO top electrode was deposited instead of the usual metal back reflector.

HOLE DRIFT MOBILITY MEASUREMENTS

The transient photocurrents measurements were done using a 4 ns illumination pulse from nitrogen laser-pumped dye laser. The devices were illuminated through their n-layers. The laser wavelengths used were 522 nm and 740 nm, corresponding to absorption depths of about 0.08 µm and 30 µm, respectively [6].

Fig. 2 (a) & (c) illustrates the transient photocurrents at 293 K at the two wavelengths for devices from US and BP1, respectively. The photocurrents are normalized as $i(t)d^2/Q_0(V+V_{bi})$ where d is the intrinsic-layer thickness, V is the applied external bias, V_{bi} is a correction for the built-in potential, and Q_0 is the total photocharge generated in the intrinsic layer. We made a correction for the built-in potential V_{bi} of the *pin* structures [7].

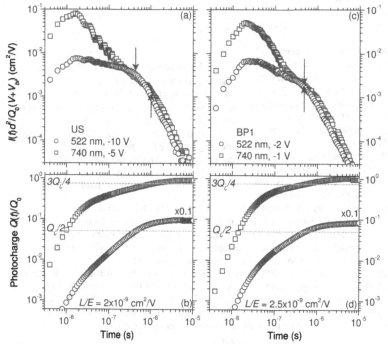

Figure 2: (a) & (c) Normalized transient photocurrents $i(t)d^2/Q_0(V+V_{bi})$ measured using two illumination wavelengths in a-Si:H *p-i-n* devices at 293 K; see text for the normalization procedure. The total photocharge Q_0 is defined as the total photocharge collected at longer times and larger bias voltages. (b) & (d) Normalized photocharge transients $Q(t)/Q_0$ obtained by time-integration of the transient photocurrent. The intersection of the transients with the horizontal lines at $Q_0/2$ (522 nm), and $3Q_0/4$ (740 nm) were used to determine the hole transit times t_T; the arrows on the upper panels indicate the resulting transit times.

For 522 nm wavelength, which is strongly absorbed near the *n/i* interface, the photocurrents are dominated by hole drift. There is a noticeable "kink" marking a transition from a shallow, power-law decay to a steeper decay that is typically identified as the hole transit time

t_T. As illustrated by the photocharge transients in the lower panels, we actually identify the transit time as the time at which half of the ultimate photocharge Q_0 has been reached. We have indicated the transit times in the upper panels with arrows, where they agree fairly well with the "kink" location. The "kink" and "half-charge" procedures are usually equivalent; we prefer the latter because we find it more reproducible [3].

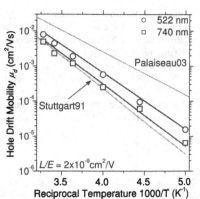

Figure 3: Hole drift mobilities μ_d as a function of reciprocal temperature $1000/T$ for several experiments. Symbols are estimates for a United Solar device measured at two illumination wavelengths; the lines through these are multiple trapping fit using the parameters $\mu_p = 0.8$ cm^2/Vs, $\Delta E_V = 43$ meV, and $v = 0.8 \times 10^{12}$ s^{-1} (522 nm) and 1.7×10^{12} s^{-1} (740 nm). Palaiseau03-ref. 10, Stuttgart91-ref. 11.

For the 740 nm illumination the photocarriers are created uniformly throughout the a-Si:H film. Half of the total photocharge Q_0 is due to the electrons; since electrons in a-Si:H have a drift mobility at least 10^2 larger than holes [8], this fraction of the photocharge is collected in less than 100 ns; the photocharge in excess of $Q_0/2$ is due to holes, and the transit time for hole collection is thus reached at $(3/4)Q_0$. As can be seen in Fig. 2, this agrees reasonable well with the "kink" in the long-time transients.

For 740 nm, we have shown the transient at half the voltage we used for 522 nm; this halving compensates for the fact that the initial, mean position of the holes for uniform absorption is already halfway across the sample, so the transit times for the 2 wavelengths should be about the same. This procedure was assumed by eq. (2) above.

In Fig. 2, for the sample BP1 there is little difference in the transit-time estimates using 540 and 722 nm wavelengths, but there is about a factor two difference for the US sample. In Fig. 3, we illustrate the temperature-dependence of the drift-mobility for this cell for both wavelengths. For the uniformly absorbed, 740 nm data we have incorporated the results of Arkhipov, et al. (cf. eq. (2)) by using the definition for the drift-mobility:

$$\mu_D^u = (4/3)^{1/2\alpha} \left(\frac{L}{Et_T^u} \right) \tag{3}$$

Fits to these data using bandtail multiple-trapping line are also illustrated (see ref. [9] for procedures); these are based on the entire set of measurements, not just those illustrated in Fig. 3. For reference, we also present two lines based on previously published, conventional time-of-flight measurements on a "polymorphous" silicon sample [10] (denoted Palaiseau03) and some measurements on a sample from Universität Stuttgart [11].

DISCUSSION

In order to address the dependence of these drift-mobility estimates on the illumination wavelength, we again consider the transit time estimates illustrated in Figure 1. The transit times measured in the BP1, BP2, and PSU samples are reasonably consistent with the Arkhipov ratio as shown; for these devices, the dispersion parameters ranged from 0.55 to 0.66, corresponding to Arkhipov ratios of 1.30 to 1.27. For these three devices the Arkhipov, *et al.* theory appears to be a better description than the naïve ratio of unity. However, the United Solar devices have distinctly larger ratios than predicted by Arkhipov, *et al.* This difference cannot be accounted for by differences in dispersion; the United Solar samples had a similar range of dispersion parameters to the other samples.

We thus believe that the measurements for the PSU and BP devices are reasonably consistent with the conventional theory of dispersive transport, which assumes that hole transport properties are constant throughout the thickness of a material. For the US samples, which were considerably thicker than the PSU and BP samples, we believe the hole drift-mobility declines for larger distances from the substrate and the n/i interface. We fitted both sets of measurements in one US sample to the bandtail multiple-trapping model; a satisfactory fit was obtained if we increased the attempt-frequency ν about threefold to fit the measurements with uniformly absorbed illumination. Changes in ν are expected due to changes in the fundamental disorder through the bandedge density-of-states N_V [12], although an equally plausible argument could be made that increased disorder should have broaden the bandtail. In either case, the present data are reversed from expectations from a changeover in structure from amorphous to microcrystalline for thicker materials [13] as the film grew thicker.

ACKNOWLEDGMENTS

We thank Christopher Wronski (Pennsylvania State University) and Gautam Ganguly (Optisolar, Inc.) for providing samples that were used in this work. This research was USDOE under the Solar American Initiative Program Contract No. DE-FC36-07 GO 17053. This project was supported in part by funding from an Empire State Development Corporation of New York State Award, granted to Syracuse University and Syracuse Center of Excellence in Environmental and Energy Systems.

APPENDIX

In this appendix we explain how we used the calculations of Arkhipov, *et al.* to obtain eq. (2). These authors assumed that the photocurrent transients are governed by multiple-trapping in an exponential bandtail, although we believe the result of eq. (2) to be a general property of anomalously dispersive transport. For strongly absorbed illumination that generates a sheet of carriers near one electrode, they obtained the following equation (eq. (34) of ref. [4]) for the transit-time with strongly absorbed illumination:

$$t_T^s = (K/\nu)\left(\frac{\nu d}{\sqrt{2}\mu E}\right)^{1/\alpha} \tag{4}$$

where K is a dimensionless function of order unity, μ is the band mobility of the carrier, ν is the rate of bandtail trapping for a mobile carrier, and d the sample thickness. As before, α is the dispersion parameter. These authors defined the transit time as the "kink" in the transient photocurrent; the half-charge method used above is expected to be equivalent [3]. Arkhipov, *et al.* subsequently published a comparable expression for the transit time for uniformly absorbed illumination (eq. (26) of ref. [5]):

$$t_T^u = (K/v)\left(\frac{vd}{\sqrt{6\mu E}}\right)^{\!\frac{1}{\alpha}} \tag{5}$$

These two expressions do not correspond to the same mean displacements of the carriers. For uniformly distributed carriers, the displacement at the transit time is half of that for strongly absorbed illumination, since the carriers initially have a mean position that is halfway across the sample. As noted earlier, to compare corresponding drift-mobilities one must use the same displacements L, or more precisely of L/E [3]. Our definition of the transit-time corresponds to $d = 2L$ for strongly absorbed illumination [3], and thus to $d = 4L$ for uniformly absorbed illumination. Substituting into (4) and (5), we obtain:

$$t_T^s = (K/v)\left(\frac{\sqrt{2}vL}{\mu E}\right)^{\!\frac{1}{\alpha}} \quad \text{and} \tag{6}$$

$$t_T^u = (K/v)\left(\frac{2\sqrt{2}vL}{\sqrt{3}\mu E}\right)^{\!\frac{1}{\alpha}} \tag{7}$$

The ratio of the transit-time for uniform illumination to that with strong illumination – at constant displacement L – is thus $(4/3)^{\frac{1}{2\alpha}}$, as used in eq. (2).

REFERENCES

1. J. Liang, E. A. Schiff, S. Guha, B. Yan, and J. Yang, *Appl. Phys. Lett.* 88, 063512 (2006).
2. One can illuminate the cell through its p-layer, and use forward bias; this approach is not generally usable because of the large dark current under these conditions.
3. Q. Wang, H. Antoniadis, E. A. Schiff, and S. Guha, *Phys. Rev. B* 47, 9435 (1993).
4. V. I. Arkhipov, M. S. Iovu, A. I. Rudenko and S. D. Shutov *Phys. Status Solidi* A 54, 67-77 (1979).
5. V. I. Arkhipov, V. A. Kolesnikov and A .I. Rudenko *J. Phys. D: Appl. Phys.*. 17, 1241-1254 (1984).
6. M. Vanecek, A. Poruba, Z. Remes, N. Beck, M. Nesladek, *J. Non-Cryst. Solids* 227–230, 967 (1998).
7. We found that the built-in-potential estimates obtained with uniform illumination are around 10% (US) to 16 % (BP1) smaller than the built-in-potential estimates obtained with surface illumination. We measured for BP1 specimen the built-in-potential to be around V_{bi}=0.5 V - uniform absorption and about V_{bi}=0.6 V -surface absorption, and for US03 sample are about V_{bi}=0.18 V -uniform absorption, respectively about V_{bi}=0.20 V -surface absorption. These US estimates are a smaller than the expected value of V_{bi} greater than 1V.
8. T. Tiedje, in *The Physics of Hydrogenated Amorphous Silicon Vol. II*, edited by J. D. Joannopoulos and G. Lucovsky (Springer-Verlag, Berlin Heidelberg, New York, Tokyo, (1984), pp. 261-300.
9. E. A. Schiff, *J. Phys.: Condens. Matter* 16, S5265 (2004).
10. M. Brinza, G.J. Adriaenssens, P. Roca i Cabarrocas, *Thin Solid Films* 427, 123 (2003).
11. C. E. Nebel, *Doctoral Dissertation* Universität Stuttgart (1991). Unpublished.
12. E. A. Schiff, *Phil. Mag. B* (in press).
13. A. Ferlauto, R. Koval C. R. Wronski, R. W. Collins, *Appl. Phys. Lett.* 80, 2666 (2002).

Poster Session:
Film Growth

Properties of Nano-Crystalline Silicon-Carbide Films Prepared Using Modulated RF-PECVD

Feng Zhu, Jian Hu, Ilvydas Matulionis, Augusto Kunrath, and Arun Madan

MVSystems, Inc., 500 Corporate Circle, Suite L, Golden, CO, 80401, USA

ABSTRACT

We report on the fabrication of nano-crystalline silicon-carbide (nc-SiC) using pulse modulated RF-PECVD technique, from silane (SiH_4) and methane (CH_4) gas mixtures which is highly diluted in hydrogen (H_2). The microstructure of nc-SiC material is nanometer-size silicon crystallites embedded in amorphous silicon-carbide (a-SiC) matrix. As carbon incorporation in nc-Si film increases, the bandgap is enlarged from 1.1eV to 1.55eV as measured by Photothermal Deflection Spectroscopy (PDS) while the crystalline volume fraction decreases from 70% to about 20%. It is found that the crystalline volume fraction, grain size and dark conductivity of nc-SiC films can be enhanced with applying a negative DC bias to substrate during deposition.

INTRODUCTION

It is well-known that the conversion efficiency of amorphous silicon (a-Si) solar cell is limited by the inherent light-induced degradation, which leads to an increase in the density of defect with prolonged illumination [1, 2]. As a result, the depletion width of the a-Si cell decreases and the recombination of the photo-generated carriers in the intrinsic layer increases, leading to a smaller collection width and hence a reduction in the open-circuit voltage (Voc), short-circuit current density (Jsc) and fill factor (FF). To reduce the instability, devices are generally made in a tandem junction configuration where the a-Si as the top cell is thin (typically 200-300nm). This limits the Jsc of the tandem cell. For instance, due to the requirement of current matching in the tandem configuration, Jsc is limited to 11-12 mA/cm^2 for the dual cells comprising of a-Si/a-Si or a-Si/nc-Si [3-5]; for a triple junction stack comprising of a-Si /a-SiGe/nc-Si or a-Si/a-SiGe/a-SiGe, Jsc is further reduced to ~9 mA/cm^2 [6-8].

In order to increase the conversion efficiency of a-Si based solar cell, it is evident that either the problem of light-induced degradation in a-Si needs to be solved or the unstable a-Si component needs to be removed from the entire structure [2, 9]. Over the last three decades, extensive studies have been performed to understand and solve the light-induced degradation effect but with limited success. In contrast, the use of nc-Si materials with an appropriate crystalline volume fraction (Xc) of 60-70% , nc-Si solar cells have exhibited less light-induced degradation [9-11]. These types of films contain fine grains whose size is in the range of 10-20 nm, embedded in an a-Si matrix. However, due to their lower bandgap (~1.1eV) the open circuit voltage of these devices is lower than that obtained in a-Si solar cell.

Incorporating carbon in nc-Si film to form nanometer sized silicon crystallites embedded in amorphous silicon-carbide matrix could lead to a promising material with higher bandgap (E_g) for application in thin film solar cells [12]. Plasma enhanced chemical vapor deposition (PECVD) technique offers advantages as it allows the manipulation of the growth surface through a control of electron temperature and its density in plasma; radical and ion flux can be manipulated on the growing surface via changes in the excitation frequency, plasma power, voltage bias, and pulse modulation, etc. In this paper, we report on the fabrication of nano-crystalline silicon-carbide (nc-SiC) materials using pulse modulated PECVD technique.

EXPERIMENT

Nc-SiC films were fabricated using pulse modulated RF-PECVD technique [13] in a cluster tool system specifically designed for thin film semiconductor market and manufactured by MVSystems, Inc. The intrinsic nc-SiC:H films were deposited using methane (CH_4), silane (SiH_4) and hydrogen (H_2) gas mixtures at a substrate temperature in the range of 100-400 °C. The RF power density used was in the range of 10-200 mW/cm^2, and the deposition pressure was in the range of 0.3-10 Torr. The CH_4 flow rate was varied to change the carbon concentration in nc-SiC film, while the SiH_4 and H_2 flow rates were kept constant. Dark (σ_d) and Photoconductivity (σ_L) were measured under AM1.5 illumination. The influence of carbon on the film structure was investigated using infrared spectroscopy (IR). The crystallinity of the films was measured by XRD and Micro-Raman spectroscopy on films deposited on silicon wafer and Corning glass (type 1737) substrate, respectively. The grain size is calculated based on Debye-Scherrer equation [14] and using the Gauss fitting technique over the three peaks (480nm, 500nm, and 520nm) in the Raman spectrum, the crystalline volume fraction (Xc) (defined as $Xc = (I_{520}+I_{500})/(I_{520}+I_{500}+I_{480})$) could be calculated. The optical bandgap (E_g) of nc-SiC films was determined by the photothermal deflection spectroscopy (PDS). For the DC bias experiment, a power supply (up to 600 V output) was used.

RESULTS AND DISCUSSION

Microstructure and crystallinity

To investigate the effect of the CH_4 flow rate on the microstructure and electrical properties of nc-SiC, nano-crystalline silicon (nc-Si) with higher Xc (70%) was first fabricated by pulse modulated RF-PECVD technique, using SiH_4 and H_2 gas mixture, and then a small CH_4 flow in a range of 0 to 0.48 sccm was added while keeping both SiH_4 and H_2 flow constant. This way, a series of SiC films with different C concentration were prepared. Fig.1(a) shows changes in the XRD pattern with CH_4 gas flow rate. (Note the orientation of <100> comes from the Si wafer substrate.) It is seen that at low CH_4 flow rate (0-0.36 sccm range), the orientations of <111> and <220> (typically related to nc-Si) are prominent which gradually vanish as the CH_4 flow rate increases further. At a CH_4 flow rate of 0.48 sccm, both orientations disappear, indicating microstructural change in the film from nano-crystalline to amorphous. The grain size of these nc-SiC films for <111> orientation, estimated using Debye-Scherrer formula, shows a decrease from 12 nm to ~ 8 nm as CH_4 flow rate increased from 0.25 to 0.4 sccm (Fig.1(b)). We note the absence of crystalline SiC peak (e.g., 3C-SiC at 2θ= ~41°). These results infer that the nc-SiC films most likely has a microstructure in which nc-Si crystals are embedded in a-SiC matrix.

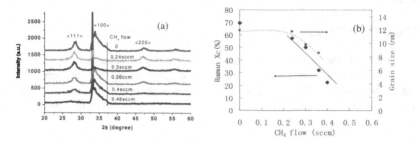

Fig.1. (a) XRD pattern and (b) Xc and grain size at <111> orientation variation with CH_4 flow rate. (The line is used as a guide)

The Xc of the nc-SiC film was determined by Micro-Raman measurement technique. The variation of Xc with CH_4 gas flow is shown in Fig.1(b). It is seen that the Xc decreases linearly from 70% to 20% as CH_4 flow rate increases from 0 to 0.4 sccm, indicating microstructure change in the film with carbon incorporation.

IR spectrum

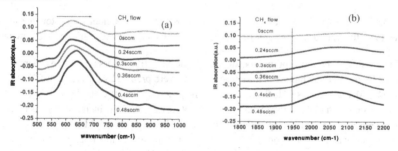

Fig.2. IR spectra at (a) 500-1000 cm^{-1} and (b) 1800-2300 cm^{-1} as function of CH_4 flow rate

Fig.2 shows the IR spectra of nc-SiC films at the 500-1000 cm^{-1} and 800-2300 cm^{-1} wave number region as function of CH_4 flow rate. It is seen from Fig.2(a) that, as CH_4 flow rate increases, the characteristic peak around 630-640 cm^{-1} (assigned to Si-H bend mode) shifts to 670 cm^{-1} (corresponding to SiC stretch mode), and the peak intensity at 670 cm^{-1} increases, suggesting more C incorporation. In addition, as shown in Fig.2(b), the characteristic peak around 2000-2100cm^{-1} increases with increasing CH_4 gas flow. Using the Gaussian fitting technique, this absorption peak can be further de-convoluted into two peaks, 2000 cm^{-1} and 2100 cm^{-1} respectively. The ratio of the peak intensity, I_{2000}/I_{2100} decreases with increasing CH_4 flow rate (CH_4 flow ≤0.36sccm) , indicating a shift of the absorption peak at 2000 cm^{-1}

(Si-H stretch) towards 2100 cm^{-1} resulting from carbon atoms bonded to Si-H (i.e., C-SiH stretch mode) and/or the presence of SiH and SiH$_2$ groups [15-17]. When CH$_4$ flow is higher than 0.4sccm the IR peak at 880 cm^{-1} becomes clear (Fig. 2(a)), attributed to wagging and bending of SiH$_2$ [17], and I$_{2000}$/I$_{2100}$ increases (Fig. 2(b)). Fig.1 shows the film becomes amorphous.

Opto-electronic properties

As CH$_4$ flow increases, both the dark (σ_d) and photoconductivity (σ_L) decreases noticeably as shown in Fig.3. It is seen that σ_d and σ_L decreases significantly, from 10^{-5} S/cm and 10^{-4} S/cm (for nc-Si) to 10^{-9} S/cm and 10^{-6} S/cm, respectively. Meanwhile, E$_g$ (deduced from Fig.4) increases with CH$_4$ flow rate due to the increase of carbon incorporation in the film.

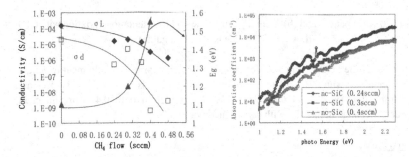

Fig.3. σ_d , σ_L and E$_g$ vs. CH$_4$ flow rate. (The lines are used for guide)

Fig.4 Absorption coefficient vs. photon energy for nc-SiC films prepared with various CH$_4$ flow rates.

Fig.4 shows the absorption coefficient of nc-SiC films versus photon energy as measured by the PDS technique (These nc-SiC films were prepared with different CH$_4$ flow, from 0.24 to 0.4 sccm.). As CH$_4$ flow increases, the absorption coefficient decreases as the E$_g$ (deduced from the PDS data) increases from ~1.2 eV to 1.55 eV. This increase in E$_g$ is caused by the increase of carbon incorporation in the film. However, it should also be noted that increasing carbon incorporation leads to a change of the film microstructure (from nano-crystalline to amorphous). In fact, for E$_g$ = 1.55 eV, Xc is decreased to as low as ~20% (Fig.1(b)).

The above results show that carbon incorporation in the film not only enlarges the E$_g$ (Fig.3), but also decreases Xc (Fig.1) and thus the conductivity in the film (Fig.3). Obviously, to maintain a high crystalline fraction while enlarging E$_g$ is a key issue in developing this material.

Effect of DC bias

In the above section, we have shown that nc-SiC films with an E$_g$ of 1.55 eV can be fabricated using the pulse modulated RF-PECVD technique. However, Xc in such nc-SiC films remains low, ~20%. In order to further increase E$_g$ of nc-SiC while maintaining a high

Xc, a negative DC bias was applied to substrate during deposition to investigate its influence on the microstructure and electrical properties of the nc-SiC film.

Fig.5 shows σ_d and σ_L as a function of negative DC bias. It is seen that, as the DC bias is changed from 0 to -100 V, σ_d increases by almost four orders of magnitude, from $\sim 10^{-9}$ S/cm to $\sim 10^{-5}$ S/cm, and saturates with further increasing bias, while σ_L decreases slightly. The film microstructure also exhibits a transition from amorphous to nano-crystalline phase as shown Fig.5(b). The XRD pattern of this series films clearly exhibits two peaks occurring at $2\theta = 28°$ and $47°$, which are associated with Si crystals in orientation of <111> and <220>, respectively. More significantly, the grain size increases from 0 to ~ 16 nm with decreasing DC bias as shown in Fig.5(a). Apparently, the increase of dark conductivity is mainly due to changes in the film microstructure and increase of the grain size. Nc-SiC prepared at -100V DC bias possessed Xc of 60%.

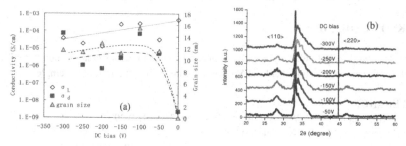

Fig.5 (a) σ_d, σ_L and grain size (The lines are used for guide), and (b) XRD pattern as function of negative DC bias.

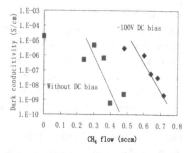

Fig.6. Comparison of the dark conductivity as function of CH_4 flow rate for a-SiC films made with and without DC bias. (The lines are used for guide)

Fig.6 shows a comparison of dark conductivity for the nc-SiC films prepared with and without DC bias. With a DC bias (-100 V in this case), σ_d increases by as much as 4 orders of magnitude at a CH_4 flow rate of 0.48 sccm. In addition, the hydrogen concentration (C_H) in the film decreases notably, i.e., from $\sim 10\%$ to $\sim 4\%$ (not shown here). This behavior mainly results from the increase of the Xc and grain size in nc-SiC film. Due to improvement of microstructure and electrical properties, nc-SiC films can be now fabricated with a higher CH_4 flow rate, i.e., up to 0.7 sccm (compared with < 0.4 sccm in the case of without DC bias).

Since the Eg of nc-SiC increases with CH_4 flow rate, we expect E_g of nc-SiC fabricated with DC bias to be further increased, i.e. >1.55 eV, while maintaining a reasonably high Xc.

CONCLUSION

Nc-SiC films were prepared at 200 °C using pulse modulated RF-PECVD technique. The nc-SiC contains nanometer-size Si crystals embedded in a-SiC matrix. Incorporation of carbon in nc-Si film widens the bandgap from 1.1 eV to 1.55 eV. However, the crystalline volume fraction is decreased from 70% to 20% and the film exhibits a phase transition from nano-crystalline to amorphous. It is found that applying a negative DC bias to substrate during deposition improves the microstructure and electrical properties of nc-SiC, which makes it possible to fabricate nc-SiC at an even higher CH_4 flow rate. Nc-SiC film with Eg of 1.55 eV and crystalline volume fraction of 60% has been achieved.

ACKNOWLEDGEMENT

This work was funded by National Science Foundation under contract number, NSF 0740359. The authors wish to thank Ed Valentich for his help in fabrication of samples.

REFERENCE:

1. D. L. Staebler and C. R. Wronski, Appl. Phys. Lett. 31, (1977) 292
2. D.E.Carlson, Appl.Phys. A 41 (1986) 305-309
3. K.Yamamoto, A. Nakajima, M.Yoshimi, et.al, Solar Energy, 77, (2004)939
4. O. Cubero, T. Söderström, F.J. Haug, et.al., Proceedings of the 23th EU-PVSEC Conference, Valencia, Spain (2008)
5. A.Shah, J.Meier, E.Vallat-Sauvain, et.al., Thin Solid Films, Vol. 403-404 (2002) 179-187
6. Y. Baojie, G. Yue, and S. Guha, Mat. Res. Soc. Proc., April (2007) A15.1
7. J.Yang, A.Banerjee, and S.Guha, Appl. Phys.Lett. 70 (1997) 2975
8. Baojie Yan, Gouzhe Yue, Yanfa Yan, et.al., Mat. Res. Soc. Proc. Vol.1066 (2008) A03.03
9. F.Meillaud, E.vallat-Sauvain, A. Shah, et.al., Appl. Phys. Lett., 103 (2008) 054504
10. Yan Wang, Xiaoyan Han, Feng Zhu, et.al., J. Non-Crystal. 352 (2006) 1909-1912
11. Guozhen Yue, Baojie Yan, Gautam Ganguly, et.al., Appl. Phys. Lett., 88, (2006) 263507
12. M. Konagai, S. Miyajima, Y. Yashiki, T. Watahiki, K. L Narayanan and A. Yamada, Proc. 31 IEEE Photovoltaic Specialists Conference, (2005) 1424
13. Scott Morrison, Jianping Xi, Arun Madan, MRS proceedings, 507 (1998) 559
14. B.D.Cullity, In Elements of X-Ray Diffraction, Addison-Wesley Publishing Company, Inc., (1978) 283-300
15. A. H. M. Smets, and M. C. M. van de Sanden, Physical Review B 76 (2007) 073202
16. G. Ambrosone, U. Coscia, S. Lettieri, et.al., Thin Solid Films 511- 512 (2006) 280-284
17. F.Demichelis, C.F.Pirri, and E.Tresso, J.Appl.Phys. 72 (4) (1992) 1327

Ion Assisted ETP-CVD a-Si:H at Well Defined Ion Energies

M. A. Wank[1], R. A. C. M. M. van Swaaij[1] and M. C. M. van de Sanden[2]
[1]Delft University of Technology, Electrical Energy Conversion Unit/DIMES, P. O. Box 5053, 2600 GB Delft, The Netherlands
[2]Eindhoven University of Technology, Department of Applied Physics, P. O. Box 513, 5600 MB Eindhoven, The Netherlands

ABSTRACT

Hydrogenated amorphous silicon (a-Si:H) was deposited with the Expanding Thermal Plasma-CVD (ETP CVD) method utilizing pulse-shaped substrate biasing to induce controlled ion bombardment during film growth. The films are analyzed with in-situ real time spectroscopic ellipsometry, FTIR spectroscopy, as well as reflection-transmission and Fourier transform photocurrent spectroscopy (FTPS) measurements. The aim of this work is to investigate the effect ion bombardment with well defined energy on the roughness evolution of the film and the material properties.

We observe two separate energy regimes with material densification and relatively constant defect density below ~ 120-130 eV and a constant material density at increasing defect density > 120-130 eV substrate bias. We discuss our results in terms of possible ion – surface atom interactions and relate our observations to reports in literature.

INTRODUCTION

Hydrogenated amorphous silicon (a-Si:H) thin films can be deposited with the expanding thermal plasma chemical vapor deposition (ETP-CVD) technique at growth rates of up to 10 nm/s [1]. At these high growth rates, substrate temperatures > 300°C are required to obtain material with good optoelectronic properties for device applications, however in solar-cell applications this imposes thermal stress on the previously deposited p-layer, leading to reduced solar-cell performance. Sinusoidal radio-frequency (RF) substrate biasing has been successfully utilized to provide the growing film with additional energy via induced ion bombardment (e. g. Smets et al. [2]). In this work, we report on a-Si:H thin film deposition utilizing a different kind of substrate biasing introduced by Wang et al. [3], refer to as pulse-shaped biasing (PSB). Utilizing a specially tailored waveform, a constant, negative potential is created on the sample holder resulting in a very narrow ion energy distribution function (IEDF) contrary to the broad and bimodal IEDF obtained from RF biasing [3]. This allows studying the effect of separate ion energies on material properties.

EXPERIMENT

The ETP-CVD technique has been described in detail before [1]: a schematic representation is shown in Fig. 1. It is based on an Ar-H_2-SiH_4 plasma that expands into a low pressure reactor chamber. The SiH_4 is dissociated mainly into SiH_3 radicals that are transported to the temperature controlled substrate holder where the film is deposited.

The PSB setup connected to our ETP-CVD reactor has been previously described by Wang et al. [3] and has been adapted for our reactor setup. The whole setup can be seen in Fig. 1b. The non-sinusoidal waveform is created by an arbitrary waveform generator (Agilent 33250A) and a broadband amplifier (Amplifier Research 150A250) with an amplification range from 10 kHz up to 250 MHz is used to control the signal amplitude. In principle, this setup enables us to operate in a frequency range for the non-sinusoidal wave from 200 kHz up to 8 MHz (the limit of the arbitrary waveform generator). Utilizing a specially tailored waveform, a constant, negative potential is created on the sample holder in between positive pulses that allow discharge of a non-conductive substrate. A constant potential on the substrate leads to a very narrow ion energy distribution function (IEDF) contrary to the broad and bimodal IEDF obtained from RF biasing [3]. In addition, the PSB setup can be used to measure the ion flux arriving at the substrate. Note that in the ETP-CVD technique the plasma potential is very low (~ 1-2 eV) and thus the applied substrate voltage alone determines the ion energy [2].

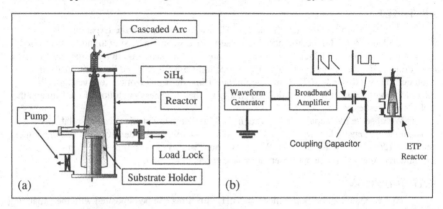

Figure 1. (a) Schematic overview of the ETP-CVD deposition setup. (b) Schematic overview of the pulse-shaped biasing setup.

All depositions were carried out under the same experimental conditions. The gas flows were 570 sccm argon and 190 sccm H_2 in the arc, 150 sccm H_2 in the nozzle and 230 sccm SiH_4 in the injection ring, at a current of 40 A in the arc. During biased depositions the heating elements had to be disconnected. The sample was heated up to 210°C prior to deposition and cooled down during deposition to about 185°C. Deposition time was 6.15 minutes. Deposition rate was about of 0.8 nm/s for unbiased and up to 1 nm/s for biased depositions.

The a-Si:H thin films have been deposited on c-Si wafers (prime wafer, 500-550 μm) with ~ 2 nm of native oxide, as determined by real-time spectroscopic ellipsometry (RTSE), and on Corning 7059 glass for optical measurements to determine the Urbach energy. Our RTSE measurements were performed using a J. A. Woollam Co., Inc M-2000F spectroscopic ellipsometer on c-Si wafers. In our RTSE data analysis we follow a procedure that was described in detail by Van den Oever et al. [4]. From the FTIR spectra obtained from films on c-Si wafers the integrated absorption was determined and analyzed to gain insight into the hydrogen bonding configuration. The Urbach energy was determined from combined reflection-transmission measurements and Fourier transform photocurrent spectroscopy (FTPS) measurements.

RESULTS

The development of surface roughness layer thickness, d_s, versus bulk film thickness, d_b, as determined from SE for different substrate voltages between 0 – 200 V can be seen in Fig. 2a.

The roughest films are obtained for unbiased depositions with a surface roughness layer thickness of about 50 Å at 3000 Å bulk film thickness. Mild ion bombardment around 10 V already leads to a strong reduction in d_s, and a further increase in substrate voltage leads to more reduction in surface roughness. At 200 V very smooth films below 20 Å d_s are obtained. Since all depositions had the same deposition time of 6.15 minutes we can also observe the increase in deposition rate above 120 V, which can be concluded from the increase in final bulk film thickness.

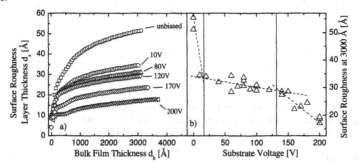

Figure 2. (a) Surface roughness layer thickness d_s as a function of bulk film thickness d_b determined from spectroscopic ellipsometry measurements. (b) The surface roughness layer thickness at a bulk film thickness of 3000 Å as a function of substrate voltage. The dashed lines are a guide for the eye.

Figure 2b shows d_s at 3000 Å bulk film thickness as a function of substrate voltage, three regions can be distinguished. Already at very low voltages < 20 V d_s strongly decreases with substrate voltage. Between 20 V and 120 V d_s only varies slightly, and above 120 V we again observe a stronger reduction in d_s. The three areas are indicated in the plot by solid lines.

FTIR results can be seen in Fig 3a and 3b. Figure 3a shows the dependence of total hydrogen concentration c_H and Fig. 3b the dependence of the LSM and HSM modes on biasing voltage. The data in Fig. 3a show some scattering, but in general the hydrogen concentration increases continuously with voltage. The same observation can be made in Fig. 3b for the LSM mode, which is hydrogen associated to vacancies. The increase appears to be stronger for voltages < 120 V and weaker for > 120 V. The HSM mode, associated to hydrogen in voids, continuously decreases until > 120V it could not be determined anymore. This transition voltage has been marked in Fig. 3b with a solid line and the regions below and above 120 V are labeled region I and II, respectively.

The infrared (IR) refractive index as a function of substrate voltage is shown in Fig. 4a and shows a similar separation into two regions. In region I (< 120 V) the refractive index continuously increases, indicating an increase in material density, in agreement with the reduction in void content observed in Fig. 3b. In region II (> 120 V) the refractive index is constant and does not change with substrate voltage. The relation between void concentration and IR refractive index indicates that for our experimental conditions the material density mainly depends on the void concentration. The Urbach energy, a good measure for the material disorder, is shown in Fig. 4b as a function of bias voltage. At low voltages the Urbach energy is constant around 50 meV and starts to increase from 100 V substrate bias onwards, with an increasingly sharp raise towards 200 V bias.

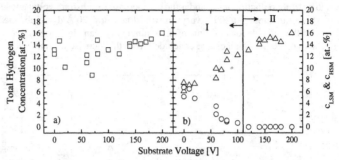

Figure 3. (a) Total hydrogen concentration as a function of substrate voltage. (b) c_{LSM} (triangles) and c_{HSM} (circles) as a function of substrate voltage. The solid line indicates the separation between material with voices (< 120 V) and without voids (> 120 V).

DISCUSSION

We observe a general trend in our results: there appears to be a transition region around 120-130 V substrate biasing. In region I we observe a reduction in void content, as concluded from FTIR data in Fig. 3, and an increase in material density as concluded from the increase in refractive index in Fig. 4a. Also the roughness layer thickness development at 3000 Å shown in Fig. 2b suggests such a transition region, where surface smoothening seems to be enhanced in region II. These observations as well as the transition in material modification mechanism will be discussed in terms of ion-surface interactions.

At the lowest substrate voltages only slight material modification at the film surface is expected. Simple heat transfer from impinging ions to surface atoms without atom displacement results in a local thermal spike and can result in local atomic rearrangement. The threshold energy for surface atom displacement of a Si surface exposed to an Ar ion beam has been estimated to be around 18 eV, while at higher energies of around 40 eV bulk atom displacement becomes more important [5]. Modelling results by Ma et al. [5] have shown how bulk atom displacement increases sharply with increasing ion energy, and more energy is deposited in subsurface layers then in the surface layer at ion energies above 100 eV. Sputtering is expected to become relevant for energies > 50 eV and the sputter yield increases drastically with increasing ion energy [6].

From the surface roughness development we conclude that local thermal spikes and surface atom displacement are already sufficient to lead to a strong decrease in surface roughness and an increase in material density without affecting the Urbach density significantly. While surface atom displacement can result in the breaking of bonds, those defects can be annihilated by the incoming growth flux particles, e. g. by sticking of an SiH_3 radical on the dangling bond, and thus do not contribute to defect creation in the material. In subsurface layers, defect annihilation requires more atomic rearrangements due to the more rigid atomic structure, and thus bulk atom displacement is more likely to lead to defect creation and deterioration of material properties. This effect can be seen in our Urbach energy data where a continuously stronger increase is observed at voltages > 100 V, which we attribute to the aforementioned mechanism of subsurface atom displacement. This increase in defect density has a detrimental effect on e. g. the photoconductivity and has been observed for RF substrate biasing before [7]. The enhanced smoothening at ion energies > 120 V observed in Fig. 2b can be related to enhanced etching or sputtering of the surface at preferred removal of material from surface hills over surface valleys.

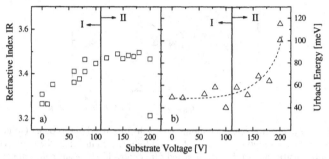

Figure 4. (a) Infrared refractive index determined from FTIR measurements. The solid line indicates the onset of void-free material for substrate biases > 120 V, as shown in Fig. 3b. (b) Urbach energy as a function of substrate bias voltage. The dashed line is a guide for the eye.

The increasing LSM mode of void-free material is attributed to the breaking of subsurface bonds due to bulk atom displacement and subsequent saturation of the created dangling bond by a hydrogen atom [2]. We observe strong similarities with results reported on RF substrate biasing below 120 V, whereas for higher voltages we do not observe the increase in void concentration reported by Smets et al. .[2] This can be related to the broad IEDF obtained in sinusoidal substrate biasing, resulting in sputtering effects due to highly energetic ions that do not occur in our well-defined ion energy distribution at similar average bias voltages.

Figure 5 shows an energy deposition map, adapted from Kaufmann et al. [7] and Smets et al. [2] In addition to data from Smets et al. [2] deposited with the ETP-CVD technique, as well as data from Hamers et al. [8] deposited with VHF-PECVD, this figure contains data from present work. It was argued by Harper et al. [9] that for ion assisted thin film deposition the range of ion energy deposition that leads to material densification without increase in defect density lies between 1 and 10 eV/atom. This range is indicated in Fig. 5 by the dashed area. Comparing our results with this range in the figure, we indeed observe that all region-I samples lie in this range, whereas samples from region II lie outside this range towards even higher energies per deposited atom. The results from Smets et al. [2] obtained for RF biasing under condition B correspond to a-Si:H films grown at 2.5 - 4.2 nm/s. They report an increase in void content at deposited energies > 10 eV/atom, which we do not observe. This difference between RF biasing and PSB is presumably related to the broad and bimodal IEDF obtained for RF substrate biasing compared to much narrower PSB IEDF [3]. More highly energetic ions can be present in an RF biasing induced ion bombardment that lead to sputtering already at much lower average substrate voltages. However, within the range < 10 eV/atom the densification of the material is comparable for both RF and PSB. Hamers et al. [8] report also material densification for ion energy depositions below < 10 eV/atom and constant material density for > 10 eV/atom.

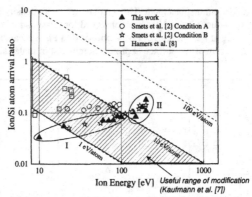

Figure 5. Energy deposition map containing the results discussed in this paper (solid triangles) as well as other data reported in literature. The highlighted area indicates the region considered useful for film densification during ion-assisted film deposition by Kaufman et al.[7] .

CONCLUSIONS

We observe an increase in material density and a reduction of void faction for ion energies below 120-130 eV, and a constant material density at a continuously increasing Urbach energy above 120-130 eV. We attribute the material densification to surface and bulk atom displacement, while the continuous increase in Urbach energy at higher energies is solely attributed to bulk atom displacement that cannot be annihilated by rearrangements of the silicon lattice. We obtain a regime of material densification in the energy deposition range of 1-10 eV/deposited atom.

ACKNOWLEDGMENTS

M. Tijssen and K. Zwetsloot are acknowledged for their skillful technical assistance. This research was financially supported by SenterNovem within the framework of EOS-LT.

REFERENCES

[1] M. C. M. van de Sanden, R. J. Severens, W. M. M. Kessels, R. F. G. Meulenbroeks and D. C. Schram, *J. Appl. Phys.* **84** (1998), p. 2426.
[2] A. H. M. Smets, W. M. M. Kessels and M. C. M. van de Sanden, *J. Appl. Phys.* **102** (2007), p. 073523.
[3] S. B. Wang and A. E. Wendt, *J. Appl. Phys.* **88** (2000), p. 643.
[4] P. J. van den Oever, M. C. M. van de Sanden and W. M. M. Kessels, *J. Appl. Phys.* **101** (2007), p. 10.
[5] Z. Q. Ma, Y. F. Zheng and B. X. Liu, *Phys. Status Solidi A-Appl. Res.* **169** (1998), p. 239.
[6] K. Wittmaack, *Phys. Rev. B* **68** (2003), p. 11.
[7] H. R. Kaufman and J. M. E. Harper, *J. Vac. Sci. Technol. A* **22** (2004), p. 221.
[8] E. A. G. Hamers, W. G. J. H. M. van Sark, J. Bezemer, H. Meiling and W. F. van der Weg, *J. Non-Cryst. Solids* **226** (1998), p. 205.
[9] J. M. E. Harper, J. J. Cuomo, R. J. Gambino and H. R. Kaufman, *Ion Bombardment Modification of Surfaces: Fundamentals and Applications*, Elsevier Science, Amsterdam (1984).

Mater. Res. Soc. Symp. Proc. Vol. 1153 © 2009 Materials Research Society   1153-A17-07

Phosphorus and Boron Doping Effects on Nanocrystalline Formation in Hydrogenated Amorphous and Nanocrystalline Mixed-Phase Silicon Thin Films

C.-S. Jiang, Y. Yan, H.R. Moutinho, and M. M. Al-Jassim
National Renewable Energy Laboratory (NREL), Golden, CO 80401

B. Yan, L. Sivec, J. Yang, and S. Guha
United Solar Ovonic LLC, Troy, MI 48084

Abstract

We report on the effects of phosphorus and boron doping on the microstructure of nanocrystallites in hydrogenated amorphous and nanocrystalline mixed-phase silicon films, using Raman spectroscopy, secondary-ion mass spectrometry, cross-sectional transmission electron microscopy, atomic force microscopy, and conductive atomic force microscopy. The characterizations revealed the following observations. First, the mixed-phase Si:H films can be heavily doped in $\sim 10^{21}/cm^3$ by adding PH_3 and BF_3 in the precursor gases. Second, the intrinsic and doped films can be made in a similar crystalline volume fraction by adjusting the hydrogen dilution ratio. The hydrogen dilution ratio is much higher for P-doped films than for the intrinsic film with the similar crystallinity. Third, the doping significantly impacts the nanostructures in the films. Nanograins aggregate to form cone-shaped clusters in the intrinsic and B-doped films, but isolate and randomly distribute in amorphous tissues in the P-doped films. The cones in the intrinsic and B-doped films are also different. The cone-angle is smaller and the nanograin density is lower in the B-doped films than in the intrinsic films. These P- and B-doping effects on the nanocrystalline formation are interpreted in terms of diffusion of Si-related radicals during film growth.

Introduction

Hydrogenated amorphous silicon (a-Si:H) and nanocrystalline silicon (nc-Si:H) thin films have attracted great interest in photovoltaic devices and large-area thin-film electronics. A common technique for fabricating a-Si:H and nc-Si:H films is plasma-enhanced chemical vapor deposition (PECVD), which decomposes SiH_4 or Si_2H_6 molecules and deposits Si radicals on foreign substrates. One of the advantages in fabricating Si:H-based thin-film devices is that the n- and p-doped layers can be produced during film growth by adding phosphorous- or boron-containing gases. High-performance solar cell devices have been successfully fabricated and manufactured by adding PH_3 in the n-layer and BF_3, $(CH)_3B$, or B_2H_6 in the p-layer. Therefore, P- and B-doping effects on the structural and electronic properties of the films are of primary importance in fundamental material study and industrial applications.

Many studies on the doping effects concentrate on macroscopic properties of the films [1–4]. In this study, we investigate the doping effects on microstructures of the films. We chose mixed-phase Si:H films because the amount of nanograins in the materials is easy to characterize [5,6]. In a-Si:H film, the nanostructured component is too low to be found; but it is too high in nc-Si:H to distinguish the changes induced by doping. In PECVD, hydrogen dilution is the main parameter controlling the amorphous/nanocrystalline transition [7]. Materials deposited in the transition regime are called mixed-phase Si:H films. Because mixed-phase Si:H films contain nanocrystallites, grain boundaries, and amorphous tissues, it provides a prototype structure for nc-Si:H study, and allow us to examine the doping effects on the amorphous/nanocrystalline transition and on the phase boundary characteristics. The understanding of mixed-phase films is expected to be useful and extended to that of fully nc-Si:H films.

Experimental

Mixed-phase Si:H films about 300 nm thick were deposited on stainless-steel substrates using a multichamber PECVD system at United Solar. PH_3 and BF_3 gases were added in H_2 and Si_2H_6 gas mixtures for P- or B-doping, respectively. The hydrogen dilution ratio ($R=H_2/Si_2H_6$) was adjusted to change the material structural properties. We found that as we added doping gases, the hydrogen dilution threshold to reach the amorphous/nanocrystalline transition is changed. In this study, we present the characterization results from three typical samples of intrinsic, P-doped, and B-doped films that were made with different hydrogen dilution ratios, but with similar crystalline volume fractions. The two doped films have similar dopant concentrations, which allow us to make a fair comparison.

The compositional, structural, and electrical properties of the films were characterized using secondary-ion mass spectrometry (SIMS), Raman spectroscopy, cross-sectional transmission electron microscopy (X-TEM), atomic force microscopy (AFM), and conductive AFM (C-AFM). Raman spectroscopy was used to measure the crystallinity of the films, where a 532-nm laser with a light spot of 1–2-μm diameter was irradiated on the samples and the Raman scattering of light was measured by a monochrometer. X-TEM was used to investigate the structures in the film bulk. The X-TEM samples were prepared by the focused ion beam (FIB) technique. A 50-nm-thick Pt layer was deposited on the sample surface to protect the thin film from damage and an 80-nm-thick slice of the sample was cut off using the FIB. Contact-mode AFM and simultaneous C-AFM measurements were used to investigate film morphology and electrical conduction. The AFM images were taken using silicon tips coated with doped diamond-like carbon to reduce the contact electrical resistance and enhance the wear resistance. When the tip was in contact with the sample surface during scanning, the tip was modified by the friction against the sample surface, which increased the tip/film contact resistance [5,6]. Therefore, we used relatively fresh tips for a fair comparison. The scan parameters were identical for all the C-AFM images.

Results

Figure 1(a) shows the Raman spectra of the intrinsic, P-doped, and B-doped films, where the mixed-phase signature is evidenced by a broad peak at 480 cm^{-1}, corresponding to amorphous phase, and a small peak at 510–520 cm^{-1}, corresponding to nanocrystalline phase. The three samples have more or less similar crystalline volume fractions. To reach similar crystallinity, the hydrogen dilution ratio was 2.3 times higher for the P-doped film than for the intrinsic film. This demonstrates that a few percent P doping significantly impacts the structure of the film, which makes the material change toward to amorphous structure. However, the crystallinity was able to change back toward nanocrystalline by increasing the hydrogen dilution. The B-doped sample was made with very different deposition conditions; thus, no comparison is presented of hydrogen dilution between the B-doped samples with others.

Because the doping efficiency is similar for P- and B-doping [3,4], the similar P and B concentrations in the films give a good comparison of doping effects on film properties. To reach a similar dopant concentration, we adjusted the PH_3 and BF_3 flow rates. Figure 1(b) shows the SIMS spectra of the doped samples, where a high doping concentration of $\sim1.5\times10^{21}$/cm^3 is measured in both P- and B-doped films. The dopant concentrations are uniform throughout the entire film thickness. At the surface of the film, the P concentration is different from that in the film bulk, which is caused by surface contamination.

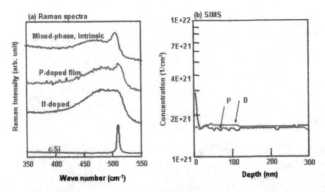

Figure 1. Raman spectra and SIMS measurements on the intrinsic, P-doped, and B-doped films.

Figure 2 shows X-TEM images of the three samples. The intrinsic Si:H film in Fig. 2(a) contains cone-shaped structures [5,8]. The high-resolution X-TEM image in Fig. 2(b) on the cone-area exhibits a high density of nanograins with sizes of several nanometers. The cones are embedded in the a-Si:H matrix. No clear nanograins were found outside the cones, which demonstrates that the intrinsic mixed-phase Si:H film has a clear amorphous and nanocrystalline phase separation. The formation of the cone structure has been explained by a faster growth rate in the nanocrystalline phase than the amorphous phase. From the cone angle of ~70°, it is estimated that the nanocrystalline phase growth rate is ~1.2 times that of the amorphous phase growth rate [9]. This faster nc-Si:H growth rate causes the nc-si:H volume fraction increases with the film growth, which is not yet reached the steady state. With the heavy doping of PH$_3$, however, the aggregations of nanograins are no longer observed, as showed in Fig. 2(c), although the Raman spectra show a similar crystalline volume fraction. The high-resolution X-TEM image of Fig. 2(d) still shows many nanograins with similar sizes of several nanometers. These nanograins are isolated and (most likely) randomly distributed in the matrix of the amorphous phase. Therefore, we conclude that the PH$_3$ doping significantly impacted the mixed-phase Si:H film structure.

With the heavy B-doping, the cone structures are still observed, as shown in Fig. 2(e). The high-resolution X-TEM image taken in the cone area, as shown in Fig. 2(f), exhibits nanograins with similar sizes to those observed in the intrinsic and P-doped films. However, the structure of the film is different from the intrinsic film. First, the cone angle (~25°) is significantly smaller than that of the intrinsic film (~70°), indicating a growth rate in the nanocrystalline phase of ~1.03 times that the growth rate in the amorphous phase [9]. Second, the cone density is significantly higher than that of the intrinsic film. The cone tops coalesce at the top of the film. Third, the nanograin density inside the cones is significantly lower than in the cones of the intrinsic film. Therefore, the crystalline volume fraction in the cones is low, which makes the X-TEM contrast of the cones much lower than the contrast in the intrinsic film. Inside the cone, besides the nc-Si:H grains, a-Si:H tissue also occupies a considerable volume fraction.

We used C-AFM to investigate the doping effects on the local electrical conduction. In C-AFM measurements, there are two possible contact barriers of tip/film and film/substrate. Because the contacts are similar to Schottky barriers, the barrier directions of the two contacts are opposite to each other. From the current images under either a positive or negative voltage between the tip and substrate, we find that the tip/film is the resistance-dominant contact. Therefore, we show here the

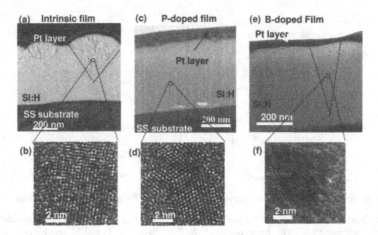

Figure 2. X-TEM images taken on: (a) and (b), intrinsic film; (c) and (d), P-doped film; (e) and (f), B-doped film. The (a), (c), and (e) images are low resolution; the (b), (d), and (f) images are high resolution.

current images under a forward bias voltage of 0.75 V to the tip/film contact. In this case, the current images mainly measure the conduction in the film, but not limited by the tip/film contact.

The AFM topographic image in Fig. 3(a) shows the cone tops of the nanocrystalline aggregations on the intrinsic film [5,6]. The cone tops disappear on the P-doped film, as shown in Fig. 3(c), which is consistent with the X-TEM observations. On the B-doped films, there are high-density particle-like features with 100–200-nm sizes, as shown in Fig. 3(e), where the coalescent small cone-tops are consistent with the X-TEM observations, as well. On the intrinsic film, the clear correspondence between the nanocrystalline aggregations in Fig. 3(a) and the high current areas in Fig. 3(b) reveals that the carrier transport paths are through the nanocrystalline aggregations [5,6]. The current on individual aggregations is relatively uniform, and the value depends on the size of the aggregations, exhibiting larger current values on larger-size aggregations. This dependency of the current on the aggregation size demonstrates that the conduction is limited by the film, but not by the tip/film contact barrier.

The current image on the P-doped film does not exhibit localized features. However, the high currents are not uniform, which is possibly caused by the complexity of the film structure. The current may depend closely on the local structure of the contact area with the probe. If the probe is just on a nanograin, the current will be higher than on an amorphous tissue area. The average current value over the image is $\overline{I} = 110$ nA, which is three orders of magnitude larger than that of the intrinsic film. On the B-doped film, no localized features are observed, although the X-TEM image shows cone-shaped nc-Si:H aggregations. This could be because of the high density of cones, and the cone tops coalesce to each other on the film surface. The reasons for the smaller average current value ($\overline{I} = 8.7$ nA) in the B-doped film than the P-doped film could be a lower mobility of the majority carriers (holes) and differences in the film structures.

348

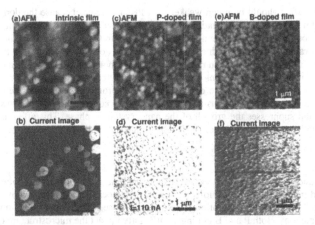

Figure 3. Topographic images (a), (c), and (e) and the corresponding current images (b), (d), and (f) were taken under a forward bias voltage of 0.75 V. The (a) and (b) images were taken on the intrinsic film; (c) and (d) images on the P-doped; and (e) and (f) images on the B-doped film.

Discussion

Based on the atomic hydrogen enhanced surface diffusion model [10], the formation of cone-shaped aggregations in the intrinsic film and the isolated nanograins in the P-doped film can be explained as follows. In the intrinsic film, once a nucleation is formed, the surface of the grain may have a higher sticking coefficient or a lower surface energy for Si radicals such as SiH_3. This leads to a continuing growth of the nc-Si:H cone. The growth rate of the nanocrystalline phase is determined by the net diffusion-in of Si radicals balanced by the diffusion-in and diffusion-out of mass [11]. Therefore, the "long-range" diffusion is the main factor determining the shape of cone structures and the nanocrystalline volume. For P-doped film, PH_3 molecules are absorbed on the growth surface with a large probability [12], preventing the "long-range" diffusion of Si radicals and thus preventing the nanocrystalline phase growth on existing Si grains. However, the increased hydrogen dilution provided more hydrogen termination of Si bonds on the growth surface. This makes the diffusion of Si radicals better in local small areas, which is not disturbed by PH_3 molecules. This enhanced local diffusion rate is beneficial for new nucleation in amorphous tissue, and such a nucleation rate is the main factor that determines the nanocrystalline volume fraction in the P-doped films.

In the case of B-doped film, the cone-shaped nc-Si:H aggregations indicate that the long-range diffusion of Si radicals is still the main mechanism for nucleation. The mechanisms for the smaller cone-angle and less dense nanograins inside cones may be explained as follows. First, the cone density in the B-doped film is much larger than that of the intrinsic film. B-related radicals may promote nc-Si:H nucleation, and this has been reported by an AFM-based study [13]. If there are dense cones, the net-diffusion mass into each cone may be reduced. Consequently, the growth rate and cone angle are reduced, because the net-diffusion-in mass would be the dominate factor for the nc-Si:H growth rate, the nc-Si:H/a-Si:H volume ratio inside the cones, and the cone angle. Second, similar to PH_3 molecules, B radicals on the surface may also reduce the long-range diffusion rates of

Si radicals. However, this effect should be much weaker than PH_3 molecules because the dominant factor of the nanocrystalline aggregation formation should still be the long-range diffusion. This reduced long-range diffusion rate can reduce the net-diffusion mass into the cones and slow the cone growth rate. Third, the B radicals may also reduce the H terminations on Si bonds in areas around the B radicals [4], and thus reduce the nucleation rate locally in amorphous tissue. The explanation is consistent with the fact that no clear nanograins were found outside the cones. Fourth, B radicals may have a catalytic effect for SiH_4 or Si_2H_6 to decompose, and to promote the growth rate of the film [1,4]. The enhanced rate of decomposing inside and outside of the cones promotes the growth of local amorphous phases and suppresses the growth of the nanocrystalline phase. This effect makes the crystalline volume fraction inside the cones of the B-doped material smaller than inside the cones in the intrinsic film.

Summary

We investigated P- and B-doping effects on the nanocrystalline formation by studying the microscopic structure and microscopic current flow in amorphous/nanocrystalline mixed-phase Si:H films with a similar crystallinity. Hydrogen dilutions need to be increased for P-doped films to reach similar nc-Si:H volume fractions. Both P and B doping significantly impact the microstructure of the films. The nanograins aggregate to cone-shaped clusters in the intrinsic and B-doped films. However, the cone angle and nanograin density in the cones in the B-doped films are much smaller than those in the intrinsic film. In the P-doped film, the nanograins are isolated and randomly distributed in a-Si:H tissues. These doping effects on the microstructures are explained based on atomic hydrogen enhanced diffusion model.

Acknowledgement

The authors thank K Alberi and R. Reedy for the Raman and SIMS measurements. This work was supported by DOE under Contract No. DE-AC36-08-GO28301 at NREL and under the Solar America Initiative Program Contract No. DE-FC36-07 GO 17053 at United Solar Ovonic LLC.

References

[1] S. Nakayama, I. Kawashima, and J. Murota, *J. Electrochem. Soc: Solid State Science & Technol.* **133**, 1721 (1986).

[2] R. Saleh and N.H. Nickel, *Thin Solid Films* **427**, 266 (2003).

[3] P. Kumar and B. Schroeder, *Thin Solid Films* **516**, 580 (2008).

[4] T. Matsui, M. Kondo, and A. Matsuda, *J. Non-Cryst. Solids* **338–340**, 646 (2004).

[5] B. Yan, C.-S. Jiang, C.W. Teplin, H.R. Moutinho, M.M. Al-Jassim, J. Yang, and S. Guha, *J. Appl. Phys.* **101**, 033712 (2007).

[6] C.-S. Jiang, B. Yan, H.R. Moutinho, M.M. Al-Jassim, J. Yang, and S. Guha, *Mater. Res. Soc. Symp. Proc.* **989**, 15 (2007).

[7] For a review, see A. V. Shah, J. Meier, E. Vallat-Sauvain, N. Wyrsch, U. Kroll, C. Droz, and U. Graf, *Sol. Energy Mater. Sol. Cells* **78**, 469 (2003).

[8] R.W. Collins, A.S. Ferlauto, G.M. Ferreira, C. Chen, J. Koh, R.J. Koval, Y. Lee, J.M. Pearce, and C.R. Wronski, *Sol. Energy Mater. Sol. Cells* **78**, 143 (2003).

[9] C.W. Teplin, E. Iwaniczko, B. To, H. Moutinho, P. Stradins, and H.M. Branz, *Phys. Rev. B* **74**, 235428 (2006).

[10] A. Matsuda, *Thin Solid Films* **337**, 1 (1999).

[11] C.W. Teplin, C.-S. Jiang, P. Stradins, and H.M. Branz, *Appl. Phys. Lett.* **92**, 093114 (2008).

[12] For a review, see T.I. Kamins, *Polycrystalline Silicon for Integrated Circuit Applications* (Kluwer Academic, Boston, 1997), p.32.

[13] T. Toyama, W. Yoshida, Y. Sobajima, and H. Okamoto, *J. Non-Cryst. Solids* **354**, 2204 (2008).

Poster Session:
Defects and Metastability

Mater. Res. Soc. Symp. Proc. Vol. 1153 © 2009 Materials Research Society   1153-A18-01

Effects of a Bias Voltage During Hydrogenation on Passivation of the Defects in Polycrystalline Silicon for Solar Cells

Yoji Saito Hayato Kohata, and Hideyuki Sano
Seikei University, 3-3-1 Kichijoji-Kitamachi, Musashino, Tokyo 180-8633, Japan

ABSTRACT

The short circuit current and conversion efficiency of the poly(multi)-crystalline solar cells are increased by the passivation process using hydrogen plasma. The passivation rate apparently increases at a reverse bias voltage near 0.6V during the hydrogenation process. The effects of the bias voltage on the passivation are large at the substrate temperatures between 200°C and 250°C. The phenomena are likely due to the existence of positively-ionized hydrogen, H^+. The H^+ ions can be accelerated from the surface into the bulk by the electric field with the negative bias. The possibility of the H^+ ions in the bulk silicon has been predicted in the previous reports. The increase of the incorporated hydrogen is confirmed by IR absorption measurements. The enhanced diffusion of hydrogen induced by the reverse bias is supported by the results of spectral response characteristics of the hydrogenated solar cells.

INTRODUCTION

Most of commercial solar cells are based on multi(poly)-crystalline silicon substrates. A large amount of lattice defects, however, are included in multi-crystalline silicon substrates, and the defects can degrade the electrical characteristics of the solar cells. Hydrogen passivation process is usually performed to reduce the electrically-active defect levels at temperatures above 300°C for hours. In this study, we found phenomena of the enhanced hydrogen diffusion into multi-crystalline silicon substrates by a bias voltage on the solar cells.

EXPERIMENTAL PROCEDURE

Fabrication Process of Solar Cells

Multi-crystalline silicon (p-type, 1.9Ωcm) substrates, which were supplied from Sharp Co., were used in this study. The substrates were cleaned with a hot alkaline solution ($NH_4OH:H_2O_2:H_2O=1:1:6$). The substrates were rinsed with deionized water, dipped HF solution, and rinsed again with deionized water. The substrates were then thermally oxidized in wet oxygen ambient to form the oxide films with thickness of about 500nm.

The solar cells were fabricated in the followings. Phosphorus was doped by thermal diffusion to form the pn junction. The front and rear aluminum electrodes were formed by evaporation. The substrates were sintered at 420°C in nitrogen ambient to form Ohmic contacts.

Hydrogenation Process

The solar cells were introduced into a conventional rf plasma system with two-paralleled electrodes, and were maintained between 200°C and 300°C. A hydrogen gas was introduced into the reaction chamber, and the pressure was maintained to be 106Pa by evacuating by a diffusion pump. The rf electric power of 10W was applied to produce atomic hydrogen. A schematic diagram of the experimental system for the hydrogenation is shown in Fig. 1. The sample was mounted onto the lower side electrode with a heater. The sample was covered by fine-meshed wire nets, made of stainless steel, to be protected from an ion impact.

During the plasma process an external voltage was applied to the solar cells. Before and after the plasma process the electrical and optical characteristics of the solar cells were measured at room temperature.

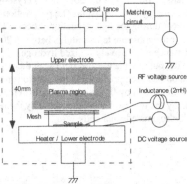

Figure 1. Schematic diagram of the hydrogenation system.

EXPERIMENTAL RESULTS AND DISCUSSION

Influence of the Bias Voltage during Hydrogenation Process

First, the dependence of electrical characteristics on the bias voltage during the hydrogenation was investigated. The hydrogenation was performed at 300°C for 60 min. The changes of the short circuit current (Isc) and the efficiency are plotted as a function of the bias voltage in Fig. 2. The initial cell efficiency (before hydrogenation) was in the range between 2 and 4%, and the cell size was 5mmX4.5mm in this experiment.

With increase of the value of the negative bias voltage, Isc and the efficiency increase as shown in Fig. 2. The increase of Isc and the efficiency is considered to be induced by the passivation of the defect levels located near the pn junction with the diffused hydrogen. The dependences of Isc and the efficiency on the bias voltage as shown in Fig. 2 are likely due to the existence of positively-ionized hydrogen, H^+. The H^+ ions can be accelerated from the surface into the bulk by the electric field with the negative bias. The possibility of the H^+ ions in the bulk silicon has been predicted in the previous reports [1]. The diffusion rate of the hydrogen is not so large at 300°C without the bias voltage, but it may be enhanced by the forward electric field.

On the other hand Isc and the efficiency decrease with the increase of the positive bias voltage as shown in Fig. 2. The degradation of the electrical characteristics is considered to be due to the defects induced by the excess hydrogen near the surfaces. [2]

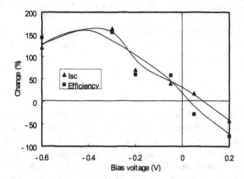

Figure 2. The changes of the short circuit current and efficiency as a function of the bias voltage during the hydrogenation at 300°C. The measurements were performed at light intensity of 15 mWcm^{-2}.

Influence of the Substrate Temperature during Hydrogenation Process

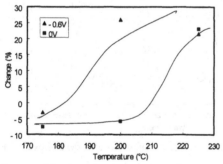

Figure 3. The changes of the short circuit current on substrate temperatures as a function of the bias voltage during the hydrogenation, where the light intensity was 100mWcm^{-2}.

The dependence of the Isc on substrate temperatures as a function of the bias voltage during the hydrogenation was investigated. The initial cell efficiency was in the range between 4 and 5%, and the cell size was 30mmX23mm in this experiment.

Figure 3 shows the changes of the short circuit current on substrate temperatures as a function of the bias voltage during the hydrogenation, where the light intensity was 100mWcm^{-2}.

When the bias voltage is 0V, the Isc increases with the increase of the temperature above about 210°C. When the bias voltage is -0.6V, the Isc increases with the temperature above 170°C. The effect of the bias voltage on the Isc becomes large around 200°C.

The Isc decreases by the hydrogenation process at low temperatures. The hydrogen hardly diffuses into the bulk at low temperatures. The decrease of the Isc is likely due to the defects induced by the plasma damage.

Hydrogen Content

We tried to estimate the depth profiles of the incorporated hydrogen by secondary ion mass spectroscopy (SIMS) measurements. In this experiment, deuterium was used instead of hydrogen to avoid the disturbance by residual hydrogen related molecules in the substrates before the hydrogenation. However, we could not observe a significant dependence of the depth profiles on the bias voltage unfortunately because of insufficient detection limit of a SIMS system.

Next, we carried out Fourier transform infrared (FTIR) spectroscopy measurements to investigate the change of hydrogen content in the substrates by the hydrogenation process. Before the FTIR measurements the Al electrodes were removed by wet etching. Figure 4 shows changes of FTIR spectra between the substrates after the hydrogenation (at the bias voltage of -0.6V and 0V) and the substrates before the hydrogenation.

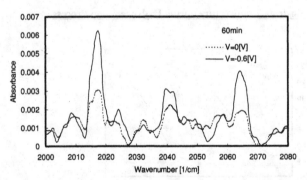

Figure 4. Changes of FTIR spectra between the substrates after the hydrogenation and the substrates before the hydrogenation.

The hydrogen content is increased by the hydrogenation, because we can see the peaks, which are located at 2017cm^{-1} and are derived from Si-H bonds. The peak for the sample, which has been hydrogenated at the bias voltage of -0.6V, is larger than that for the sample hydrogenated at 0V. The reverse bias voltage, therefore, enhances the hydrogen incorporation.

356

Next, we measured spectral response as a clue of the depth profiles of passivated defects. Figure 5(a) and 5(b) show spectral response characteristics of the solar cells hydrogenated at the bias voltage of 0V and -0.6V, respectively. At the bias voltage of 0V, the spectral response increases for all wavelengths of the measurement range as shown in Fig. 5(a). This result indicates that the defects around the depletion layer would be passivated by the incorporated hydrogen. On the other hand, the spectral response of the sample, hydrogenated at the bias voltage of -0.6V, increases largely at the long wavelengths around 700nm as shown in Fig. 5(b). Comparing the spectral response in Fig. 5(b) to those in Fig. 5(a), passivated region for V=-0.6V is deeper than that for V=0V, because the light with long wavelength is absorbed in the deep region from the surfaces. These results supports that the diffusion rate of hydrogen would be enhanced by the reverse bias voltage during the hydrogenation process.

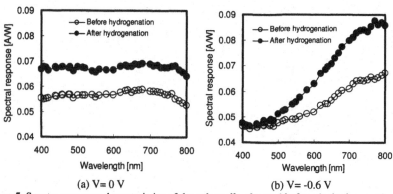

(a) V= 0 V   (b) V= -0.6 V

Figure 5. Spectra response characteristics of the solar cells after and before the hydrogenation,

CONCLUSIONS

The short circuit current and conversion efficiency of the solar cells are increased by the passivation process using hydrogen plasma. The hydrogenation rate apparently increases at a reverse bias voltage near 0.6V during the hydrogenation process. The effects of the bias voltage on the hydrogenation are large at the substrate temperatures between 200°C and 250°C, are considered to be caused by H^+ ions, which would exist in the substrates. The results of FTIR spectra and spectral response characteristics support the enhanced diffusion of hydrogen by the reverse bias voltage.

ACKNOWLEDGMENTS

The authors would like to thank T. Momma, S. Tomita, and M. Kondo of Seikei Univ. for their technical assistance. The authors would like to thank Sharp Corp. for the providing the multi-crystalline silicon substrates. This research was financially supported partially by a Grant-in-Aid for Scientific Research (C) from JSPS, Japan.

REFERENCES

1. S. Darwiche, M. Nikravech, D. Morvan, J. Amouroux and D. Ballutaud, *Solar Energy Materials and Solar Cells*, **91**, Issues 2-3, 23, 195 (2007).
2. S. Martinuzzi, I. Perichaud, *Solar Energy Materials and Solar Cells*, **72**, 343 (2003)

Mater. Res. Soc. Symp. Proc. Vol. 1153 © 2009 Materials Research Society   1153-A18-02

Structural Properties of a-Si:H Films With Improved Stability
Against Light Induced Degradation

G. van Elzakker[1], P. Šutta [2], M. Zeman [1]
[1] Delft University of Technology, DIMES, P.O. Box 5053, 2600 GB Delft, The Netherlands
[2] University of West Bohemia, NTRC, Univerzitní 8, 306 14 Plzeň, Czech Republic

ABSTRACT
X-ray diffraction (XRD) analysis of thin silicon films was carried out using both the symmetric Bragg-Brentano and asymmetric thin-film attachment geometries. The asymmetric configuration allows quantitative phase analysis of the films and reveals that amorphous silicon films deposited from silane diluted with hydrogen have the strongest peak in the XRD patterns located around 27.5 degrees. This peak corresponds to the signal from ordered domains of tetragonal silicon hydride and not from cubic silicon crystallites. The full width at half maximum (FWHM) of this peak narrows from 5.1 to 4.8 degrees as the ratio of hydrogen to silane flow (R) increases to 20 and does not change significantly for higher hydrogen dilutions. The amorphous silicon films fabricated at different hydrogen dilution were applied as absorber layers in single-junction solar cells. Degradation experiments confirm a substantial reduction of the degradation when the dilution ratio is increased from R=0 to R=20. The light induced degradation of solar cells with absorber layers prepared at R > 20 is not further reduced by increasing R.

INTRODUCTION
Hydrogenated amorphous silicon (a-Si:H) is a promising material for low-cost solar cells. Unfortunately, a-Si:H suffers from light-induced degradation known as Staebler-Wronski effect [1]. It has been demonstrated that solar cells with a-Si:H absorber layers prepared from silane diluted with hydrogen in plasma-enhanced chemical vapor deposition (PECVD) showed less degradation during light exposure than solar cells with undiluted absorbers [2,3]. This a-Si:H material is referred to as protocrystalline silicon (pc-Si:H) due to the fact that it will eventually evolve from the amorphous to microcrystalline phase when grown to a sufficient thickness [4,5].
In this article the structural properties of pc-Si:H deposited by PECVD are analyzed. The X-ray diffraction (XRD) technique was used to study the influence of hydrogen dilution on the structural properties of pc-Si:H films. In particular the narrowing of the first scattering peak (FSP) in the XRD spectra was investigated since this parameter is used as an indication of an improved medium range order (MRO). The pc-Si:H films fabricated at different hydrogen dilutions were applied as absorber layers in single-junction solar cells. The degradation behavior of these solar cells was investigated and related to the structural properties of the films.

EXPERIMENTAL DETAILS
Individual films and solar-cell absorber layers were deposited under the following deposition conditions; an rf-power of 4 W, a silane (SiH$_4$) flow of 5 sccm, a substrate temperature of 180°C. The hydrogen dilution is expressed by the dilution ratio R, which is defined as R=[H$_2$]/[SiH$_4$]. The dilution was varied between R > 0 and R = 40. An undiluted reference film (R = 0) was grown at a different pressure of 0.7 mbar and a SiH$_4$ flow of 40 sccm. The thickness of all films was ~ 300 nm, which corresponds to the thickness of the absorber layer in solar cells. In order to

obtained 300 nm thick pc-Si:H films at all R the chamber pressure of 2.6 mbar was used. The individual films were deposited on Corning Eagle 2000 type glass substrates and on crystalline silicon (c-Si) substrates coated with a 20 nm thick a-Si:H layer grown from pure silane. Single junction p-i-n solar cells were deposited on Asahi U-type substrates using the above described films as the absorber layers. The solar cells have the following structure: p-type a-SiC:H layer (10 nm)/a-SiC:H buffer layer/intrinsic absorber layer (300 nm)/n-type a-Si:H layer (20 nm). The back contact consists of 300 nm aluminum. The external parameters of the solar cells (efficiency η, fill factor ff, short-circuit current density J_{SC}, open-circuit voltage V_{OC}) were determined from I-V measurements using an Oriel Corporation solar simulator. The solar cells were degraded at a constant temperature of 50 °C with halogen lamps using a power density of 100 mW/cm^2.

The structural properties of the films were studied by X-ray diffraction analysis using an automatic powder diffractometer X'pert Pro in both symmetric and asymmetric goniometer configurations with CuK-alpha radiation (lambda = 0.154 nm).

X-RAY DIFFRACTION ANALYSIS

X-ray diffraction analysis is a very useful tool in the field of thin films since it allows us to probe only a few micrometers of the thickness of the films. There are several configurations how to carry out the X-ray measurement in order to obtain diffraction patterns of the films deposited on a substrate. The most suitable configuration for the analysis of a film depends on its value of the X-ray linear absorption coefficient and its thickness. A widely-used goniometer with a symmetric Bragg-Brentano geometry is suitable for thicker (poly-) crystalline materials with a high value of the linear absorption coefficient. In this case the X-ray beam usually does not reach the substrate and only the diffraction patterns of the film are detected. On the other hand, for materials with a low value of the linear absorption coefficient the penetration of the X-ray beam can be larger than the film thickness and the intensities of diffraction lines of the film are very low in comparison with stronger intensities of diffraction lines originating from the substrate material. In such cases the asymmetric geometry configuration of the X-ray measurement is more convenient for the structural analysis. We analysed Si:H films using both configurations.

The main feature of the Bragg-Brentano symmetric geometry is that both the incident and diffracted beams are at the same angle to the sample surface. For a given family of lattice planes this angle is equal to the Bragg's angle. It means that only those lattice planes $\{h\,k\,l\}$ of crystallites can be detected, which are parallel to the film surface and fulfil the Bragg's condition. The main disadvantage of the Bragg-Brentano geometry is that the irradiated volume of the film decreases with increasing diffraction angle that results in a weak intensity of diffraction line recorded at higher diffraction angles. This can be solved by using an automatic divergence slit, which keeps the irradiated volume of the film constant.

In the case of very thin films low (glancing) incident angles are used to prevent that the radiation penetrates far into the substrate. As a result, diffraction lines of the film exhibit higher intensities in comparison to the substrate lines. In this configuration the (pseudo-) parallel beam optics is used, the incident angle (ω) is kept fixed and the diffraction angle (2ϑ) changes. The geometry of the experiment is therefore asymmetric. Due to the fixed incident angle the irradiated volume is constant, which results in higher intensities for diffraction lines observed at higher diffraction angles. In our experiment the incident angle ω was fixed to 0.5 deg and the diffraction angle 2ϑ varied from 15 to 40 deg (short scan) or from 15 to 65 deg (long scan). In the short scan only the first broad amorphous peak of the films can be obtained and in the long scan both the first and also the second broad amorphous peaks can be detected.

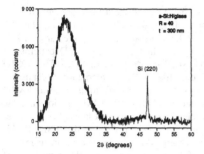 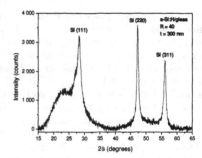

Figure 1: The XRD pattern of a Si:H film on glass recorded using the symmetric geometry.

Figure 2: The XRD pattern of a Si:H film on glass recorded using the asymmetric geometry.

RESULTS AND DISCUSSION
Analysis of structural properties of pc-Si:H films with XRD

In order to clearly illustrate the difference of using both XRD geometries for the analysis of thin silicon films we measured a hydrogenated silicon (Si:H) film containing both amorphous and crystalline phase components which was deposited on glass substrate at R = 40. Figure 1 and 2 show the XRD patterns recorded for the Si:H film on the symmetric Bragg-Brentano geometry and on the asymmetric geometry with the thin-film attachment, respectively. Figure 1 gives a qualitative analysis of the film, i.e. the presence of both amorphous and crystalline fractions are observed, but it is impossible to carry out the quantitative XRD analysis of the film. The fraction of amorphous phase cannot be determined since the contribution of the amorphous phase to the signal is completely screened by the signal from the glass substrate. In addition, the symmetric geometry cannot detect those crystallites that do not have their planes parallel to the sample's surface. As figure 2 demonstrates the asymmetric geometry provides more information about the film. More diffraction patterns from silicon crystallites were detected and the signal from the substrate was weaker which allowed us to separate the signal from the amorphous phase of the film from the signal from the glass substrate. In this case also the quantitative XRD analysis is possible because the amorphous and polycrystalline phases in the film could be separated as shown in figure 6. The quantitative analysis of this Si:H film becomes clearer after presenting the results of the XRD analysis on pc-Si:H films that contain only amorphous phase.

The asymmetric geometry configuration was used to investigate the structural differences between the pc-Si:H films deposited at different R. Figure 3 shows an example of the raw XRD pattern measured on a pc-Si:H film deposited on a c-Si substrate at R = 20. The sharp peaks on the black line belong to the Laue diffraction from the c-Si substrate created by the continuous X-ray radiation, which was not fully attenuated by the beta filter. The initial processing of the XRD patterns that includes background determination and subtraction was done using the X'pert HighScore plus software. For determining the average size of the ordered domains and micro-strains we used a procedure utilizing the integral breadth of a diffraction line. This procedure is based on a Voigt function applied to the breadths of the diffraction line as proposed by Langford [6,7]. This procedure can only be used for symmetric line profiles. In case of an asymmetric profile, prior to the line profile analysis, it was fitted using the least squares method with two symmetric Pearson VII functions. Figure 4 shows the FSP of the film, corrected for the background, and the fit of the data with two Pearson VII functions. It was found that the FSP peak of all films of the dilution series could be fitted with a large-area peak centered around 27.5 degrees and a smaller peak centered around 32.5 degrees. In figure 4 the XRD patterns for crystalline silicon (Si) and crystalline silicon hydride (Si_4H) standards are included. These XRD standards were constructed from the ICDD PDF 2 data file (ICDD – International Centre for Diffraction Data, PDF – powder diffraction file). Comparing the standards to the measured XRD patterns we find a striking match between the position of the two Pearson VII functions and the silicon hydride lines, while the silicon (111) line clearly does not correspond to the center position at 27.5 degrees. Given the controversy regarding the origin of the a-Si:H FSP [8], this result provides an interesting new interpretation of the ordered domains in the amorphous matrix. A good agreement between the position of the peaks of the measured lines and the standard XRD patterns of the silicon hydride provides evidence that the diffracting signal comes mainly from ordered domains of a tetragonal silicon-hydride [9] and not from ordered domains of silicon.

For all films, the 2θ positions of the strongest peak in our XRD patterns, corresponding to the silicon hydride (001) line in figure 4, are between 27.49 and 27.59 deg. These positions result in inter-planar spacings of 0.3245 to 0.3233 nm. For the tetragonal silicon hydride lattice the reference values of the 2θ peak position and inter-planar spacing of the (001) line are 27.53 deg and 0.3238 nm, respectively [9]. The results from our measurements match well with the reference values of the (001) line of silicon hydride lattice. The lattice spacings represent distances between the lattice planes occupied mostly by silicon atoms. Hydrogen atoms, when they are incorporated in the lattice structure (for example as hydrides), can influence the lattice parameters of the structure and cause the observed deviations.

The dominant peak at 27.5 degrees was used for the further analysis of the structural properties. Figure 5 shows the FWHM values of this peak as a function of the hydrogen dilution. The figure demonstrates that the FWHM decreases with increasing R from 0 to 20, and remains nearly constant when R is further increased from R = 20 to R = 40. This confirms previous reports about the narrowing of the FSP [10], but shows for the first time that this effect saturates for a certain value of R.

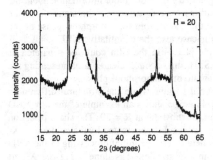

Figure 3: Raw XRD pattern measured on a 300 nm thick pc-Si:H film deposited at R=20 on a c-Si substrate.

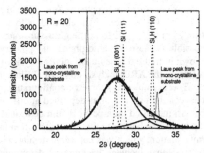

Figure 4: XRD patterns of the FSP after background correction. Also shown are the XRD lines for crystalline Si and Si_4H standards.

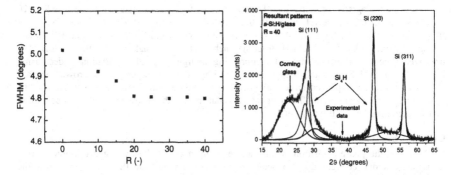

Figure 5: FWHM values of the FSP as a function of R of the pc-Si:H films deposited on a c-Si substrate.

Figure 6: XRD patterns of the FSP after background correction for a Si:H film deposited on a glass substrate at R = 40.

Degradation characteristics of solar cells

The a-Si:H films prepared from hydrogen diluted silane were implemented as absorber layers in p-i-n solar cells. The solar cells were subjected to a degradation experiment. Figure 7 shows the evolution of the efficiency and the fill factor of the solar cells with exposure time. In both figures the parameters are normalized to their initial values before degradation. The degradation experiment confirms that the cells with absorber layers deposited using hydrogen dilution are more stable to light exposure. A clear reduction of the degradation is already observed when the dilution ratio is increased from R = 0 to R = 10 and the degradation is further suppressed when R is increased to 20. However, solar cells with absorber layers prepared at R > 20 exhibit similar degradation behavior as the cell with the absorber layer prepared at R = 20. The efficiency of the solar cells with absorber layers prepared at R ≥ 20 stabilizes at around 88% of their initial efficiency.

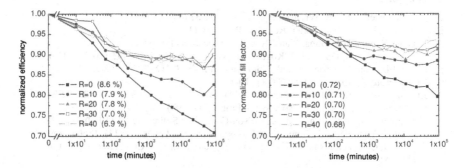

Figure 7: Normalized η and ff as a function of illumination time of solar cells with absorber layers deposited at different R. The initial values of η and ff are given in brackets.

The similar degradation behavior of solar cells with absorber layers deposited at R $\geq$ 20 indicates that there is a correlation with the structural properties observed for the individual films. The saturation of the FWHM value of the FSP for pc-Si:H films prepared at R $\geq$ 20 is a strong indication that the structural order has achieved an optimal state in these films. The improved structural order contributes to suppressing the light induced degradation by decreasing the amount of weak Si-Si bonds which are removed by hydrogen etching of the growing surface [11]. However, the improved structural order in amorphous films still does not prevent the occurrence of the degradation. The mechanisms that initiate the degradation have to be further investigated in order to fully prevent it.

CONCLUSIONS

Although the phase transition is the most evident effect of the hydrogen dilution in Si:H films, more subtle changes do also occur in the amorphous silicon that is grown below the threshold thickness before the phase transition. The detailed XRD analysis of the pc-Si:H films revealed a narrowing of the FSP of the XRD patterns with increasing hydrogen dilution. For the first time it is demonstrated that this effect saturates and an increase of the hydrogen dilution above a certain value does not result in a further narrowing of the FSP. The most important result from the XRD analysis of the pc-Si:H films is the presence of ordered domains of tetragonal silicon hydride in the films. The XRD pattern is obtained from diffracting domains of tetragonal silicon hydride and not from cubic silicon.

A degradation experiment demonstrated the improved stability of the cells with absorber layers prepared at higher hydrogen dilution. However, the degradation of solar cells with absorber layers prepared at R > 20 is not further reduced with a further increase of R. This result indicates that there is a strong link between the degradation behavior of the solar cells and the medium range order of the individual films (estimated from the FSP width of the XRD pattern), which also saturates for films prepared at R > 20.

ACKNOWLEDGEMENTS

This work was supported by Delft University of Technology and by the Ministry of Education, Youth and Sports of the Czech Republic under the Project No. 1M06031.

REFERENCES

1. D. L. Staebler and C. R. Wronski, Appl.Phys. Lett. **31** (4), 292 (1977).
2. L. Yang and L. F. Chen, Mater. Res. Soc. Proc. **336** (1994) 669.
3. J. Yang, X. Xu and S. Guha, Mater. Res. Soc. Proc. **336** (1994) 687.
4. C.R. Wronski, J.M. Pearce, R.J. Koval et al., Mater. Res. Soc. Proc. **715** (2002) A13.4.
5. G. van Elzakker, V. Nádaždy et al., Thin Solid Films **511-512** (2006) 252-257.
6. J.I. Langford: J. Appl. Cryst. (1978) 11, 10-14.
7. J.I. Langford, A. Boultif, J.P. Auffrédic and D. Louër: J. Appl. Cryst. (1993) 26, 22-33.
8. D.L. Williamson, Mater. Res. Soc. Symp. Proc. **559** (1999) 251.
9. S. Minomura, K. Tsuji et al., Journal of Non-Crystalline Solids **35&36** (1980) 513-518.
10. S. Guha, J. Yang, D.L. Williamson et al., Applied Physics Letters **74** (13), 1860 (1999).
11. C. C. Tsai, G. B. Anderson, R. Thompson et al., Journal of Non-Crystalline Solids **114** (1989) 151.

Mater. Res. Soc. Symp. Proc. Vol. 1153 © 2009 Materials Research Society   1153-A18-03

First-Principles Modeling of Structure, Vibrations, Electronic Properties and Bond Dynamics in Hydrogenated Amorphous Silicon: Theory Versus Experiment

A.I. Shkrebtii[1,*], I.M. Kupchak[1,2], and F. Gaspari[1]

[1]University of Ontario Institute of Technology, Oshawa, ON, L1H 7L7, Canada
[2]V. Lashkarev Institute of Semiconductor Physics NAS, Kiev, 03028, Ukraine

ABSTRACT

We carried out extensive first-principles modeling of microscopic structural, vibrational, electronic properties and chemical bonding in hydrogenated amorphous silicon (a-Si:H) in a wide range of hydrogen concentration and preparation conditions. The theory has been compared with experimental results to comprehensively characterize this semiconductor material. The computer modeling includes *ab-initio* Molecular Dynamics (MD), atomic structure optimization, advanced signal processing and computer visualization of dynamics. We extracted parameters of hydrogen and silicon bonding, electron charge density and calculated electron density of states (EDOS) and hydrogen diffusion. A good agreement of the theory with various experiments allowed us to correlate microscopic processes at the atomic level with macroscopic properties. Here we focus on correlation of the amorphous structure of the material, atom dynamics and electronic properties. These results are of increasing interest due to extensive application of a-Si:H in modern research and technology and to the significance of detailed understanding of the material structure, bonding, disordering mechanisms and stability.

INTRODUCTION

The utilization of sophisticated computer modelling to analyze structure and dynamical processes in solid state, including amorphous materials, allows the uncovering of many of their characteristics which are difficult or even impossible to isolate and analyze experimentally[1,2]. Hydrogenated amorphous silicon (a-Si:H) has been the subject of intensive investigation for over 30 years; there exists extensive literature covering all of its most important properties and applications[3,4]. The main role of hydrogen in amorphous silicon is to passivate the Si dangling bonds (DBs) to restore a proper energy gap and semiconducting properties, thus enabling application of a-Si:H in microelectronics and photovoltaics. Due to the importance of hydrogen, many experimental methods have been used to characterize DB passivation, H-Si bonding and related mechanisms of degradation. Among the numerous experimental techniques to investigate a-Si:H and the role of hydrogen, Fourier Transform Infrared Spectroscopy (FTIR) is used extensively to analyze vibrational spectra of a-Si:H. Although FTIR represents one of the most common and powerful characterization techniques, no microscopic links between the observed vibrational spectra and the microscopic properties of a-Si:H have been established. Other important experimental techniques such as Isothermal Capacitance Transient Spectroscopy (ICTS) and Constant Photocurrent method (CPM) are widely used for the determination of EDOS, (for a review of these techniques, see [4]). It is important, however, to theoretically model such electron properties to be able to correlate preparation conditions with the material's quality and stability.

On the other hand, a variety of experimental techniques and numerical methods were used not only to characterize the material, but to prove that a computer simulation produced a system that corresponds to the experimental reality. For instance, we have proved recently[5,6] by extensive ab-initio simulation of hydrogenated amorphous silicon that the commonly used radial distribution function cannot ensure reproducibility of macroscopic experimental features of a-Si:H. Instead, the derivation of realistic vibrational spectra is necessary for the validation of a particular numerical model. To this aim we have developed a comprehensive numerical algorithm that not only extracts the vibrational spectra[7], but more importantly, gives unprecedented access to microscopic processes in non-crystalline materials and interpretation of the experimental results. In this paper, by applying this and other numerical formalisms to hydrogenated amorphous silicon, we have decoded the EDOS in terms of chemical bonding, structure, its stability, and defect formation, which are crucial to the understanding of the complex nature of this important material. The calculated results are in a good agreement with experiment.

THEORY

We have used Car-Parrinello ab initio molecular dynamics[8] to simulate a-Si:H with different hydrogen concentrations within the Density Functional Theory (DFT) as implemented in the software package Quantum-Espresso[9]. The 64 Si atom supercell and hydrogen atoms added has been used throughout the MD runs. Kohn-Sham orbitals have been expanded in a plane wave basis set using an energy cutoff of 12 Ry, and the Brillouin zone of the supercell lattice has been sampled by the Γ-point. A few runs with higher cutoff up to 20 Ry have been performed to confirm that the simulations at 12 Ry are sufficiently accurate to describe the dynamics of both H and Si atoms. In addition a 128 Si atom supercell with 12 Ry cutoff have been stimulated as well. We have found that the vibrational frequencies, atomic distribution and DOS for 64 and 128 Si atom supercells are practically identical, thus verifying that a 64 atom supercell is sufficient to produce accurate results. The equations of motion were integrated using Verlet algorithm with a discrete time step of 0.24 fs. The Nosé [9] thermostat has been used to control the temperature of the supercells during the melting and slow cooling down. Hydrogenated amorphous silicon was obtained from crystalline 64 Si atom supercell by randomly adding different number of hydrogen atoms and further melting that followed by slow cooling to room temperature and subsequent annealing. Given numbers of H atoms correspond to the following atomic concentrations: 4 – 6%, 6 – 9%, 8 – 11%, 10 – 14% and 12 – 16%, respectively. For each concentration a few different samples were prepared with different initial configurations of hydrogen. The samples were melted at 3000 K or 5000 K typically for 5 ps. After equilibration of the molten system for 5-10 ps, the samples were cooled down to zero temperature with different cooling rates: 5 K per each 300 or 500 MD steps, corresponding to $(4 - 7) \times 10^{14}$ K/sec. As we have found, the melting time is not very important because the molten system at 3000 K or 5000 K thermalized very rapidly (within 2 – 5 ps) and being further thermalized during a slow cooling until the temperature drops to 1200K. The choice of cooling rate is defined by requirements to obtain a "quality" amorphous sample, which produced different a-Si:H characteristics comparable to those experimental ones. If cooling rate is too high, the system did not have time enough to solidify properly, and such very rapid cooling is inconsistent with technological process of amorphous silicon fabrication.

Both melting and cooling processes were carried out with temperature control via ionic Nose thermostat[9]. However, we carried out MD simulations necessary to collect statistical data for calculation of vibrational spectra, radial distribution functions, *etc.*, without temperature control. After the cooling finished, the system was geometrically optimized to zero temperature, and then the atom positions in these structures were slightly distorted from the initial geometry to initiate the MD at the given temperature. The MD temperature of the systems was stable in most simulations. However, if structural changes occurred in the system, changes in the temperature up to ±150°C were observed in some cases. Such changes happened usually when we annealed the supercell to higher temperature, which agrees with experimental finding. After the system was structurally changed during the annealing, the temperature keeps stable, or slowly decreases.

The following methods of MD trajectories analysis have been used: *Visual analysis.* To understand changes of atomic structure for the numerically generated samples and the processes, which take place during the MD simulation, computer animation has been successfully used. This allowed us to draw important conclusion about the quality of the amorphous structure, the short range symmetry, the hydrogen complexes present in the system, their modifications, *etc.* Also computer animation made it possible to track structural changes such as those due to diffusion, motion of particular atoms or complexes with time. *Structural analysis.* For comparison of the numerical structures with experimental ones, the radial distribution functions (RDF), the coordination numbers and the mean square displacement of each sort of atoms have been calculated. For instance, a pair-correlation function $g(r)$ is a probability to find one atoms on the distance r from the second one with respect to the corresponding probability for the same density of ideal gas.

RESULTS AND DISCUSSION

We were able to generate for the first time all possible complexes, such as monohydride SiH, dihydride SiH_2, trihydride SiH_3, and "hydrogen molecule" H_2. (This is not a real molecule, since H_2 complex has different interatomic separation and vibrational frequencies compared to free H_2). In most cases such complexes are stable at room temperature during the MD runs for 10 ps or even longer. We have to stress that these complexes were created in the "natural way" as the result of random atomic motion during the melting, in contrast to other way of creating amorphous hydrogenated Si artificially (see, for instance, [10]). In our numerical experiment the complexes were created in the amorphous silicon after the samples were slowly cooled down to room temperature from the melt and all of them are demonstrated in figure 1. The "hydrogen molecule" H_2, dihydride SiH_2 and trihydride SiH_3 complexes have been proposed experimentally (see, for instance, [11]) but previously were not numerically obtained using computer simulations. In agreement with experimental conclusions, SiH complexes were created most often and these monohydrates were present in the all samples, shown in figure 1.

 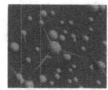 

Figure1. Examples of hydrogen complexes in the amorphous silicon, created numerically. Red balls depict hydrogen atoms, cyan balls are silicon atoms, and only SiH bonds are shown. The

left panel shows two dihydrides SiH_2 as well as monohydrides SiH; in the center a trihydride bond and monohydrates are shown; the hydrogen "molecule" is shown in the right panel.

In the table below we summarize all types of hydrogen complexes created numerically in our computer simulations for the *same* hydrogen concentration, namely $Si_{64}H_{10}$ supercells. These different complexes were obtained at different initial atomic configurations and preparation conditions. This is one important confirmation that the creation of such complexes is a random process, and complexes stability directly indicates their realistic surroundings.

Table 1. Various complexes, obtained theoretically using $Si_{64}H_n$ supercells (n=4 – 12).

H amount (%)	SiH	SiH_2	SiH_3	H_2	Si_2H
$Si_{64}H_4$ (6%)	x	-	-	-	-
$Si_{64}H_6$ (9%)	x	-	x		x
$Si_{64}H_8$ (11%)	x	-	-	x	-
$Si_{64}H_{10}$ (14%)	x	x	x	x	-
$Si_{64}H_{12}$ (16%)	x	-	-	-	x

In addition to standard complexes, table 1 shows a new Si_2H complex, the so-called "jumping atom" – a hydrogen atom that is shared between two neighboring silicon atoms. Si-H-Si was proposed in [3], and such complexes have been found in two different numerical samples. Being stable during simulations, the jumping atom complex closely resembles a bond centered hydrogen state [3]. However, detailed analysis of the H atom trajectory and the electron charge density [6] indicates that the jumping H atom actually forms Si-H bond with either first or second Si atoms one at a time, rather than sharing the charge density between these two Si atoms. This H atom spends most of the time in the vicinity of one or another Si, while regularly switching between them. Being very important from the point of view of light-soaking degradation [12], such unstable (jumping atom) complex poses a serious numerical challenge to extract vibrational density of stated (VDOS) due to the hanging frequency of the hydrogen vibration. Nevertheless, our new VDOS approach allowed for a successfully overcoming of this problem [6,7].

Figure 2. Sample $Si_{64}H_{10}$ before (a) and after (b) annealing. One SiH_2 complex, seen as the left in (a), breaks and forms two monohydrates SiH (b).

After samples have been cooled down, some of them were annealed at 150 – 500°C for a long time (10-40 ps). The annealing influenced the sample structure, *e.g.*, resulting in hydrogen or silicon diffusion in some cases, while in other cases no changes occurred. A diffusion process has been observed in the $Si_{64}H_{12}$ structure, where the silicon atom "passed" its bonded hydrogen atom to another silicon atom, that is, the H was shifted by a distance comparable with interatomic distance and such configuration remained stable during the following 30 ps. In other cases processes of dissociation of SiH_3 onto SiH_2 and SiH, dissociation of SiH_2 onto two SiH occurred randomly. One of such processes is demonstrated in figure 2.

To characterize the modification of the electronic properties, the EDOS of the amorphous structure was calculated before and after annealing. To optimize the method, two samples were created using the standard approach (from melt): $Si_{128}H_{16}$ and $Si_{64}H_8$, corresponding to the hydrogen atomic concentration of 11%. We have found that the EDOS for both of these samples were very similar: therefore we show here our results obtained for the 64-atom unit cells.

To simulate conditions responsible for the Staebler-Wronski[12] effect, we started with an amorphous silicon sample $Si_{64}H_{10}$, containing two dihydrides and six monohydrates (sample A). Then, two different samples were created by removing two hydrogen atoms, so that one sample contained 8 monohydrates (sample B), and the second contained 4 monohydrates and 2 dihydrides (sample C). These samples were geometrically optimized to zero temperature and MD runs at room temperature were performed to collect statistical data. At the same time, all of these samples were annealed at a temperature between 600K and 800K for approximately10 ps. Finally, geometry optimization and MD runs were performed once again. All characteristics calculated for the system before annealing are indicated as "ba" and as "aa" for annealed samples. Details of the simulations for all three samples will be discussed elsewhere[6]; here we demonstrate the angle distribution function (ADF) (figure 3) and EDOS (figure 4) modifications in the process of annealing only for sample (A). This sample contained six monohydrides and two dihydrides. During the annealing, one of the dihydrides has been split into two monohydrates, and one monohydrate changed its Si host atom.

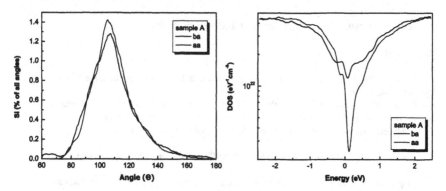

Figure 3. ADF for Si for sample A   **Figure 4**. The EDOS for sample A.

While modification of the ADF is not significant, EDOS and VDOS[7] are modified essentially during the annealing in very good agreement with the experimental finding.

CONCLUSIONS

We simulated amorphous Si in a range of hydrogen content up to 20%, with different cooling and annealing rates and temperatures. The technologically important correlations of the hydrogen vibrations with macroscopic properties were established in a-Si:H at different temperatures. In particular, using computer animation and by comparing our theory with infrared spectroscopy experiments we interrelated for the first time the vibrational spectra of a-Si:H, hydrogen migration and radial distribution function with the preparation conditions, and annealing. We have identified vibrational signatures of hydrogen instability and its rebonding in an amorphous network, and formation of mono-, di- and tri-hydride Si-H complexes. Finally, a comparison with the experimental data provided a wealth of information regarding the hydrogen diffusion, and the quality of a-Si:H. We also demonstrated that this method offers the possibility of accessing other important a-Si:H macroscopic characteristics and its stability in general, including the correlation of a-Si:H dynamics with the detrimental Staebler-Wronski effect. Finally, this novel parameter-free modeling technique allows an extension to an analysis of the structural, dynamical and electronic properties of a wider class of amorphous semiconductors and non-crystalline materials.

ACKNOWLEDGMENTS

The research was supported by the Centre for Materials and Manufacturing/Ontario Centres of Excellence (OCE/CMM) "Sonus/PV Photovoltaic Highway Traffic Noise Barrier" project, Discovery Grants from the Natural Sciences and Engineering Research Council of Canada (NSERC) and the Shared Hierarchical Academic Research Computing Network (SHARCNET).

---

* Corresponding author, e-mail address: Anatoli.Chkrebtii@uoit.ca

REFERENCES

1 M. C. Payne, M. P. Teter, D. C. Allan, T. A. Arias, and J. D. Joannopoulos, Rev. Mod. Phys. **64**, 1045 (1992).
2 M. Griebel, S. Knapek, and G. Zumbusch, *Numerical Simulation in Molecular Dynamics: Numerics, Algorithms, Parallelization, Applications* (Springer Verlag, Berlin, 2007).
3 R. A. Street, *Hydrogenated Amorphous Silicon* (Cambridge Univ. Press, Cambridge, UK, 1991).
4 T. Searle, *Amorphous Silicon and its Alloys* (INSPEC, London, 1998), chapters VI-VII, p. 271-337.
5 F. Gaspari, I. M. Kupchak, A. I. Shkrebtii, and J. M. Perz, Phys. Rev. B, accepted (2009).
6 I. M. Kupchak, A. I. Shkrebtii, and F. Gaspari, private communication.
7 I. M. Kupchak, F. Gaspari, A. I. Shkrebtii, and J. M. Perz, J. Appl. Phys. **103**, 123525-1 (2008).
8 R. Car and M. Parrinello, Phys. Rev. Lett. **55**, 2471 (1985).
9 S. Nose, J. Phys.: Condens. Matter **2**, SA115 (1990).
10 T. A. Abtew and D. A. Drabold, Phys. Rev. B **74**, 085201 (2006).
11 G. Lucovsky, R. J. Nemanich, and J. C. Knights, Phys. Rev. B **19**, 2064 (1979).
12 D. L. Staebler and C. R. Wronski, Appl. Phys. Lett. **31**, 292 (1977).

Mater. Res. Soc. Symp. Proc. Vol. 1153 © 2009 Materials Research Society   1153-A18-05

Defects in hydrogenated amorphous silicon carbide alloys using electron spin resonance and photothermal deflection spectroscopy

Brian J. Simonds,[1] Feng Zhu[2], Josh Gallon[1,2], Jian Hu[2], Arun Madan[2], P. Craig Taylor[1]
[1] Physics Department, Colorado School of Mines, Golden, CO 80401, U.S.A.
[2] MV Systems, Inc., Golden, CO 80401, U.S.A.

ABSTRACT

Hydrogenated amorphous silicon carbide alloys are being investigated as a possible top photoelectrode in photoelectrochemical cells used for hydrogen production through water splitting. In order to be used as such, it is important that the effects of carbon concentration on bonding, and thus on the electronic and optical properties, is well understood. Electron spin resonance experiments were performed under varying experimental conditions to study the defect concentrations. The dominant defects are silicon dangling bonds. At room temperature, the spin densities varied between 10^{16} and 10^{18} spins/cm^3 depending on the carbon concentration. Photothermal deflection spectroscopy, which is an extremely sensitive measurement of low levels of absorption in thin films, was performed to investigate the slope of the Urbach tail. These slopes are 78 meV for films containing the lowest carbon concentration and 98 meV for those containing the highest carbon concentration.

INTRODUCTION

Amorphous silicon carbide (a-SiC:H) has long been studied as a promising material for use in solar cells as a result of its tunable bandgap. Recently, interest has been extended to include use as a photoelectrode in photoelectrochemical cells for hydrogen production by splitting water [1-4]. For amorphous silicon (a-Si:H), several generally held principles have been developed relating film properties and eventual device performance made from such films. However, in the case of a-SiC:H, relationships between basic material characterization techniques and device performance have not been made explicit. In order to do this, we must first identify the parameters in the films we wish to eventually use as absorber layers in a *pin* solar cell. In this work, we investigate the opto-electronic properties of the films with photothermal deflection spectroscopy (PDS) and electron spin resonance (ESR). PDS will allow us to measure sub-bandgap absorption as well as give us an indication of the density of states from disorder and defect levels. ESR measurements will allow us to ascertain bonding structure resulting from the incorporation of carbon.

EXPERIMENT

A series of a-SiC:H films were made at MVSystems in Golden, CO by plasma enhanced chemical vapor deposition (PECVD) in a mixture of SiH$_4$, CH$_4$, and H$_2$. The amount of carbon was controlled by varying the amount of methane in the gas mixture with methane gas ratios

[CH$_4$/(SiH$_4$ + CH$_4$)] varying from 0.20 to 0.55. The bandgap, conductivity, and the parameter gamma (γ) were tested on Corning glass (type 1737) substrate. Gamma (γ) is related to the photoconductivity by $\sigma_{ph}=\sigma_0F^\gamma$ where σ_0 is a constant of proportionality and F is the intensity of the illumination [8]. Out of this series of films, we chose three to characterize as they make reasonable absorber layers in *pin* solar cell devices. For these samples, the SiH$_4$ flow was 20 sccm and the H$_2$ flow was 100 sccm. The methane was diluted 20% in H$_2$ which meant that the actual amount of H$_2$ varied from 120 sccm to 140 sccm. It was shown separately that this amount of change in the H$_2$ flow rate did not appreciably alter the bandgap. During PECVD deposition the pressure was 550 mTorr, the power 20 W and the substrate temperature 200 °C. These films were deposited on thin (~1mm), high quality quartz chosen so that PDS and ESR could be performed. The concentration of carbon in these films was measured with energy dispersive x-ray spectroscopy on a JEOL JSM-7000F field emission scanning electron microscope with an EDAX Genesis energy dispersive x-ray spectrometer. The atomic percentage of carbon was 6%, 9%, and 11% corresponding to methane gas ratios of 0.20, 0.29 and 0.33 respectively.

To measure absorption in these films, we employed a technique known as photothermal deflection spectroscopy (PDS) [5]. This is an extremely sensitive technique that allows one to measure absorption several orders of magnitude below that of standard reflection and transmission techniques. Subsequently one can measure sub-bandgap absorption resulting from states created by structural disorder or defect levels within the bandgap. Thus, an estimate of defect density can be calculated by the method introduced by Jackson and Amer [6]. According to this method the absorption resulting from defect bands is singled out and then integrated and related to a defect density by a constant.

The final characterization tool employed was electron spin resonance (ESR) in order to quantify the spin densities in these films. The data were taken on a Bruker 9.5 MHz X-band spectrometer at room temperature. ESR is sensitive only to paramagnetic defects which, in these films, are commonly accepted to be the neutral silicon dangling bonds [13]. Spin density calculations were made by double integration of the signal and then comparing this to a spin standard of known defects.

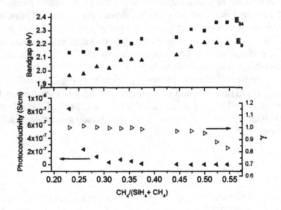

Figure 1. The Bandgap, photoconductivity, and gamma variation with methane gas ratio. In the bottom plot, the points of the symbols point to the appropriate ordinate.

RESULTS AND DISCUSSION

In the upper plot of Figure 1, we show the optical bandgaps for the series of a-SiC:H films. The E_{04} values are defined as the energy at which the absorption coefficient $\alpha = 10^4$ cm^{-1}, which is a standard metric for disordered films. The curve labeled E_g is the value obtained based on the method prescribed by Tauc [7] where simple, parabolic energy bands are assumed. This assumption is generally only believed to be valid in the region $\alpha > 10^4$ cm^{-1}. By fitting the data in this region E_g is extracted by using the equation $(\alpha E)^{1/2} = C(E - E_g)$ [8]. As is typical, these values are approximately 0.15-0.20eV lower than the E_{04} values.

As shown in the lower plots in Figure 1, σ_{ph} decreases with increasing methane gas ratio implying an increase of the recombination centers in the material which is in contrast with the interpretation when $\gamma > 0.5$ [8]. However σ_{ph}, as measured under white light illumination , does not take into account the increase of E_g and a concomitant reduction in the absorption coefficient. In order to evaluate this further, an appropriate way is to measure the photocurrent, I_p, at a long wavelength λ, (550nm), when there should be uniform bulk absorption. The photocurrent can be expressed as,

$$I_p = e\, N_{ph}(\lambda)\, (1-R_\lambda)\, [1-\exp(-\alpha_\lambda\, d)]\, \eta \bar{\tau}/t_t, \tag{1}$$

where $N_{ph}(\lambda)$ is the photon flux, R_λ is the reflection coefficient, α_λ is the absorption coefficient, d is the film thickness, η is the quantum efficiency of photo generation, $\bar{\tau}$ is the recombination lifetime and t_t is the transit time. Assuming that η, $\bar{\tau}$, t_t, and $(1-R_\lambda)$ are constant for different films (i. e. different E_g), then to a first order approximation, $X = I_p/[1-\exp(-\alpha_\lambda\, d)]$, can account for the changes in the absorption coefficient as E_g increases [8]. This is shown in reference [1] and it is noted that the normalized current, X, does not change significantly as E_g increases. This is in contrast to the decrease in σ_{ph} with E_g shown in Figure 1. This then suggests that the DOS has remained low and is consistent with $\gamma > 0.9$ (low DOS) throughout methane gas ratio up to 50%.

Photothermal Deflection Spectroscopy

Figure 2 shows the absorption coefficient of each of the three chosen films with carbon concentrations of 6, 9, and 11 atomic %. The signal seen here is a convolution of optical absorption from every possible electronic region including extended, localized and deep defect states. In the linear region between about 1.7eV – 2.1eV, the absorption coefficient primarily results from localized to extended state transitions and is known as the Urbach tail. This region can be described by $\alpha = \alpha_0 \exp(E/E_0)$ where E is the excitation energy and E_0 is the Urbach energy which is the inverse slope of the data when plotted versus $\ln(\alpha)$. Since the absorption coefficient here directly depends on the density of localized states, E_0 is considered to be a measure of the amount of disorder [9]. Their bandgap values are presented in Figure 1 as previously discussed and E_0 is 78, 85, and 98 meV for carbon concentrations of 6, 9, and 11%, respectively. For comparison, a typical value for device grade a-Si:H is ~50 meV [8]. As the carbon concentration increases, so too does the value of E_0. This is expected as the density of localized states is increasing with more disorder created by introducing more carbon. Also, there

373

is an increase in the bandgap from E_{04} = 2.06 eV to 2.18 eV with carbon concentration. This is known to be a result of at least some of the carbon being incorporated in the form of sp^3 carbon which is essentially an insulator [10]. The feature at 0.88eV in Figure 2 is an overtone of an O-H vibrational stretch mode from the quartz substrate.

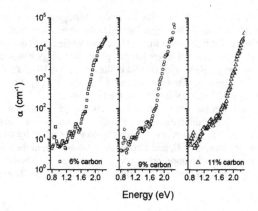

Figure 2. Absorption coefficient curves of three a-SiC films measured by PDS.

Electron Spin Resonance

In Figure 3 the ESR data of these films are shown. These were taken from single films

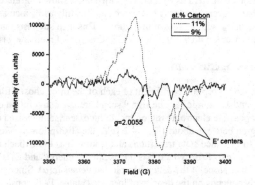

Figure 3. ESR spectra shown as the first derivative of the absorption. The main feature is a silicon dangling bond at g=2.0055. The feature around 3385G is a an E' center resulting from the quartz substrate.

~1μm thick on quartz substrates. The main feature at around 3375G, corresponding to a g-value of 2.0055, is attributed to an unpaired electron in a silicon dangling bond and is the main defect in these films. From comparison with a standard sample, we obtain defect densities of $8x10^{17}$ cm^{-3} and $2x10^{17}$ cm^{-3} for the samples with the most and second most carbon, respectively. The 6% carbon sample data is not shown as there was no signal above the noise. Thus we can only place an upper bound on the spin density of $9x10^{16}$ cm^{-3}. The feature at a slightly higher field value than the main peak is an E' center which results from the interface at the quartz substrate where the silicon atom is attached to 3 oxygen atoms.

What these spectra show is that as the carbon concentration increases so too does the defect density in the films. Based on comparisons to other research [10], we believe that the carbon is incorporated into our films as a mixture of sp^2 and sp^3 carbon. As a result, some silicon bonds are left unpassivated, thus paramagnetic and detectable with ESR. As more carbon is introduced, the number of dangling bonds increases which is clearly shown in Figure 3. For comparison, device quality amorphous silicon has typical defect densities less than or approximately equal to 10^{16} defects per cm^3 [13].

Defect density comparison

From both the PDS and the ESR data we have estimated an absolute number of defects. It has been shown previously for a-Si:H that these two measurements show a one-to-one correspondence [6]. However, as our data show in Figure 4 there is a diverging disagreement as the methane gas flow ratio increases. Also plotted is the data from Demichelis *et al.* [11] on a-SiC which were had much higher methane flow ratios and thus presumably more carbon. Their data show up to an order of magnitude disagreement between the two methods. Discrepancies between these two methods of defect calculation has been seen before in other a-Si:H alloys as well as in microcrystalline silicon. In general, it is believed that the method of integrating the excess absorption to estimate defect densities from PDS is over simplistic for the more complicated structures created when alloys are added. It has been found that for materials outside a-Si:H that the PDS calculations no longer correlate one-to-one with the ESR data using the same constant of proportionality. As for a-SiC at least one group speculates that this disagreement is a result of the complicated bonding structure resulting form the incorporation of the carbon bond [12].

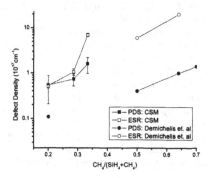

Figure 4. Defect density comparison of data from PDS and ESR. Data for the films in this paper

are given by the squares and data given by Demichelis *et al.* [11] is given by circles.

CONCLUSIONS

By characterizing these a-SiC films we make several observations. First, as evidenced from the PDS data, the Urbach energies are 50% to 100% higher than is typically seen in device grade a-Si:H. This is typically interpreted as an increase in localized states within the bandgap region just above the valence band and below the conduction band resulting from structural disorder. Also, there is an increase in the bandgap with carbon incorporation. In comparison with other research, our results are consistent with the interpretation that carbon is being incorporated as a mixture of sp^2 and sp^3 carbon. As a result of this mixture, increasing densities of silicon dangling bonds are detected with increasing carbon concentration by our ESR measurements. The comparison of these results for spin densities with the defect density calculations from PDS shows a disagreement that is consistent with what other groups have found for a-Si:H alloys. In future work, it will be important to correlate these results of shifting absorption edge and increasing defect densities with solar cell devices made for each film studied here.

ACKNOWLEDGEMENTS

Research supported in part by NREL KXEA-3-33607-36, MVSystems, and the Renewable Energy Materials Research Science and Engineering Center NSF DMR-0820518.

REFERENCES

1. F. Zhu, J. Hu, A. Kunrath, I. Matulionis, B. Marsen, B. Cole, E.L. Miller, and A. Madan, Proc. SPIE, Vol. 6650, 66500S (2007).
2. J. Hu, F. Zhu, I. Matulionis, A. Kunrath, T. Deutsch, L. Kuritzky, E. Miller, and A. Madan, Proc. 23rd European Photovoltaic Solar Energy Conference, 69 (2008)
3. J. Hu, F. Zhu, I. Matulionis, T. Deutsch, N. Gaillard, and A. Madan, MRS Spring meeting (2009) S.3.5
4. F. Zhu, J. Hu, I. Matulionis, Todd Deutsch, Nicolas Gaillard, A. Kunrath, E. Miller and A. Madan, Philosophic Magazine, (to be published)
5. Nabil M. Amer and Warren B. Jackson, *Semiconductors and Semimetals*, vol. 21 part B, edited by Jacques I. Pankove (Academic Press, Inc., Florida, 1984)
6. Warren B. Jackson and Nabil M. Amer, Phys. Rev. B **25**, (1982) 5559
7. J. Tauc et al., Phys. Stat. Sol. **15**, (1966) 627
8. Arun Madan and Melvin P. Shaw, *The Physics and Applications of Amorphous Semiconductors* (Academic Press, Inc., California, 1988)
9. G.D. Cody, et al., Journal de Physique C4, (1981) 301
10. I. Solomon, Applied Surface Science **184**, (2001) 3-7
11. F. Demichelis et al., Applied Surface Science 70/71, (1993) 664-668
12. F. Demichelis et al., Mod. Phys. Lett. **5**, (1991) 285-292
13. Robert A. Street, *Hydrogenated Amorphous Silicon* (Cambridge University Press, Cambridge, 1991)

Poster Session:
Novel Device Applications

Mater. Res. Soc. Symp. Proc. Vol. 1153 © 2009 Materials Research Society   1153-A19-01

Optical Processing Devices for Optical Communications: Multilayered a-SiC:H Architectures

P. Louro[1,2], M. Vieira[1], M. A. Vieira[1], M. Fernandes[1], A. Fantoni[1], G. Lavareda[2], C.N. Carvalho[3]. [1]Electronics Telecommunication and Computer Dept. ISEL, R. Conselheiro Emídio Navarro, 1949-014 Lisboa, Portugal Tel: +351218317290, Fax:+351218317114, plouro@deetc.isel.ipl.pt, [2] CTS-UNINOVA, Monte da Caparica, 2829-516, Portugal. [3] ICEMS, IST, Av. Rovisco Pais, 1049-001, Lisboa, Portugal

ABSTRACT

In this paper three multilayered architectures based on a-SiC:H with voltage controlled spectral selectivity in the visible spectrum range are analyzed. Multiple simultaneous modulated communication channels (red, green and blue or their polychromatic mixtures) were transmitted together at different frequencies. The combined optical signal was analyzed by reading out the photocurrent signal generated by the devices, under different applied voltages. Results show that the multiplexed signal depends on the device architecture and is balanced by the wavelength and transmission speed of each input channel, keeping the memory of the incoming optical carriers. In the single graded p-i'i-n configuration the device acts mainly as an optical switch while in two stacked p-i'-n-(ITO)-p-i-n configurations, the input channels are selectively tuned by shifting between forward and reverse bias. An electrical model, supported by a numerical simulation gives insight into the device operation.

INTRODUCTION

Wavelength division multiplexer (WDM) is a standard technique used in optical communications to enlarge the bandwidth of the transmission channel. For its implementation it is necessary to use a multiplexer device to combine multiple signals into the optical fiber and a demultiplexer device to perform the splitting of the combined signals into the original ones. In long distance communications the infrared window is used in the WDM technique [1]. In short range communications the transmission window lays within the visible spectrum, and thus the multiplexer and demultiplexer devices must operate in this range which demands for the design of adequate mux/demux devices [2, 3]. Some of the applications related to short range optical communication [4] include short range networking, e.g. indoor and LAN applications, the auto industry, consumer electronics, and the medical imaging for image-transfer applications.
In this paper we present a series of wavelength sensitive transducers working as voltage controlled optical filters and optimized to take advantage of the whole visible spectrum.

DEVICE ARCHITECTURE

 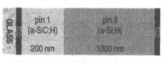

Figure 1 - WDM device architecture: a) pi'in stucture with graded intrinsic layer (NC11); b) stacked pi'n/pin structure (NC10).

The semiconductor sensor element is based on single or stacked a-SiC:H p-i-n structures using different architectures, as depicted in Fig. 1. All devices were produced by PE-CVD on a glass substrate. The semiconductor structures are sandwiched between two transparent conductive oxides (TCO) that constitute front and back electrical contacts.

The common characteristic in all the architectures lays on the composition and geometry of the absorber layers. The a-SiC:H absorbers were optimized for high collection of the blue and high transmittance of the red lights, while the a-Si:H layers were designed to achieve high collection of the red light. In the first configuration (NC11, Fig. 1a) is used a single pi'in photodiode with two different absorbers: a thin layer (200 nm) based on a-SiC:H (i') and a thicker one (1000 nm) based on a-Si:H (i). In the two other configurations (Fig. 1b) it is used a multilayered structure based on two stacked pi'n/pin photodiodes, one with a thin intrinsic absorber based on a-SiC:H and the other with a thicker made of a-Si:H (NC10). In configuration NC12 an additional TCO is placed in-between both photodiodes. In all the configurations the incident light comes from the glass side. Thus, photo generation occurs firstly in the a-SiC:H absorber, and the remaining non-absorbed light goes through the a-Si:H layer.

OPTOELECTRONIC CHARACTERIZATION

The optoelectronic characterization of the devices was performed through spectral response (400-800 nm) measurements under different electrical bias (-10V to +4V) and ac I-V characteristics at different wavelength. In Fig. 2, for NC11 device, it is displayed the spectral photocurrent (a) and the ac current-voltage characteristics under different wavelengths (b). In Fig 3 the same dependence is displayed for NC10 and NC12 devices.

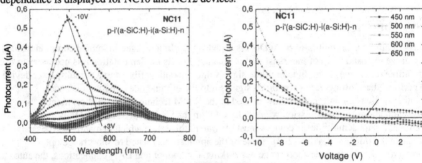

Figure 2 - Spectral response at different applied bias (a) and photocurrent voltage characteristics under different light wavelengths (b) for NC11 device.

Results show that the spectral sensitivity depends on the applied bias, mainly under reverse bias, and is distributed asymmetrically along the visible spectrum. For the p-i'i-n graded cell (NC11) and wavelengths higher than 620 nm the dependence is weak, while for shorter wavelengths an enlargement of the spectrum is observed with an additional peak that emerges from the spectrum, centered at 470 nm. A change in the photocurrent regime is observed, in the short wavelength range, even under negative applied voltages. In the stacked NC10 and NC12 devices, and short wavelengths, there is an enlargement of the spectrum with the increase of the reverse bias. Here, for longer wavelengths the photocurrent becomes independent on the applied bias. The ac I-V curves, as expected, show that at 600 nm and 650 nm the collection efficiency is independent of the applied voltage for the stacked devices, exhibiting a slight variation for the graded pi'in structure. Under shorter wavelengths, in both, the collection increases as the applied voltage changes from forward to reverse, at a rate dependent on the light wavelength. It is interesting to notice that the presence of the transparent ITO layer in-between both diodes do not affect the spectral response and I-V characteristics.

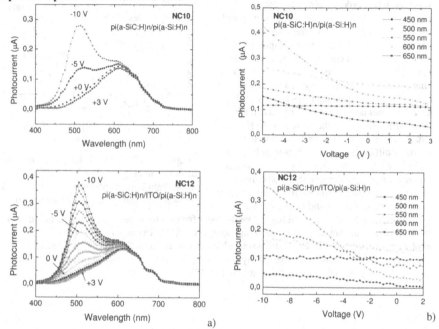

Figure 3 - Spectral response at different applied bias (a) and photocurrent voltage characteristics under different light wavelengths (b) for the stacked architectures.

Data show different behaviors when single and double structures are compared. In the single configuration the electrical field is asymmetrically distributed across the graded i'i -layer leading to a collection efficiency that is dependent on the light depth penetration across it. In the double, the contribution of both front and back diodes is evident (Fig. 3) and the effect of the internal p-n junction crucial on the device functioning and future applications. In the pi'in configuration is possible to reject selectively the different wavelengths by changing the applied voltages (arrows in Fig. 2) while in the double ones, under forward bias the device becomes sensitive to the reddish region and under reverse bias to the blue one working as an selective optical in the visible range.

WAVELENGTH DIVISION MULTIPLEXING

In Fig. 4 and in Fig. 5 are displayed, respectively, the multiplexed signals acquired by using the single pi'in and the stacked pi'n-pin devices, at reverse and forward bias, under monochromatic light bias (R: 650 nm, G: 550 nm, B: 450 nm) and under triple wavelengths combination (RGB: 650 nm/550 nm/450 nm). The light modulation frequency of each light bias was chosen to be multiple of the others in order to ensure a synchronous relation of ON-OFF states of the optical bias along each cycle. The corresponding optical bias states along the cycle are displayed.

The light intensity of each diode (around 15 µW) was adjusted in order to generate similar photocurrents. From Fig. 4 and 5, it is observed that the photocurrent, under the multiple combinations and reverse bias, exhibits eight different levels that correspond each to different optical bias states. Under forward bias the signal amplitude decreases as suggested by the results of the photocurrent-voltage characteristics (Fig. 2). It becomes negative in the graded cell (Fig. 4) due to the inversion of the current flow. Here, simultaneous color and speed transmission becomes difficult; the device acts to only as an optical switch. Results from Fig. 5 show that for the stacked pi'n/pin devices the photocurrent signal at reverse bias exhibits also eight different threshold values corresponding each to the

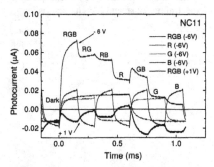

Figure 4 - Multiplexed signals obtained by device pi'in at -6 V and + 1 V under different input light bias: R (650 nm); G (550 nm), B (450 nm) and RGB (650/550/450 nm).

specific conditions of the optical biasing while under forward they are reduced to one half due to the blindness of the device to the blue component (Fig. 3). The extinction of the blue photocurrent signal component under positive bias allows the identification of the three input channels (color and transmission speed) by comparing the different threshold magnitudes of the multiplexed signals under forward and reverse bias [3].

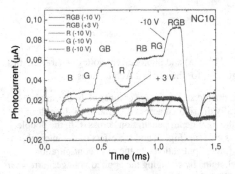 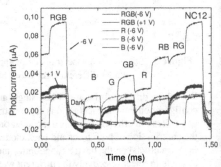

Figure 5 - Multiplexed signals obtained by the stacked pi'n-pin devices at reverse and forward biases under different input light bias: R (650 nm); G (550 nm), B (450 nm) and RGB (650/550/450 nm).

The use of the internal terminal between both photodiodes (NC12) indicates the establishment of better defined photocurrent thresholds and consequently it forecasts an enlargement of the channel spacing, which is prone to reduce the inter-wavelength cross talk and to increase the S/N.

ELECTRICAL MODEL

Based on the experimental results and device configuration an electrical model was developed and supported by a SPICE simulation. To better understand the transient effects due to the time-varying irradiation, the linear state equation [5] was analyzed and used to design the equivalent electric circuit as shown on Fig. 6a. It was taken into account that in the stacked geometry both front and back structures are optically and electrically in series. Under steady state conditions the current (I) and the potential drop across each diode ($V_{1,2}$) will be given, respectively, by eq. 1 [6].

$$I = 0.5 \times \left(-(I_1 + I_2) + \sqrt{(I_1 - I_2)^2 + 4 I_{o1} I_{o2} \exp\left(\frac{V}{\eta V_T}\right)} \right) \qquad V_{1,2} = \eta V_T \ln\left(\frac{(I + I_{1,2})}{Io_{1,2}}\right) \qquad (1)$$

Under transient conditions the circuit becomes a bucket-brigade device. The transfer charge between C1 and C2 is analogous to filling and emptying water buckets. Assuming that the flow of carriers across the internal p-n junction is proportional to the difference in the emitter-base voltages (v_1, v_2) of both transistors, and applying the Kirchhoff laws, the time periodic linearized state equation description is given by:

$$\frac{dv_{1,2}}{dt} = \begin{bmatrix} -\dfrac{1}{r_1 c_1} & \dfrac{1}{r_1 c_1} \\ \dfrac{1}{r_1 c_2} & -\dfrac{1}{r_1 c_2} - \dfrac{1}{r_2 c_2} \end{bmatrix} v_{1,2}(t) + \begin{bmatrix} \dfrac{1}{c_1} \\ \dfrac{1}{c_2} \end{bmatrix} i_{1,2}(t) \qquad i(t) = \begin{bmatrix} \dfrac{1}{r_1} & \dfrac{1}{r_2} \end{bmatrix} v_{1,2}(t) \qquad (2)$$

Where $r_{1,2}$ and $c_{1,2}$ are respectively the dynamic emitter resistances and capacitances of Q1 and Q2. It is a two phase system since two separate pulses, $i_{1,2}(t)$ are needed to provide the input signals. Two square wave current sources with different frequencies, I1 and I2, are used to simulate the input blue and red channels, respectively. The green channel is simulated by two pulsed sources with the same frequencies, I3 and I4, since the green photons are absorbed across both front and back junctions.

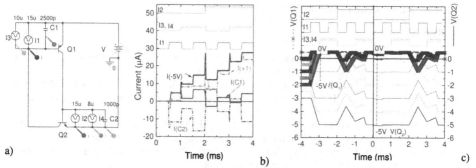

a)   Time (ms)   b)   Time (ms)   c)

Figure 6 - a) Equivalent electric circuit used for simulation proposes. SPICE simulation under red (I2=15uA), blue (I1=15uA) and green (I3=10uA, I4=8uA) pulsed lights and negative and positive applied voltages: b) current; c) voltage drop across Q1 (symbols) and Q2 (lines).

In Fig. 6b) it is shown the transient input channels, the current across the capacitors, IC1, IC2, under negative bias and the multiplexed signals, I, at -5V and +1 V. In Fig. 6c) it is displayed the base-emitter voltage, across the Q1 (symbols) and Q2 (lines) between -5 V and 0 V during the first (t<0) and the second cycles (t>0).

Good agreement between experimental (Fig. 5) and simulated (Fig. 6b) data was observed. Results show that the device stores and transports the minority carriers generated by different current pulses, through the C1 and C2 capacitors keeping the memory of the input channels (color and frequency) acting as a multiplexer device. Under negative bias, once the blue channel is ON, the emitter-base of Q1 becomes optically forward biased and C2 is rapidly charged in inverse polarity of C1 ($i_{C1}(t)C_2 = -i_{C2}(t)C_1$; Fig. 6b) with an input voltage in which a threshold value is inserted for clamping (arrows in Fig. 6c) resulting in a reinforcement of the reverse bias at Q2. The current source keeps filling the capacitors for the duration of the pulse, Δt, of the input channel and the transferred charge between C1 and C2 will reach the output terminal as a capacitive charging current. The presence of the red channel changes in an opposite way the charge of both capacitors. With the green channel ON both front and back contributions are considered in opposite way and the current is the balance between the blue- and the red-like contributions (Fig. 1, Fig. 2).

CONCLUSIONS

Optical processing devices with different multilayered a-SiC:H architectures were compared. Experimental results show that in the single p-i-n architecture the device acts mainly as color sensor while in the stacked ones, the multiplexed signal contains the complete information (color and transmission speed) of each input channel. An electrical model, based on the analysis of the linear state equations, was developed and supports the working functioning of the stacked architectures.

More work has to be done in the optimization of the device configuration in order to enlarge the number of input channels and to improve the frequency response, which are crucial parameters to enhance the bandwidth of the optical transmission system.

ACKNOWLEDGEMENTS

This work supported by POCTI/FIS/70843/2006.

REFERENCES

1. Michael Bas, Fiber Optics Handbook, Fiber, Devices and Systems for Optical Communication, Chap, 13, Mc Graw-Hill, Inc. 2002.
2. S. Randel, A.M.J. Koonen, S.C.J. Lee, F. Breyer, M. Garcia Larrode, J. Yang, A. Ng'Oma, G.J Rijckenberg, H.P.A. Boom. "Advanced modulation techniques for polymer optical fiber transmission". proc. ECOC 07 (Th 4.1.4). Berlin, Germany (2007) 1-4.
3. P. Louro, M. Vieira, M A Vieira, M. Fernandes, A. Fantoni, C. Francisco, M. Barata "Optical multiplexer for short range application", Physica E 41 (2009) 1082-1085.
4. O. Ziemann, J. Krauser, P.E. Zamzow, W. Daum, POF Handbook, "Optical Short Range Transmission Systems", Springer, 2nd Ed., 2008.
5. Wilson J. Rugh, "Linear System Theory" (2nd Edition) Prentice-Hall Information and Systems Science Series, Serie E (1995).
6. M. Vieira, A. Fantoni, M. Fernandes, P. Louro, G. lavareda, C. N. Carvalho "pinpín and pinpii'n multilayer devices with voltage controlled optical readout" Journal of Nanoscience and Nanotechnology, Vol 9, In Press, Corrected Prof, Available online, February 2009.

Mater. Res. Soc. Symp. Proc. Vol. 1153 © 2009 Materials Research Society   1153-A19-02

Fine Tuning of the Spectral Sensitivity in a-SiC:H Stacked p-i'i-n Graded Cells

A. Fantoni[1,2], M. Fernandes[1], P.Louro[1,2], Guilherme Lavareda[2,3], Carlos N Carvalho[3] M. Vieira[1,2]
[1] ISEL- DEETC, Rua Conselheiro Emídio Navarro 1, 1949-014 Lisbon, Portugal
[2] CTS, UNINOVA, Monte da Caparica, Portugal
[3] DCM, FCT-UNL, Monte da Caparica, Portugal

ABSTRACT

It is presented in this work a p(a-SiC:H)/i(a-SiC:H)/i((a-Si:H)/ n(a-Si:H) single junction for application in color sensing domain produced with the PECVD technique. The interest of this device resides in its simplicity of realization and utilization, as it takes advantage from the well known properties of the a-Si:H p-i-n junctions together with the possibility of color filtering by tailoring the optical gap of a-Si:H with the introduction of a small percentage of Carbon. The thicknesses of the i(200 nm) and i (1000 nm) layers are optimized for light absorption in the blue and red ranges, respectively. Measurements of the spectral response under forward and reverse polarization show a dependence of the wavelength of the maximum absorption on the intensity of the applied bias. A comparison of the photocurrent reading with and without a 650 nm background DC optical bias permits a complete separation of blue and red color under reverse and forward applied bias, respectively. The application of the LSP technique (AC regime of the optical bias) permits a complete RGB reading of the incoming light.
Simulation results obtained with the program ASCA will support and explain the measurements about spectral response and photocurrent reading under DC and AC regimes.

INTRODUCTION

Following a great effort in the development of amorphous Si/SiC phototransistors and photodiodes as photo-sensing devices [1], color sensor devices based on a-Si:H technology have been widely studied and presented in literature. Different configurations of the device structure have been presented, ranging from PINIP [2] to NIPIN [3] or PINPIN [4] configuration or to structures based on multiple staked junction [5]. All of them take advantage of the possibility for the optical gap modulation by means of dilution of a small percentage of carbon into a-Si:H for controlling light penetration and absorption into the different active layers of the device. All of those devices have been successfully used as bias-controlled two of three color detectors. While separation of red and blue color is easily achieved through a calibrated application of the external bias (generally with opposite polarity), detection of the green color is often problematic.

In order to achieve a complete RGB mapping we present here a simple p(a-SiC:H)/i(a-SiC:H)/i((a-Si:H)/ n(a-Si:H) single junction prepared by plasma enhanced chemical vapor deposition (PECVD) technology. The color detection is achieved by photocurrent measurement under three different values of the externally applied bias. The results here presented have been obtained under standard DC regime and through the application of the Laser Scanner Photodiode (LSP) technique [6] (AC regime of the optical bias) which lead to a complete three color detection of the incoming light.

DEVICE STRUCTURE, SIMULATION AND CHARACTERIZATION

The device structure is depicted in Figure 1. It is a ITO/p (a-SiC:H)/i'(a-SiC:H)/i((a-Si:H)/ n(a-Si:H)/ITO. The thicknesses of the i' (200 nm) and i (1000 nm) layers are optimized for light absorption in the blue and red ranges, respectively. Transparent contacts have been deposited on both surfaces of the device in order to permit the light to enter from both sides.

Figure 1. Device structure: ITO/p (a-SiC:H)/i'(a-SiC:H - 200 nm)/ i(a-Si:H – 1000 nm)/ n(a-Si:H)/ITO.

In Figure 2 and Figure 3 it is reported the spectral response (SR) as measured experimentally and simulated with the ASCA program [7], obtained by illuminating the structure through the p-layer. A good correlation between the experimental data and the simulation result have been obtained on SR and I-V characteristic, so to permit an insight of the device functioning based on the analysis of the internal electrical configuration obtained with the ASCA simulator. The device responsivity is bias dependent, showing an increasing value in the blue region under the application of the reverse bias. Low responsivity is observed in the red region for both forward and reverse bias. Improvement of the SR in the red region has been observed under large values (10 V) of reverse polarization on the junction.

The signal produced by the device (Is) is obtained by comparing the photocurrent produced by the background light shining through the p-layer (IL) and the photocurrent produced by the same background in superposition with a lower intensity light probe (650 nm) shining through the n-layer (IP). Conceptually it can be reduced to the measure of the photocurrent variation produced by the light probe:

$$I_S = I_L - I_P \tag{1}$$

All the results presented in this section are obtained under stationary condition, so to exclude any capacitive effect of the junction. Results extracted from the simulation and obtained by photocurrent measurement are reported in Figure 4 and 5, respectively. The obtained signal is restricted to the blue region of the background spectrum. The device behaves as a very selective detector of blue light, being completely blind to green/red light. The signal obtained in the blue region enhances with the increasing reverse bias. The explanation for such a behavior resides in the low conductivity of the a-SiC:H layer which results in a very high value of the serial resistance. The red radiation (from the background as well from the light probe) only generates hole-electron pairs into the a-Si:H layer. Even under forward bias, the photogenerated carriers do not arrive to cross the a-SiC:H, where they recombine, and the resulting photocurrent is very

386

low. On the other hand, photogeneration under blue radiation is mainly restricted into the a-SiC:H layer, and the device SR under reverse bias is similar to the one we observed with an a-SiC:H pin photodiode [8]. The superposition of a blue background and a red probe results in a complete depletion of both the i-layers, in a great improvement of the photocurrent and finally in a good value of the produced signal.

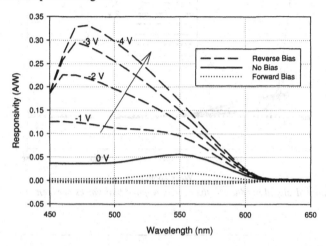

Figure 2. Simulated spectral response of the device. Applied bias varies between -4 and 4 V

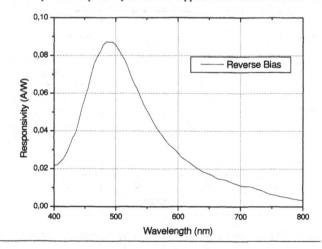

Figure3. Measured spectral response of the device. Reverse bias = - 4 V

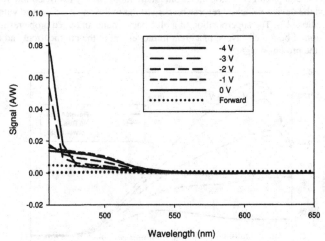

Figure 4. Simulated signal produce by the device. Applied bias varies between -4 and 4

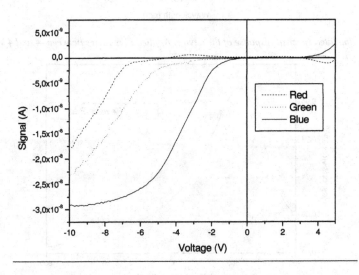

Figure 5.Measured signal (as defined by eq. 1) produce by the device for different background light conditions.

THREE COLOR DETECTION

The measurement we performed under steady state condition, as well as the simulation results, show as this device can be applied, when reverse biased, as a selective detector for radiation in the blue region. The application of the LSP technique (AC regime of the light probe) permits to extend the device performance to a complete RGB mapping of the background light. The form for extracting the signal from the devices is still represented by eq. 1 , but the light probe shining on the back surface of the structure becomes pulsed.

While the behavior under reverse bias is similar to the one observed under stationary condition (i.e. large signal for blue radiation), when the signal is extracted with the application of the LSP technique we observe a selective bias controlled color tuning of the output. Figure 6 reports the photocurrent produced by the light probe obtained with blue (450 nm) green (550 nm) and red (650) background light and compared by the output obtained with no background at all. It can be observed here that by tuning the applied bias it is possible to selectively reduce the output obtained with each one of the background to the one obtained without background (-3 V for green, 0 V for red and 4V for blue). Defining the Iph(-3), Iph(0)and Iph(4) the signal measured for these three values of the applied bias, the RGB mapping can be extracted through the resolution of the linear system:

$$\begin{cases} \alpha_R + \alpha_B = I_{ph}^{(-3)} \\ \alpha_G + \alpha_B = I_{ph}^{(0)} \\ \alpha_R + \alpha_G = I_{ph}^{(4)} \end{cases}$$

(2)

Where α_R, α_G and α_B are the RGB component of the background light.

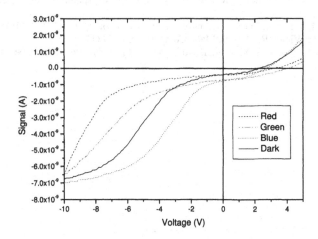

Figure 6. Signal (as defined by eq. 1)extracted under AC condition through the LSP technique

CONCLUSIONS

The presented device is a p(a-SiC:H)/i(a-SiC:H)/i((a-Si:H)/ n(a-Si:H) junction for application in color sensing domain. The presented results show that such device can be applied for selective blue light detection under stationary conditions and as a full three color sensor through the application of the LSP technique. Future work on this promising structure will focus on color image sensor application and on Wavelength Domain Multiplexing application for optical communication in the visible range through Plastic Optical Fibers.

ACKNOWLEDGMENTS

We are grateful for financial support by the Polytechnic of Lisbon (Project 5822/2004: PASTA) and by the Fundação para a Ciência e a Tecnologia (project PTDC/FIS/70843/2006

REFERENCES

[1] C.Y. Chang, J.W. Hong, Y.K. Fang, IEE Proceedings-J, Vol. 138, No. 3 (1991) 226
[2] G. De Cesare, F. Irrera, F. Lemmi and F. Palma, Appl. Phys. Lett. 66 (1995) 1178.
[3] H. Stiebig, J. Giehl, D. Knipp, P. Rieve and M. Bohm, Mater.Res. Soc. Symp. Proc. 377 (1995) 815.
[4] P. Louro , M. Vieira , A. Fantoni ,M. Fernandes, C. N. Carvalho , G. Lavareda, Sensors and Actuators A, 123–124, (2005) 326–330
[5] M. Topic, F. Smole, J. Furlan, W. Kusian, Journal of Non-Crystalline Solids 198-200 (1996) 1180-1184
[6] Fernandes M, Vieira M, Martins R, Journal of Non-Crystalline Solids 352 (9-20) 1801-1804.
[7] A. Fantoni, M. Vieira, R. Martins, Math. Comput. Simul. 49 (1999) 381
[8] A. Fantoni, M. Fernandes, Y. Vygranenko M. Vieira, Phys. Stat. Sol. (a) 205, No. 8, (2008) 2069–2074

Mater. Res. Soc. Symp. Proc. Vol. 1153 © 2009 Materials Research Society   1153-A19-03

Laser-Induced Crystallization of SiGe MEMS Structural Layers Deposited at Temperatures Bbelow 250°C

Joumana El-Rifai[1,3,4], Sherif Sedky[1,2], Rami Wasfi[1], Chris Van Hoof[3,4] and Ann Witvrouw[3]

[1]Youssef Jameel Science and Technology Research Center, The American University in Cairo, Cairo, Egypt
[2]Physics Department, The American University in Cairo, Cairo, Egypt
[3]IMEC, Leuven, Belgium
[4]Katholieke Universiteit Leuven, Leuven, Belgium

ABSTRACT

This work is a step towards a viable process for poly-SiGe MEMS structural layers deposited at substrate temperatures below 250°C. Laser annealing was used for post-deposition layer treatment to realize poly-SiGe structural layers with the desired electrical and mechanical properties at low substrate temperatures. The technique uses a pulsed excimer laser beam for the local thermal treatment of a SiGe layer deposited by Plasma Enhanced Chemical Vapor Deposition (PECVD) at 210°C. By tuning the laser treatment and the film deposition conditions, 1-1.8 μm thick films having an electrical resistivity as low as 14.1 mΩ·cm and optimal strain gradient in the range of -4.3×10^{-6} to $+6.8\times10^{-6}$ μm^{-1} were realized.

INTRODUCTION

Over the past decades, microelectromechanical systems (MEMS) have been implemented in a variety of applications. Integrating such systems with the pre-fabricated CMOS electronics allows for the production of compact devices which have an improved signal/noise ratio as compared to hybrid approaches [1-4]. A number of monolithic integration techniques are possible [1-5]. The most promising technique to integrate MEMS and CMOS is the post-processing method. Not only does it make use of standard CMOS processing techniques, but constructing the MEMS structures directly on top of the pre-fabricated CMOS has proven to be the most compact and cost effective [1, 2]. A major limitation to such an approach is that the processing temperature for the MEMS layers must be compatible with the underlying CMOS layers. Such a temperature limitation may restrict the choice for the employed MEMS material.

The acceptable thermal budget for the MEMS process depends on the CMOS process used. In the literature thermal budgets of 450-520°C for 90 minutes to deposit on pre-fabricated 0.35 μm CMOS and 450°C for 60 minutes for 0.25 μm CMOS were reported [6, 7].

Apart from CMOS integration, fabrication of MEMS on passive and flexible substrates is an emerging interest. This implies a further reduction in temperature budget since the substrate materials (e.g. Benzocyclobutene (BCB), Silicone and Polyimide) have a maximum processing temperature in the range of 230 to 300°C [8].

Poly-SiGe layers have been proven to be good MEMS structural layers for above-IC integration [1, 2]. However, lowering the deposition temperature of such materials below 250°C for post-processing on temperature sensitive substrates will degrade the electrical and mechanical properties of the films. High stress, strain gradient and electrical resistivity are the main challenges encountered with such temperature reduction [9]. Deposition temperature reduction should always be accompanied by a technique that improves the electrical and mechanical properties of the material.

Prior research conducted to lower the Si and SiGe deposition temperature has involved alloying Si with Ge [10,11], multilayer depositions [12], high hydrogen dilution [12], metal induced crystallization [1, 13, 14] and laser induced crystallization [15]. Combining Chemical Vapor Deposition (CVD) and PECVD depositions in a multilayer approach produced poly-SiGe with good electrical and mechanical properties when deposited at 450°C [12]. Using PECVD and a high hydrogen dilution, good MEMS structural layers, consisting of micro-crystalline SiGe, can even be obtained in-situ at temperatures around 350 °C [12]. Metal-induced crystallization was used to crystallize LPCVD films at 400°C [1] and PECVD films at 300-370°C [13]. Laser induced crystallization or laser annealing (LA) has produced good PECVD SiGe structural layers at deposition temperatures as low as 210°C with a stress gradient as low as $1\times10^{-6}\mu m^{-1}$ for an 0.3μm thick film [15]. However, a high-energy laser treatment may result in severe stress changes, cracks in the film and the formation of voids in these hydrogen-containing layers. Therefore, only a mild laser treatment was used in [15], which resulted in a shallow surface crystallization of the thin films and a minimum resistivity of 80 mΩ·cm.

In this work a laser treatment of thick SiGe layers (1.0 to 1.8 μm) was performed with the aim of having at the same time a low resistivity and a low strain gradient. For this purpose, a thorough study of the impact of the initial deposition conditions, the laser pulse number, energy and repetition rate on the electrical and mechanical properties of the films under consideration has been done. The effect of using a single pulse treatment in comparison to a multiple pulse treatment as in [15] was also examined.

EXPERIMENT

Amorphous SiGe layers were deposited by PECVD in an Oxford Plasma Lab 100 System at a constant substrate temperature of 210°C. A 200nm Si layer was used in between the SiGe and the 2μm thick sacrificial SiO_2 layers for improved adhesion. The initial stress of blanket wafers was measured using an FSM film stress measurement tool. Deposited films were subsequently patterned with cantilever and other MEMS test structures to evaluate the strain gradient. The effect of the Ge content and total SiGe layer thickness on the properties of the film was monitored by depositing layers with diverse thicknesses (from 0.1 to 1.8 μm) and Ge concentrations (from 11% to 69%).

Prior to the laser treatment, the electrical and mechanical properties of the as-grown amorphous layers and the effect of varying the Ge content and SiGe thickness were examined. The sheet resistance was measured using a four-point probe and the resistivity of the layer was quantitatively calculated from the measured values. Moreover, an optical interferometer was used to determine the layer's strain gradient from cantilever out of plane deflection (OPD) data after the cantilever structures were released in a 10% HF solution and dried using a Critical Point Dryer (CPD) tool. Released 650 and 750 μm long cantilever structures were used for this purpose. In addition, SEM images of the deflected cantilevers were collected.

After sufficient data has been gathered regarding the as-grown layers, unreleased patterned SiGe films were subjected to a laser annealing (LA) treatment. Local thermal treatment of the amorphous layers was performed using a 248 nm KrF excimer laser with a 0.48×0.48 cm^2 spot size and a 24 ns pulse duration. The impact of varying the pulse energy, number of pulses and pulse frequency on electrical and mechanical properties of the films under consideration has been investigated. Employed variations were in the range of 20 to 360 mJ/cm^2, 1 to 1000 pulses and 1 to 50 Hz respectively. The electrical and mechanical properties of the treated layers were examined using the same methods and tools as for the amorphous layers and a comparison between pre-treatment and post-treatment resistivity and strain gradient values was made.

For all treatments, the laser fluence has been restricted to low values that allow a good control of stress and strain gradient and at the same time prevent void formation. As a result only shallow crystallization is expected. The crystallization depth was examined using Spreading Resistance Profile (SRP) measurements and by inspecting TEM cross sections of the treated layers.

DISCUSSION

Laser annealing conditions will depend on the strain gradient and mean stress of the as-grown films [2]. Since we are interested in using the technique to eliminate the strain gradient and at the same time produce reasonable resistivity values only mild laser treatments were effective. The as-grown film properties, treatment conditions and properties of the treated films will be discussed in the following subsections.

As Grown Film Properties

All depositions for the SiGe films were performed at 210 °C and the as-grown films were amorphous with initial resistivity measurements indicating either an open circuit or very large values in the range of 1.4×10^4 $\Omega \cdot cm$. It has been demonstrated before that the laser treatment will produce an overall increase in the layer's mean tensile stress [15, 16]. Therefore, it is necessary to start with films possessing a (strong) compressive mean stress. The mean stress and strain gradient of the layers can be tuned by controlling the deposition conditions such as the Ge concentration, deposition pressure and total SiGe layer thickness [15].

For a Ge concentration of 11%, an increase in the film thickness will result in an increase in the film's compressive stress as illustrated in Fig. 1a. On the other hand, for a fixed layer thickness of 250 nm, an increase in the Ge content will produce more compressive layers as illustrated in Fig. 1b. Films selected to be laser annealed were insured to have an initial compressive stress as outlined in Table 1. In addition, a few experiments were done on an 0.7 μm thick SiGe layer with 69% of Ge and a mean tensile stress of 20 MPa.

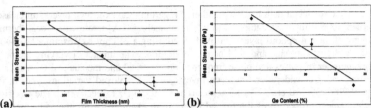

Figure 1 - Tuning the properties of deposited films. (a) Effect of film thickness on mean stress with fixed Ge content of 11%. (b) Effect of Ge content on the mean stress of SiGe with fixed thickness of 250 nm.

Ge%	Layer Thickness (μm)	Mean Stress (MPa)	As-grown Strain Gradient (μm⁻¹)	As-grown Resistivity (Ω.cm)	Treatment SP – Single Pulse MP – Multiple Pulse	Strain Gradient After Treatment (μm⁻¹)	Resistivity After Treatment (Ω.cm)
28	1.0	-43	8.3×10⁻⁵	open	MP-120 mJ/cm², 100 pulses, 10 Hz	6.8×10⁻⁶	1.55×10⁻²
28	1.8	-117	-1.11×10⁻⁵	1.39×10⁴	SP-240 mJ/cm²	-4.2×10⁻⁶	1.69×10⁻²
28	1.8	-117	-1.11×10⁻⁵	1.39×10⁴	MP-120 mJ/cm², 500 pulses, 10 Hz	-2.5×10⁻⁶	3.58×10⁻²
11	1.6	-130	-1.44×10⁻⁵	open	SP-200 mJ/cm²	-4.3×10⁻⁶	1.41×10⁻²
11	1.6	-130	-1.44×10⁻⁵	open	MP-100 mJ/cm², 50 pulses, 1 Hz	-7.5×10⁻⁷	2.97×10⁻²

Table 1 – Layer conditions pre and post optimized treatment.

Another noticeable fact is that prior to laser treatment, the released cantilever structures have already an OPD. The OPD is an indication of the strain gradient in the layer and the amount of the deflection can be controlled by the Ge concentration and layer thickness. Figure 2a indicates that for the same Ge concentration, as the thickness increases, the OPD of the cantilever decreases. Similarly for the same thickness, as the Ge concentration increases, the OPD decreases as seen in Fig. 2b.

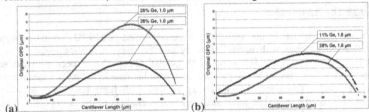

Figure 2 - Tuning the OPD of the as-deposited films (cantilever anchors are at the origin). (a) The as-grown OPD for 650μm long cantilevers for 28% Ge films with different thicknesses. (b) The as-grown OPD for 650μm long cantilevers for films with similar thicknesses and different Ge content.

Laser Annealing Treatment Conditions and Properties of Treated Films

A proper laser annealing condition would decrease the film's resistivity and strain gradient, without damage to the hydrogen containing material. Either a single or multiple pulse technique can be used. Let us first consider the effect of the single pulse and multiple pulse treatment on the material's resistivity. The resistivity of a 1.8 μm thick 28% Ge layer was reduced from 1.39×10^4 Ω·cm prior to laser annealing to 5.9×10^{-2} Ω·cm and 1.69×10^{-2} Ω·cm after a single pulse of 120 and 240 mJ/cm², respectively. In the case of a multiple pulse treatment, a laser treatment of 500 pulses at 120 mJ/cm² and 10Hz was sufficient to reduce the resistivity of the previous 28% Ge sample to 3.58×10^{-2} Ω·cm.

Increasing the total supplied laser energy to the layer, whether an increase in the number of pulses or the single pulse energy, is associated with a significant decrease in electrical resistivity. In addition, as the Ge content is increased, the structural changes become more pronounced for the same laser energy. For example, single pulse laser energy of 160 mJ/cm^2 produced a resistivity of 1.80×10^{-1} $\Omega \cdot$cm for an 11% Ge layer and an even lower resistivity of 2.97×10^{-2} $\Omega \cdot$cm for a 28% Ge layer.

With regards to the effect of the laser treatment on the strain gradient, a comparison between pre and post-annealing OPD of the 650 μm long cantilever of the 28% Ge sample was made. SEM images of pre and post-treatment cantilevers for the 1.8 μm thick layer are displayed in Fig. 3. As the laser annealing densifies the top surface of the layer, it is an advantage to have a negative strain gradient for the as-grown material, (cf. Figs. 2 and 3.a). An optimized laser treatment can then result in flat cantilevers. In Fig 3b 500 pulses at 120 mJ/cm^2 and 10Hz were used to produce a 28% Ge 1.8 μm film layers with a strain gradient as low as -2.5×10^{-6} μm^{-1}. Single pulse treatment can also be used successfully. A pulse of 180 mJ/cm^2 was sufficient to produce a 1.6 μm SiGe film with 11% Ge with a strain gradient of -2.7×10^{-6} μm^{-1}. The range of optimized strain gradient values for tests performed was between -4.3×10^{-6} and $+6.8 \times 10^{-6}$ μm^{-1} after laser annealing. Generally the absolute strain gradient value of the layer will decrease with increasing laser energy until a maximum energy value is reached, after which the strain gradient will start to increase again and the material properties will degrade.

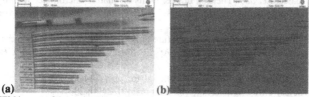

(a)   **(b)**

Figure 3 - SEM images of the suspended cantilever structures of 1.8 μm SiGe films with 28% of Ge. (a) As grown (b) After laser treatment with 500 pulses at 120 mJ/cm^2 and 10Hz.

To achieve proper strain gradient values it is necessary to keep the total supplied laser energy for a single pulse treatment below 240 mJ/cm^2 for a 28% Ge sample and 60 mJ/cm^2 for a 69% Ge sample. This restraint will ultimately affect the minimum achievable resistivity and the maximum crystallization depth. TEM and SRP analysis showed that only a maximum crystallization depth of 200 nm is achieved for a 1.8 μm SiGe film with 28% Ge annealed at either a single pulse of 280 mJ/cm^2 or 500 pulses at 10Hz and 120 mJ/cm^2.

The best combination of electrical and mechanical properties for the films after laser annealing is listed in Table 1.

CONCLUSIONS

We have demonstrated a correlation between the layer thickness, Ge concentration, mean stress and optimal laser annealing conditions. The deposited layers should be (slightly) compressive to allow larger laser annealing energy values. Using a higher energy will directly influence the annealed layer, resulting in a significant decrease in electrical resistivity. By tuning factors such as the laser annealing energy, number and frequency of pulses, Ge content and layer thickness, it is possible to realize thin films having an electrical resistivity as low as 14.1 mΩ·cm and a strain gradient in the range from -4.3×10⁻⁶ to +6.8×10⁻⁶ μm⁻¹. These films are clearly promising structural layers for MEMS applications that require a post-processing temperature of 210°C.

REFERENCES

1. T. J. King et al., "Recent Progress in Modularly Integrated MEMS Technologies," *Proc. IEDM*, pp. 199-202, 2002.
2. S. Sedky, *Post-Processing Techniques for Integrated MEMS*. Boston, MA: Artech House, 2006.
3. A. E. Franke, T. J. King et al., "Integrated MEMS Technologies," *Mat. Res. Soc. Bulletin*, pp. 291-295, Apr. 2001.
4. L. Haspeslagh et al., "Highly reliable CMOS-integrated 11MPixel SiGe-based micro-mirror arrays for high-end industrial applications," *IEDM 2008, IEEE*, pp.1-4, 2008.
5. J. H. Smith et al., "Embedded micromechanical devices for the monolithic integration of MEMS with CMOS," *Elec. Dev. Meeting*, pp. 609-612, Dec. 1995.
6. S. Sedky, A. Witvrouw et al., "Experimental Determination of the Maximum Post-Process Annealing Temperature for Standard CMOS Wafers," *IEEE Trans. Electron Devices*, vol. 48, no. 2, pp. 377-385, Feb. 2001.
7. H. Takeuchi et al., "Thermal budget limits of quarter-micrometer foundry CMOS for post-processing MEMS devices," *IEEE Trans. on Electron Dev.*, vol. 52, pp. 2081-2086, Sept. 2005.
8. N. D. Young et al., "Low Temperature Poly-Si on Flexible Polymer Substrates for Active Matrix Displays and Other Applications," *Mat. Res. Soc. Symp. Proc.*, vol. 769, 2003.
9. S. Sedky, "SiGe: An attractive material for post-CMOS processing of MEMS," *Microelectronic Engineering*, vol. 84, pp. 2491-2500, 2007.
10. S. Sedky et al., "Thermally insulated structures for IR bolometers, made of polycrystalline silicon germanium alloys," *International Conference on Solid-State Sensors and Actuators Proceedings*, vol. 1525, pp. 237-240, 1997.
11. R. T. Howe and T. J. King, "Low Temperature LPCVD MEMS Technologies," *Mat. Res. Soc. Symp. Proc.*, vol. 729, 2002.
12. M. Gromova et al., "Characterization and strain gradient optimization of PECVD poly-SiGe layers for MEMS applications," *Sensors and Actuators A*, vol. 130-131, pp. 403-410, 2006.
13. S. Sedky et al., "Low tensile stress SiGe deposited at 370 ° C for monolithically integrated MEMS applications", *Mat. Res. Soc. Symp. Proc.*, vol. 808, A.4.19, Spring 2004.
14. S. F. Chen et al., "Low temperature grown poly-SiGe thin film by Au metal-induced lateral crystallization (MILC) with fast MILC growth rate," *Elec. Lett.*, vol. 39, no. 22, Oct. 2003.
15. S. Sedky et al., "Optimal Conditions for Micromachining Si₁₋ₓGeₓ at 210°C", *JMEMS*, vol. 16, no. 3, pp. 581-588, June 2007.
16. S. Sedky et al., "Pulsed laser annealing, a low thermal budget technique for eliminating stress gradient in poly-SiGe MEMS structures," *J. Microelectromech. Syst.*, vol. 13, no.4, pp. 669-675, Aug. 2004.

Mater. Res. Soc. Symp. Proc. Vol. 1153 © 2009 Materials Research Society   1153-A19-04

Evaluation of the Electrical Properties, Piezoresistivity and Noise of poly-SiGe for MEMS-above-CMOS applications

P. Gonzalez[1,2], S. Lenci[1], L. Haspeslagh[1], K. De Meyer[1,2], and A. Witvrouw[1]
[1]IMEC, Kapeldreef 75, 3001 Leuven (Belgium)
[2]K.U. Leuven, Kasteelpark Arenberg 10, 3001 Leuven (Belgium)

ABSTRACT

In this work, the electrical properties of heavily doped poly-SiGe deposited at temperatures compatible with MEMS integration on top of standard CMOS are reported. The properties studied are resistivity, temperature coefficient of resistance, noise, piezoresistivity, Hall mobility and effective carrier concentration. The obtained results prove the potential of using poly-SiGe as a sensing layer for MEMS-above-CMOS applications.

INTRODUCTION

The monolithic integration of the micro-electro-mechanical systems (MEMS) with the driving, controlling and signal processing electronics on the same CMOS substrate can improve performance, yield and reliability as well as lower the manufacturing, packaging and instrumentation costs [1]. The post-processing route (fabricating MEMS directly on top of CMOS) is the most promising approach for CMOS-MEMS monolithic integration as it enables integrating MEMS without introducing any changes in standard foundry CMOS processes. However post-processing limits the maximum fabrication temperature of MEMS to 450°C-520°C [2], to avoid introducing any damage or degradation in the performance of the existing electronics or interconnects. This temperature constraint is quite restrictive for post-processing surface micromachined MEMS, as it might affect relevant physical properties such as crystallinity, growth rate, mechanical properties, doping activation, etc. Polycrystalline silicon germanium (poly-SiGe) has emerged as an attractive alternative to polycrystalline silicon (poly-Si) for MEMS-above-CMOS applications thanks to its lower electrical resistivity [3] and lower transition temperature from amorphous to polycrystalline [4,5], which can be as low as 400°C (with the appropriate germanium content).

The main objective of this work is to investigate the resistivity, temperature coefficient of resistance, piezoresistivity and noise of poly-SiGe deposited at temperatures compatible with MEMS post-processing on top of standard CMOS wafers with Al interconnects. For this study, boron in situ doped (with a chemical concentration of $3.6 \cdot 10^{21} \mathrm{cm}^{-3}$) poly-$Si_{24}Ge_{76}$ was deposited at a wafer temperature of 450°C using chemical vapor deposition (CVD). The resistivity is measured in the temperature range of -10 to 200°C using a four point probe. The active carrier concentration and mobility are determined from Hall measurements. The piezoresistivity is estimated by measuring the resistance variation when a uniform and uniaxial stress provided by a four point bending fixture is applied to the films. To allow for comparison, boron-doped poly-Si layers were also deposited at 550°C with an estimated chemical doping concentration of $1 \cdot 10^{21} \mathrm{cm}^{-3}$.

EXPERIMENTAL

The samples were fabricated using a two mask-process. Each sample measures 8x80 mm^2 and consists of a 361±11 nm thick (determined using a Dektak profiler) boron-doped poly-SiGe (76% Ge) film deposited on top of (100) silicon wafers covered with 100nm-thick SiO₂. Poly-SiGe has been deposited using chemical vapor deposition (CVD) at 460°C chuck temperature (450°C wafer temperature) and 4.3 Torr in an Applied Materials (AMAT) PECVD CxZ chamber, mounted on an Applied Materials Centura Giga-Fill SACVD platform. The silicon gas source is pure silane, whereas 10% germane in hydrogen has been used as the germanium gas source. 1% diborane in hydrogen has been used as the boron gas source. The silane/germanium flow ratio equals 0.9 and the diborane flow is 40 sccm for all depositions. Rutherford Back Scattering (RBS) and Secondary Ion Mass Spectrometry (SIMS) analyses were used to determine the germanium and boron concentrations [6]. The values measured on similar samples are $3.6 \cdot 10^{21}$ at/cm^{-3} for boron concentration and 76% for Ge content.

The resistors are created by etching the poly-SiGe in the central region of each sample. Each resistor is contacted by 1μm-thick aluminum copper (AlCu 0.1%) traces to four bonding pads in order to measure the resistance by the four-point probe method, eliminating any spurious contribution of contacts and metal traces. Resistors of different lengths (from 0.5 mm to 18mm) and widths (from 25 μm to 100 μm) have been patterned. Figure 1 shows microscope pictures of typical samples used for piezoresistive and Hall measurements, respectively.

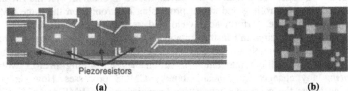

(a) (b)

Figure 1 *Microscope picture of a) typical sample used for piezoresistive measurements with 4 differently orientated poly-SiGe resistors and b) four cross-shaped resistors of different sizes used for Hall measurements*

To allow for a direct comparison between the electrical properties of poly-SiGe and poly-Si, *in situ* boron doped poly-Si layers were also deposited by low pressure CVD at 550°C and 300 mTorr (in a vertical furnace, batchtool) with an estimated chemical doping concentration of $1 \cdot 10^{21}$ cm^{-3}. The poly Si layers, which showed a thickness (determined using a Dektak profiler) of 378±25nm, were patterned following the same two-mask process used for the poly SiGe layers.

DISCUSSION

Electrical properties

The resistivity is measured in the temperature range of -10 to 200°C by performing four point resistance measurements of resistors of different dimensions. The measured resistivities for poly-SiGe and poly-Si are plotted as a function of temperature in figure 2. As can be seen from this figure, the resistivity is more or less a linear function of temperature in the temperature interval -10 to 200°C. Based on this observation, the TCR can be defined in the following way [7]:

$$TCR = \frac{\rho_2 - \rho_1}{\rho_1 \cdot (T_2 - T_1)} \qquad (1)$$

where ρ_2 and ρ_1 represent the resistivity at temperature T_2 and T_1, respectively. In this work T_1 is -10°C and T_2 is 200°C. Eq. (1) was used to calculate the TCR for all samples. The obtained results for resistivity and TCR of poly-SiGe and poly-Si (table 1) are compatible with previously reported data for heavily doped films [3, 7- 9]. The poly SiGe film showed very good electrical properties with a resistivity of only 0.86 mΩ·cm and a low TCR of 270 ppm/°C. The resistivity of poly-SiGe was lower than that of poly-Si (2.6 mΩ·cm), what was predictable as, according to [3], in boron doped films the resistance decreases with increased Ge content due to both an increased effective carrier concentration and a higher Hall mobility. As expected for heavily-doped polycrystalline films, the resistivity of both poly-SiGe and poly-Si increases with temperature leading to a positive TCR. This is because, as stated in [9], at high doping levels the barrier height is small and the effect of grain boundaries on electrical conduction is negligible compared to the resistivity of the grains. Within the grains, the current transport is by carrier drift and the resistivity of the grain region behaves essentially like that of the single crystal material [10]. Therefore, as the mobility decreases with temperature (due to the thermal vibrations of the atoms), the resistivity of the grain region increases with temperature, translating into a positive TCR. We can observe that poly- SiGe (table 1) exhibits lower TCR than poly-Si [7,9] at similar doping levels; this property can be very useful for the design of some devices as for example pressure sensors with piezoresistive sensing elements. Moreover, similar as in poly-Si, the TCR is expected to modulate with doping concentration [9, 10] and therefore it could be possible to obtain lower TCR or make it even zero for a lower doping level.

The active carrier concentration and mobility are determined from Hall measurements using cross-shaped resistors (fig. 1) at a current of 30mA and a varying magnetic field from -6 to 6T. The Hall factor was found to be a linear function of the magnetic field for the applied range and therefore the Hall scattering factor (r) can be set to unity [7]. The obtained results are summarized in table 1. The relation between active carrier concentration (N_{act}) and chemical concentration (N_{che}) is also shown in table 1. From the results we conclude that even at temperatures as low as 450°C, a lot of the dopants are activated in poly-SiGe. By comparing the ratio N_{act}/N_{che} we see that the relative number of activated dopants in poly-SiGe is similar to that of poly-Si deposited at a temperature 100°C higher. This is in agreement with previously reported results [3], which show that poly-SiGe films with boron doping require lower temperatures than equivalently doped poly-Si films to activate the dopants.

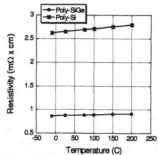

Figure 2 *Resistivity as a function of measurement temperature for poly $Si_{24}Ge_{76}$ and poly Si*

	Poly-Si$_{24}$Ge$_{76}$	Poly-Si
Resistivity (mΩ·cm)	0.86 (±4%)	2.6 (±8%)
TCR (°C^{-1})	2.7·10^{-4} (±4%)	3·10^{-4} (±3%)
N$_{act}$ (cm^{-3})	1·10^{21} (±8%)	3.3·10^{20} (±7%)
Mobility (cm^2/Vs)	7.8 (±8%)	7.1 (±13%)
N$_{act}$/N$_{che}$ (%)	28	33

Table 1 *Measured electrical properties of poly SiGe and poly Si*

Piezoresistivity

The longitudinal, transversal and shear piezoresistive coefficients (π_l, π_t and π_{lt}, respectively) of poly-SiGe are determined using the procedure described in [11]. The relative resistance variation of a polycrystalline piezoresistor subjected to a uniform and unidirectional stress σ and orientated at an angle φ with respect to the stress direction is given by the formula [11]:

$$\Delta R / R = \sigma(\pi_l \cos^2 \varphi + \pi_t \sin^2 \varphi - \pi_{lt} \cos\varphi\sin\varphi) \quad (2)$$

where R and ΔR are the zero-stress resistance and the resistance variation due to σ, respectively.

The samples were stressed in a four point bending fixture (figure 3), which creates a pure bending condition. Four blades exert a force F resulting in a uniform and unidirectional stress on top of the region in between the inner blades, where the piezoresistors are placed.

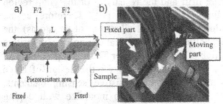

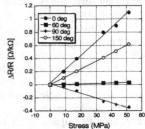

Figure 3 *a) General sketch of a sample stressed in a four-point bending fixture. b) Picture of the sample in the four-point bending setup. The wires that connect the resistors to the parameter analyzer are also visible*

Figure 4 *Relative resistance variation vs. applied stress for Poly-SiGe resistors with 4 different orientations (in degrees)*

Figure 5 *relative resistance variation vs. applied stress for Poly-Si resistors with 4 different orientations (in degrees)*

	Poly-Si₂₄Ge₇₆ (this work)	Poly-Si (this work)	Poly-Si₄₀Ge₆₀ [12]*	Poly-Si [14]**	c-Si [15]
G	2.8	4.2	7	6-10	20-30

Table 2 *Comparison between longitudinal gauge factors for different materials with similar doping levels*
* Stack of 4× (200nm CVD+800nm PECVD) Poly-Si₄₀Ge₆₀
** Value obtained by extrapolation of the theoretical curve proposed in [14] for the gauge factor as a function of doping concentration

To determine the value of the stress σ induced in the piezoresistors, finite element simulations (COMSOL3.3) were performed, using an assumed Young's modulus E=140GPa [12] for poly-SiGe and E=163GPa for poly-Si [14]. Fig. 4 (fig. 5) shows the relative resistance variation of several poly-SiGe (poly-Si) resistors versus applied force for φ=0, 60, 90 and 150 degrees. By substituting the measured relative resistance variation and the simulated value of the stress in (2), the piezoresistive coefficients were found. Results show $\pi_l = 2 \cdot 10^{-11}$, $\pi_t = -0.7 \cdot 10^{-11}$ and $\pi_{lt} = 0.8 \cdot 10^{-11}$ [Pa⁻¹] for poly SiGe and $\pi_l = 2.6 \cdot 10^{-11}$, $\pi_t = -1.11 \cdot 10^{-11}$ and $\pi_{lt} = -0.7 \cdot 10^{-11}$ [Pa⁻¹] for poly Si. The maximum variation in the piezoresistive coefficients from sample to sample is ±20% (8 samples measured) in the case of poly-SiGe and ±13.3% for poly-Si (7 samples measured). These results translate into a longitudinal gauge factor (defined as the ratio between relative resistance change and strain, $\Delta R/\varepsilon R$) of 2.8 (4.2) for poly SiGe (poly Si), which is comparable to previously reported values [12, 14]. Table 2 offers a comparison with gauge factors of other materials with similar doping level found in literature. The doping dependence of the poly-Si gauge factor, which tails off for high and low doping levels, reaching a maximum at a doping concentration of around $1 \cdot 10^{19} \text{cm}^{-3}$ [14], is well known. Therefore the low gauge factors found both for poly-

SiGe and poly-Si are not surprising taking into account the high doping concentrations of the studied films. Moreover the relatively lower gauge factor exhibited by the poly-SiGe samples compared to the poly-Si ones could also be explained by the higher effective carrier concentration found in poly-SiGe (table 1). Finally, assuming poly-SiGe piezoresistive properties will exhibit similar modulation with doping concentration as poly-Si, we can expect to improve the gauge factor by decreasing the doping concentration. But even if the maximum achievable gauge factor of poly-SiGe is lower than that of poly-Si, the possibility to post-process on top of CMOS would still make it an interesting material for MEMS piezoresistive sensors as monolithic integration leads to a higher signal/noise ratio which might offset a hypothetical lower gauge factor.

Noise

In resistors, the fundamental noise sources may be divided into frequency independent thermal noise and frequency dependent conductance fluctuation noise or 1/f noise. The total noise spectrum can be expressed like in (3), where α is the so-called Hooge factor (or simply 1/f factor), f is the frequency, V_B, the bias voltage, k_B represents the Boltzman constant and R the resistance value. N is the total number of carriers and is given by $N=p \cdot L \cdot W \cdot T$, being p the boron-doping dose, and L, W and T the length, width and thickness of the resistor.

$$S_{vf} = \underbrace{\frac{\alpha}{N \cdot f} \cdot V_B^2}_{\substack{\text{Hooge's formula} \\ \text{for 1/f noise}}} + \underbrace{4 \cdot K_B \cdot T \cdot R}_{\text{Johnson noise}} \qquad (3)$$

Although the previous empirical relation for the 1/f noise was originally proposed for homogeneous materials [16], it has been used in the past for the study of the noise in heavily doped polysilicon films [17]. And, even though, a complete analysis of the noise in any polycrystalline material cannot be carried out using only the value of α, it gives a good measure of the strength of the noise in the material [18]. Two point noise measurements were carried out at room temperature using a BTA9812B noise analyzer in conjunction with an HP35665A dynamic signal analyzer. Similar setup has already been successfully used to measure noise of poly SiGe resistors in the past [19]. In this work poly-$Si_{24}Ge_{76}$ resistors of different lengths (0.5, 1, 2 and 3mm) and widths (25, 50 and 100 μm) were studied. In all resistors only thermal noise could be measured; not even at high bias voltages (up to 9V) 1/f noise appeared. According to (3) this would mean a 1/f noise factor α lower than $1 \cdot 10^{-5}$. This would be in agreement with results reported in [18] for the noise in boron-doped poly-$Si_{30}Ge_{70}$, if the data is extrapolated to higher doping levels. This value is more than one order of magnitude lower than the one reported in our previous work [12] for 4 μm-thick Poly-$Si_{40}Ge_{60}$ (4× (200nm CVD + 800nm PECVD). This low noise could be very interesting for many applications as for example piezoresistive pressure sensors since lower noise translates into lower minimum detectable pressure [12].

CONCLUSIONS

The properties of heavily doped CVD poly-$Si_{24}Ge_{76}$, deposited at temperatures compatible with MEMS integration on top of standard CMOS, have been experimentally evaluated. Poly-$Si_{24}Ge_{76}$ proved to have very good electrical properties, with a resistivity lower than 1 mΩ·cm and a TCR of only $2.7 \cdot 10^{-4}$ °C^{-1}. Moreover, the layer showed piezoresistive properties similar to those of similarly doped poly-Si and very low 1/f noise with a Hooge factor below 10^{-5}. These results confirm that poly-SiGe is a very attractive structural material for MEMS-above-CMOS applications.

REFERENCES

1. A. Witvrouw *et al*, *"Poly SiGe: a Superb Material for MEMS"*, proc. MRS, Vol. 782, pp. 25-36, 2004.
2. S Sedky *et al*, *"Experimetal Determination of the maximum post-processing temperature of MEMS on top of standard CMOS wafers"*, IEEE Trans. Electron Devices, Vol.48 (2), pp. 1-9, 2001.
3. D. Bang *et al*, *"Resistivity of boron and phosphorus doped polycrystalline $Si_{1-x}Ge_x$ films"*, Applied Physics Letters, 66 (2), pp. 195-197, (1995).
4. T.J. King *et al*, *"Deposition and properties of low-pressure chemical vapor deposited polycrystalline silicon-germanium films"*, Journal of The Electrochemical Society, 141, 2235-2241 (1994).
5. T.J. King *et al*, *"A low temperature ($\leq 550°C$) silicon germanium MOS thin-film transistor technology for large area electronics"*, IEDM '91 technical digest,567-570.
6. G. Bryce *et al*, *'Simultaneous optimization of the material properties, uniformity and deposition rate of polycrystalline CVD and PECVD Silicon-Germanium layers for MEMS applications'*, Proc. ECS '08, 16 (10), pp. 353-364 (2008).
7. M. Rydberg *et al*, *"Temperature Coefficient of Resistivity in Heavily Doped Oxigen-Rich Polysilicon"*, Journal of The Electrochemical Society, 148 (12) pp 725-733 (2001)
8. S. Sedky *et al*, *"Electrical Properties and Noise of Poly SiGe deposited at Temperatures Compatible With MEMS Integration on Top of Standard CMOS"*, MRS Proceedings, Vol. 729 (2002)
9. M. Boutchich *et al*, *"Characterization of Phosphorus and Boron Heavily Doped LPCVD Polysilicon Films in the temperature Range 293-373 K"*, IEEE Electron Device Letters, 23 (3), pp. 139-141 (2002)
10. K. N. Bhat, *"Silicon Micromachined Pressure Sensors"*, Journal of the Indian Institute of Science, Vol. 87:1, 2007
11. S. Lenci *et al*, *"Determination of the piezoresistivity of microcrystalline Silicon-Germanium and application to a pressure sensor"*, Proc. MEMS '08, 427-430
12. P. Gonzalez *et al*, *"Evaluation of Piezoresistivity and 1/f Noise of Polycrystalline SiGe for MEMS sensors applications"*, Proceedings Eurosensors '08, pp. 881-884
13. W.N. Sharpe *et al*, *"A New Technique for Measuring Properties of Thin Films"*, Journal of micromechanical systems, 6 (3) (1997)
14. P.J. French, A.G.R Evans,*"Piezoresistance of Polysilicon"*, Electronics Letters, Vol 20 (24),1984
15. S.M Sze, *Semiconductor Sensors*, John Wiley & Sons, Inc. (1994)
16. F.N Hooge, *"1/f noise is no surface effect"*, Phys Letters A, Vol.29, pp. 139-140, 1969
17. M.J. Deen, *"Low frequency noise in heavily doped polysilicon thin film resistors"*, Journal Vacuum Science Technology B, 16 (4), pp. 1881-1884 (1998)
18. X.Y. Chen *et al*, *"1/f noise in poly-SiGe analyzed in term of mobility fluctuations"*, Solid-State Electronics 43, pp. 1715-1724 (1999)
19. K. Chen et al, *"Analysis of low-frequency noise in boron-doped polycrystalline silicon-germanium resistors"*, Applied physics letters, 81 (14), pp.2578-2580 (2002)

Mater. Res. Soc. Symp. Proc. Vol. 1153 © 2009 Materials Research Society   1153-A19-05

Noise Spectra of Si$_x$ Ge$_y$ B$_z$:H Films for Micro-Bolometers

A.Kosarev, I.Cosme, A.Torres

Electronics, National Institute for Astrophysics, Optics and Electronics, L.E. Erro 1, col. Tonantzintla, Puebla, 72840, Mexico

ABSTRACT

Noise spectra in plasma deposited Si$_x$Ge$_y$B$_z$:H thermo-sensing films for micro-bolometers have been studied. The samples were characterized by SIMS (composition) and conductivity (room temperature conductivity, activation energy) measurements. The noise spectra were measured in the temperature range from T= 300 K to T=400 K and in the frequency range from f=2 Hz to f=2x10^4 Hz. The noise spectra $S_I(f)$ for the samples Si$_{0.11}$Ge$_{0.88}$:H and Si$_{0.04}$Ge$_{0.71}$B$_{0.23}$ can be described by $S_I(f) \sim f^{-\beta}$ with $\beta = 1$ and $\beta = 0.4$, respectively. For the sample Si$_{0.05}$Ge$_{0.67}$B$_{0.26}$ two slopes were observed: in low frequency region f$\leq$ 10^3 Hz β_1= 0.7 and at higher frequencies f>10^3 Hz β_2= 0.13. Increasing temperature resulted in an increase of noise magnitude and a change of β values. The latter depended on film composition. The correlation observed between noise and conductivity activation energies suggests that noise is due to bulk rather than interface processes. Noise spectrum of the thermo-sensing film Si$_{0.11}$Ge$_{0.88}$:H was compared with that for micro-bolometer structure with the same thermo-sensing film. The micro-bolometer structure showed higher noise value in entire frequency range that assumed additional processes inducing noise.

INTRODUCTION

Thermal detectors (bolometers) based on plasma deposited films have demonstrated many attractive advantages such as high responsivity, moderate resistance, compatibility of the fabrication process with standard silicon CMOS technology. Pioneering works with boron doped amorphous silicon in this field have been reported by "LETI" [1] and "Raytheon" [2] In our previous works we have reported on the study of fabrication and characterization of micro-bolometers with non doped silicon-germanium thermo-sensing films [3-5]. Noise measurements are of principal importance for performance characterization of the devices, because noise and responsivity determine detectivity that is a figure of merit based on signal-to-noise ratio enabling a comparison of different detectors (not only thermal ones). Another reason to study noise is related to its high sensitivity to fabrication and material parameters. So far noise in thermo-sensing materials and micro-bolometers has been poorly reported in literature. Only a few papers considered noise in plasma deposited non-crystalline materials [6 and references therein]. Noise related data for micro-bolometers can be found in refs. [7, 8].

The goal of this work is to study experimentally noise spectra in several plasma deposited thermo-sensing films used in micro-bolometers.

EXPERIMENT

The Si$_x$Ge$_y$B$_z$: H films were deposited by low frequency (LF) plasma enhanced chemical vapor deposition (PE CVD). Deposition parameters were as follows: substrate temperature

T_s= 300 °C, power W = 450 W, the discharge frequency f=110 kHz, pressure P= 0.6 Torr, a gas mixture consisting of silane (SiH_4), germane (GeH_4) and diborane (B_2H_6) diluted with hydrogen. Composition of the films was changed by means of the variation of flows of the aforementioned gases (details see e.g. in ref. [9]). Composition of the films was determined by SIMS profiling. The films were deposited on glass ("Corning 1737") substrates for electrical and optical characterization and on Si substrates for IR and SIMS analysis. The characteristics of the thermo-sensing films studied are listed in Table I.

Table I. Characteristics of the thermo-sensing films.

		Thermo-sensing films		
		Process 479	Process 480	Process 443
Film Thickness (μm)		0.42	0.51	0.5
Deposition rate ($\overset{o}{A}$ /s)		7	9.5	2.8
Solid content	x (Si)	0.06	0.04	0.11
obtained from	y(Ge)	0.67	0.71	0.88
SIMS	Z(B)	0.26	0.23	Traces ($2x10^{-5}$)
Film	E_a (eV)	0.21	0.18	0.345
	TCR (K^{-1})	-0.027	-0.023	-0.044
properties	σ_{RT} (Ωcm)$^{-1}$	$1x10^{-2}$	$2.5x10^{-2}$	$6x10^{-5}$
	σ_0 (Ωcm)$^{-1}$	36.46	24.55	34.85

Conductivity and noise measurements were conducted with planar structures with Ti bottom electrodes deposited on glass. The current-voltage I(U) characteristics of the structures were measured in vacuum (P = 10 mTorr) thermostat ("MMR Inc.") with electrometer ("Keithley" – 6517-A). The noise spectral density (NSD) measurements were performed with lock-in amplifier ("Stanford Research Systems" – SR 530) in the frequency range of 4 to $2x10^4$ Hz and Δf= 1Hz.

RESULTS AND DISCUSSION

Noise spectral density (NSD) is defined as $S_I(f) = [I_{noise}(f)/(\sqrt{\Delta f})]^2$, where $I_{noise}(f)$ is r.m.s. current noise, Δf = 1 Hz is a bandwidth. Noise spectra measured at room temperature are shown in Figure 1. It can be seen that these films showed different $S_I(f)$ behavior. The sample $Si_{0.11}Ge_{0.88}$:H (443) without boron showed the lowest noise while the sample $Si_{0.04}Ge_{0.72}B_{0.23}$ (480) showed the highest noise in the entire range of frequency. All samples revealed increase of noise with reducing frequency, but they have different slopes extracted from $S_I(f) \sim f^\beta$ approximations. The sample $Si_{0.11}Ge_{0.88}$:H (443) showed the highest slope β = 1 i.e. "classical" 1/f noise behavior, the sample $Si_{0.04}Ge_{0.72}B_{0.23}$ (480) had the lower slope β =0.4 and the sample 479 showed 2 slopes: β_1 = 0.74 in low frequencies (f< 2 $x10^2$ Hz) and lower value β_2 =0.13 in the range of frequency f > $2x10^2$ Hz.

Interesting data were obtained from noise-temperature measurements. NSD for $Si_{0.11}Ge_{0.88}$:H sample (process 443) is shown in Figures 2 . At room temperature this sample has one slope β =

1 but increasing temperature results in more complex noise spectra, where two regions with different slopes β_1 and β_2 can be distinguished. The insert in Figure 2 shows dependence of

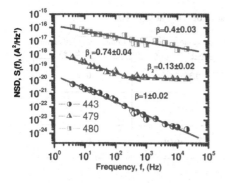

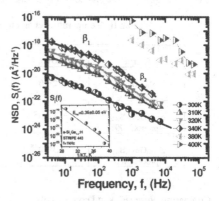

Figure 1. Noise spectra for different thermo-sensing films at T=300K : a) $Si_{0.11}Ge_{0.88}$:H (443), b) $Si_{0.06}Ge_{0.67}B_{0.26}$:H (479), and c) $Si_{0.04}Ge_{0.71}B_{0.23}$:H (480).

Figure 2. Noise spectra at different temperatures for the sample $Si_{0.11}Ge_{0.88}$:H (443).

noise at f=1 kHz as a function of temperature. Solid line is the best fit of the experimental points that reveals temperature activation behavior characterized by noise activation energy E_{na}= 0.36 eV. Noise spectra measured at different temperatures for other samples studied are presented in Figure 3 a) (sample $Si_{0.06}Ge_{0.67}B_{0.26}$: H, process 479) and Figure 3 b) (sample $Si_{0.04}Ge_{0.71}B_{0.23}$: H,

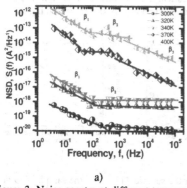

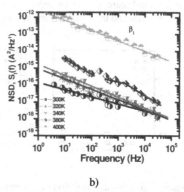

a)

b)

Figure 3. Noise spectra at different temperatures for the samples: a) $Si_{0.06}Ge_{0.67}B_{0.26}$:H (479), and b) $Si_{0.04}Ge_{0.71}B_{0.23}$:H (480).

process 480). In all samples studied increasing temperature causes more complex and different, depending on composition of the film, behavior of noise spectra. Thermally activated character of noise change with temperature was observed in all samples, but they demonstrated different room temperature noise and activation energy values. The values obtained are listed in Table II.

Table II. Temperature characteristics of noise at f=1 kHz.

Sample	S_I(f=1 kHz), A^2/Hz	Noise activation energy, E_{na}, eV	Conductivity activation energy, E_a, eV
$Si_{0.11}Ge_{0.88}$ (process 443)	2.2×10^{-23}	0.36 ± 0.05	0.35
$Si_{0.06}Ge_{0.67}B_{0.26}$ (process 479)	2.1×10^{-20}	0.7 ± 0.1	0.21
$Si_{0.04}Ge_{0.71}B_{0.23}$ (process 480)	6.3×10^{-18}	0.16 ± 0.01	0.18

Comparing noise (E_{na}) and conductivity (E_a) activation energies presented in Table II it can be seen a close correlation between these values for the sample $Si_{0.11}Ge_{0.88}$:H (443) and $Si_{0.04}Ge_{0.72}B_{0.23}$:H (480). Although for the sample $Si_{0.06}Ge_{0.67}B_{0.26}$:H (479) noise activation energy determined in the same way as for the previous samples at $f=10^3$ Hz is remarkably larger than conductivity activation energy. It should be noted, however, that for this sample it was observed a change of noise mechanism with temperature: in the temperature range 300 K $\leq T \leq$ 340 K noise at $f = 10^3$ Hz corresponded practically to "white" noise, while at higher temperature $T > 340$ K it became "1/f" type noise with $S_I \sim f^{-\beta}$. The correlation between E_{na} and E_a observed in two samples allows us to think that noise is due to bulk rather than interface processes. Additionally changing character of noise spectra with temperature can be revealed from temperature dependence of slopes in noise spectra. In two samples (443 and 479) noise spectra demonstrated two-three regions with different slopes. Temperature dependence of slope (s) is presented in Figure 4. Low frequency slope β_1 reduces with temperature in the sample 443 without boron and increases in the samples with boron 479 and 480. β_3 determined at higher

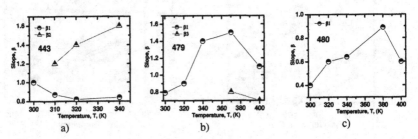

Figure 4. β factor as a function of temperature: a) for the sample 443, b) for the sample 479 and c) for the sample 480. Solid connecting lines are guides for the eyes.

406

frequency than β_l rises with temperature in the sample 443 without boron and reduces in the sample with boron 479. Low frequency slope in the samples with boron 479 and 480 show similar trend with temperature (increases, reaches maximum and decreases) differing from behavior in the sample without boron (continuous reducing and saturation with temperature).

Comparing noise spectra (normalized by current cross section area) presented in Figure 5 for the thermo-sensing film (sample 443) and for micro-bolometer fabricated with the same film it can

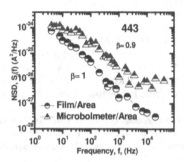

Figure 5. Comparison of noise spectra for thermo-sensing film and micro-bolometer with the same thermo-sensing film (NSD is normalized by current cross section area).

be seen that noise within the entire frequency range studied is higher for micro-bolometer than that in the same thermo-sensing film. Slopes in low frequencies $f < 10^3$ Hz have very close values. For higher frequencies $f > 10^3$ Hz micro-bolometer has higher noise by about 2 orders of value. Moreover device in this region shows "white" noise, while the film shows "1/f " noise. The observed difference in noise spectra for the thermo-sensing film and the micro-bolometer with the same thermo-sensing film means an existence of additional processes inducing noise.

CONCLUSIONS

Noise spectra $S_I(f)$ have been studied for $Si_xGe_yB_z$:H thermo-sensing films used in micro-bolometers. These films were deposited by LF PECVD with variation of gas mixture composition. Strong dependence of noise spectra on composition was observed. This dependence can not be related only to conductivity difference. The measurements of noise spectra at different temperatures have revealed a thermal activated increase of noise with temperature. The correlation observed for two samples between noise and conductivity activation energies suggests that noise is due to bulk rather than interface processes. Noise observed in micro-bolometer was higher than that for the thermo-sensing film in the entire frequency region. $S_I(f)$ for micro-bolometer structure showed difference in slopes (β factors) in comparison with that for the thermo-sensing film. Both observations allow us to think that additional processes induce noise in the device structure.

ACKNOWLEDGMENTS

Authors acknowledge CONACyT project No.48454F for financial support, Dr. Yu. Kudriavtsev (CINVESTAV, Mexico) for SIMS analysis, Dr. S.Rumiantsev (Rensselaer Polytecnic Institute, USA) for useful discussions of the results. I.Cosme acknowledges CONACyT scholarship No. 204372.

REFERENCES:

1. J.L.Tissot and F.Rothan, *Proc. SPIE*, **3379**, 139 (1998).
2. T.Schimert and D.Ratcliff, *Proc. SPIE*, **3577**, 96 (1999).
3. A.Torres, A.Kosarev, M.L.Garcia Cruz, R.Ambrosio. *J.Non-Cryst. Solids*,**329**, 179 (2003).
4. M.Garcia, R.Ambrosio, A.Torres, A.Kosarev. *J.Non-Cryst. Solids,* **338-340**, 744 (2004).
5. A.Kosarev, M.Moreno, A.Torres, R.Ambrosio. *Mater., Res., Symp., Proc.,* **910**, 0910-A17-05 (2006).
6. R.E.Johanson, M.Guenes, S.O.Kasap. *IEE Proc.-Circuits Devices Syst.,* **149 (1)**, 68 (2002).
7. A.Ahmed and R.N.Tait. *Infrared Physics and Technology*, **46**, 468 (2005).
8. M. M. Moreno, A. Kosarev, A.J. Torres, and I. Cosme. *Mater. Res. Soc. Symp. Proc.*, **1066** 1066-A18-05 (2008).
9. M.Moreno, A.Kosarev, A.Torres, R.Ambrosio. *Int. J. High Speed Electronics and Systems*, **18(4)** 1045 (2008).

Mater. Res. Soc. Symp. Proc. Vol. 1153 © 2009 Materials Research Society 1153-A19-06

Electronic detection and quantification of ions in solution using an a-Si:H field-effect device

J. Costa[1], M. Fernandes[1], M. Vieira[1], G. Lavareda[2], C. N. Carvalho[3], A. Karmali[4]
[1]Electronics Telecommunication and Computer Dept. ISEL, R.Conselheiro Emídio Navarro, 1949-014 Lisboa, Portugal. E-mail: jcosta@deetc.isel.ipl.pt
[2]CTS-UNINOVA, Monte da Caparica 2829-516, Portugal
[3]ICEMS-Av. Rovisco Pais, Lisbon
[4]Chemical Engineering and Biotechnology Research Center, Instituto Superior Engenharia de Lisboa R.Conselheiro Emídio Navarro, 1949-014 Lisboa, Portugal

ABSTRACT

Progress in microelectronics and semiconductor technology has enable new capabilities in the field of sensor construction, particularly of pH sensors based on field-effect transistors (FETs) as transducers of chemical signal. While crystalline devices present a higher sensitivity, their amorphous counterpart present a much lower fabrication cost, thus enabling the production of cheap disposable sensors for use in the food industry. Interest in biosensors consisting of a semiconductor transducer and a functionalized surface with biomolecule receptors (BioFET) continues to grow as they hold the promise for highly selective, label-free, real-time sensing as an alternative to conventional optical detection techniques.

We have been involved in the development of a biosensor where the enzymatic activity of recombinant amidase from *Escherichia coli* is coupled to a semiconductor transducer for the detection of toxic amides in food and industrial effluents. The devices were fabricated on glass substrates by the PECVD technique in the top gate configuration, where the metallic gate is replaced by an electrolytic solution with an immersed Ag/AgCl reference electrode. Silicon nitride is used as gate dielectric enhancing the sensitivity and passivation layer used to avoid leakage and electrochemical reactions. In this article we report on the semiconductor unit, showing that the sensor displays the desired current-voltage characteristics. In addition we present an electrical model of the device, in agreement with the experimental data, that is sensitive to the pH of the solution.

INTRODUCTION

This paper reports our efforts in the development of an ion sensitive field effect transistor (ISFET) sensor to detect toxic amides in food. The scientific interest arises from the fact that there are no biosensors to assay for either aliphatic or aromatic amides. On the other hand, there is great interest as far as general public health is concerned because significant levels of acrylamide were detected in food (eg. chips and breakfast cereals). Among various requirements, the sensor should be inexpensive and compact so that it can be used *in situ*, thus the interest in using amorphous silicon.

The hydrogenated amorphous silicon (a-Si:H) technology allows the fabrication of a broad range of devices, such as thin film transistors (TFT) and photodiodes, on a variety of low cost rigid or flexible substrates, like glass metal foil or polymeric films. Additionally, the low-temperature processing and the good optoelectronic properties makes a-Si:H an attractive material for the development of low cost sensors for the detection of small quantities of biological molecules. In the past, these devices have been used for the detection of hydrogen in the gas phase using a MISFET structure with a palladium gate, which acts as a catalyzer for the dissociation of the molecular hydrogen [1]. Later Gotoh et al. [2] reported the use of a-Si:H as the active material in an ISFET for biosensing applications in an electrolyte medium. Since then, the changes in the gate voltage at fixed source–drain voltage and source–drain current, caused by pH variations in the electrolyte, have been thoroughly investigated for a variety of dielectric gate materials [3].

This paper is mainly concerned with the ISFET unit which will be described in the next section. In order to obtain a sensor that is sensitive to toxic amides, such as acrylamide, the ISFET unit must be integrated with a membrane containing immobilized recombinant amidase from *Escherichia coli*. The role of the amidase is to catalyze the hydrolysis of acrylamide into ammonia. The acrylamide is detected indirectly by measuring the presence of the ammonium ions. A specific zeolite, such as clinoptilolite, is deposited on the ISFET surface making the device selective to the ammonium ions.

One interesting aspect concerning the ISFET unit we suggest here is the fact that is totally based on amorphous silicon technology, leading to low production costs and possibly a disposable sensor. Nevertheless its performance should be comparable to its crystalline counterpart thus making it an interesting solution.

DEVICE FABRICATION

Due to the operation principle of the ISFET the top gate configuration was chosen, the structure of the device is sketched in Figure 1. The source and drain contacts where patterned on an Indium Thin Oxide (ITO) covered glass, defining a channel whose dimensions are: Width (W) =3500μm and Length (L)=100 μm thus with a geometric factor W/L=35. Next, the semiconductor and insulator layers were fabricated sequentially by plasma-enhanced chemical vapour deposition (PECVD) under the deposition conditions shown in Table 1. The resulting thicknesses are 1200 Å and 2000 Å for the a-Si:H and a-SiN layers respectively. The final step, prior to encapsulation, was the patterning of the dual layer in order to expose the source and drain contacts. The device was then encapsulated using epoxy resin leaving exposed only the a-SiN pH-sensitive surface over the channel region.

Table 1- Fabrication conditions

Material	a-Si:H	a-SiN
Substrate temperature (°C)	300	350
Gas pressure (mTorr)	163	172
RF power (W)	20	20
Silane flow (SCCM)	20	16
Ammonia flow (SCCM)	-	4

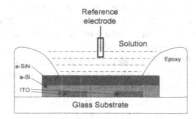

Figure 1- a-Si:H ISFET structure

I-V Characteristics

After assembling, the devices where tested by measuring the I_D-V_{DS} and I_D-V_{GS} as in conventional MOSFET's. The operation of the two devices is very similar, except for the fact that metal gate is replaced by a buffer solution and a reference electrode. By dipping the ISFET and the reference electrode (Ag/AgCl) in a buffer solution with known pH the curves can be obtained, by stepping the voltage applied to the drain and reference electrode, as presented in Figure 5. Apart from the low current levels observed, and the comparatively high off currents, the devices shows good field effect behaviour with a threshold voltage (V_t) around -1V.

ELECTRICAL MODEL

Based on the experimental results and device configuration an electrical model is presented and supported by a SPICE simulation. We have used a methodological approach inspired in the behavioral model of Martinoia and Massobrio for crystalline ISFETs, which was adapted to the a-Si:H transistor studied here. The SPICE simulation was performed using the software AIM-SPICE [4].

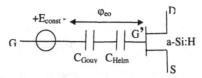

Figure 2-Equivalent circuit of the ISFET device, showing the capacitors and potentials described in the text.

In ISFETs the threshold voltage provides a mean to detect the ion concentration in solution. The gate voltage is the voltage of the reference electrode, so the threshold voltage V_{th}^{ISFET} contains terms which reflect the presence of the electrolyte [5]:

$$V_{th}^{ISFET} = V_{th} + E_{ref} + \varphi_{lj} + \chi_e - \varphi_{eo} - \frac{\phi_m}{q} \tag{1}$$

In the previous expression E_{ref} is the potential of the reference electrode relative to the hydrogen electrode, χ_e is the electrolyte-insulator surface dipole potential, ϕ_m is the workfunction of the metal contact, φ_{lj} the liquid-junction potential difference between the reference solution and the electrolyte and V_{th} is the usual threshold voltage. All terms are constant and can be represented by a single term E_{const}, except φ_{eo} which depends on the ion concentration. It is this term that makes the ISFET sensitive to the electrolyte pH.

We have considered the ISFET divided in two stages: an electronic stage, the a-Si:H transistor, and an electrochemical stage which includes the capacitances of the double layer and the interface potentials of the reference electrode (see Figure 2). From the site-binding theory we have a relation between the proton concentration H_s and the surface charge density σ_0, at the electrolyte-insulator interface [6]. For the Si_3N_4 insulator the expression is the following:

$$\sigma_0 = qN_{Sil}\left(\frac{H_S^2 - K_A K_B}{H_S^2 + K_A H_S + K_A K_B}\right) + qN_{Nit}\left(\frac{H_S}{H_S + K_N}\right) \qquad (2)$$

where q is the electron charge; K_N the dissociation constant for the amine sites; K_A, K_B the two dissociation constants for silanol sites; N_{Nit} and N_{Sil} the surface densities of primary amine and silanol sites, respectively. The Boltzmann factor provides the relation between the bulk proton concentration (H_b) and the interface proton concentration (H_s):

$$H_s = H_b \exp(-\varphi_{eo}/V_T) \qquad (3)$$

where φ_{eo} is the electrolyte-insulator interface potential.

On the other hand, from the electrical double layer theory, and assuming, as usual, that the charge density in the semiconductor is much smaller than σ_0 [7,8], we have the following relation:

$$\sigma_0 = C_{eq}\varphi_{eo} \qquad (4)$$

where C_{eq} is the series capacitance formed by the Gouy-Chapman and Helmholtz capacitances [6,7]:

$$C_{Helm} = \frac{\varepsilon_{IHP}\varepsilon_{OHP}}{\varepsilon_{OHP}d_{IHP} + \varepsilon_{IHP}d_{OHP}}WL \qquad\qquad C_{Gouy} \cong \frac{\sqrt{8\varepsilon_w kT}}{2V_T} \qquad (5)$$

In the expressions, W and L are the ISFET channel width and length; ε_{OHP} and ε_{IHP} the dielectric constants of the outer and inner Helmoholtz layers, d_{OHP} and d_{IHP} the length of the inner and outer Helmholtz layers, ε_w the dielectric constant of the electrolyte, T the temperature, k the Boltzmann constant and $V_T=kT/q$ the thermal voltage.

When equations (3) and (4) are inserted into equation (2) the result is a non-linear equation in φ_{eo} which is dependent on proton concentration (pH) and density of binding sites N_{Nit} and N_{Sil}. The equation can be solved numerically to determine the value of φ_{eo} to be used in the equivalent circuit of Figure 2. The electrochemical parameters, such as number of binding sites have been taken from the literature [5]. The derivative of φ_{eo} with respect to the pH was obtained and gives the maximum sensitivity of the device, which is shown in Figure 3. One can conclude that for an acidic solution the sensitivity is highly dependent on the density of amine binding sites and thus there is advantage in using silicon nitride as gate dielectric.

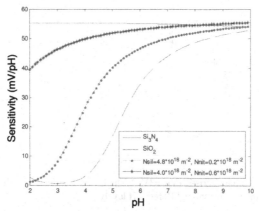

Figure 3- ISFET sensitivity ($-d\varphi_{eo}/dpH$) as a function of the pH for different densities of binding sites. Four situations are displayed including Si_3N_4 ($N_{Sil}=3x10^{18}$ m^{-2}, $N_{Nit}=2x10^{18}$ m^{-2}) and SiO_2 (absence of amine sites).

For the a-Si:H stage we have used the model developed by Schur et. al [9], with the physical dimensions set according to our device and the remaining parameters extracted from the experimental curves using the tool AIM-extract. The fitting was reasonable successfully as can be seen in Figure 4. The SPICE simulation was performed using the software AIM-SPICE [4].

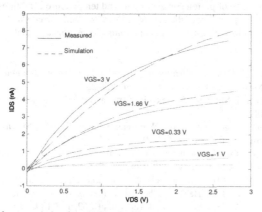

Figure 4- I_{DS}-V_{DS} charactreristics for pH 7 and different V_{GS} values

Finally, in Figure 5 we compare the simulated $I_{DS} - V_{GS}$ characteristic for three different pH levels. The sensitivity calculated from the curves is close to the expected Nernstian behavior.

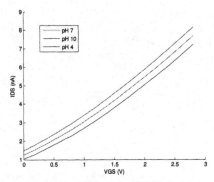

Figure 5- The I_{DS}-V_{GS} characteristic ($V_{DS}=3V$) of the ISFET obtained with the electrical model described in the text for three different pH levels.

CONCLUSIONS

The overall aim of our research is to produce a low cost ion sensitive transistor that can be used in the food industry to detect toxic amides. The ISFET prototype that we report on this paper, an a-Si:H device in the top gate configuration, displays the desired current-voltage characteristics obtained experimentally and can be used to detect small changes in the pH of the solution according to the simulated results. Further studies will be necessary to determine the response of the sensor in a wide range of proton concentrations and temperature conditions. Future work will focus on improving the geometry of the device to increase drain currents and also on its integration with the biological sensing unit containing the enzyme and zeolite.

REFERENCES
[1] A. D'Amico and G. Fortunato, Ambient sensors, in Semiconductors and Semimetals, vol. 21, J. I. Pankove, Ed. New York: Academic, 1984, pp. 209–237.
[2] M. Gotoh, S. Oda, I. Shimizu, A. Seki, E. Tamiya and I. Karube, Sensors and Actuators, 16, 1/2, pp. 55–65, (1989).
[3] Hung-Kwei Liao, Jung-Chuan Chou, Wen-Yaw Chung, Tai-Ping Sun and Shen-Kan Hsiung, Sensors and Actuators B, 50, 104-109, (1998).
[4]T. A. Fjeldly, T. Ytterdal, and M. Shur, Introduction to Device Modeling and Circuit Simulation, John Wiley & Sons, New York, 1998.
[5] S. Martinoia, G. Massobrio, L. Lorenzelli, Sensors and Actuators B, 105, 14-27, (2005).
[6] D.E. Yates, S. Levine, T.W. Healy, J. Chem. Soc. Faraday Trans. 70, 1807–1818, (1974).
[7] S. Martinoia, G. Massobrio, Sensors and Actuators B, 62, 182-189, (2000).
[8] L. Bousse, N.F. de Rooij, P. Bergveld, Operation, IEEE Trans. Electron Devices ED, 30, 10, 1263–1270, (1983).
[9] M. Shur and M. Hack, J. Appl. Phys. 55, 10, 3831, (1984).

Poster Session:
Large Area and Flexible Processing

Effects of electro-mechanical stressing on the electrical characterization of on-plastic a-Si:H thin film transistors

Jian Z. Chen[1,*], Chih-Yong Yeh[2], I-Chung Chiu[2], I-Chun Cheng[2,3], Jung-Jie Huang[4], Yung-Pei Chen[4]

[1]Institute of Applied Mechanics, National Taiwan University, Taipei 10617, Taiwan
[2]Graduate Institute of Photonics and Optoelectronics, National Taiwan University, Taipei 10617, Taiwan
[3]Dept. of Electrical Engineering, National Taiwan University, Taipei 10617, Taiwan
[4]Display Technology Center, Industrial Technology Research Institute, Chutung, HsinChu County 310, Taiwan

ABSTRACT

We analyzed the effect of electromechanical stressing on the electrical characteristics of hydrogenated amorphous silicon thin-film transistors. It had been shown that the TFTs, fabricated at 150 °C, respond to tension/compression by a rise/fall in electron mobility. In TFTs fabricated using the same process, a slight shift of threshold voltage was observed under prolonged high compressive strain and the gate leakage current slightly increases after ~2% compressive strain. In general, the change of TFT performance due to pure mechanical straining is small in comparison to electrical gate-bias stressing. From the comparison among Maxwell stress (induced by electrical gate-bias stressing), mechanical stress (applied by bending), and drifting electrical force for passivated hydrogen atom, the most significant cause for the change of electrical characterization of a-Si:H TFTs should be the trapping charges inside the dielectric, under combined electrical and mechanical stressing. The mechanical stress does not act on Si-H bonds to drift hydrogen atoms, while it is mainly balanced by the rigid Si-Si networks in a-Si:H or a-SiN$_x$. Therefore, mechanical stress has very little effect on the instability of low temperature processed a-Si:H TFTs.

INTRODUCTION

Flexible electronics requires devices working during and after mechanical straining. a-Si:H TFTs, one of the most promising flexible electronic components, are more defective when fabricated at low process temperature (100-200°C) than those made in the standard range of process temperatures of 250-350°C [1]. The stability of a-Si:H TFTs under mechanical stress has been studied by several groups [2-6]. Gleskova *et al.* studied the effect of short-term mechanical strain on TFTs processed on Kapton at 150°C [2-4]. They observed a rise of the electron field effect mobility in tension and a drop in compression, and a very small change in subthreshold slope in the opposite direction [2-4,6]. In our previous study, we found small but measurable changes in the electrical characteristics when a very large compressive mechanical strain was applied for a long duration. The gate leakage current slightly rises when the applied compressive strain exceeds 2% [5]. Won *et al.* experimented on the performance of a-Si:H TFTs under the influence of combined electrical and mechanical stresses. Their results agree with the findings from other research groups: (1) the shift of threshold voltage (instability) is mainly due to electrical gate-bias stressing; (2) the mechanical stress (applied by bending) causes the change of mobility [6]. Over all, the change of TFT electrical characteristics due to pure mechanical straining is small in comparison to that caused by electrical gate-bias stressing.

Under electrical gate-bias stressing, there are two mechanisms identified as being responsible for the instability of TFTs – charge trapping in the gate dielectric and defect creation in the a-Si:H channel layer [7-13]. The defect creation is due to the breaking of Si-H bonds and the drifting of passivated hydrogen atoms [18]. More recent studies have shown that Si-Si bond breaking is the rate limiting step for defect creation in a-Si:H TFTs under these circumstances [26]. There is no clear cut for so-called high and low electric field, for it is also determined by the defect nature of the dielectrics [1].

In this paper, we re-examined the electrical characteristics of PECVD SiN_x dielectric for low temperature processed a-Si:H TFTs. The quality of SiN_x is affected by the deposition power [23-24]. Combined with studies regarding defects of SiN_x:H, we provided more information for the abnormal temperature dependent stability of low temperature processed a-Si:H TFTs using SiN_x:H as gate dielectric [14], which is quite different from TFTs made at conventional temperatures ($250^{\circ}C\sim350^{\circ}C$) [15-18]. We then compare the orders of magnitude of the Maxwell stress (induced by gate-bias stressing), the mechanical stress (applied by bending), and the drifting electrical force for passivated hydrogen atoms. According to these estimations, for low-temperature processed a-Si:H TFTs, the instability of TFTs (shift of threshold voltage) should be most plausibly caused by the injected charges in the gate dielectric during the gate-bias electrical stressing. The mechanical stress, mainly balanced by Si-Si rigid network in a-Si:H material [25], is not the direct cause for the drifting of mobile hydrogen atoms, however, it does lower the activation energy for the breakage of Si-Si bonds [26]. Therefore, a-Si:H TFTs are subjected to a larger threshold voltage shift when mechanical stresses are applied in addition to electrical gate-bias stressing.

EXPERIMENTAL RESULTS AND DISCUSSION

I. Effect of deposition power on the quality of SiN_x:H dielectric

Figure 1 shows the schematics of the sample for the leakage current measurement. A 200 nm MoW layer (metal 1) was deposited by sputtering, followed by photolithography to define the pattern. Then a 100 nm SiN_x dielectric layer was deposited by plasma enhanced chemical vapor deposition (PECVD) with four different powers, 250W, 1000W, 1500W, and 2000W. Via holes were opened to make contact to metal 1. A 200 nm aluminum layer (metal 2) was then deposited and patterned as electrodes to measure the leakage currents. I-V characteristics of these samples were then measured by HP 4156 semiconductor parameter analyzer at various temperatures.

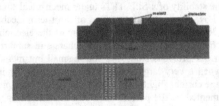

Metal 1 : 200 nm MoW by sputter
Dielectric : 100 nm SiN_x by PECVD
Metal 2 : 200 nm Al by sputter

Figure 1 Schematics of samples for SiN_x dielectric leakage current measurement.

It is well accepted that carrier conduction in dielectric is due to three major contributions: Poole-Frenkel (PF) mechanism, Fowler-Nordheim (FN) mechanism, and hopping process of thermally excited electrons [15,19]. PF is usually dominant at high fields and high temperatures; FN is dominant at high fields and low temperatures; and hopping process is dominant at relative low fields and moderate temperatures. In real dielectrics, there is no clear-cut in field strength and temperature between these processes [15,19,22].

Figure 2 and 3 show the PF and FN plots at temperatures of 30°C and 70°C, respectively. In a specified material system, the characteristics of PF is that $ln(I/E)\sim E^{1/2}$ at fixed temperature and $ln(I)\sim 1/T$ at fixed electric field, in which I is the current and E is the electric field. In FN mechanism, the characteristics is $ln(I/E^2)\sim -1/E$, and temperature is not a dependent parameter in FN process.

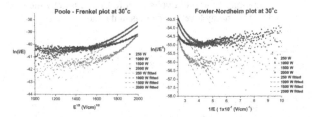

Figure 2 SiN$_x$ dielectric leakage current measurement at temperature of 30°C.

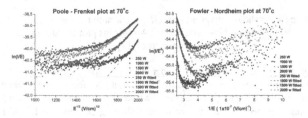

Figure 3 SiN$_x$ dielectric leakage current measurement at temperature of 70°C.

From Fig. 2 and Fig. 3, both FN and PF linearly well fit to the data at high electric field. This indicates that the conduction mechanism in the SiN$_x$ dielectric is mixed at temperatures of 30 °C and 70 °C. At the operation temperature of 30 °C, among the four deposition powers for SiN$_x$, the SiN$_x$ dielectric is least leaky when deposited at a power of 1500 W. For the SiN$_x$ deposited at a power of 250 W, it performs better at high electric field comparing to those deposited at a power of 1000 W and 2000 W. On the other hand, when the SiN$_x$ dielectric is tested at a temperature of 70 °C, 250 W deposited SiN$_x$ exhibits the best dielectric property. This temperature dependence of I-V characteristics confirms the presence of the PF mechanism. This also indicates that the increase of PECVD SiN$_x$ deposition power may improve the dielectric property [23-24], but too large deposition power may render poor quality SiN$_x$, resulted from the ion bombardment in plasma during deposition.

II. Abnormal temperature-dependent stability of low-temperature processed a-Si:H TFTs

Defects are more prone to exist in a-Si:H TFTs fabricated at low process temperatures (100-200°C) [1]. The nature of defects in a gate dielectric significantly affects the electrical response of low temperature processed TFTs. For instance, low temperature processed a-Si:H

TFTs exhibit abnormal temperature-dependent stability in response to the gate-bias stressing [14], in comparison to those made at conventional temperature (250-350°C) [15-18].

In ref.[14], the amorphous silicon thin-film transistors (a-Si:H TFTs) were fabricated at 150°C. In that experiment, at ~50°C, release of trapped electrons in shallow defects of these nitrogen-rich PECVD SiN$_x$ dielectrics causes a saturation of threshold voltage shift under gate-bias electrical stressing (Fig. 4(b)), as well as the abrupt increase of gate leakage current (Fig. 4(c)) and off current (Fig. 4 (a)). In comparison to PECVD processed silicon-rich SiN$_x$, in which the defects are deeply situated near the mid-bandgap, nitrogen-rich SiN$_x$ usually consists of defects located shallowly from the conduction band edge [21-22]. Therefore, the temperature for the occurrence of the saturation of threshold voltage shift ΔV_t, caused by the electrical gate-bias stressing, is determined by the energy levels of defects in the dielectric.

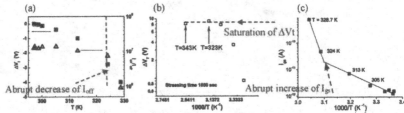

Figure 4 (a) Drop of threshold voltage and on-off current ratio temperature. (b) Arrhenius plot of ΔV_T after 1850 sec of gate-bias stressing. (c) Arrehenius plot of gate leakage current vs temperature. [14]

III. Transient breakdown stage of dielectric under electromechanical stressing

A transient breakdown stage of the dielectric was also found after alternating electrical and mechanical stressing [5]. In this experiment, we stepwisely increased the compressive strain and straining time, applied to the TFT with the following sequence – 1.15% 10min, 1.15% 20min, 1.37% 10min, 1.37% 20min, 1.37% 30min, 2.0% 10min, 2.0% 20min, 2.0% 30min, 2.0% 60min, 2.34% 10min and 2.34% 20min. No apparent instability (increase of threshold voltage) was found and TFTs are well-functional before completely breakdown at mechanical strain-time of 2.34%-20min, confirming the results from literatures [2,3]. However, for TFTs with exceptional high quality (low gate leakage current, low defects) a-SiN$_x$:H dielectric (low gate leakage current of ~1.88A/μm^2, TFT dimension W/L = 400μm/40μm), a transition region of dielectric degradation was observed before complete breakdown of TFTs, as shown in Fig. 5. Figure 5(a) is the characteristics of TFTs prior to straining. The gate leakage current before straining is de facto the measurement limit of our instrumentation ~ 0.1pA. When the applied strain-time was increased to 2.0%-20min, there is a rise of gate leakage current, as shown in Fig. 5(b), indicating a degradation of gate dielectric. This gate leakage current increases exponentially in V_g, which is the typical I-V relation for dielectrics [19]. This risen gate leakage current stays through 2.0%-30min, 2.0%-60min, 2.34%-10min strain-time until the application of 2.34%-20min stain-time, at which the gate dielectric completely breaks down such that TFTs can not be turned on anymore, as shown in Fig. 5(c).

Repeated electrical and mechanical stressing on SiN$_x$ can accelerate the growth of the defects or the micro-cracks, especially for those with trapped charges [20]. The slight increase of gate leakage current could be a sign before complete breakdown of the gate dielectric.

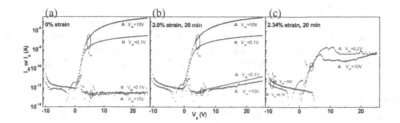

(a) TFT characteristics without straining. (b) A rise of gate leakage current occurs when the strain-time was increased to 2.0%-20min. (c) After the application of 2.34%-20min stain-time, the gate dielectric is completely breakdown. [5]

Without considering the stress concentration for electrical fields and mechanical stresses due to device geometry, the corresponding mechanical stress induced in a-Si:H or SiN$_x$ layer is ~2 GPa for mechanical strains of ~2%, which is the typical maximal allowable compressive strain without devastating failure of a-Si:H TFTs [2-5]. With the typical gate-bias stressing voltage of 20 V, the resultant Maxwell stress $\sigma_{maxwell}$ at the a-Si:H and SiN$_x$:H interface is estimated to be ~50 kPa by $(\varepsilon_{Si} - \varepsilon_{SiNx}) E^2$. The electric drifting force on a single valence charged atom is ~ 6 MPa, estimated by qE divided by atomic cross section area. The latter two types of stresses are at least two orders of magnitude smaller than the stress corresponding to ~2% mechanical strain.

From experimental investigations, although the applied mechanical stress is much greater than electrical stress, surprisingly, electrical stressing has a more significant effect on the instability of a-Si:H TFTs [2-6]. It is highly plausible that the defect level configuration in the dielectric dominates in the stability behavior of low temperature processed a-Si:H TFTs. On the other hand, because the rigidity of a-Si:H is supported by Si-Si networks [25], the applied mechanical stresses are mainly balanced by the rigid Si-Si networks in a-Si:H. As such, the mechanical stresses do not directly drift mobile hydrogen atoms, yet the mechanical stresses could lower the activation energy for the creation of the dangling bonds [26], implying that the Si-Si bonds are weaken under externally applied mechanical stresses. In this regard, pure mechanical stress contributes very little to the shift of threshold voltage, but extra amount of threshold voltage shift reveals when TFTs are subjected to the mechanical stressing in addition to the electrical gate-bias stressing.

Conclusion

Due to the complicated material nature of low temperature processed a-Si:H TFTs, the characterization of a-Si:H TFTs under the influence of electro-mechanical stressing can not be theoretically analyzed merely with macroscopic mechanics. The defect levels in dielectric plays dominant role in the stability of low temperature processed a-Si:H TFTs under the electro-mechanical stressing. The applied mechanical stresses are mainly balanced by the rigid Si-Si networks in a-Si:H or a-SiN$_x$, to lower the activation energy for the creation of the dangling bonds. Pure mechanical stresses contribute very little to the shift of threshold voltage. When TFTs are subjected to mechanical stressing, extra threshold voltage shift reveals due to the mechanical stresses applied in addition to electrical gate-bias stressing.

ACKNOWLEDGMENTS
The authors would like to thank Display Technology Center, Industrial Technology Research Institute, Taiwan, for the preparation of dielectric leakage current measurement samples. J.Z.C. and I.C.C gratefully acknowledge partial financial support from National Science Council, Taiwan, under Grand nos. NSC97-2221-E-002-052 (JZC) and NSC97-2221-E-002-236 (ICC).

REFERENCES
1. C. -S. Yang, L. L. Smith, C. B. Arthur, and G. N. Parsons, J. Vac. Sci. Technol. B **18**, 683 (2000).
2. H. Gleskova, S. Wagner, W. Soboyejo, and Z. Suo, J. Appl. Phys. **92**, 6224 (2002).
3. H. Gleskova, S. Wagner, and Z. Suo, Appl. Phys. Lett. **75**, 3011 (1999).
4. H. Gleskova and S. Wagner, Appl. Phys. Lett. **79**, 3347 (2001).
5. J. Z. Chen, I.-C. Cheng, S. Wagner, W. Jackson, C. Perlov, C. Taussig, Mater. Res. Soc. Symp. Proc. **989**, 0989-A09-04 (2007); J. Z. Chen, C. R. Tsay, I-C. Cheng, H. Gleskova, S. Wagner, W. B. Jackson, C. Perlov, and C. Taussig, presentation in 7th Annual USDC Flexible Electronics and Displays Conference and Exhibition '07, Phoenix, AZ, USA.
6. S. H. Won, J. K. Chung, C. B. Lee, H. C. Nam, J. H. Hur, and J. Jang, J. Electrochem. Soc. **151**, G167 (2004).
7. C. van Berkel and M. J. Powell, Appl. Phys. Lett. **51**, 1094 (1987).
8. M. J. Powell, C. van Berkel, and J. R. Hughes, Appl. Phys. Lett. **54**, 1323 (1989).
9. W. B. Jackson and M. D. Moyer, Phys. Rev. B **36**, 6217 (1987).
10. W. B. Jackson, J. M. Marshall, and M. D. Moyer, Phys. Rev. B **39**, 1164 (1989).
11. Y. Kaneko, A. Sasano, and T. Tsukada, J. Appl. Phys. **69**, 7301 (1991).
12. S. C. Deane, R. B. Wehrspohn, and M. J. Powell, Phys. Rev. B **58**, 12625 (1998).
13. R. B. Wehrspohn, S. C. Deane, I. D. French, I. Gale, and M. J. Powell, J. Appl. Phys. **87**, 144 (2000).
14. J. Z. Chen and I.-C. Cheng, J. Appl. Phys. **104**, 044508 (2008).
15. M. J. Powell, Appl. Phys. Lett. **43**, 597 (1983).
16. F. R. Libsch and Kanicki, Appl. Phys. Lett. **62**, 1286 (1993).
17. M. J. Powell, C. van Berkel, and J. R. Hughes, Appl. Phys. Lett. **54**, 1323 (1989).
18. W. B. Jackson, J. M. Marshall, and M. D. Moyer, Phys. Rev. B **39**, 1164 (1989).
19. S. M. Sze, *Physics of Semiconductor Devices*, 2nd ed. (1983).
20. R. M. McMeeking, J. Appl. Phys. **62**, 3116 (1987).
21. R. Kärcher, L. Ley, and R. L. Johnson, Phys. Rev. B **30**, 1896 (1984).
22. M. H. W. M. van Delden, and P. J. van der Wel, IEEE 41st Annual International Reliability Physics Symposium, pp.293 (2003).
23. A. Z. Kattamis, K. H. Cherenack, B. Hekmatshoar, I.-C. Cheng, H. Gleskova, J. C. Sturm, and S. Wagner, IEEE Electron Dev. Lett. **28**, 606 (2007).
24. J.-J. Huang, C.-J. Liu, H.-C. Lin, C.-J. Tsai, Y.-P. Chen, G.-R. Hu, and C.-C. Lee, J. Phys. D: Appl. Phys. **41**, 245502 (2008).
25. R. A. Street, Hydrogenated amorphous silicon (1991).
26. R. B. Wehrspohn, S. C. Deane, I. D. French, I. Gale, J. Hewett, and M. J. Powell, and J. Robertson, *J. Appl. Phys.* **87**, 144 (2000).

Mater. Res. Soc. Symp. Proc. Vol. 1153 © 2009 Materials Research Society   1153-A20-03

Ehsanollah Fathi and Andrei Sazonov
G2N Laboratory, Electrical and Computer Engineering Department, University of Waterloo, 200
University Ave. W., Waterloo, Ontario, Canada

ABSTRACT

In this work, we optimized different thin film silicon layers in a single junction p-i-n solar cell at deposition temperature of 150 °C. Using the optimized doped and undoped layers, 0.5 cm^2 test cells fabricated both on glass and polyethylene naphthalate (PEN) substrates. The cells show identical open circuit voltages and fill factors, whereas the short circuit current and consequently the efficiency of the cell fabricated on glass is higher than the efficiency of 3.99% of the cell fabricated on PEN substrate.

INTRODUCTION

Increasing demand for mobile electronic devices requires adequate power supply. Existing technologies in energy conversion and storage need to keep up with ever increasing capabilities of portable information systems. One possible solution is in integrating thin film silicon solar cells with rechargeable thin film batteries on unbreakable and light-weight plastic foil substrates to provide an uninterrupted power supply compatible with mobile electronics. These self-rechargeable power supply units would eliminate bulky rechargeable batteries and make portable electronics truly mobile [1].

Thermal budget of thin film battery fabrication is lower than that of thin film solar cell. Thus, the battery in an integrated system can either be on top of the solar cell or on the opposite side of the substrate. Due to difficulty in mechanical integrity maintenance on flexible substrates, laminating the battery on the opposite side of the substrate would not be a suitable choice. As a result, one is limited to superstrate configuration, where the substrate must be transparent.

Nearly 85% transparency of plastic foils like polyethylene terephthalate (PET) and polyethylene naphthalate (PEN) in visible and near infrared regions makes them suitable substrates for our target application. To comply with thermal budget of these plastic substrates, fabrication process should be optimized at deposition temperatures not exceeding 150 °C. Several authors reported on amorphous silicon (a-Si:H) and microcrystalline silicon (µc-Si) solar cells on PET and PEN substrates [2, 3]. However, most researchers focused on solar cells with substrate (n-i-p) configuration. In order to reach and exceed performance the substrate cells demonstrate, all layers in superstrate cells made on plastic foil have to be optimized.. Here, we optimized doped and undoped thin film silicon layers at deposition temperature of 150 °C, and employed these layers to fabricate single junction p-i-n solar cells both on glass and PEN substrates.

EXPERIMENT

All silicon films and solar cells investigated in this paper were grown using a conventional multichamber (13.56 MHz) plasma-enhanced vapor deposition (PECVD) system.

This system consists of separate PECVD chambers for undoped and doped film deposition, and a load-lock for handling. A robotic arm transfers substrates between different chambers without breaking the vacuum. Silane (SiH_4) diluted in hydrogen was used for i-layer deposition, and hydrogen diluted trimethylboron (TMB) and phosphine (PH_3) gases were used for doping. Process pressure of 900 mTorr and the substrate temperature of 150 °C were maintained throughout every film deposition. All doped and undoped silicon films were deposited and optimized on 0.5 mm thick 3" Corning Eagle 2000 glass wafers, and single junction solar cells were fabricated on both Corning Eagle 2000 and 125 μm thick Teonex PEN substrates.

Undoped silicon films were deposited by varying silane ratio from 16.6% to 25% while the RF power and deposition time were set to 5 W and 60 min for all depositions. The evolution of structural and electronic properties of n- and p-type hydrogenated silicon films was investigated by gradually varying the phosphine-to-silane flow ratio ($[PH_3] / [SiH_4]$) from 0.5 to 2% and the TMB-to-silane flow ratio ($[TMB] / [SiH_4]$) from 0.25 to 1.5%. Here, 1% phosphine and 1% TMB both diluted in hydrogen were used. The RF power, deposition time and hydrogen dilution ratio ($H_2 / (H_2 + SiH_4)) \times 100\%$ were maintained at 2 W, 60 min, and 99%, respectively. After film deposition, coplanar Al electrodes of 2 mm width, 18 mm length and, 2 mm apart were sputtered on top of the film through a shadow mask to measure the dark- and photoconductivity.

The film crystallinity was estimated from Raman spectra measured in the back-scattering geometry using the Renishaw micro-Raman spectrometer with a 488nm Ar ion laser line focused on ~5 μm^2 spot. The power of the incident beam was kept at 10 mW to avoid the re-crystallization of silicon films.

Using optimized doped and undoped silicon layers, single junction p-i-n solar cells were fabricated on Corning Eagle 2000 and Teonex PEN substrates. The cells consisted of Corning Eagle 2000 glass (or PEN) substrate / sputtered-ZnO (700 nm) / nc-Si p-layer (25 nm) / a-Si:H (340 nm) / nc-Si n-layer (30 nm) / Al. Sputtered Al films approximately 200 nm thick defined different cells with area of 0.5 cm^2 on 3" round substrates. Figure 1 shows the photograph of fabricated cells on the PEN substrate.

Open circuit voltage (Voc) and fill factor (FF) of solar cells were obtained from current-voltage (J-V) measurements performed at 25 °C under AM1.5 solar spectrum with a dual-lamp solar simulator and the external quantum efficiency (EQE) curves were given at short circuit condition.

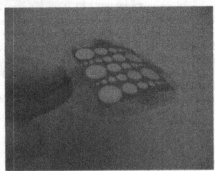

Figure 1: Photograph of fabricated cells on the PEN substrate.

RESULTS AND DISCUSSION

Optimization of doped films

For deposition of doped silicon films we started with hydrogenated nanocrystalline silicon (nc-Si:H) film previously developed at deposition temperature of 300 °C [4], and deposited undoped nc-Si:H at the same process conditions at 150 °C. Then hydrogen diluted phosphine or TMB were added in order to obtain n-type or p-type nc-Si:H, respectively. Figure 2 shows the Raman spectra of undoped nc-Si:H films deposited at 150 °C with thicknesses of 40nm and 15 nm, respectively. This figure shows that as the film becomes thinner, the peak at 520 cm^{-1} ,characteristic of crystalline Si phase, decreases. However, even at 15 nm, the film shows higher crystalline peak at 520 cm^{-1} compare to the peak due to amorphous phase at 480 cm^{-1}. This result is in agreement with the results presented in [4] for the films deposited at 300 °C. Accordingly, even at the substrate temperature of 150 °C, nanocrystalline grains can be formed at early stages of film growth.

Figure 3 shows Raman spectra of p-doped films deposited at different TMB-to-silane flow ratios in the range of 0 - 1.3%. As TMB-to-silane ratio increases, local deformations induced by the impurity atoms lead to a more profound amorphous peak at 480 cm^{-1}. It is concluded that by increasing the concentration of dopants, the film experiences a gradual transition from nanocrystalline to amorphous structure. Similar behavior was observed in Raman spectra of the n-doped films with different phosphine-to-silane ratios and was reported elsewhere [5].

Figure 4 presents the dark conductivities of p-doped and n-doped nc-Si films as a function of the doping gas-to-silane ratio. The conductivities of both p-type and n-type films increase as the dopant gas ratios increased from 0.35% to 1%. This increase is believed to be the consequence of the rise in the number of active dopants in the film as it does not correlate with decreasing crystal fraction (see Figure 3). However, further increase in the ratio causes a rapid drop in dark conductivities due to the film amorphization. Therefore, at the optimum doping gas-to-silane ratio of 1% ([TMB] / [SiH$_4$]) and 0.75% ([PH$_3$] / [SiH$_4$]), maximum dark conductivity was achieved.

Optimization of undoped film

At low deposition temperatures (< 300 °C), hydrogen dilution of silane source gas plays a critical role in tuning thin film silicon microstructure. It has been shown that by increasing hydrogen dilution, the bandgap can be increased and defect density reduced. However, further increase of hydrogen dilution results in reduction of the open-circuit voltage due to transition to mixed amorphous-microcrystalline silicon films. Protocrystalline silicon (pc Si:H) represents the material right at this threshold [7]. Employing this material as an absorber layer, results in better light soaking stability of short circuit current while the open circuit voltage is still high as in amorphous silicon cells. In order to determine experimentally if we have protocrystalline silicon, an analysis of variations in the film optoelectronic properties has to be carried out [3]. Here, we deposited undoped silicon films with various silane-to-hydrogen ratios ([SiH$_4$] / ([SiH$_4$] + [H$_2$])) and used the photosensitivity σ_{photo} / σ_{dark} as the quality factor.

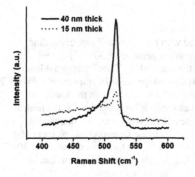

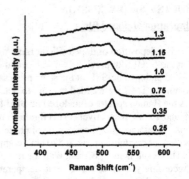

Figure 2: Raman spectra of undoped nc-Si:H films with 40 nm and 15 nm thickness deposited at 150 °C substrate temperature.

Figure 3: Raman spectra of p-doped nc-Si:H films deposited at different TMB-to-silane flow ratios in the range of 0.25-1.3%.

Figure 5 shows the photosensitivity $\sigma_{photo} / \sigma_{dark}$ of the undoped films as a function of the hydrogen-to-silane ratio. One can see that the photosensitivity reaches a maximum of 1.5×10^6 at the hydrogen-to-silane ratio of 18%. As it was mentioned in introduction section, this point indicates the amorphous-to-microcrystalline phase transition. The value of the photosensitivity of the deposited protocrystalline film is quite comparable with the state of the art absorber material requirement ($>10^5$) [6].

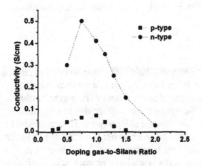

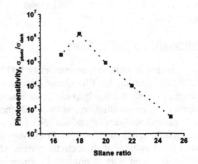

Figure 4: Dark conductivity of p- and n-type nc-Si:H films vs. doping gas-to-silane percentage ratio in the range of 0.25-1.5% for p-type and 0.5 to 2% for n-type films.

Figure 5: Photosensitivity ($\sigma_{photo} / \sigma_{dark}$) of undoped a-Si:H films vs the ratio ([SiH$_4$] / ([SiH$_4$] + [H$_2$] × 100%)).

Solar cell fabrication

Figure 6 shows the J (V) characteristics of solar cells fabricated in the same deposition run on PEN and glass substrates. There were no light trapping or antireflection coatings applied to the fabricated cells. The table inset in Figure 6 shows the performance characteristics of the cells. While V_{oc} and FF of both cells are identical, the higher short circuit current density of the cell fabricated on glass results in a higher efficiency of 4.7% compared to that of the cell fabricated on PEN (3.99%). Figure 7 shows spectral dependencies of transmittance for PEN and glass substrates used in this work. One can see that both substrates have almost the same transmittance for the wavelengths above 400 nm. Furthermore, most photons with wavelengths below 400 nm are either absorbed in ZnO or p-doped nc-Si layer. As a result, the intensity of the light that reaches the intrinsic layer is the same for PEN and glass substrates. Therefore, the lower short circuit current density of the cell fabricated on PEN substrate is attributed to reduced absorption and/or transport properties of undoped layer. We believe that in case of PEN substrate, the actual substrate temperature might have been lower than that for the glass substrate as the specific heat capacity of PEN is about 3 times higher than that of the glass substrates ($2.24 \ J \ g^{-1} \ K^{-1}$ against $0.84 \ J \ g^{-1} \ K^{-1}$).

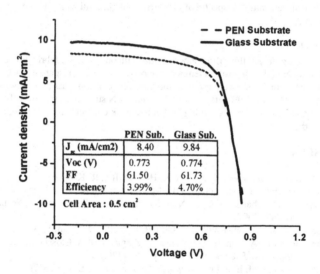

	PEN Sub.	Glass Sub.
J (mA/cm2)	8.40	9.84
Voc (V)	0.773	0.774
FF	61.50	61.73
Efficiency	3.99%	4.70%

Cell Area : 0.5 cm^2

Figure 6: Current density vs. voltage characteristics of p-i-n single junction cells under 1.5 AM illumination fabricated on glass and PEN substrates. The table inset shows the performance parameters.

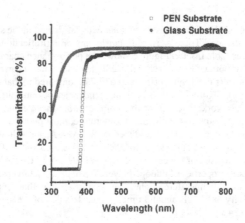

Figure 7: Optical transmittance spectra of bare PEN and glass substrates in the wavelength range of 300-800 nm.

CONCLUSIONS

We have optimized thin film nanocrystalline silicon layers for fabrication of a single junction p-i-n solar cell at deposition temperature of 150 °C on a transparent plastic foil. The optimized process conditions were used to fabricate solar cells on glass and PEN substrates. The efficiencies of 4.7% and 3.99% attained for glass and PEN substrates, respectively. Incorporation of the light management schemes and high reflective back contact can further improve the efficiency of these cells.

REFERENCES

[1]   G. Dennler, S. Bereznev, D. Fichou, K. Holl, D. Ilic, R. Koeppe, M. Krebs, A. Labouret, C. Lungenschmied, A. Marchenko, D. Meissner, E. Mellikov, J. Meot, A. Meyer, T. Meyer, H. Neugebauer, A. Opik, N. S. Sariciftci, S. Taillemite, and T. Wohrle, Solar Energy, **81** 947 (2007).

[2]   J. Bailat, V. Terrazzoni-Daudrix, J. Guillet, F. Freitas, X. Niquille, A. Shah, C. Ballif, T. Scharf, R. Morf, A. Hansen, D. Fischer, Y. Ziegler, and A. Closset, in *20th EU Photovoltaic Solar Energy Conference*, p. 1529 (2005).

[3]   Y. Ishikawa and M. B. Schubert, *Jpn. J. Appl. Phys.* **45**, 6812 (2006).

[4]   M. R. Esmaeili-Rad, F. Li, A. Sazonov, and A. Nathan, *J. Appl. Phys.* **102**, 064512 (2007).

[5]   H. J. Lee, PhD Thesis, University of Waterloo, 2008.

[6]   J. Poortmans and V. Arkhipov, in *Thin Film Solar Cells: Fabrication, Characterization and Applications* (John Wiley & Sons, 2006), p. 191.

[7]   R. W. Collins and A. S. Ferlauto, Current Opinion in Solid State and Materials Science, **6**, 425 (2002).

Poster Session:
Thin-Film Transistors

Mater. Res. Soc. Symp. Proc. Vol. 1153 © 2009 Materials Research Society   1153-A21-04

The effect of active-layer thickness on the characteristic of nanocrystalline silicon thin film transistor

Sun-Jae Kim, Sang-Myeon Han, Seung-Hee Kuk, Jeong-Soo Lee and Min-Koo Han

School of Electrical Engineering, Seoul National University, Gwanak-gu, Seoul 151-742, Korea

ABSTRACT

We fabricated nc-Si TFTs in order to investigate the effect of the active-layer thickness on the characteristic of the nc-Si TFT. Bottom gate nc-Si TFTs were fabricated at 350°C using ICP-CVD. The thicknesses of the nc-Si layer were remained to 700, 1200 and 1700 Å. As the active-layer thickness increases, the mobility and the on-current level were not altered. However, the off-current level increased considerably and on/off ratio decreased. It may be attributed to highly doped characteristic of thick nc-Si film. As the nc-Si film thicker, the conductivity increases considerably and the Fermi level approaches to the conduction band minimum, which indicates the increases of doping level. The oxygen concentration shows high level of unintentional doping. Also, columnar growth of nc-Si film makes that the crystallinity of top region is much higher than that of bottom region. So, the conductivity of thick nc-Si film becomes high compared to that of thin nc-Si film. The structure of the nc-Si TFT with thick nc-Si film can be similar to the serial connection of N+, N- and N+ resistance, so that it suffers difficulty to suppress the off current and to secure high on/off ratio. Therefore, the off current can be suppressed by thinning of the high conducting active nc-Si layer and nc-Si TFT with channel thickness of 700 Å shows good on/off characteristic. It is deduced that bottom gate nc-Si TFT is necessary to have intrinsic channel layer as well as thin channel layer to reduce the leakage current.

INTRODUCTION

Nanocrystalline silicon (nc-Si) thin film transistor (TFT) has attracted considerable attention [1]. It can be fabricated with rather simple process as compared with polycrystalline silicon (poly-Si) TFT since it doesn't need additional recrystallization process. And it also shows better uniformity while recrystallization process causes severe non-uniformity problem in poly-Si TFT [2,3]. nc-Si TFT also has advantage compared to hydrogenated amorphous silicon (a-Si:H) TFT due to better electrical stability [4].

nc-Si TFT can be fabricated by simple process compatible to widely used a-Si:H TFT fabrication process. However, off-current level is rather high in nc-Si TFT as compared with a-Si:H TFT. Several efforts were done for the suppression of leakage current level of nc-Si TFT [5-7]. However, since nc-Si structure has special feature of columnar crystalline growth, which may attribute the conduction mechanism, intrinsic properties of nc-Si film with various thicknesses need to be concretely.

The purpose of our work is to report the effect of active-layer thickness on the off-current characteristic of nc-Si TFT. Using etch-back process, nc-Si TFTs with various active-layer thicknesses were fabricated employing inductively coupled plasma chemical vapor deposition (ICP-CVD). The property of nc-Si film was analyzed from conductivity and activation energy.

EXPERIMENT

We have fabricated nc-Si TFTs of an inverted staggered structure. All the process was carried out based on conventional a-Si:H TFT process. Molybdenum was sputtered with thickness of 2000 Å on the glass substrate and patterned as gate metal. SiN_X of 4000 Å was deposited and nc-Si film of 2000 Å was deposited.

ICP-CVD was employed for the deposition of nc-Si film. ICP-CVD provide certain advantages such as high deposition rate and improved crystallinity of nc-Si film, can be used for fabrication of high quality nc-Si TFTs. ICP-CVD employs remote plasma, which reduces troublesome ion bombardment problem issues [8]. It is also important to note that ICP-CVD generates high density plasma compared with conventional plasma enhanced chemical vapor deposition (PECVD) [9].

The process temperature of the deposition was set to 350 °C. Helium gas was used for the dilution of the reactant gas. The flow rate of Helium : Silane (SiH_4) was 40:3 [sccm]. The base pressure in the reactor was less than 1 mT and the process pressure during the deposition was set to 50 mT and the ICP power was 700W.

The 500Å-thick n+ layer was deposited and the active patterning was carried out. 2000Å-thick molybdenum as the source/drain metal was sputtered and patterned. The n+ layer on the channel region was etched. Fig. 1 shows the cross-sectional TEM images of the fabricated nc-Si TFT. It shows the gate, gate insulator, nc-Si, n+ layer and source/drain metal. The thicknesses of channel region were remained about 700, 1200 and 1700 Å after etch-back process. The high resolution cross-sectional TEM image shows the interface between nc-Si and SiN_X. Clear silicon lattice patterns are observed right on the SiN_X layer, which implies that the incubation layer of amorphous phase is not formed during the nc-Si deposition. I-V characteristics of the TFTs with W=90 μm and L=5 μm were measured using an Agilent 4156 semiconductor parameter analyzer.

Fig. 1 The cross-sectional TEM images of the fabricated nc-Si TFT and high resolution cross-sectional TEM image showing the interface between nc-Si and SiN_X.

For the investigation of property of the nc-Si film, the electrical conductivity in the dark was measured with coplanar aluminum contact, the activation energy was calculated from temperature dependence of dark conductivity. And Raman spectrum was obtained using Jobin Yvon LabRam HR Raman spectrometer.

RESULT

The channel thickness in the bottom-gate nc-Si TFT was varied by controlling etch-back process time. Fig. 2 shows the transfer characteristics of the nc-Si TFTs which have active layers of various thicknesses. The active layer of each TFT is 700, 1200 and 1700 Å. As the thickness of nc-Si film as an active layer increases, the off-current elevates dramatically, while the on-current is not varied. The 31 pA of 700Å-thick nc-Si film increases to 900 nA by almost 30,000 times when the thickness active layer becomes 1700 Å. Table 1 summarizes the mobility, threshold voltage, on/off ratio and off current of the nc-Si TFTs. The mobility is not varied with the change of active layer's thickness, but on/off ratio decreases abruptly and switching characteristic of nc-Si TFT disappears as active layer becomes thick.

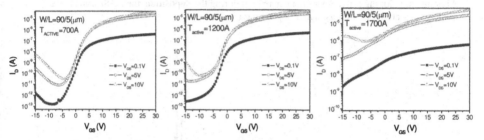

Fig. 2 The transfer characteristics of the nc-Si TFT with various thickness of nc-Si active layer. (a) Active layer thickness = 700 Å. (b) Active layer thickness = 1200 Å. (c) Active layer thickness = 1700 Å.

Table 1 Summarized parameters of nc-Si TFTs with various active layer thicknesses.

nc-Si thickness	700Å	1200Å	1700Å
Mobility (cm²/Vsec)	0.64	0.63	0.67
V_{TH} (V)	1.43	1.43	-2.1
On/Off ratio (@V_{DS}=10V)	1.5×10^6	8.3×10^5	6.7×10^1
Off current	31 pA	61 pA	900 nA

DISCUSSION

The dependence of characteristic of the nc-Si on the thickness of channel layer may be attributed to the conductivity of the nc-Si film. The dark-conductivities of 400, 700, 1200 and 2000 Å-thick nc-Si films are 1.30E-7, 3.40E-6, 2.83E-5 and 1.44E-3 S/cm at room temperature, respectively. The dark-conductivity vs. 1000/T plot is shown in Fig. 3(a). The dark conductivity value increases rapidly as the nc-Si film thicker.

We obtain the value of dark-conductivity activation energy of $\sigma = \sigma_0 \exp(\dfrac{-E_a}{kT})$ by exponential fitting . The activation energy of 400, 700, 1200 and 2000 Å-thick nc-Si films are 0.69, 0.61, 0.45 and 0.11 (eV), respectively. The activation energy vs. the nc-Si film thickness is shown in Fig. 3(b). The value of activation energy indicates how much the position of the Fermi level is below the conduction band edge. It can be clearly observed that the activation energy decreases from 0.69 to 0.11 eV as the nc-Si film thicker. This result is deduced from the narrowing of energy difference between the effective Fermi level and the conduction band edge [10]. Thus, the Fermi level position increases rapidly.

It is generally regarded that a-Si or nc-Si film shows a slightly n-type characteristic [11]. Although any doping process was not carried out during the deposition, the nc-Si film shows higher conductivity as well as higher doped characteristic.

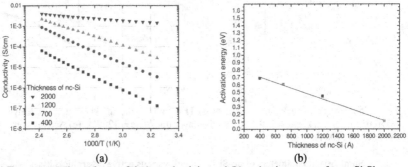

(a)   (b)

Fig. 3 (a) Temperature dependence of dark conductivity and (b) activation energy for nc-Si film with various thicknesses.

The n-type characteristic of the as-deposited nc-Si film may be attributed to the oxygen impurity concentration in the nc-Si film, native defects near conduction band in the nc-Si film and so on. Our SIMS data indicates that the oxygen concentration of the 2000 Å-thick nc-Si films fabricated 350°C is about 10^{20} atoms/cm^2, which is enough for reported level of unintentional doping [12]. The incorporation of oxygen into the nc-Si film can be attributed to the out-diffusion from the SiO$_2$ at the process chamber wall.

Due to the unintentional doping by oxygen incorporation, thick nc-Si film is likely to act as a resistance [13]. In the top of nc-Si, grain size is large and the crystallinity is high due to columnar crystalline growth. So, the conductivity in the top region becomes high compared to that in the bottom region. The structure of the nc-Si TFT with thick nc-Si film can be similar to the serial connection of N+, N- and N+ resistance, as depicted in Fig. 4, so that it suffers difficulty to suppress the off current and to secure high on/off ratio. Therefore, the off current can be suppressed by thinning of the high conducting active nc-Si layer during etch back process and nc-Si TFT with channel thickness of 700 Å shows good on/off characteristic.

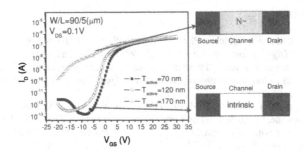

Fig. 4 The schematic of the lateral doping structure of nc-Si TFT with various nc-Si film thicknesses.

CONCLUSION

We have fabricated 350°C nc-Si TFT with bottom gate structure using ICP-CVD. And the leakage current of nc-Si TFT was analyzed with nc-Si TFT with various active-layer thicknesses. The thickness of the nc-Si layer was controlled during etch-back process and remained 700, 1200 and 1700 Å. The nc-Si TFT with 700Å-thick channel showed good electrical characteristic, such as mobility of 0.64 cm^2/V·sec and V_{TH} of 3.18V. As the channel thickness increases, the mobility and the on-current level were not altered.

However, the off-current level elevates from 31pA to 900 nA and on/off ratio decreases from 1.5×10^6 to 6.7×10^1, as active-layer thickness increases from 700 Å to 1700 Å. It may be attributed to highly doped characteristic of thick nc-Si film. As the nc-Si film thicker, the conductivity increases considerably and the Fermi level approaches to the conduction band minimum, which indicates the increases of doping level. The oxygen concentration shows high level of unintentional doping. Also, columnar growth of nc-Si film makes that the crystallinity of top region is much higher than that of bottom region. So, the conductivity of thick nc-Si film becomes high compared to that of thin nc-Si film.

The structure of the nc-Si TFT with thick nc-Si film can be similar to the serial connection of N+, N- and N+ resistance, so that it suffers difficulty to suppress the off current and to secure high on/off ratio. Therefore, the off current can be suppressed by thinning of the high conducting active nc-Si layer during etch back process and nc-Si TFT with channel thickness of 700 Å shows good on/off characteristic.

Our results imply that bottom gate nc-Si TFT is necessary to have intrinsic channel layer as well as thin channel layer to reduce the leakage current.

ACKNOWLEDGMENT

This work was supported by the Korea Science and Engineering Foundation (KOSEF) grant funded by the Korea government (MEST). (No. R0A-2007-000-20117-0)

435

REFERENCES

1. J. H. Lan and J. Kanichi, "Planarization of a-Si TFT for High-Aperture-Ratio AMLCDs", *Tech. Digest of Int. Display Res. Conf. 1997*, pp. L58-L61, (1997).
2. J. S. Im and H. J. Kim, "Phase transformation mechanisms involved in excimer laser crystallization of amorphous silicon films", *Appt. Phys. Lett*. Vol. 63 (14), pp. 1969-1971, (1993).
3. M. Itoh, Y. Yamamoto, T. Morita, H. Yoneda, Y. Yamane, S. Tsuchimoto, F. Funada and K. Awane, "High-Resolution Low-Temperature Poly-Si TFT-LCDs Using a Novel Structure with TFT Capacitors", *Tech. Digest of SID 1996*, pp17~20, (1996).
4. K. Ichikawa, S. Suzuki, H. Maino, T. Aoki, T. Higuchi and Y. Oana, "14.3 in.-Diagonal 16-Color TFT-LCD Panel Using a-Si:H TFTs", *Tech. Digest of SID 1999*, pp. 226-229, (1989).
5. A. T. Hatzopoulos, N. Arpatzanis, D.H. Tassis, C.A. Dimitriadis, M. Oudwan, F. Templier and G. Kamarinos, "Study of the Drain Leakage Current in Bottom-Gated Nanocrystalline Silicon Thin-Film Transistors by Conduction and Low-Frequency Noise Measurements Electron Devices", *IEEE Transactions Electron Devices*,Vol. 54, (2007).
6. H. Lee, A. Sazonov and A. Nathan, "Leakage current mechanisms in top-gate nanocrystalline silicon thin film transistors", *Appl. Phys. Lett.*, Vol.92, (2008).
7. M. R. Esmaeili-Rad, A. Sazonov and A. Nathan, "High Stability, Low Leakage Nanocrystalline Silicon Bottom Gate Thin Film Transistors for AMOLED Displays", *IEDM 2006*,pp. 1-4, (2006).
8. M. Goto, H. Toyoda, M. Kitagawa, T. Hirao, and H. Sugai, "Low Temperature Growth of Amorphous and Polycrystalline Silicon Films from a Modified Inductively Coupled Plasma", *Jpn. J. Appl. Phys*. 36, pp. 3714-3720, (1997).
9. J. Hopwood, "Review of inductively coupled", *Plasma Sources Sci. Technol*. 1, pp.109-116, (1992).
10. L. H. Teng and W. A. Anderson, "Thin film transistors on nanocrystalline silicon directly deposited by a microwave plasma CVD", *Solid-State Electronics*, Vol 48, pp.309~314, (2004).
11. R. Platz and S. Wagner, "Intrinsic microcrystalline silicon by plasma-enhanced chemical vapor deposition from dichlorosilane", *Appl. Phys. Lett.*, Vol. 73, pp.1236~1238, (1998).
12. P. Torres, J. Meier, R. Fluckiger, U. Krol, J. A. Anna Selvan, H. Keppner, A, Shah, S. D. Littlewood, I. E. Kelly and P. Giannoules, "Device grade microcrystalline silicon owing to reduced oxygem contamination", *Appl. Phys. Lett.*, Vol. 69, pp. 1373-1375, (1996).
13. C. Lee, A. Sazonov and A. Nathan, "High-mobility nanocrystalline silicon thin-film transistors fabricated by plasma-enhanced chemical vapor deposition", *Appl. Phys. Lett.*, Vol.86, 222106, (2005).

Mater. Res. Soc. Symp. Proc. Vol. 1153 © 2009 Materials Research Society   1153-A21-05

Effects of Hydrogen Plasma Treatment on Hysteresis Phenomenon and Electrical Properties for Solid Phase Crystallized Silicon Thin Film Transistors

Sung-Hwan Choi, Sang-Geun Park, Chang-Yeon Kim, and Min-Koo Han

School of Electrical Engineering and Computer Science, Seoul National University, 599 Gwanangno, Gwanak-gu, Seoul, 151-744, Korea

ABSTRACT

We have investigated the effects of hydrogen plasma treatment on the hysteresis phenomenon and electrical properties of solid phase crystallized silicon thin film transistors (SPC-Si TFTs) employing alternating magnetic field crystallization (AMFC). We employed H_2 plasma treatment on the SPC-Si active layer before SiO_2 gate insulator deposition. By increasing the power and time duration of H_2 plasma treatment, it was observed that the hysteresis phenomenon of SPC-Si TFT was effectively suppressed and electrical properties such as threshold voltage, field effect mobility was improved considerably. This is due to the role of hydrogen atoms by passivating the defects and grain boundary trap states in SPC-Si film. However, relatively high power and long hydrogen plasma treatment (100W, 5 minutes) could degrade the electrical characteristics of the device. SPC-Si TFT for 100W power of PECVD and 3 minutes with the H_2 plasma treatment exhibit the significant improvement of electrical characterics (V_{TH} = -3.85V, μ_{FE} = 21.16cm^2/Vs), and a smaller hysteresis phenomenon (ΔV_{TH} = -0.30V) which is suitable for high quality AMOLED Display.

INTRODUCTION

Active matrix organic light emitting diode (AMOLED) employing thin film transistor (TFT) pixel circuits have been considered as a future display technology due to its high brightness, compactness, fast response time and wide viewing angle [1]. Therefore the quality of TFTs as well as a life time of electroluminescence should be improved. Recently, solid phase crystallized silicon TFT (SPC-Si TFT) on the glass substrate has gained a considerable attention as a pixel element of the driving circuit in the AMOLED technology because it shows superior current stability than hydrogenated amorphous silicon (a-Si:H) TFTs, and it has better current uniformity than low temperature polycrystalline silicon (LTPS) TFTs. However the threshold voltage of SPC-Si TFT is relatively high and leakage current of device is also large. They also suffer from a hysteresis phenomenon similar to other devices, especially in p-type devices. Residual image sticking has still served as a critical problem in the AMOLED display. Residual image sticking might be observed for a few minutes due to the hysteresis phenomenon of the driving transistor in the AMOLED display [2].

The purpose of this work is to improve the electrical characteristics of SPC-Si TFTs and reduce the hysteresis phenomenon of p-type devices on glass substrates by employing H_2 plasma treatment on the solid phase crystallized silicon before gate insulator deposition. The threshold voltage of SPC-Si TFT was decreased and leakage current level was also reduced with the H_2 plasma treatment. It was also found that the hysteresis phenomenon of the device was also suppressed for driving a high quality AMOLED display based on SPC-Si TFT.

FABRICATION OF SPC-SI TFT

We have fabricated the p-type SPC-Si TFT on a glass substrate employing the conventional coplanar top-gate structure. First, we deposited a 50-nm-thick amorphous silicon (a-Si) active layer by plasma enhanced chemical vapor deposition (PECVD) on the buffer layer. And it was rapidly crystallized at 700°C for 15 minutes employing alternating magnetic field crystallization (AMFC) method [3]. After the thermal annealing process, the a-Si thin film was transformed to solid phase crystallized film with a grain size of 30nm as shown in Fig 1 (a). Hydrogen plasma treatments were performed on the solid phase crystallized silicon at 420°C before the gate oxide was deposited. The poly-Si film was patterned by active mask, and then a 59nm thick SiO_2 (single gate insulator) was deposited as a gate insulator by PECVD at 420°C. Finally, the gate electrode was patterned and source-drain regions were formed by ion implantation. The structure was fabricated to function as the drain and source electrodes by the self-aligned doping process without using an additional mask.

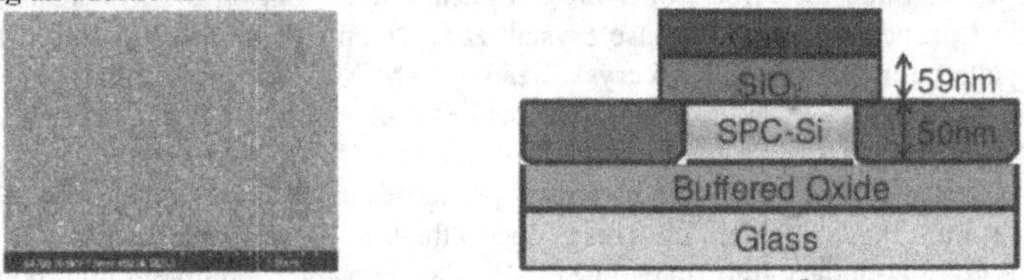

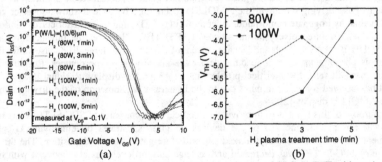

(a)   (b)

Figure 1 (a) SEM micrograph of fabricated SPC-Si thin film, (b) Top-gate p-channel SPC-Si TFT structure fabricated on SiO_2 (300nm) stacked as a buffer layer on glass substrate.

RESULTS AND DISCUSSION
Enhancement of electrical characteristics and reliability in SPC-Si TFT

SPC-Si thin film consists of randomly oriented grain boundaries, containing a high- density of intragrain defects [4]. Trap states in the grain boundary could mainly influence the electrical properties of TFTs [5]. Hydrogen plasma treatment has been widely used to terminate defects, such as dangling bonds [5]. In order to improve the electric characteristics of SPC-Si TFTs, the various H_2 plasma treatments were applied to the solid phase crystallized silicon after the crystallization of amorphous silicon and before the SiO_2 gate insulator deposition at 420°C.

Figure 2 (a) Transfer characteristics and (b) threshold voltages of SPC-Si TFT with various H_2 plasma treatments. All devices were measured with the dimension of (W/L) = (10/6)μm.

The applied plasma (RF) power of PECVD was varied between 80W and 100W. The time required for hydrogen plasma treatment was 1, 3 and 5 minutes (No.1 - No.6), each other. The transfer characteristics of SPC Si-TFTs with various H_2 plasma treatments are shown in Fig 2 (a). As the RF power of PECVD during the H_2 plasma treatment was increased, the threshold voltage of the SPC-Si TFTs was decreased, except that of No. 6 which was treated for 5 minutes with 100W PECVD power as shown in Figure 2 (b).

The improvement of electrical characteristics (such as V_{TH} and field effect mobility) with H_2 plasma treatment could be explained by passivating the trap states and grain boundaries in the solid phase crystallized silicon (SPC-Si) active layer [6]. The SPC-Si active layer has many grain boundaries compared to low temperature poly-Si active layer (Grain size of SPC-Si film is about 20-50 nm) [4]. It is known that trap states are mainly located in the grain boundaries. Because the trap states could trap free carriers in the active layer, they could play a role in degrading the carrier transport in SPC-Si active layer. When the H_2 plasma treatment was applied in the SPC-Si active layer, the trap states were decreased by binding between hydrogen atoms and dangling bonds. This is because the hydrogen atoms could terminate the defect states in the grain boundary [7]. By reducing the number of defect states, free carriers could be effectively activated to the conduction band edge by the applied electrical bias, which contributes to the electrical conductivity of SPC-Si thin film.

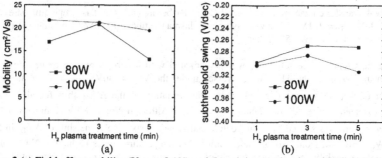

Figure 3 (a) Field effect mobility (V_{DS} = -0.1V) and (b) subthreshold swing of SPC-Si TFT with various H_2 plasma treatments.

The interface between the active layer (SPC-Si) and the silicon dioxide (gate insulator) was also improved by H_2 plasma treatment. However when the H_2 plasma treatment was applied for 5 minutes with 100W power of PECVD, the threshold voltage of SPC-Si TFT was increased and the mobility of SPC-Si TFT was decreased compared to that of lower power (PECVD) or shorter time of H_2 plasma treatment. In the case of high power (100W) with relatively long plasma treatment (5 min), the poor surface of the silicon layer was induced by the excessive hydrogen plasma damage. Due to the existence of a damaged interface between active layer and gate insulator, the threshold voltage of No. 6 (100W, 5 min) was increased compared to other hydrogen plasma treated samples. Excessive hydrogen plasma treatment would have a rather deleterious effect upon the gate insulator and SPC-Si thin film interface region.

The field effect mobility (μ_{FE}) was also increased as the power of PECVD was increased from 80W to 100W for 1 minute and 3 minutes of hydrogenation plasma treatment duration as shown in Figure 3(a). This improvement in field effect mobility could be explained by reducing the defect state density with reconstruction of dangling bonds and redistribution of hydrogen atoms into the interface regions. Because the existence of the trap levels in the grain structure contribute to the potential barrier height, the electrical conductivity of the SPC-Si TFT is principally influenced by the grain boundaries [4]. I_{ON}/I_{OFF} current ratio also increases and the subthreshold swing improves due to the termination of grain boundary and reduction of intragrain trap states by hydrogen plasma treatment as shown in Figure 3 (b). However, there is no longer improvement of electrical characteristic such as mobility and subthreshold swing in the case of hydrogen plasma treatment time (5 min). Actually, some degradation was even observed if the SPC-Si TFT were exposed to hydrogen plasma for longer than 5 minutes with 100W power of PECVD. This is due to excessive etching effect of SPC-Si surface by hydrogen atoms [8].

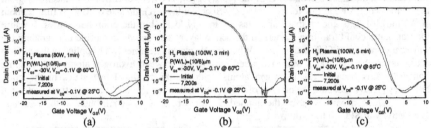

(a)   (b)   (c)

Figure 4 Transfer characteristics of the SPC-Si TFT before and after bias temperature stress at V_{GS}= -30V, V_{DS}= -0.1V, 7200s with hydrogen plasma treatment (a) 80W / 1 minute, (b) 100W / 3 minutes and (c) 100W / 5 minutes.

The bias temperature reliability was investigated with hydrogen plasma treated devices as shown in Figure 4. The TFTs were stressed under the V_{GS} = -30V, V_{DS} = -0.1V at elevated temperatures (60°C) respectively. It shows the variation of the ΔV_{TH} shift for all hydrogen plasma treated devices during 2 hours stress time. Although No. 1 (80W, 1 min) and No. 6 (100W, 5 min) was shifted by about -0.96V and -1.34V respectively, the shift of ΔV_{TH} in No. 5 (100W, 3 min) was only about -0.48V after bias temperature stress (2 hours). No. 5 plasma treated device exhibit less degradation, indicating that appropriate H_2 plasma treatment (100W, 3 min) leads to better bias temperature reliability.

Suppression of hysteresis phenomenon in SPC-Si TFT

In order to investigate the hysteresis phenomenon with various hydrogen plasma treatments, we have measured I_{DS}-V_{GS} transfer characteristics by double sweep mode. The image sticking caused by hysteresis phenomenon could be evaluated by the threshold voltage variation (ΔV_{TH}) with gate voltage sweep direction. Figure 5(a) shows that hysteresis in the transfer characteristics (measured at V_{DS} = -0.1V) was observed between forward (V_{GS} = 10V to -20V, gate voltage sweep step = 0.2V) and reverse gate voltage sweeps (V_{GS} = -20V to 10V).

To reduce the hysteresis phenomenon of SPC-Si TFT, we passivated the silicon dangling bond by using hydrogen plasma treatment. When applied drain bias was -0.1V, the threshold voltage of forward voltage sweep direction (from +10V to -20V) in the sample No. 1 (80W, 1 min) was -6.91V while that of reverse voltage sweep direction (from -20V to +10V) was -7.44V due to the different quantities of initially trapped charges. And in the sample No. 5 (100W, 3 min), the threshold voltage of forward and reverse gate voltage sweep direction was -3.85V and -4.15V. For sample No. 6 (100W, 5 min), it was measured that the threshold voltage with forward and reverse gate bias sweep direction was -5.42V and -5.97V respectively.

The variation of V_{TH} with gate voltage sweep direction was decreased considerably with increasing power of hydrogen plasma treatment except that of 5 minutes (applied time). As the applied time of the hydrogen plasma treatment was increased to 5 minutes, the variation of the threshold voltage between forward and reverse gate sweep direction was increased for the 80W and 100W power of PECVD as shown in Figure 5 (b).

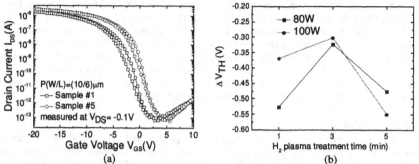

(a) (b)

Figure 5 (a) Transfer Characteristics (double sweep mode) of SPC-Si TFT, and (b) the variation of threshold voltages of SPC-Si TFT between forward sweep and reverse sweep mode with various H_2 plasma treatments

The hysteresis phenomenon of sample No.1 (80W, 1 min) ($\Delta V_{TH} = 0.53V$) is significantly larger than that of No. 5 (100W, 3 min) ($\Delta V_{TH} = 0.3V$), which shows the smallest value amongst all of the samples as shown in Figure 5 (b). A quantity of initially interface trapped charge between SPC-Si film and the gate oxide is dominantly affected by trap state density and intragrain defect, which is strongly related with the grain structure of active layer.

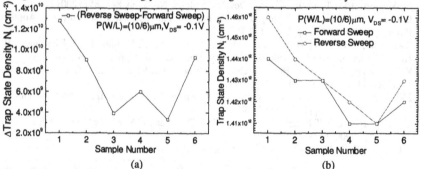

(a) (b)

Figure 6 (a) Trap state density with various H_2 plasma treatments in SPC-Si TFT, (b) variation of trap state density between forward sweep and reverse sweep mode.

To analyze the difference in the hysteresis phenomenon from No. 1 (80W, 1 min) to No. 6 (100W, 5 min), we have extracted the trap state density with various H_2 plasma treatments by Levinson's model [9]. Especially, No.1 exhibits a relatively high ΔN_t value of $1.28 \times 10^{10} cm^{-2}$, compared with $3.34 \times 10^9 cm^{-2}$ of No.5. Because sample No.1 shows a relatively high trap state density, it exhibits a large hysteresis phenomenon compared with other samples.

Although the H_2 plasma treatment was also performed on No.1, there was not enough time to decrease the trap states by passivating the grain boundary trap states as hydrogen atoms. By increasing the power and applied time duration for hydrogen plasma treatment, a relatively small ΔN_t value which related with hysteresis phenomenon in the SPC-Si TFT was shown. For 5 minutes with 100W power with PECVD (No. 6), on the contrary, the ΔN_t value was increased. This is because the hydrogen atoms no longer passivate the dangling bonds. Instead, the hydrogen plasma has a negative effect on the interface by decomposing the Si-H bonding of the surface region. Immoderate hydrogen plasma treatment may have destructive effects between the gate oxide and SPC-Si interface region [8]. Our experimental results showed that the trap state density of TFT effectively decreased after 100W, 3 minutes hydrogen plasma treatment, which suppresses the hysteresis phenomenon for achieving a high quality AMOLED panel.

CONCLUSIONS

In summary, we have improved electrical properties and suppressed the hysteresis phenomenon of SPC-Si TFT by employing a hydrogen plasma treatment prior to SiO_2 gate insulator deposition. It was also verified that bias temperature stability enhances with appropriate hydrogen plasma treated SPC-Si TFT (100W power of PECVD and 3 minutes) by reducing the defects and grain boundary trap states. This is due to the role of hydrogen atoms by improving interface between SPC-Si film and gate oxide. Therefore it could be possible to reduce a trap state density by hydrogen plasma treatment, which suppress the hysteresis phenomenon and enhance the electrical characteristics for high quality AMOLED panel composed of SPC-Si TFTs.

REFERENCES

1. M. Stewart, IEEE Trans. on Electron Devices, Vol. 48, No. 5, pp. 845–851 (2001).
2. B.K. Kim, O.H. Kim, H.J. Chung, J.W. Chang, and Y.M. Ha, Japanese Journal of Applied Physics, vol.43, no.4A, pp. L482-L485 (2004).
3. C.-Y. Kim, S.-G. Park, M.-K. Han, H.-K. Lee, S.-W. Lee, S.-H. Jung, C.-D. Kim, and I. B. Kang, Journal of The Electrochemical Society, Vol. 155, No. 4, pp. H224-H227 (2008).
4. Y. Morimoto, Y. Jinno, K. Hirai, H. gata, T. Yamada, and K. Yoneda, J. Electrochem. Soc., Volume 144, Issue 7, pp. 2495-2501 (1997).
5. I.W. Wu, T.-Y. Huang, W.B. Jackson, A.G. Lewis, and A. Chiang, IEEE Electron Device Lett., Vol. 12, No. 4, pp. 181-183 (1991).
6. Y.S. Lee, H.Y. Lin, T.F. Lei, T.Y. Huang, T.C. Chang and C.Y. Chang, Jpn. J. Appl. Phys., Vol. 37, pp. 3900-3903 (1998).
7. S. Luan and G.W. Neudeck, J. Appl.Phys. Vol. 68 (1990).
8. J.W. Tsai, C.Y. Huang, Y.H. Tai, and H.C. Cheng, F.C. Su, F.C. Luo, and H. C. Tuan, Appl. Phys. Lett. / Vol. 71, Issue 9, pp. 1237-1239 (1997).
9. J. Levinson, F. R. Shepherd, P. J. Scanlon, W. D. Westwood, G. Este, and M. Rider, J. Appl. Phys. Vol. 53, Issue 2, pp. 1193-1202 (1982).

Film Growth II

Mater. Res. Soc. Symp. Proc. Vol. 1153 © 2009 Materials Research Society   1153-A22-01

Potential of Hot Wire CVD for Active Matrix TFT Manufacturing

R.E.I. Schropp, Z.S. Houweling, V. Verlaan[1]
Utrecht University, Faculty of Science, Debye Institute for Nanomaterials Science,
Department of Physics and Astronomy, Section Nanophotonics – Physics of Devices,
P.O. Box 80.000, 3508 TA Utrecht, The Netherlands
[1]Present address: Eindhoven University of Technology, Eindhoven, The Netherlands

ABSTRACT

Hot Wire Chemical Vapor Deposition (HWCVD) is a fast deposition technique with high potential for homogeneous deposition of thin films on large area panels or on continuously moving substrates in an in-line manufacturing system. As there are no high-frequency electromagnetic fields, scaling up is not hampered by finite wavelength effects or the requirement to avoid inhomogeneous electrical fields. Since 1996 we have been investigating the application of the HWCVD process for thin film transistor manufacturing. It already appeared then that these Thin Film Transistors (TFTs) were electronically far more stable than those with Plasma Enhanced (PE) CVD amorphous silicon. Recently, we demonstrated that very compact SiN_x layers can be deposited at high deposition rates, up to 7 nm/s. The utilization of source gases in HWCVD of $a-Si_3N_4$ films deposited at 3 nm/s is 75 % and 7 % for SiH_4 and NH_3, respectively. Thin films of stoichiometric $a-Si_3N_4$ deposited at this rate have a high mass-density of 3.0 g/cm^3. The dielectric properties have been evaluated further in order to establish their suitability for incorporation in TFTs. Now that all TFT layers, namely, the SiN_x insulator, the a-Si:H or μc-Si:H layers, and the n-type doped thin film silicon can easily be manufactured by HWCVD, the prospect of "all HWCVD" TFTs for active matrix production is within reach. We tested the 3 nm/s SiN_x material combined with our protocrystalline-Si:H layers deposited at 1 nm/s in 'all HW' TFTs. Results show that the TFTs are state of the art with a field-effect mobility of 0.4 cm^2/Vs. In order to assess the feasibility of large area deposition we are investigating in-line HWCVD for displays and solar cells.

INTRODUCTION

High deposition rates are required to reduce production costs of large-area applications (e.g., displays and solar cells) in mass fabrication. The common approach in the development of plasma sources for high deposition rate is to aim for high ionization rates and high gas utilization. Fundamentally, in plasma enhanced chemical vapor deposition (PECVD) high electron energy is required to fragment molecules while the probability of occurrence is low, since point collisions are required in a 3D space. Hot Wire Chemical Vapor Deposition (HWCVD) is fundamentally different since molecules are cracked on a 2D catalyst material. Because the source gases are catalytically decomposed, the method is often referred to as Thermo-Catalytic Chemical Vapor Deposition or Catalytic Chemical Vapor Deposition (Cat-CVD). A schematic view of our HWCVD reactors is given in figure 1.

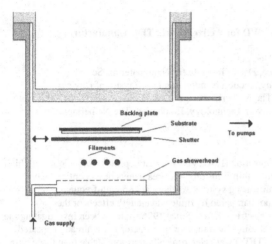

Figure 1. Schematic drawing of the experimental HWCVD reactor.

The technology of HWCVD has made great progress during the last couple of years. The quality of thin film materials can be accurately controlled and there is an increasing number of examples showing that large area continuous coating is feasible.

Thin film silicon nitride is a Hot-Wire deposited material that is likely the first to be commercially applied. Many applications are possible for SiN_x. This material has been demonstrated as encapsulation barrier against H_2O and O_2 (even on top of sensitive organic layers), as passivating dielectric in AlGaN/GaN high mobility field effect transistors, as a mechanically strong material for microelectromechanical structures (MEMS), as the gate dielectric in thin film transistors (TFTs), and as a passivating antireflective layer on polycrystalline solar cells.

In this paper we show recent progress on the application of SiN_x :H and a-Si:H in Hot-Wire deposited thin film transistors (TFTs) as well as results of our research on utilizing the HWCVD technology in an in-line manufacturing approach.

RESULTS

Silicon nitride by Hot Wire CVD

Hot Wire CVD has some significant advantages over techniques that utilize plasmas for decomposition of the source gasses. The first requirement for obtaining high deposition rate is to obtain a high dissociation rate of the source gasses. Fundamentally, high electron energy is required to fragment molecules while the probability of occurrence is low, since point collisions are required in a 3D space. Hot Wire CVD is fundamentally different since molecules are cracked on a 2D catalyst material, which makes it far more probable for each source gas molecule to be decomposed.

The decomposition probability for SiH_4 in a single collision with the filament is 40% [1]. As the number of collisions at commonly used pressures (0.01 mbar) on a catalytic surface of typically 50 cm^2 is close to 10, the overall decomposition probability of SiH_4 molecules in a stagnant gas is close to unity. In practice, taking into account that also the reaction products can be pumped out, the total gas utilization efficiency for a-Si:H deposition from SiH_4 can be as high as ~80%, as is observed experimentally [2]. This utilization rate is ~5 times larger than in PECVD.

For a SiH_4/NH_3 mixture at relatively low total flow rate, and thus low deposition rate, utilization rates of 98% and 52% for SiH_4 and NH_3 have been obtained, respectively [3]. At a high deposition rate (3.0 nm/s) we still have a SiH_4 utilization rate of > 70% [4]. Good gas utilization and high deposition rate are highly desirable from a cost point of view. The efficient catalytic decomposition of feedstock gasses has led to ultrahigh deposition rates of up to 7.3 nm/s for device quality, transparent, near-stoichiometric films [5]. The application of HW-SiN_x at a deposition rate of 3 nm/s to polycrystalline Si wafer solar cells has led to cells with 15.7% efficiency in collaboration with the Energy research Center of the Netherlands (ECN) [6]. The 7.3 nm/s material has also been tested as passivating layer on polycrystalline silicon cells in collaboration with Centrotherm Photovoltaics Technology GmbH, Germany, leading to poly-Si cells with 14.9% efficiency [7].

Another advantage is that the technique is free of dust formation, free of ion-bombardment issues and can easily be scaled up to offer large area capability [4]. The low H concentration of 9 at.-%, specifically the low density of Si-H bonds [8], is an advantage for the use as sidewall and liner material in ultra large scale integration p-type metal-oxide-semiconductor transistors. This is the case because H, coming from the SiN_x at the sidewalls, can create defects at the gate/gate-insulator interface [9]. We have applied HW-SiN_x as the gate dielectric in TFTs [10].

Thin Film Transistors by Hot Wire CVD

The HWCVD technique has been shown to be a viable method for the deposition of silicon-based thin films and solar cells [6,11-13]. It can be considered to be a "remote" technique, as the substrate itself has no active role in generating the active precursors, unlike the case of PECVD where it usually has at least a role as the grounded electrode. As there is no requirement to achieve an equipotential plane at the substrate, it is easier in HWCVD to transport either rigid or foil type substrates while deposition is taking place and thus it is easier to scale up to large areas. Needless to say, the morphology and the structure of the substrate surface still influence the film formation. Also the substrate temperature is still of importance. Therefore, when scaling up the substrate size and/or transporting the substrate during deposition, the substrate temperature should be kept homogeneous and at the desired value.

The HWCVD technique for thin film deposition of a-Si and SiN_x is very well suitable for thin film transistor (TFT) fabrication because it is highly desirable to have large deposition rates and good gas utilization for economic production of TFTs for active matrix applications. It has been reported earlier that TFTs with thermally grown SiO_2 on c-Si as gate dielectric with HWCVD a-Si as conducting channel, are electronically more stable than when the a-Si is made with PECVD [14,15]. The thin-film alternative for the thermally oxidized wafer is SiN_x on glass, made with HWCVD, because then both the dielectric film and the a-Si conducting channel can be made with HWCVD. The application of this very fast and easy scalable technique for all layers decreases processing time greatly. A disadvantage of PECVD is that it may lead to SiN_x materials that are too porous as well as too H-rich at high N concentration [16].

TFTs exclusively made with HWCVD have been investigated in the past and showed good performance and stability [17,18]. Here, we report on results of "all hot wire TFTs" with amorphous hydrogenated SiN_x as the gate dielectric, made at very fast rates of up to 3 nm/s and a-Si:H as conductive channel made at a very fast rate of 1 nm/s.

Nishizaki *et al.* [19] showed that a-Si TFTs made with HWCVD, show high stability upon bias stressing and that the threshold voltage shift is due to charge trapping in the SiN_x only. To quantify trapped charges and to optimize our HWCVD SiN_x deposited at high deposition rates (~3 nm/s), we determined the mechanical stress, the root-mean-square (*rms*) roughness and the dielectric properties including fixed and trapped charges of our HWCVD SiN_x for various compositions.

Thin SiN_x films deposited using HWCVD have a maximum in mass density and a minimum in H concentration at an atomic N/Si ratio of 1.3 [5]. For this reason we selected this material for testing in TFTs. Trilayer TFT structures in bottom gate staggered configuration have been made, the structure is shown in figure 2. The structure consists of HWCVD $SiN_{1.3}$ (made at 3 nm/s), HWCVD a-Si:H (made at 1 nm/s) [20] and PECVD μc-Si:H highly doped n-layers. For the HWCVD SiN_x/a-Si:H stack this leads to a total deposition time of less than 4 minutes. We maintained a high SiH_4 utilization efficiency of 75%, despite the much higher flow rates that are necessary to obtain the high SiN_x deposition rate [4]. The utilization efficiency for the less expensive ammonia under these conditions is 7%. The depositions were performed at Utrecht Solar Energy Laboratory (USEL) on Corning 1737 glass with pre-patterned 100 nm thick Cr gates as provided by Japan Institute of Advanced Science and Technology (JAIST). After deposition of the HWCVD films, the photolithographic structuring was performed at JAIST. Unfortunately, it was necessary to introduce an air break between the SiN_x deposition and the a-Si:H deposition, which most likely posed an upper limit to the achievable mobility. In these first experiments, the SiN_x gate insulation was made thick (400 nm), to avoid electrical breakdown of SiN_x. The result is that no pinholes have been found within the matrix made (86 TFTs) but the S-value is rather large (1 V/decade). Analysis shows that these "all hot wire" TFTs have an on/off ratio of 10^6 and a mobility of 0.4 cm^2/Vs after a forming gas anneal, the transfer characteristics of the TFTs were measured three times [10].

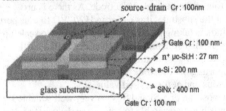

Figure 2. Schematic diagram of the trilayer structure made by USEL and JAIST.

The transfer characteristics are shown in figure 3. The subsequent curves correspond to the first, second, and third measurement. The threshold voltage is slightly increased in the direction of more positive gate voltage after each measurement.

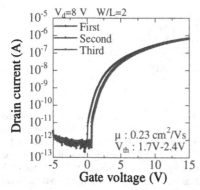

Figure 3. Transfer characteristics of the HWCVD SiN$_x$/a-Si:H TFT, before forming gas anneal.

It is known that the V$_{th}$ shift caused by charge trapping is dependent on the composition of the SiN$_x$ dielectric in the case of plasma-deposited films [21] and a study of the defect density and bond structure specifically for hot-wire deposited SiN$_x$ was done at JAIST [19]. It is shown that V$_{th}$ shift can be minimized by carefully optimizing the composition of the SiN$_x$. The present TFT data are consistent with a composition of the SiN$_x$ film that is neither too Si rich nor too N rich, since the initial V$_{th}$ shift is very small.

The S-value is rather large, but this is due to the large thickness (400 nm) of the SiN$_x$ film. The series resistance of the source and drain contacts is very low because of the use of microcrystalline n-type source and drain contact layers. The mobility (in Table 1) is estimated from the gradient of the square root of I$_d$ as a function of V$_g$ in the saturation region (V$_d$ = 8 V).

Although the characteristics of the present TFTs are already suitable for applications, the as-deposited mobility is only moderately high. We annealed the TFTs in forming gas (H$_2$/N$_2$) at 200°C for two hours. This time, the mobility was measured in three different ways. The first method uses the I$_d$ –V$_d$ data in the linear regime. The second method uses the transconductance g$_m$, at constant V$_g$, The third method is based on the output characteristics around V$_d$ = 0. This method could be used because the contacts had very low Ohmic resistance. Table I shows the mobilities obtained by all three methods, before and after forming gas anneal. The three methods show consistent results and it is seen that the highest measured mobility is 0.4 cm^2/Vs.

These TFTs show that HWCVD SiN$_x$, deposited at 3 nm/s is already suitable for application in TFTs. To determine whether x = 1.3 (as used in the TFTs described above) is the optimal N/Si ratio, stress, *rms* roughness and electrical measurements are performed.

In a second round of tests of all-HWCVD TFTs, we used a slightly more N-rich SiN$_x$:H layer (x = 1.35) with only half the thickness (200 nm) and a HWCVD deposited *microcrystalline* silicon (μc-Si:H) layer, deposited at 0.2 nm/s. The SiN$_x$ layer was deposited at 2.8 nm/s. The manufacturing of the TFTs was undertaken in collaboration with G. Fortunato at IMM-CNR Sez. Roma, Italy.

Table 1. Mobility (cm²/Vs) determined by three different methods, both before and after forming gas anneal.

Mobility	Method 1		Method 2		Method 3	
Anneal:	before	after	before	after	before	after
W/L=2	0.29	0.34	0.32	**0.38**	0.31	0.34
W/L=50	0.17	0.29	0.18	0.32	0.18	0.31

During the fabrication no aggressive procedures were used (no ultrasound wet etch, no diluted HF wet etch, etc.) to avoid cracking or delamination of the deposited layers. The devices show a behavior that is similar to amorphous Si TFTs, with mobility values ~0.8 cm²/Vs, threshold voltage V_{th} ~ 3 V. Although the mobility of these μc-Si:H TFTs is amorphous-like, it is twice as high as our previous a-Si:H TFTs. As can be seen in figure 4, in the unannealed state the OFF current was relatively high, probably due to residual water, but we found that it could be reduced substantially by thermal annealing at 120°C for 1 h in nitrogen.

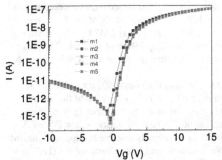

Figure 4. Transfer characteristics of the unannealed HWCVD SiN$_x$/μc-Si:H TFT. The subsequent curves correspond to the first through the fifth measurement. The threshold voltage is slightly increased (by ~0.6 V) in the direction of more positive gate voltage after each measurement (labeled as m1 through m5).

These are the first reported μc-Si:H all-HW TFTs. Top gate TFTs usually have better mobility, because the incubation layer (in which the channel should develop, see figure 2) can be avoided as the crystallinity develops to higher volume ratio towards the top of the μc-Si:H layer. Saboundji et al. [22] have shown HWCVD top gate TFTs with a higher mobility (25 cm²/Vs). The gate insulator in those TFTs was a silicon oxide layer (SiO₂) made by rf sputtering, so these TFTs are not "all HW" TFTs.

Top gate TFTs were beyond the scope of the present study. Further, the bottom gate TFTs still suffer from some drift of the threshold voltage, both in the as fabricated state and in the annealed state.

Mechanical Stress and rms roughness in HW-SiN$_x$

For our HWCVD SiN$_x$ films the intrinsic stress is *tensile* for SiN$_x$ depositions with $1.2 < x < 1.6$. We assume that the thermal stress in a SiN$_x$ film does not depend on composition in this range and amounts to -240 MPa (compressive) for films deposited at a substrate temperature of 450°C. A linear dependence of the stress with increasing x is found, with the largest stress for the most N-rich samples. For layers with N/Si = 1.2 the tensile stress becomes as low as 16 MPa. Mechanical stress is an important issue. SiN$_x$ films deposited with conventional methods like PECVD tend to have high stress values in the range of 100–1000 MPa [23]. Passivating layers for organic light emitting diodes for example, require very low stress in the order of 10 MPa [24,25], in addition to low deposition temperatures. A low stress in these films is also important in MEMS applications and plastic electronics.
Low stress in thin films is also important in micro-electro-mechanical (MEMS) applications, plastic electronics and when TFTs are deposited on thin polymer foil.

The root-mean-square (rms) roughness (as determined by atomic force microscopy (AFM)) measured on 300 nm thick SiN$_{1.3}$ layers is about 1 nm. A low rms value is necessary for high field-effect mobility in thin film transistors, for more N-rich layers (x >1.4) lower *rms* values of 0.5 nm are found.

Dielectric Properties of HW-SiN$_x$

In figure 5 the results of the C-V measurements as a function of the layer composition are shown for a range of films with $1.1 < x < 1.6$. In the top graph the static dielectric constant (ε) is seen not to change significantly over this range. It amounts to an average value of 6.3, as deduced from the accumulation capacitance of the metal-insulator-semiconducting (MIS) structures.

The middle graph shows the behavior of N$_{traps}$, which is the number of shallow interface traps (trapped charges) and is determined from the hysteresis in the backward sweep, as observed in the capacitance-voltage characteristics. The bottom graph shows N$_{fix}$, which is the number of fixed charges, for simplicity including all fixed, mobile and surface state charges. All films had comparable thickness. N$_{fix}$ is determined from the flat band voltage of the MIS structure and is assumed to be homogeneously distributed in the SiN$_x$ film. We deduce 6×10^{16} cm^{-3} < N$_{fix}$ < 8×10^{16} cm^{-3} and 1.3×10^{10} cm^{-2} < N$_{traps}$ < 1.7×10^{10} cm^{-2}, both for N-rich samples with x $\geq$ 1.45. For x = 1.3 the fixed charges and trapped charges appear to have a maximum and for x < 1.3 the charges decrease again with increasing x, but are still ten times as large as for the N-rich films.

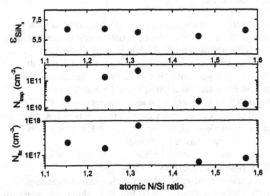

Figure 5. Static dielectric constant, number of trapped charges and total fixed charges as a function of N/Si ratio.

Current-Voltage (I-V) measurements were performed on the MIS structures, where the breakdown voltage is defined as $J = 10^{-6}$ A/cm^2. The MIS structures show no breakdown for SiN$_x$ films with $x \geq 1.45$ and $d > 174$ nm. Also for MIS structures with $x = 1.25$ and $d = 400$ nm there is no electrical breakdown observed. Consequently, the breakdown fields obtained exceed 5.9 MV/cm. Other groups also report that SiN$_x$ films with higher N/Si have higher breakdown fields but also report higher static dielectric constants as the layers become more N-rich [26,27], while we report a roughly constant value for the static dielectric constant. Breakdown fields for PECVD or HWCVD SiN$_x$ of 2-6 MV/cm have been reported [26,28,29]. In order to meet the requirements for a good gate dielectric, the SiN$_x$ material has to withstand electric fields of ≥ 2 MV/cm [26], and thus the HWCVD SiN$_x$ films also qualify for TFT applications in this respect.

In line Hot Wire CVD

We have conducted preliminary investigations on the feasibility of performing a HWCVD process uniformly on *moving* Corning 2000 glass (for optoelectrical characterization and displays) or TCO-coated glass substrates (for solar cell fabrication). As the creation of charged dust particles in the gas phase is avoided in HWCVD, deposition could be undertaken with the substrates facing *upward* (this is at variance with figure 1), thus further simplifying the mounting of the substrates. The deposition experiments have been conducted in our PILOT system, a two-chamber plus load lock system, originally installed to produce homogeneous thin film silicon layers on a 30 cm x 40 cm substrate PECVD. The transport system utilizes rails to guide the substrate holder between chambers or within one of the chambers. The holder can be pushed or pulled along the rail system by a transport arm, thus allowing the movement of substrates *during* deposition, by which in-line manufacturing can be mimicked. The "intrinsic" chamber (or i-chamber) was adapted to accommodate a hot-wire assembly while keeping the original rf electrode in place. For the purpose of hot-wire deposition, multiple, parallel, 99.9 % pure Ta filaments with a diameter of 0.50 mm were mounted above the substrate transport system.

Figure 6. An a-Si:H thin film with device-quality properties made on moving glass substrate (~20 cm x ~20 cm; a one Euro coin is added for reference).

Above the filament assembly, a shower head gas supply is mounted to achieve an evenly distributed feedstock gas flow.

Figure 6 shows an amorphous silicon thin film deposited on a plain glass substrate while it was moving under the filaments. The dark and photoconductivity of amorphous silicon on moving glass substrates were $\sigma_d = 8 \cdot 10^{-12}$ Ω^{-1}cm^{-1} and $\sigma_{ph} = 1.9 \cdot 10^{-6}$ Ω^{-1}cm^{-1} (showing an appropriate photoresponse for solar cell applications, in excess of 10^5) and the dark activation energy was $E_a = 0.92$ eV, which indicates the intrinsic property of this material, having a band gap of 1.85 eV [30]. Microcrystalline silicon on moving glass substrates had dark and photoconductivities, respectively, of $\sigma_d = 2.2 \cdot 10^{-7}$ Ω^{-1}cm^{-1} and $\sigma_{ph} = 4.0 \cdot 10^{-5}$ Ω^{-1}cm^{-1} (photoresponse of ~200). The dark activation energy was $E_a = 0.6$ eV, again indicating a Fermi level position close to mid gap. The Raman crystalline fraction was 0.68.

To test these films, we ventured the demonstration of p-i-n solar cells. We used standard Asahi U-type TCO-coated glass substrates. The p- and n-layer were made by (stationary) PECVD in another chamber of this system. For the intrinsic layers we used the processing conditions for amorphous silicon on moving substrates. Despite the two air breaks that were needed because the PECVD is performed with the substrates facing down while the HWCVD is done with the substrate facing up, the solar cells performed very well already with our first attempt.

The efficiency was 7.1%, the short circuit current density (J_{sc}) was 15.0 mA/cm^2, the open circuit voltage (V_{oc}) was 0.71 V, and the fill factor was 0.67. The thickness of the i-layer was ~300 nm. While the V_{oc} is rather low, which is attributed to the need to have air breaks at the p/i and i/n interfaces, the fill factor is not influenced by the two air breaks.

CONCLUSIONS

We have shown that HWCVD is suitable for TFTs in displays. 'All-HW' TFTs have been made, using HWCVD SiN$_x$ deposited at 2.8 – 3.0 nm/s. The a-Si:H active layers have been deposited at 1.0 nm/s and the μc-Si:H layers at 0.2 nm/s. To our knowledge, we have produced the first 'all-HW' μc-Si:H TFTs. The mobility is 0.8 cm^2/Vs, twice as high as that of the a-Si:H TFTs.

453

Further, we have shown that it is relatively straightforward to scale up the HWCVD technique. We demonstrated deposition of device-quality amorphous and microcrystalline silicon on moving substrates, showing that in-line manufacturing on moving glass substrates or on a moving web in roll to roll mode is perfectly feasible.

ACKNOWLEDGMENTS

The authors thank the entire Physics of Devices group at Utrecht University for their excellent work. We also thank Prof. Hideki Matsumura, Mr. Shogo Nishizaki, and Mr. Yohei Endo at JAIST for the fabrication of the a-Si:H TFTs mentioned in this paper. We further thank Prof. G. Fortunato and Dr L. Maiola at IMM-CNR Sez. Roma, Italy for the fabrication and characterization of the μc-Si:H TFTs mentioned in this paper.

REFERENCES

1. N. Honda, A. Masuda, H. Matsumura, *J. Non-Cryst. Solids* **266–269**, 100 (2000).
2. B.P. Nelson, Y. Xu, A.H. Mahan, D.L. Williamson, R.S. Crandall, *Mater. Res. Soc. Symp. Proc.* **609**, A22.8.1 (2000).
3. S.G. Ansari, H. Umemoto, T. Morimoto, K. Yoneyame, A. Izumi, A. Masuda, H. Matsumura, *Thin Solid Films* **501**, 31 (2006).
4. R.E.I. Schropp, *Jpn. J. Appl. Phys.* **45**, 4309 (2006).
5. V. Verlaan, A.D. Verkerk, W.M. Arnoldbik, C.H.M. van der Werf, R. Bakker, Z.S. Houweling, I.G. Romijn, D.M. Borsa, A.W. Weeber, S.L. Luxembourg, M. Zeman, H.F.W. Dekkers, and R. E. I. Schropp, *Thin Solid Films* **517**, 3499 (2009).
6. V. Verlaan, C.H.M. van der Werf, Z.S. Houweling, H.F.W. Dekkers, I.G. Romijn, A.W. Weeber, H.D. Goldbach, and R.E.I. Schropp, *Prog. Photovolt. Appl.* **15**, 563 (2007).
7. R.E.I. Schropp, C.H.M. van der Werf, V. Verlaan, J.K. Rath, and H.Li, *Thin Solid Films* **517**, 3039 (2009).
8. V. Verlaan, C.H.M. van der Werf, W.M. Arnoldbik, H.D. Goldbach, R.E.I. Schropp, *Phys. Rev. B* **73**, 195333 (2006).
9. A. Akasaka, *Extended Abstracts of the 4th Int'l. Conf. on Hot wire CVD Process*, Takayama, Japan (2006).
10. R.E.I. Schropp, S. Nishizaki, Z.S. Houweling, V. Verlaan, C.H.M. van der Werf, and H. Matsumura, *Solid St. Electron.* **52**, 427 (2008).
11. H. Matsumura, *J. Appl. Phys.* **65**, 4396 (1989).
12. A.H. Mahan, J. Carapella, B.P. Nelson, R.S. Crandall and I. Balberg, *J. Appl. Phys.* **69**, 6728 (1991).
13. K.F. Feenstra, R.E.I. Schropp and W.F. van der Weg, *J. Appl. Phys.* **85**, 6843 (1999).
14. J. K. Rath, F.D. Tichelaar, H. Meiling and R.E.I. Schropp, *Mat. Res. Soc. Symp. Proc.* **507**, 879 (1998).
15. H. Meiling and R.E.I. Schropp, *Appl. Phys. Lett.* **70**, 2681 (1997).
16. S. Hasegawa, L.He, Y.Amano and T. Inokuma, *Physical Review B* **48**, 5315 (1993).
17. M. Sakai, T. Tsutsumi, T. Yoshioka, A. Masuda, and H. Matsumura, *Thin Solid Films* **395**, 330 (2001).
18. B. Stannowski, J.K. Rath, and R.E.I. Schropp, *Thin Solid Films* **395**, 339 (2001).
19. S. Nishizaki, K. Ohdaira, and H. Matsumura, *13th Int. Display Workshop*, Vol. **2**, Dec. 7;

AMDp-3 (2006).

20. R.E.I. Schropp, M.K. van Veen, C.H.M. van der Werf, D.L. Williamson, A.H. Mahan, *Mat. Res. Soc. Symp. Proc.* **808**, A8.4.1 (2004).
21. M.J. Powell, C. van Berkel, and J.R. Hughes, *Appl. Phys. Lett.* **54**, 1323 (1989).
22. A. Saboundji, N. Coulon, A. Gorin, H. Lhermite, T. Mohammed-Brahim, M. Fonorodana, J. Bertomeu, J. Andreu, *Thin Solid Films* **487**, 227 (2005).
23. W.A.P. Claassen, W.G.J.M. Valkenburg, W.M. v.d. Wijgert, and M.F.C. Willemsen, *Thin Solid Films* **129**, 239 (1985).
24. A. Masuda, M. Totsuka, T.Oku, R. Hattori, and H. Matsumura, *Vacuum* **74**, 525 (2004).
25. M. Takano, T. Niki, A. Heya, T. Osono, Y. Yonezawa, T. Minamikawa, S. Muroi, S. Minami, A. Masuda, H. Umemoto, and H. Matsumura, *Jpn. J. Appl. Phys.* **44**, 6A, 4098 (2005).
26. A. Sazonov, D. Stryahilev, A. Nathan, and L. D. Bogomolova, *J. Non-Cryst. Solids* **299– 302**, 1360 (2002).
27. L.J. Quinn, S.J.N. Mitchell, B.M. Armstrong, and H.S. Gamble, J. Non-Cryst. Solids **187**, 347 (1995).
28. B. Stannowski, *Silicon-based thin-film transistors with a high stability*, Ph.D. thesis, Universiteit Utrecht (2002).
29. F. Liu, S. Ward, L. Gedvilas, B. Keyes, B. To, Qi Wanga. *J. Appl. Phys.* **96**, 2973 (2004).
30. R.E.I. Schropp and M. Zeman, *Amorphous and Microcrystalline Solar Cells: Modeling, Materials and Devices Technology* (Springer, 1998).

Mater. Res. Soc. Symp. Proc. Vol. 1153 © 2009 Materials Research Society   1153-A22-02

Comparative Study of MVHF and RF Deposited Large Area Multi-Junction Solar Cells Incorporating Hydrogenated Nano-Crystalline Silicon

Xixiang Xu[1], Yang Li[1], Scott Ehlert[1], Tining Su[1], Dave Beglau[1], David Bobela[2], Guozhen Yue[1], Baojie Yan[1], Jinyan Zhang[1], Arindam Banerjee[1], Jeff Yang[1], and Subhendu Guha[1]
[1]United Solar Ovonic LLC, 1100 West Maple Road, Troy, MI, 48084, U.S.A.
[2]National Renewable Energy Laboratory, 1617 Cole Blvd, Golden, CO, 80401, U.S.A.

ABSTRACT

We report our investigations of large area multi-junction solar cells based on hydrogenated nano-crystalline silicon (nc-Si:H). We compared results from cells deposited by RF (13.56 MHz) at lower deposition rate (~3 Å/s) and by Modified Very High Frequency (MVHF) at higher rate ($\geq$ 10 Å/s). With optimized process conditions and cell structures, we have obtained ~12% initial small active-area (~0.25 cm^2) efficiency for both RF and MVHF cells and 10~11% large aperture-area (~400 cm^2) encapsulated MVHF cell efficiency for both a-Si:H/nc-Si:H double-junction and a-Si:H/nc-Si:H/nc-Si:H triple-junction structures on Ag/ZnO coated stainless steel substrate.

INTRODUCTION

Hydrogenated nano-crystalline silicon (nc-Si:H) thin films have been widely used in multi-junction silicon based thin film solar cells. Compared with hydrogenated amorphous silicon (a-Si:H) thin films, nc-Si:H thin films have superior long wavelength response and low light-induced degradation [1-3]. These films are materials of choice for middle and bottom cells in multi-junction solar cells. In order to achieve higher light absorption at longer wavelengths, nc-Si:H cells must be much thicker than the a-Si:H based cells. Typical thickness of the nc-Si:H based multi-junction cell structure ranges from 2 to 4 µm, while the a-Si:H based cells are typically ~0.5 µm thick. Such a thick cell requires higher deposition rate for nc-Si:H to be practical for manufacturing. Currently, the most commonly used method to increase deposition rate for plasma enhanced chemical vapor deposition (PECVD) is to use MVHF (30 to 150 MHz) excited plasma instead of RF (13.56 MHz) plasma.

Key issues of using MVHF are: 1) whether one can achieve good spatial uniformity over a large area at higher rate; and 2) whether the performance of MVHF deposited high rate solar cells is comparable to RF deposited lower rate solar cells. Previously, we have reported achieving good spatial uniformity over large areas using MVHF PECVD process [2]. In this work, we present results of achieving high deposition rate ($\geq$ 10Å/s) and high quality nc-Si:H over a large area using the MVHF technique. We also present results from a comparative study on multi-junction cells incorporating nc-Si:H component cells deposited by RF and MVHF techniques.

EXPERIMENTAL

Large area multi-junction solar cells were deposited in batch deposition systems with the substrate dimension of 15" x 14". Both RF (13.56 MHz) and MVHF (40-100 MHz) techniques were used for the comparative study. Details of the deposition system can be found elsewhere [2]. Typical deposition rate is about ~3 Å/s using RF, and ≥10 Å/s using MVHF.

Three groups of nc-Si:H based multi-junction solar cells were deposited for the comparative study. Group A contains double-junction cells (a-Si:H/nc-Si:H) made with RF technique. Group B contains double-junction cells (a-Si:H/nc-Si:H) made with MVHF technique, and group C are MVHF deposited triple-junction cells (a-Si:H/nc-Si:H/nc-Si:H).

For small-area cells, indium-tin-oxide (ITO) and metal contacts were deposited to form cells with an active area of 0.25 cm^2. For large-area (~400 cm^2) encapsulated cells, grid wires are used to collect current. Current-density versus voltage (J-V) and quantum efficiency (QE) measurements were performed for solar cell characterization.

RESULTS AND DISCUSSION

Initial cell performance

Table I summarizes the typical initial J-V characteristics of the small-area cells. From Table I, it can be seen that the cells deposited with RF have relatively higher open-circuit voltage (V_{oc}) and fill factor (FF) than those made by MVHF. On the other hand, the short-circuit current density (J_{sc}) for the MVHF deposited cells is larger than that for the RF deposited cells. This can be mainly attributed to that RF deposited nc-Si:H cells are ~30% thinner than the MVHF deposited nc-Si:H cells in this study. Despite these differences, the initial efficiencies of the cells deposited at higher rate with MVHF are quite comparable to those deposited at lower rate using RF. This demonstrates that MVHF deposition can achieve high deposition rate, while maintaining high quality intrinsic material and cell performance.

Table I. J-V characteristics of small-area solar cells made with RF and MVHF. The thicknesses of intrinsic nc-Si:H layer in bottom cells are 1.0~1.3 μm for group A, 1.5-2.0 μm for groups B and C.

Group	Run No.	Type	Dep. Rate (Å/s)	Dep. Method	J_{sc} (mA/cm^2)	V_{oc} (V)	Fill Factor	Eff. (%)
A	12088	DBL	~ 3	RF	10.45	1.50	0.76	11.80
	12186	DBL	~ 3	RF	10.64	1.49	0.75	11.97
B	14358	DBL	> 10	MVHF	11.97	1.37	0.71	11.60
	13588	DBL	> 10	MVHF	11.72	1.41	0.71	11.73
C	14368	TRI	> 10	MVHF	8.77	1.81	0.72	11.40
	14370	TRI	>10	MVHF	8.90	1.80	0.72	11.50

Figure 1 shows the quantum efficiency (QE) of a double-junction cell (a) and a triple-junction cell (b) made with MVHF. The total current density is 24.6 mA/cm^2 for the double-junction cell and 27.7 mA/cm^2 for the triple-junction cell. The current densities from each component cells are also labeled in Fig.1. The double-junction cell current is limited by the top cell, and the current of the triple cell is intentionally made to be limited by the nc-Si:H component cells, as such structure can reduce the light-induced degradation, mostly due to the degradation of the top a-Si:H cell.

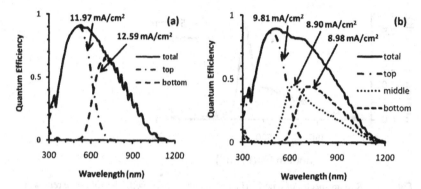

Figure 1. The quantum efficiency (QE) of a double-junction cell (a) and a triple-junction cell (b) deposited by MVHF.

We also fabricated large-area nc-Si:H based multi-junction solar cells. Table II shows the initial characteristics of 2 encapsulated a-Si:H/nc-Si:H double-junction and 2 a-Si:H/nc-Si:H/nc-Si:H triple-junction cells. The aperture area is 400 cm^2. The efficiency for the double-junction cells is 10-10.6%, while the efficiency for the triple-junction cells is only 10%, indicating that the triple-junction cells have yet to be optimized.

Table II. J-V characteristics of large-area encapsulated nc-Si:H based multi-junction solar cells made using MVHF plasma deposition. The aperture area is 400 cm^2.

Cell #	Cell Structure	V_{oc} (V)	J_{sc} (mA/cm^2)	FF	P_{max} (W)	Efficiency (%)
14334	a-Si:H/nc-Si:H	1.41	11.5	0.66	4.25	10.6
14337	a-Si:H/nc-Si:H	1.38	11.6	0.63	4.00	10.0
14371	a-Si:H/nc-Si:H/nc-Si:H	1.84	7.6	0.72	4.01	10.0
14393	a-Si:H/nc-Si:H/nc-Si:H	1.83	8.0	0.68	4.00	10.0

Material characterization of nc-Si:H films

Figure 2 shows the Raman spectra of a single-junction nc-Si:H cell deposited under similar conditions for the bottom cells of those listed in Table I. The solid line represents data using excitation light of 632 nm wavelength, and the dashed line shows data using 532 nm excitation light. The 532 nm Raman spectrum is indicative of the crystalline volume fraction near the surface, while the 632 nm Raman spectrum reflects the crystalline volume fraction for the entire film thickness. The two curves are almost identical, indicating that the crystalline volume fraction is nearly constant as the nc-Si:H film grows. The crystalline volume fraction is close to 50%. This is consistent with the previous result that the best cell performance is obtained with a crystalline volume fraction close to 50% [2].

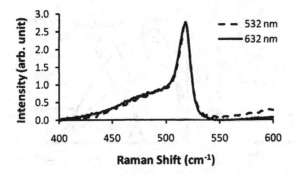

Figure 2. Raman spectra of a nc-Si:H single-junction solar cell. The solid line represents spectrum using 632 nm excitation light, and the dashed line represents spectrum using 532 nm excitation light.

Figure 3 shows the XRD spectrum from the same sample. The crystalline orientations in crystalline silicon are indicated by the arrows. The other large peaks are due to signals from the substrate. The relative intensities of the peaks in the diffraction pattern indicate a random distribution of crystalline orientations.

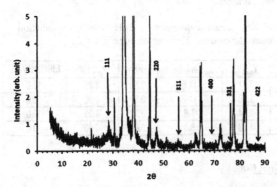

Figure 3. X-ray diffraction spectrum of the single- junction nc-Si:H cell. The positions of different silicon crystalline orientations are labeled. The strong peaks are contributed by SS/Ag/ZnO substrate and back reflector. The nc-Si:H layer is ~2 μm thick.

Light-soaking stability

In order to aim at more stable cell structure, we started light soaking the large-area encapsulated nc-Si:H based multi-junction cells. As illustrated in Fig.4, the MVHF deposited a-Si:H/nc-Si:H/nc-Si:H triple-junction cells undergo less light-induced degradation than the a-Si:H/nc-Si:H double-junction cells, ~5% versus ~19% after 450 hours of light soaking. One large-area encapsulated RF a-Si:H/a-SiGe:H/a-SiGe:H triple-junction cell was deposited on SS/Al/ZnO as a reference sample, and is also being light-soaked side-by-side with the MVHF deposited nc-Si:H based multi-junction cells. The RF cell shows ~10% degradation after 450 hours of light soaking, which is a typical value of degradation for the cell type. The greater than expected degradation for the MVHF a-Si:H/nc-Si:H double-junction cells can be attributed to the thick a-Si:H layer and the top cell limited current mis-match.

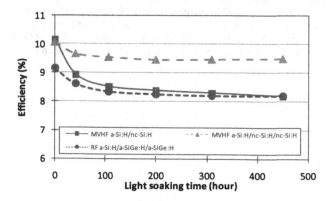

Figure 4. Efficiency of large-area (~400 cm^2) encapsulated cells versus light-soaking time. All cells are light soaked with 100mW/cm^2 white light at 50°C under open-circuit condition.

SUMMARY

We have successfully deposited large-area nc-Si:H based multi-junction solar cells with MVHF at high deposition rates ($\geq$10Å/s). The initial performance of the MVHF deposited high-rate cells is comparable to that of the RF deposited lower-rate cells. We have obtained $\geq$10% initial large-aperture-area efficiency for both the a-Si:H/nc-Si:H double-junction and a-Si:H/nc-Si:H/nc-Si:H triple-junction encapsulated cell structures. Preliminary light soaking results indicate that the a-Si:H/nc-Si:H/nc-Si:H triple-junction cell structure is more stable than the a-Si:H/nc-Si:H double-junction cell structure.

ACKNOWLEDGMENTS

The authors thank K. Younan, D. Wolf, T. Palmer, N. Jackett, L. Sivec, B. Hang, R. Capangpangan, J. Piner, G. St. John, A. Webster, J. Wrobel, B. Seiler, G. Pietka, C. Worrel, D. Tran, Y. Zhou, S. Liu, and E. Chen for sample preparation and measurements. The work was supported by US DOE under the Solar America Initiative Program Contract No. DE-FC36-07 GO 17053.

REFERENCES

[1] B. Yan, G. Yue, and S. Guha, *Mater. Res. Soc. Symp. Proc.* Vol. 989, 2007, p. 335.
[2] X. Xu, B. Yan, D. Beglau, Y. Li, G. DeMaggio, G. Yue, A. Banerjee, J. Yang, S. Guha, P. Hugger, and D. Cohen, *Mater. Res. Soc. Symp. Proc.* Vol. 1066, 2008, p. 325.
[3] G. Ganguly, G. Yue, B. Yan, J. Yang, S. Guha, *Conf. Record of the 2006 IEEE 4th World Conf. on Photovoltaic Energy Conversion*, Hawaii, USA, May 7-12, 2006, p.1712.

Mater. Res. Soc. Symp. Proc. Vol. 1153 © 2009 Materials Research Society   1153-A22-04

Gas phase conditions for obtaining device quality amorphous silicon at low temperature and high deposition rate

J.K. Rath, M.M. de Jong, A. Verkerk, M. Brinza, R.E.I. Schropp
Utrecht University, Faculty of Science, Debye Institute for Nanomaterial Science, Nanophotonics – Physics of Devices, P.O. Box 80000, 3508 TA Utrecht, the Netherlands. J.K.Rath@uu.nl

ABSTRACT

The aim of this paper is to find a parameter space for deposition of amorphous silicon films at low substrate temperature by VHF PECVD process for application in solar cell fabrication on cheap plastics. Our studies show that at lower substrate temperature, keeping the pressure constant, the ion energy flux reaching the growth surface decreases, which we partly attribute to increasing gas phase collisions arising from an increase in gas density. The role of hydrogen is twofold: (1) higher hydrogen dilution increases the ion energy and restores it to its required value at low temperatures; (2) a normal to dusty plasma transition occurs at lower hydrogen to silane flow ratios and this transition regime shifts to higher dilution ratios for lower substrate temperatures. Thus the role of high hydrogen dilution at low temperature is to avoid the dusty regime. The ion energy flux at low substrate temperature can also be restored to the value obtained at high substrate temperature, without increasing hydrogen dilution, by simply lowering the chamber pressure or increasing the delivered plasma power, though the ion energy distribution functions (IEDFs) in these cases differ substantially from the IEDF at high temperature conditions. We propose that a low pressure or high power in combination with a modest hydrogen dilution (high enough to avoid dusty regime) will deliver silicon films at low temperature without sacrificing deposition rate.

INTRODUCTION

Hydrogenated amorphous silicon based solar cells on cheap plastic substrates will pave the way for achieving low cost solar energy. This will also boost the roll-to-roll manufacturing process for thin film silicon solar cells, which are at present mostly fabricated by inline or batch process on glass substrates. However, the consideration of throughput and belt size needs a fast fabrication. Amorphous silicon (a-Si) or nanocrystalline silicon (nc-Si) show the best electronic properties and device quality at a temperature around 200 °C [1] and most of the processing steps, including the choice of substrates are chosen to meet this demand. A decrease of substrate temperature (Ts) from its optimum value of 200 °C leads to a decrease in structural order and compactness [2]. However, high electronic quality can be maintained even for films made at low temperatures by depositing a-Si at the amorphous to nanocrystalline transition regime (the so called protocrystalline or edge materials) and the hydrogen dilution at the transition regime monotonically shifts to higher values with decrease of substrate temperature [3]. A systematic study by ellipsometry revealed that the compactness of the film suffers if the substrate temperature is decreased from 200 °C to 100 °C, however, with an optimum dilution the compactness of the film made at 100 °C can be restored to the high substrate temperature value, in case of plasma CVD process [4]. Using such optimized a-Si materials p-i-n and n-i-p type solar cells have been made at 100 °C with efficiencies of 7.3% [3] and 5.3% [5] respectively. Moreover, at comparable gas phase conditions at low substrate temperature, the films made by HWCVD are less compact than the ones made by the plasma CVD process [4]. The inference from the above studies is that both hydrogen dilution and ion energy support the growth of device quality materials at low substrate temperature. The former has already been suggested as

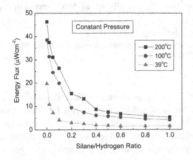 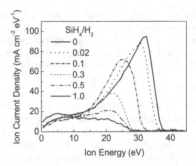

Fig. 1: (a) Ion energy flux as a function of silane to hydrogen ratio at three substrate temperatures. (b) IEDF at various silane to hydrogen ratios. Power=12.3W, p=0.10 mbar, Ts=200 °C.

part of the growth model that helps the diffusion of precursor radicals at the growth surface [1]. However, the respective contribution from these two parameters, namely, density of atomic hydrogen and the ion energy to the improvement of the material quality and their interdependence are still unclear. Moreover, we have observed that a high photosensitivity of ~ 10^6 can be achieved for materials made at substrate temperatures of 200, 100 and 39 °C, but the optimum hydrogen to silane ratio in the gas phase is at 1, 5 and 40 respectively, leading to a lower deposition rate for device-quality material at lower temperature. To further understand the effect of ion energy on the structure of materials made at low substrate temperatures, the IEDFs have been measured by means of a retarding field electrostatic ion energy analyzer [6]. The aim of this paper is to find the role of hydrogen dilution in manipulating the structure of the material at low temperature and whether alternative methods can be employed to obtain similar compact device quality material without sacrificing deposition rate.

EXPERIMENTAL

The IEDF of the ionic contribution to the deposition species was analyzed using a 4-grid electrostatic energy analyzer with retarding field. Experiments have been done at two frequencies 50MHz and 60MHz at inter-electrode distance, d, of 27mm and comparable results have been obtained for both the frequencies. The in-situ ion energy diagnosis was carried out in the ATLAS high vacuum system. The substrate temperature was controlled by the temperature of the heater block on the back of the substrate holder. The retarding field ion energy analyzer with a 4-grid stack was developed, fabricated and installed in the middle part of the substrate holder in ATLAS system, ensuring minimal disturbance of the plasma (by leveling the surface of the analyzer with the substrate holder's top surface). A detail of the energy analyzer assembly and the measurement method is given elsewhere [6]. The ion current values have been recorded simultaneously with the analyzing voltage applied to the third grid. The resulting retarding curves have been differentiated with respect to voltage and normalized in mA.cm^{-2}.eV^{-1} to obtain the IEDF. Consequently, integral of the IEDF in the form,

$$\int_0^\infty f_i(\varepsilon)d\varepsilon$$

gives the ion current density to the grounded electrode surface.

464

RESULTS

The ion energy distribution function was found to be varying with temperature and silane/hydrogen ratio in the gas phase. Fig. 1a shows the calculated energy from IEDF as a function of silane to hydrogen ratio at different substrate temperatures. From this figure two observations can be made; 1. at any silane to hydrogen ratio, the energy flux decreases at lower substrate temperature, 2. at any substrate temperature energy flux increases sharply below certain threshold silane to hydrogen ratio and this threshold shifts towards lower silane to hydrogen ratio at lower substrate temperature, a behaviour reminiscent of the shift of transition regime with decrease of substrate temperature [3]. This result already suggests that a relation between the ion energy and the optimum photosensitive material exists. Hence, the loss of ion energy due to lower substrate temperature has to be mitigated by adaptation of the deposition regime. From Fig. 1a it is clear that this can be done by a change of hydrogen dilution. This partly explains why a high hydrogen dilution is needed to obtain device quality material at low substrate temperature.

From the above argument a question arises; is this the only way to recover the ion energy? This question is important because hydrogen dilution has a negative consequence in the form of lowering of deposition rate due to the combination of reduced silicon species flux and enhanced etching. We look then at other scenarios that can mitigate the loss of ion energy and we arrive at two possibilities;

(a) Increase of power. Experiment has shown that peak of ion energy increases with increasing delivered power and this dependence is similar at all the three substrate temperatures studied [6]. A detailed study however showed that the average ion energy does not change significantly which will become clear later in Fig. 4. On the other hand, the ion flux showed a monotonic increase with increase in power. The combined effect is an increase of ion energy flux reaching the growing surface (Fig.2a). Fig. 2a shows two series of samples made at silane fractions of 0.1 and 0.3 respectively for the deposition at substrate temperature of 39 °C. The average ion energy and energy flux at the growing surface are shown for a pressure of 0.10 mbar and a power of 12.3W at the substrate temperatures of 100 and 200 °C respectively for two cases of silane fraction: 0.1 (top box) and 0.3 (lower box) respectively. As expected from our previous

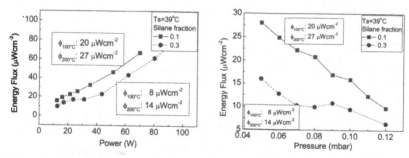

Fig. 2: Ion energy flux as a function of power (a) and pressure (b) for Ts=39 °C. Average ion energy and energy flux for Ts=100 and 200 °C for silane fraction of 0.1 and 0.3 are given as reference.

discussion, the ion energy flux for the 0.1 fraction is higher than for the 0.3 fraction. It is evident from the figure that the ion energy flux at 39 °C is lower than that of 100 °C and much lower than that of 200 °C at both these silane fraction conditions for the same power and pressure gas phase conditions. The good message that comes from this study is that the ion energy characteristics of the high temperature conditions can be attained even at 39 °C by increasing the power adequately. Moreover, this adaption process becomes easier when a combination of high dilution and high power is used, for example using silane fraction of 0.1, instead of silane fraction of 0.3.

(b) Decrease of pressure : Fig. 3a shows the IEDF of the plasma at different pressure conditions. The IEDFs move towards the high energy region with decrease in pressure and the peak shows a monotonic shift towards high energies. This could be attributed to reduced collision in the sheaths at lower pressures. This characteristic that occurs at all process temperatures can be effectively utilized to compensate the ion energy that is lost at low substrate temperature. The calculated average ion energy also follows this trend. In addition, the ion flux increases at low pressures. This could be attributed to a higher mobilty and larger mean free path of gas species in the reactor at low pressure condition. The combined effect of these two contributions is that the energy flux reaching the substrate increases with lower pressure conditions. Fig.2b shows the ion energy flux as a function of chamber pressure for two series of samples under two cases of silane fraction of 0.1 and 0.3 respectively. As expected, the ion energy flux is higher at the lower silane fraction. The average ion energy and ion flux at 12.3W, 0.13 mbar for the substrate temperature of 100 °C and 200 °C for two cases of silane fraction of 0.1 and 0.3 respectively are given in Fig 2b. It is clear that for Ts of 39 °C, the required ion energies can be reached by lowering of the pressure, and with a combination of lower silane fraction and lower pressure the required energies are even easier to reach. An explanation for the lowering of ion energy flux at low temperature may be attributed to higher gas density that follows from the simple gas equation (pV =nRT, where V is the volume of the reactor, T is the gas temperature, R is the molar gas constant and n is the molar gas density) when the pressure is kept constant. If this is the case, then lowering the pressure can restore the gas density, which roughly could explain the aforementioned technique of adaptation of pressure to restore ion energy flux. To verify this hypothesis depositions were performed at three temperatures under the same gas density conditions (Fig 3b). We see that the hypothesis is not completely valid, as the curves in Fig 3b do

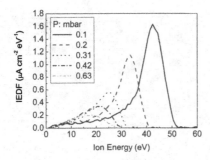 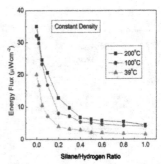

Fig. 3. (a) IEDF at various pressures. Power=12.0W, Ts=200 °C, (b) Ion energy flux as a function of silane to hydrogen ratio at constant gas density.

466

not merge, though a substantial change of ion energy flux takes place compared to Fig.1a. However, as can be seen from Fig. 2b, there exists a pressure regime where the required ion energy can be restored. This data poses two questions; (1) why is the ion energy not restored at low temperature at constant density? And (2) are the plasma conditions reproducible with restoring the ion energy flux? The answer to the first question needs a related question as to whether any other change in plasma condition is taking place at low substrate temperature. One of the hypotheses is that the effective dilution is lower at low substrate temperature. This can arise if less silicon species are incorporated into the film, leading to building up of silane concentration in the gas phase and lowering hydrogen density,

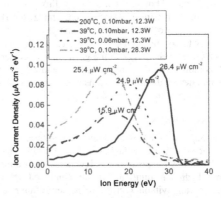

Fig. 4 : IEDFs of plasma condition at high temperature (200 °C) compared to IEDF with and without restored ion energy flux at low substrate temperature (39 °C)

as each incorporation of a silicon atom releases two hydrogen atoms to the gas phase. The former can occur if the sticking coefficient drops at lower substrate temperature. For SiH₃ type of precursors, earlier reports suggest against such a behavior at the temperature range of our study and it happens only at high temperatures > 300 °C. To verify this we looked at the temperature dependence of the deposition rate at a silane to hydrogen ratio of 0.25, which revealed that the deposition rate indeed decreases by 13% when the temperature is decreased from 200 °C to 39 °C [7]. Considering SiH₃ as the main precursor, a sticking coefficient of 10% [8] will lead to a decrease of effective hydrogen to silane ratio of 2.1%. This is much smaller than the discrepancy that we see in Fig. 3b However, this temperature effect on effective hydrogen dilution will be significant if other types of radicals with high sticking coefficients (such as SiH₂) are considered as growth precursors. Moreover, the refractive index at 632nm also decreased from 4.2 to 3.83 with this lowering of temperature, suggesting porous/less compact material at low substrate temperature. This will further lead to reduced incorporation of silicon into the films. However, its effect on the effective silane to hydrogen ratio will be marginal considering the rather small change of refractive index. From the literature [2] also it is evident that the silicon density changes marginally with lowering of temperature. The second hypothesis on the change in plasma condition is about the so called normal to dusty regime transition (the so-called α to γ' transition). We indeed observed by a V-I probe that a normal to dusty plasma transition occurs at lower hydrogen to silane flow ratios and this transition regime shifts to higher dilution ratios for lower substrate temperatures. This is consistent with the reports that the incubation time for the dusty regime decreases with lowering of gas temperature [9]. We speculate that at low temperature in a dusty regime, a significant loss of ion density as well as ion energy occurs due to neutralization collisions between the negative clusters and positive ions. Future studies will confirm which of these hypotheses is true. .

As far as the second question is concerned, we looked into the IEDFs of unrestored and restored ion energy flux plasma conditions at low substrate temperature (39 °C) and compared these to the IEDF at high substrate temperature (200 °C). Figure 4 shows that lowering the substrate temperature from 200 °C to 39°C decreased both the peak of the ion energy and also the total ion energy flux. Though the ion energy flux can indeed be restored by either increasing power by 16W or decreasing pressure by 0.04 mbar, the IEDFs are not restored to the high temperature shape. Much of the restored ion energy flux for the low temperature case comes from the low energy tail of the IEDFs. Future studies will reveal how efficiently the low energy ions are able to transfer their energy to the growing surface and this will depend on the type of ionic species. A combination of hydrogen dilution and a low pressure or high power regime can be found at which device quality a-Si films can be deposited at low substrate temperature without too much sacrificing of the deposition rate.

CONCLUSION

For deposition at low substrate temperature, the role of hydrogen is more complex than simple surface reactions; the higher hydrogen dilution increases the ion energy and restores its required value at low temperatures. A low pressure or a high power in combination with a comparatively low hydrogen dilution can be employed to achieve device quality silicon films at low temperature without sacrificing the deposition rate.

ACKNOWLEDGEMENTS

The authors thank Caspar van Bommel for the thin film depositions, Ruurd Lof for technical support. This research was financially supported by the Netherlands Agency for Energy and the Environment (SenterNovem) of the Ministry of Economic Affairs: regeling Energie Onderzoek Subsidie: lange termijn' (EOS LT).

REFERENCES

[1] A. Matsuda, Jap. J. Appl. Phys. **43** (2004) 7909.
[2] A. Fontcuberta i Morral, P. Roca i Cabarrocas, C. Clerc, Phys. Rev. B, **69** (2004) 125307.
[3] P.C.P. Bronsveld, J.K. Rath, R.E.I. Schropp, Proceedings of the 20th EUPVSEC, Barcelona (2005) p.1675.
[4] J. K. Rath, R. E.I.Schropp, and Pere Roca i Cabarocas, Phys. Stat. Sol.(c) **5**, No.5, (2008) 1346–1349. J.K.Rath, R.E.I.Schropp, Pere Roca i Cabarocas, F.D. Tichelaar, J.Non.Cryst. Solids, **354** (19-25), (2008) pp. 2652-2656.
[5] M.Brinza, J.K. Rath and R.E.I. Schropp, Sol. Energy Mater. Sol. Cells **93** (2009) 680-683
[6] J.K. Rath, A. D. Verkerk, M. Brinza, R.E.I. Schropp, W.J. Goedheer, V.V. Krzhizhanovskaya, Y.E. Gorbachev, K.E. Orlov, E.M. Khilkevitch, A.S. Smirnov, Proc. 33rd IEEE PVSC, San Diego (2008).
[7] A. Verkerk, J.K. Rath, M. Brinza, R.E.I. Schropp, W.J. Goedheer, V.V. Krzhizhanovskaya, Y.E. Gorbachev, K.E. Orlov, E.M.Khilkevitch, A.S. Smirnov, Mater. Sci. Eng. B **159-160** (2009) 53-56.
[8] A Matsuda, K.Nomoto, Y.Takeuchi, A.Suzuki, A.Yuuki, J.Perrin, Surface Science, **227** (1990) 50.
[9] K. de Bleecker, A. Bogaerts, W. Goedheer, Phys. Rev. B **70** (2004) 056407.

Hydrogen in Silicon

Mater. Res. Soc. Symp. Proc. Vol. 1153 © 2009 Materials Research Society   1153-A23-04

^{1}H NMR Study of Hydrogenated Nanocrystalline Silicon Thin Films

K.G. Kiriluk[1], D. C. Bobela[2], T. Su[3], B. Yan[3], J. Yang[3], S. Guha[3], A. Reyes[4], P. Kuhns[4], P.C. Taylor[1]

[1]Colorado School of Mines, Golden, CO 80401
[2]NREL, Golden, CO 80401
[3]United Solar Ovonic LLC, Troy, MI 48084
[4]National High Magnetic Field Lab, Tallahassee, FL 32310

ABSTRACT

Hydrogenated nano-crystalline silicon (nc-Si:H) is a promising material for multi-junction solar cells. We investigated the local hydrogen environments in nano-crystalline silicon thin films by nuclear-magnetic-resonance (NMR). At room temperature, ^{1}H NMR spectra have broader components than those observed in standard device grade hydrogenated amorphous silicon (a-Si:H). As the temperature decreases, the ^{1}H NMR exhibits a broadening of the line shape attributed to hydrogen atoms at the interfaces between the amorphous silicon (a-Si) and the crystalline silicon (c-Si) regions. These results suggest that the local hydrogen structure in nc-Si:H is very different from that in a-Si:H. In particular, the hydrogen clusters contributing to broadened spectra may exist on the surfaces of the a-Si/c-Si interfaces which do not exist in the more dense matrix of a-Si:H and may contribute to certain unique optoelectronic properties of these nc-Si:H thin films. The dependence of the spin-lattice relaxation time (T_1) on temperature, however, is very similar to that in a-Si:H, which indicates the spin-lattice relaxation mechanism, i.e. via spin diffusion through molecular hydrogen, is common to both systems.

INTRODUCTION

Advances in solar cell technology lies in nanotechnology. The most well-understood of these technologies is hydrogenated nanocrystalline silicon (nc-Si:H). nc-Si:H absorbs photons at a lower energy than hydrogenated amorphous silicon (a-Si:H), which is why it can be used as the bottom cell absorption layer in multi-junction solar cells. nc-Si:H also has the advantage that it degrades less over time than a-Si:H [1].

During the growth process for nc-Si:H, silane gas is diluted with hydrogen to promote the growth of nanocrystals inside an amorphous matrix. The dilution ratio is varied to maintain constant density and size of the nanocrystals. Isolated ^{1}H atoms passivate dangling bonds in the amorphous region, while others are believed to cluster together at the interfaces between the amorphous region and the crystallites.

We have used ^{1}H NMR to study the relative hydrogen content at different sites. To gain insight into the molecular hydrogen content of these systems, we have also investigated the temperature dependence of the spin-lattice relaxation in a nc-Si:H sample.

EXPERIMENT

^{1}H NMR experiments were conducted on a-Si:H and nc-Si:H samples deposited at United Solar Ovonics, LLC using very high frequency (VHF) plasma enhanced chemical vapor deposition (PECVD). A sample of thickness 1μm was deposited on aluminum foil. The foil was etched using a diluted 3% HCL solution. The remaining sample, now a powder, was then collected in an NMR tube and dried in ambient air. X-ray diffraction experiments on a similar sample indicated a grain size of approximately 20 nm. Raman scattering indicated a 50% crystalline fraction that was relatively independent of depth in the film.

The NMR experiments were completed at the National High Magnetic Field Lab (NHMFL) in Tallahassee, FL and the Colorado School of Mines in Golden, CO. The experiments at the NHMFL were conducted using a 17 T magnet and MagRes2000 spectrometer. The experiments at CSM were conducted using an 8 T Oxford magnet and a TecMag spectrometer. At the NHMFL, the ^{1}H signal was studied at 3.3 T and 141.4 MHz using a 5 MHz sampling rate. The sample was studied at various temperatures: 8, 10, 15, 20, 30, 40, 60, 90, 200, and 300K. For each temperature, the line shape was probed using a 90° solid echo pulse sequence [2]. At CSM, the spin-lattice relaxation time was probed at temperatures: 10, 20, 40, 80, and 300K.

RESULTS AND DISCUSSION

Lineshape

Figure 1 shows the Fourier transformed lineshapes for both a-Si:H and nc-Si:H at 7K and 10K respectively. The width of the a-Si:H lineshape is temperature independent.

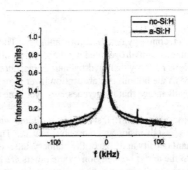

Figure 1. Fourier tranformed lineshapes for a-Si:H at 7K and nc-Si:H at 10K. The broad line is much broader for nc-Si:H than it is for a-Si:H. The width of the a-Si:H lineshape is temperature independent.

In both samples, the lineshape is composed of two parts. The first is a narrow peak which is best represented by a Lorentzian curve and is therefore attributed to the isolated ^{1}H atoms that have passivated the dangling bonds in the amorphous bulk. The broad peak is best represented by a Gaussian curve and therefore has the interpretation of arising from H atoms that are somehow clustered together.

Figure 2 shows the full width at half maximum (FWHM) for the nc-Si:H narrow and broad lines in plots a) and b) respectively. The full lineshape was fit with a Gaussian and a Lorentzian both characterized by a full width at half maximum (FWHM) parameter. In both plots a) and b), the lineshape remains constant until 90K. At temperatures below 90K, the width broadens rapidly.

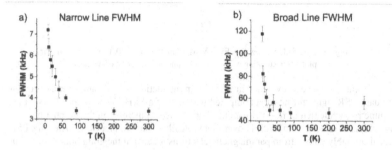

Figure 2. nc-Si:H lineshape FWHM as a function of temperature. a) is the FWHM for the narrow line, b) is the FWHM for the broad line.

In a-Si:H, the lineshape remains constant as a function of temperature. From previous studies, it is known that the lineshape for a-Si:H is governed by unresolved H-H dipole-dipole interactions [3]. The width of the lineshape depends on the number of atoms, as well as the distance between atoms. Because the nc-Si:H lineshape is dependent upon temperature, it cannot be governed purely by the nuclear dipole-dipole interaction.

A possible explanation for the observed temperature dependence is the interaction with paramagnetic centers. Because paramagnetic electrons are governed by the Boltzmann equation, one might expect that the lineshape would be linear as a function of T^{-1}. Using the approximation that $e^x \approx 1 + x$ for small x, Figure 3 shows the FWHM for the lineshape as a function of 1000/T on a linear-linear plot. The narrow line FWHM is linear within the error.

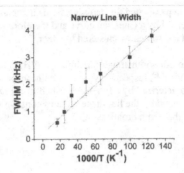

Figure 3. nc-Si:H narrow line FWHM as a function of 1000/T on a linear-linear
plot approximating $e^x \approx 1+x$. The line is a linear fit of the data.

Another piece of evidence in favor of an interaction with paramagnetic centers is
the dark ESR signal for nc-Si:H compared with that of a-Si:H. The signal for a-Si:H is
symmetric, while that of nc-Si:H is not. There is also an increase in the number of
paramagnetic electrons in nc-Si:H over that typically observed in a-Si:H. The dark ESR
signal probably comes from paramagnetic electrons located at the grain boundaries in nc-
Si:H. [1]

Spin-Lattice Relaxation Time

The spin-lattice relaxation time was studied as a function of temperature. At each
temperature, a free induction decay (FID) was taken using varying repetition rates for the
pulse. The intensity of the FID was determined and plotted as a function of the repetition
rate. T_1 was determined using Equation 1 to fit the curve,

$$I(t) = I_\infty(1 - e^{\frac{-\tau_1}{T_1}})$$

(1)

where I(t) is the intensity, I_∞ is the saturated intensity, τ_1 is the pulse repetition rate, and
T_1 is the spin-lattice relaxation time [4].

Figure 4 shows the spin-lattice relaxation time for each temperature. The open
circles represent T_1 for the broad line and the black squares represent T_1 for the narrow
line. The black triangles represent the a-Si:H T_1 for the same temperatures [3].

474

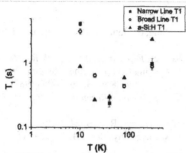

Figure 4. nc-Si:H T_1 measurement for temperatures 300, 80, 40, 20, and 10 K.
Black squares represent the narrow line, open circles represent the broad line. The
black triangles represent T_1 for a-Si:H at the same temperatures [3].

T_1 is the same for the broad and narrow lines within the error bars for all
temperatures except 10K. The T_1 curve is very similar in shape to the T_1 curve for a-Si:H
[3]. It is known for a-Si:H that the shape of the T_1 curve comes from relaxation of the
hydrogen via transferring energy to the motion of rotating molecular hydrogen. [5-6].
Because the shape of the T_1 curve for nc-Si:H is similar, the spin-lattice relaxation is
most likely driven by molecular hydrogen in the sample.

CONCLUSIONS

Both the lineshape and the spin-lattice relaxation time change as a function of
temperature. The broadening of the lineshape as a function of decreasing temperature is
not yet understood but suggests that the local environment in nc-Si:H is different from
that of a-Si:H. The similiarity of the behavior of the spin-lattice relaxation time as a
function of temperature in both a-Si:H and nc-Si:H suggests that the nc-Si:H behavior is
driven by molecular hydrogen.

ACKNOWLEDGMENTS

The work performed at the NHMFL and CSM is partially supported by NSF under
grants numbered DMR-0702351 and DMR-0073004 and by the NSF REMRSEC
program under grant number DMR-08-20518. The work performed at United Solar
Ovonic LLC and at CSM is also partially supported by the DOE under the Solar America
Initiative Program Contract Number DE-FC36-07 GO 17053.

REFERENCES

1. T. Su, T. Ju, B. Yan, J. Yang, S. Guha, and P. C. Taylor. Mater. Res. Soc. Symp. Proc. **1066**, 273 (2008)
2. E. Fukushima and S. Roeder, *Experimental Pulse NMR, A Nuts and Bolts Approach*, (Addison Publishing Company, Inc., Reading, MA, 1982), pp252.
3. W.E. Carlos and P.C. Taylor, Phys. Rev. B, **26**, 3605 (1982)
4. C. P. Slichter, *Principles of Magnetic Resonance*, (Springer-Verlag, New York), pp8.
5. W.E. Carlos and P.C. Taylor, Phys. Rev. B, **24**, 1435 (1982)
6. M.S. Conradi and R.E. Norberg, Phys. Rev. B, **24**, 2285 (1982)

AUTHOR INDEX

478

SUBJECT INDEX

Printed in the United States
By Bookmasters